Elasticity and Plasticity of Large Deformations

Albrecht Bertram

Elasticity and Plasticity of Large Deformations

Including Gradient Materials

Fourth Edition

Springer

Albrecht Bertram
Institut für Mechanik
Technische Universität Berlin
Berlin, Germany

ISBN 978-3-030-72327-9 ISBN 978-3-030-72328-6 (eBook)
https://doi.org/10.1007/978-3-030-72328-6

This Springer imprint is published by the registered company Springer Nature Switzerland AG
The registered company address is: Gewerbestrasse 11, 6330 Cham, Switzerland

Preface to the First Edition

This book is based on the lecture notes of courses given by the author over the last decade at the Otto-von-Guericke University of Magdeburg and the Technical University of Berlin. Since the author is concerned with researching material theory and, in particular, elasto-plasticity, these courses were intended to bring the students close to the frontiers of today's knowledge in this particular field, an opportunity now offered also to the reader.

The reader should be familiar with vectors and matrices, and with the basics of calculus and analysis. Concerning mechanics, the book starts right from the beginning without assuming much knowledge of the subject. Hence, the text should be generally comprehensible to all engineers, physicists, mathematicians, and others.

At the beginning of each new section, a brief *Comment on the Literature* contains recommendations for further reading. Throughout the text we quote only the important contributions to the subject matter. We are far from being complete or exhaustive in our references, and we apologise to any colleagues not mentioned in spite of their important contributions to the particular items.

It is intended to indicate any corrections to this text on our website

http://www.uni-magdeburg.de/ifme/l-festigkeit/elastoplasti.html

along with remarks from the readers, who are encouraged to send their frank criticisms, comments and suggestions to

albrecht.bertram@ovgu.de.

All the author's royalties from this issue will be donated to charitable organisations like *Terres des Hommes*.

Acknowledgment. The author would like to thank his teachers RUDOLF TROSTEL, ARNOLD KRAWIETZ, and PETER HAUPT who taught him *Continuum Mechanics* in the early seventies, and since then have continued to give much helpful advice.

Many colleagues and friends also made useful comments and suggestions to improve this book, including ENRICO BROSCHE, CARINA BRÜGGEMANN, SAMUEL FOREST, SVEN KASSBOHM, THOMAS KLETSCHKOWSKI, JOHN KINGSTON, WOLFGANG LENZ, GERRIT RISY, MANUELA SCHILDT, MICHAEL SCHURIG, GABRIELE SCHUSTER, BOB SVENDSEN and, most of all, THOMAS BÖHLKE and ARNOLD KRAWIETZ, who gave countless valuable comments. The author is grateful to all of them.

Preface to the First Edition

Preface to the Second Edition

The new edition of such a textbook provides the chance to eliminate any errors or misprints identified in the first edition, to correct any misleading expressions, and to add new references. This has all been done here.

The valuable input of students and colleagues alike has enabled me to make many small improvements to the first edition, without, however, changing the main structure of the book.

I would also like to express my gratitude to everyone who helped me with this task, in particular to Thorsten Hoffmann, Alexander Lacher, Jan Kalisch, Sven Kaßbohm, and Uwe Herbrich, to mention just a few.

Magdeburg, May 2008

Preface to the Third Edition

Nonlinear Continuum Mechanics is a rapidly growing field of research. Since the last edition of this book, many important results in this field have been published. In this new edition we have done our best to refer to the most important such results.

We have enlarged the part on hyperelastic models and anisotropic yield criteria, included an outlook on *Material Plasticity*, and tried to improve the text wherever possible.

I would like once again to express my gratitude to everyone who helped me with this task, in particular to Rainer Glüge and Jan Kalisch. And I would also sincerely like to thank Jim Casey for the many stimulating discussions that I enjoyed during my stay at the UC Berkeley in spring 2010.

Magdeburg, June 2011

Preface to the Fourth Edition

This edition of the book includes two new chapters.

Nonlinear Rheology

For many applications involving large deformations of materials like polymers, one finds that a hyperelastic material law would be too restrictive since it is not able to describe observed effects like relaxation or creep. For this purpose, an extension to viscoelasticity is desirable. Within the context of small deformations this is common practice, while its formulation for finite deformations involves many problems, which are the object of current research work. We found out that we can learn a lot for this purpose from plasticity. Therefore, this chapter is placed after the chapter on plasticity, with direct reference to the concepts developed there.

Gradient Materials

In this essay a relatively novel field of material modelling is presented, to which the author has dedicated most of his research work during the last decade. In this short introduction to these models we will mainly focus on the fundamental issues like balance laws, boundary conditions, elasticity and plasticity rather than applications to specific problems.

The author would like to thank all those friends and colleagues who helped with discussions, comments, and suggestions, in particular ARNOLD KRAWIETZ and JAN KALISCH.

Berlin, December 2020

Contents

List of Frequently Used Symbols

Sets

$\mathscr{B}$	body (-manifold)
$\mathscr{B}_0$	domain of the body in the reference placement
$\mathscr{B}_t$	domain of the body in the current placement
$\mathscr{Conf}$	configurations space in Sect. 12.2
$\mathscr{E}$	EUCLIDean point space
$\mathscr{Elap}$	elastic range
$\mathscr{Inv}$	set of invertible tensors (general linear group)
$\mathscr{Lin}$	space of linear mappings from $\mathscr{V}$ to $\mathscr{V}$ (*2*nd-order tensors)
$\underline{\mathscr{Lin}}$	space of hardening variables in Chap. 10
$\mathscr{Orth}$	set of orthogonal *2*nd-order tensors (general orthogonal group)
$\mathscr{Psym}$	set of symmetric and positive-definite *2*nd-order tensors
$\mathscr{R}$	space of real numbers
$\mathscr{Sym}$	space of symmetric *2*nd-order tensors
$\mathscr{Skw}$	space of antisymmetric or skew *2*nd-order tensors
$\mathscr{Unim}$	set of *2*nd-order tensors with determinant ± 1 (general unimodular group)
$\mathscr{V}$	space of vectors (EUCLIDean shifters)

A superimposed + such as $\mathscr{Inv}^+$ means: with positive determinant. $\mathscr{R}^+$ denotes the positive reals. $\mathscr{Orth}^+$ denotes the special orthogonal group.

Variables and Abbreviations

$\mathbf{a} = \mathbf{r}^{\bullet\bullet}$	$\in \mathscr{V}$	acceleration
$\mathbf{A}$	$\in \mathscr{Unim}^+$	symmetry transformation
$\mathbf{B} = \mathbf{F}\,\mathbf{F}^T$	$\in \mathscr{Psym}$	left CAUCHY-GREEN tensor
$\mathbf{b}$	$\in \mathscr{V}$	specific body force
$\mathbf{c}$	$\in \mathscr{V}$	translational vector in the EUCLIDean transform.
$\mathbf{c}_i$	$\in \mathscr{V}$	lattice vector in Chap. 10.5
$\mathbf{C} = \mathbf{F}^T\,\mathbf{F}$	$\in \mathscr{Psym}$	right CAUCHY-GREEN tensor
$\mathbf{C}_e = \mathbf{P}^T\,\mathbf{C}\,\mathbf{P}$	$\in \mathscr{Psym}$	transformed right CAUCHY-GREEN tensor in Chap. 10

$\boldsymbol{\mathsf{C}} = \mathbf{F}^{-1}\, Grad\, \mathbf{F}$		configuration tensor
COOS		coordinate system
$\mathbf{d}_O = \int_{\mathcal{B}_t} \mathbf{r}_O \times \mathbf{v}\, dm$	$\in \mathcal{V}$	moment of momentum with respect to O
$\mathbf{d}_\alpha$		directional vector of a slip system in Chap. 10.5
$\mathbf{D} = sym(\mathbf{L})$	$\in \mathcal{S}ym$	stretching tensor, rate of deformation tensor
da		element of area in current placement
da_0		element of area in reference placement
$\mathbf{da}$		vectorial element of area in current placement
$\mathbf{da}_0$		vectorial element of area in reference placement
Div, div		material and spatial divergence operator
dm		element of mass
dv		element of volume in current placement
dv_0		element of volume in reference placement
$\mathbf{dx}$		vectorial line element in current placement
$\mathbf{dx}_0$		vectorial line element in reference placement
$\mathbf{e}_i$	$\in \mathcal{V}$	basis vector of ONB
$\mathbf{E} = sym(\mathbf{H})$	$\in \mathcal{S}ym$	linear strain tensor
$\mathbf{E}^a = \frac{1}{2}\,(\mathbf{I} - \mathbf{B}^{-1})$	$\in \mathcal{S}ym$	ALMANSI´s strain tensor
$\mathbf{E}^b = \mathbf{I} - \mathbf{V}^{-1}$	$\in \mathcal{S}ym$	spatial BIOT´s strain tensor
$\mathbf{E}^B = \mathbf{U} - \mathbf{I}$	$\in \mathcal{S}ym$	material BIOT´s strain tensor
$\mathbf{E}^F := det(\mathbf{F})\, \mathbf{F}^{-T}$	$\in \mathcal{I}nv^+$	area placement tensor
$\mathbf{E}^G = \frac{1}{2}\,(\mathbf{C} - \mathbf{I})$	$\in \mathcal{S}ym$	GREEN´s strain tensor
$\mathbf{E}^{gen}$	$\in \mathcal{S}ym$	spatial generalised strain tensor
$\mathbf{E}^{Gen}$	$\in \mathcal{S}ym$	material generalised strain tensor
$\mathbf{E}^h = ln\, \mathbf{V}$	$\in \mathcal{S}ym$	spatial HENCKY´s strain tensor
$\mathbf{E}^H = ln\, \mathbf{U}$	$\in \mathcal{S}ym$	material HENCKY´s strain tensor
E	$\in \mathcal{R}$	internal energy
$\mathbf{f}$	$\in \mathcal{V}$	resulting force acting on the material body
$\mathbf{F} = Grad\, \boldsymbol{\chi}$	$\in \mathcal{I}nv^+$	deformation gradient
$\mathbf{g}_i$, $\mathbf{g}^i$	$\in \mathcal{V}$	basis vectors of dual bases
$\mathbf{g} = grad\, \theta$	$\in \mathcal{V}$	spatial temperature gradient
$\mathbf{g}_0 = Grad\, \theta$	$\in \mathcal{V}$	material temperature gradient
$Grad$, $grad$		material and spatial gradient operator
G_{ik}, G^{ik}	$\in \mathcal{R}$	metric coefficients

$\mathbf{H} = Grad\, \mathbf{u}$	$\in \mathcal{Lin}$	displacement gradient
$\mathbf{I}$	$\in \mathcal{Psym}$	unit tensor of 2nd-order
$\boldsymbol{I}$		unit tensor of 4th-order
$\boldsymbol{I}^S$		symmetriser of 4th-order
$\boldsymbol{I}^A$		antimetriser of 4th-order
$J = det\, \mathbf{F}$	$\in \mathcal{R}^+$	JACOBIan, determinant of the deformation gradient
$K = \frac{1}{2} \int_{\mathcal{B}_t} \mathbf{v}^2\, dm$	$\in \mathcal{R}$	kinetic energy
$\mathbf{K} = Grad(\kappa_0\ \underline{\kappa}_0^{-1})$	$\in \mathcal{Inv}^+$	local change of the reference placement
L	$\in \mathcal{R}$	(global) stress power
L_e	$\in \mathcal{R}$	(global) power of the forces
$\mathbf{L} = grad\, \mathbf{v} = \mathbf{F}^{\bullet}\, \mathbf{F}^{-1}$	$\in \mathcal{Lin}$	velocity gradient
$l = \rho^{-1}\, \mathbf{T} \cdot \mathbf{L}$	$\in \mathcal{R}$	specific stress power
$\mathbf{m}_O$	$\in \mathcal{V}$	moment with respect to O
m	$\in \mathcal{R}^+$	mass
$\mathbf{n}$	$\in \mathcal{V}$	normal vector
$\mathbf{o}$	$\in \mathcal{V}$	zero vector
$\mathbf{0}$	$\in \mathcal{Sym}$	zero tensor
ONB		orthonormal basis
p	$\in \mathcal{R}$	pressure
$\mathbf{p} = \int_{\mathcal{B}_t} \mathbf{v}\, dm$	$\in \mathcal{V}$	linear momentum
$\mathbf{P}$	$\in \mathcal{Inv}$	plastic transformation in Chap. 10
PFI	Principle of form invariance	
PISM	Principle of invariance under superimposed rigid body motions	
PMO	Principle of material objectivity	
$\mathbf{q}$	$\in \mathcal{V}$	spatial heat flux vector
$\mathbf{q}_0 = J\, \mathbf{F}^{-1}\, \mathbf{q}$	$\in \mathcal{V}$	material heat flux vector
Q	$\in \mathcal{R}$	heat supply
$\mathbf{Q}$	$\in \mathcal{Orth}^+$	versor in EUCLIDean transformation
$\mathbf{r}$	$\in \mathcal{V}$	position vector
r	$\in \mathcal{R}$	specific heat source
$\mathbf{R}$	$\in \mathcal{Orth}^+$	rotation tensor
$\mathbf{S} = \mathbf{F}^{-1}\, \mathbf{T}\, \mathbf{F}^{-T}$	$\in \mathcal{Sym}$	material stress tensor
$\mathbf{S}^P = -\, \mathbf{C}\, \mathbf{S}\, \mathbf{P}^{-T}$	$\in \mathcal{Lin}$	plastic stress tensor in Chap. 10

$\boldsymbol{S} = \mathbf{F}^{-1} \circ \boldsymbol{T}$		material hyperstress tensor (triad)
$\mathbf{T}$	$\in \mathcal{S}ym$	CAUCHY´s stress tensor
$\mathbf{T}^{B} = sym(\mathbf{T}^{2PK}\,\mathbf{U})$	$\in \mathcal{S}ym$	BIOT´s stress tensor
$\mathbf{T}^{K} = J\,\mathbf{T}$	$\in \mathcal{S}ym$	KIRCHHOFF´s stress tensor
$\mathbf{T}_{k} = \mathbf{F}^{T}\,\mathbf{T}\,\mathbf{F}$	$\in \mathcal{S}ym$	convected stress tensor
$\mathbf{T}^{M} = \mathbf{C}\,\mathbf{T}^{2PK}$	$\in \mathcal{L}in$	MANDEL´s stress tensor
$\mathbf{T}_{r} = \mathbf{R}^{T}\,\mathbf{T}\,\mathbf{R}$	$\in \mathcal{S}ym$	relative stress tensor
$\mathbf{T}^{1PK} = J\,\mathbf{T}\,\mathbf{F}^{-T}$	$\in \mathcal{L}in$	*1*st PIOLA-KIRCHHOFF stress tensor
$\mathbf{T}^{2PK} = J\,\mathbf{F}^{-1}\,\mathbf{T}\,\mathbf{F}^{-T} = J\,\mathbf{S}$	$\in \mathcal{S}ym$	2nd PIOLA-KIRCHHOFF stress tensor
$\mathbf{t} = \mathbf{T}\,\mathbf{n}$	$\in \mathcal{V}$	stress vector on a surface
t	$\in \mathcal{R}$	time
$\mathbf{U} = \sqrt{\mathbf{C}}$	$\in \mathcal{P}sym$	right stretch tensor
$\mathbf{u}$	$\in \mathcal{V}$	displacement vector
$\mathbf{V} = \sqrt{\mathbf{B}}$	$\in \mathcal{P}sym$	left stretch tensor
V	$\in \mathcal{R}^{+}$	volume in the current placement
V_0	$\in \mathcal{R}^{+}$	volume in the reference placement
$\mathbf{v} = \mathbf{r}^{\bullet}$	$\in \mathcal{V}$	velocity
$\mathbf{w}$	$\in \mathcal{V}$	spin vector (axial vector of $\mathbf{W}$)
$\mathbf{W} = skw(\mathbf{L})$	$\in \mathcal{S}kw$	spin tensor
w	$\in \mathcal{R}$	specific elastic energy
$\mathbf{x}$	$\in \mathcal{V}$	position vector in space
$\mathbf{x}_0$	$\in \mathcal{V}$	position vector in the reference placement
$\mathbf{Y}_p = \mathbf{C}_e^{\bullet} - \mathbf{P}^{T}\,\mathbf{C}^{\bullet}\,\mathbf{P}$	$\in \mathcal{S}ym$	incremental plastic variable in Chap. 10
$\mathbf{Z}$	$\in \underline{\mathcal{L}in}$	hardening variables in Chap. 10

Greek Letters

χ		motion
δ		variation, virtual
$\delta_i^{\,j},\ \delta^j_{\ i},\ \delta_{ij},\ \delta^{ij}$		KRONECKER symbols
δ	$\in \mathcal{R}$	specific dissipation
ε	$\in \mathcal{R}$	specific internal energy
ε_{ij}		linear strains (components of $\mathbf{E}$)
ε_{ijk}		permutation or LEVI-CIVITA-symbol
η	$\in \mathcal{R}$	specific entropy
φ^{i}		spatial coordinates
φ		yield criterion in Chap. 10
Φ_p		yield criterion for an elastic range in Chap. 10

κ		placement
κ_0		reference placement
λ	$\in \mathscr{R}^+$	plastic consistency parameter in Chap. 10
$\lambda_i = \mu_i^2$	$\in \mathscr{R}^+$	eigenvalue of **C** and **B**
μ_i	$\in \mathscr{R}^+$	eigenvalue of **U** and **V**
θ	$\in \mathscr{R}^+$	temperature
ρ	$\in \mathscr{R}^+$	mass density in the current placement
ρ_0	$\in \mathscr{R}^+$	mass density in the reference placement
τ	$\in \mathscr{R}$	process time
ψ	$\in \mathscr{R}$	specific free energy
Ψ^i		material coordinates
$I_{\mathbf{T}}, II_{\mathbf{T}}, III_{\mathbf{T}}$	$\in \mathscr{R}$	principal invariants of a tensor **T**
$[A_{pj}]$		matrix with components A_{pj}

Compositions

$\cdot$	inner or scalar product
$\times$	between vectors: vector or cross product
$\times$	between sets: Cartesian product
$\otimes$	tensor product or dyadic product
$*$	RAYLEIGH product defined in (1.35)
$\circ$	tensor transformation defined in (12.1)

Logical Symbols

$\in$	"element of"
$\forall$	"for all"
$\wedge$	"and"
$=$	equality
$:=$	equal by definition
$\equiv$	identification ("is set equal in this place")
$\approx$	approximate equality ("almost the same")
$\Leftrightarrow$	equivalence ("if and only if")
$\Rightarrow$	implication ("thus")

List of Selected Scientists

Emilio **Almansi** (1869-1948)
Johann **Bauschinger** (1834-1893)
Maurice Anthony **Biot** (1905-1985)
Ludwig **Boltzmann** (1844-1904)
Augustin Louis **Cauchy** (1789-1857)
Arthur **Cayley** (1821-1895)
Elwin Bruno **Christoffel** (1829-1900)
Rudolf Julius Emmanuel **Clausius** (1822-1888)
Gaspard Gustave de **Coriolis** (1792-1843)
Eugène Maurice Pierre **Cosserat** (1866-1931)
Francois **Cosserat** (1852-1914)
Maurice **Couette** (1858-1943)
Daniel Charles **Drucker** (1918-2001)
Pierre Maurice Marie **Duhem** (1861-1916)
Jerald Laverne **Ericksen** (1924-)
Euclid (ca. 320-260 a. c.)
Leonhard **Euler** (1707-1783)
Joseph **Finger** (1841-1925)
Jean Baptiste Joseph de **Fourier** (1768-1830)
Maurice Rene **Fréchet** (1878-1973)
Galileo **Galilei** (1564-1642)
René Eugène **Gâteaux** (1887-1914)
Carl Friedrich **Gauß** (1777-1855)
Paul **Germain** (1920-2009)
Josiah Willard **Gibbs** (1839-1903)
George **Green** (1793-1841)
Georg **Hamel** (1877-1954)
William Rowan **Hamilton** (1805-1865)
Hermann Ludwig Ferdinand von **Helmholtz** (1821-1894)
Heinrich **Hencky** (1885-1951)
Rodney **Hill** (1921-2011)
Robert **Hooke** (1635-1703)
Maksymilian Tytus **Huber** (1872-1950)
Carl Gustav Jacob **Jacobi** (1804-1851)
James Prescott **Joule** (1818-1889)
Robert **Kappus** (1904-1973)
Gustav Robert **Kirchhoff** (1824-1887)
Joseph Louis **Lagrange** (1736-1813)
Gabriel **Lamé** (1795-1870)
Edmund **Landau** (1877-1938)
Erastus Henry **Lee** (1916-2006)

Gottfried Wilhelm **Leibniz** (1646-1716)
Tullio **Levi-Civita** (1873-1941)
Jean **Mandel** (1907-1982)
Julius Robert von **Mayer** (1814-1878)
Raymond D. **Mindlin** (1906-1987)
Richard von **Mises** (1883-1953)
Paul Mansour **Naghdi** (1924-1994)
Claude Louis Marie Henri **Navier** (1785-1836)
Carl Gottfried **Neumann** (1832-1925)
Isaac **Newton** (1643-1727)
Walter **Noll** (1925-2017)
James Gardner **Oldroyd** (1921-1982)
Gabrio **Piola** (1794-1850)
Max Karl Ernst Ludwig **Planck** (1858-1947)
William **Prager** (1903-1980)
Ludwig **Prandtl** (1875-1953)
Lord **Rayleigh** (John William Strutt) (1842-1919)
Markus **Reiner** (1886-1976)
Endre (A.) **Reuss** (1900-1968)
Osborne **Reynolds** (1842-1912)
Hans **Richter** (1912-1978)
Bernhard Georg Friedrich **Riemann** (1826-1866)
Ronald Samuel **Rivlin** (1915-2005)
Adhémar Jean Claude Barré **de Saint-Venant** (1797-1886)
Erich **Schmid** (1896-1983)
Hermann Amandus **Schwarz** (1843-1921)
George Gabriel **Stokes** (1829-1903)
Geoffry Ingram **Taylor** (1886-1975)
Henri Edouard **Tresca** (1814-1885)
Rudolf **Trostel** (1928-2016)
Clifford Ambrosius **Truesdell** (1919-2000)
Richard A. **Toupin** (1926-2017)
Woldemar **Voigt** (1850-1919)

Introduction

While several textbooks on non-linear elasticity have already been published, the subject of finite plasticity appears not yet to have gained the maturity of a textbook science, in spite of its paramount importance for practical applications in, *e.g.*, metal-forming simulations. One of the reasons for this surprising fact is that even the fundamental concepts of plasticity still lack a rational introduction based on clear physical and mathematical reasoning. One of the aims of this book is to at least partially reduce such shortcomings of plasticity theory (see the last chapter).

We intend to introduce plasticity in a *rational* way, *i.e.*, based on axiomatic assumptions with clear physical and mathematical meanings. We are far from believing that this has already been successfully completed. Too many questions are still to be answered and need further concretisation. So the part of this book on plasticity can be considered as an essay which will stimulate and encourage further research work.

In our opinion, it takes a solid grounding in finite elasticity to properly understand finite plasticity. Therefore, we will introduce the fundamental concepts of elasticity in some detail. Before we do so, however, a careful description of non-linear continuum mechanics shall be given. As almost all quantities in continuum mechanics are described by tensors or tensor fields of different order, the book starts with an introduction to tensor algebra and analysis. This is necessary because many different notations of tensor calculus are used in the literature.

In mechanics, like any other precise science, it is impossible in principal to *define* all concepts. Certain concepts and laws have to be introduced as primitive ones, relying on an *a priori* empirical understanding of the reader. Based on these primitive concepts, we can then derive other concepts *by definition*. And only if a certain structure of the theory has been constructed, can we try and assess the validity of our axioms. Such an axiomatic approach is aimed at in the present context. We try to give all the concepts used a certain mathematical rigour, without revelling in pure formalisms.

The author considers himself as following in the tradition of the *Berlin school of continuum mechanics*, which was made famous by the likes of GEORG HAMEL, ISTVAN SZABO, WALTER NOLL, RUDOLF TROSTEL, HUBERTUS WEINITSCHKE, ARNOLD KRAWIETZ, and PETER HAUPT. It seems that this school no longer exists in Berlin, most of its members have already passed away, while others left to spread these ideas across the world.

A word on notation. It should be mentioned in passing that one of the early habits of this school was to use a direct notation for vectors and tensors, long before it became fashionable. Of course, we will adopt this elegant notation here whenever it is feasible.

It is practically impossible to give each symbol a unique meaning, without drowning in indices, tildes, primes, etc. The reader will therefore be confronted with some double meanings in the book, but we have always tried to avoid confusion. For this purpose, a list of the symbols repeatedly used in different parts of the book has been included at the beginning of the book. We have always tried to use notations that are common in today's literature, which has been hugely influenced by the masterpiece of TRUESDELL/ NOLL (1964), which included elasticity, but unfortunately not plasticity. Only on a few occasions do we prefer different notations. As an example, all scalar products in this text are denoted by a dot, regardless of the order of tensors involved.

1 Mathematical Preparation

A Comment on the Literature. Although this mathematical preparation is rather detailed in the field of tensor calculus, basic knowledge of mathematics is required. For further reading we recommend the books by LOOMIS/STERNBERG (1968), CHOQUET-BRUHAT et al. (1977), and FLEMING (1977).

In continuum mechanics physical quantities can be

- **scalars** or **reals**, like time, energy, power,
- **vectors**, like position vectors, velocities, or forces,
- **tensors**, like deformation and stress measures.

Since we can also interpret scalars as *0*th-order tensors, and vectors as *1*st-order tensors, all continuum mechanical quantities can generally be considered as tensors of different orders.

They can either be defined for the whole body as a **global variable** (like the resulting force acting on a body), or as **local** or **field variables**, *i.e.* defined in every point of a body (like its velocity).

Our short mathematical preparation shall make us familiar with the mathematical concept of tensors. We will start with the *algebra of tensors*. Thereafter, we will consider tensor-functions or tensor fields, on which we can perform calculus or *tensor analysis*. Finally, we will consider integrals over such fields, which will give us resulting tensors or mean values of fields.

Notations. The standard notation of scalars (reals) is by italic letters like α , β , ... or a , b , ... For vectors we will use small Latin letters in bold like $\mathbf{a}$, $\mathbf{b}$, ... *2*nd-order tensors are notated by large Latin letters in bold like $\mathbf{A}$, $\mathbf{B}$, … . For *4*th-order tensors we use larger and italic letters like $\boldsymbol{\mathcal{A}}$, $\boldsymbol{\mathcal{B}}$ … . Sets and spaces are notated in cursive, like

$\mathscr{R}$	the set of real numbers
$\mathscr{R}^+$	the set of positive real numbers
$\mathscr{R}^n$	the set of n ordered real numbers (n-tuples).

If $\mathscr{V}$ and $\mathscr{W}$ are sets, a **mapping** or **function**

$$f : \mathscr{V} \to \mathscr{W}$$

assigns to each element of its **domain** $\mathscr{V}$ uniquely one element of its **range** $\mathscr{W}$. If we want to give the variables names, we write

A. Bertram, *Elasticity and Plasticity of Large Deformations*,
https://doi.org/10.1007/978-3-030-72328-6_1

$$f : v \mapsto w\,.$$

Clearly, the **argument** is $v \in \mathscr{V}$ and its **value** or **image** under f is $w = f(v) \in \mathscr{W}$.

The **composition** of two functions such as

$$f : \mathscr{V} \to \mathscr{W} \qquad \text{and} \qquad g : \mathscr{W} \to \mathscr{U}$$

is a function

$$gf : \mathscr{V} \to \mathscr{U}$$

defined by

$$gf(v) := g(f(v)) \qquad \forall\, v \in \mathscr{V}.$$

The latter exists if the range of f is included in the domain of g.

If $\underline{\mathscr{V}}$ is a subset of $\mathscr{V}$, thus $\underline{\mathscr{V}} \subset \mathscr{V}$, then the image of $\underline{\mathscr{V}}$ under f is the set

$$f(\underline{\mathscr{V}}) := \bigcup_{v \in \underline{\mathscr{V}}} f(v) \subset \mathscr{W}.$$

The **restriction** of f to $\underline{\mathscr{V}}$ is denoted by $f|_{\underline{\mathscr{V}}}$.

If the image of the domain of a function coincides with its range

$$f(\mathscr{V}) \equiv \mathscr{W}$$

we will call it **surjective** or **onto**. Surjectivity of a function can always be achieved by appropriately restricting its range.

A function is called **injective** if each element of its range corresponds to not more than one argument. Hence we have the implication

$$f(v_1) = f(v_2) \qquad \Rightarrow \qquad v_1 = v_2\,.$$

A function which is both surjective and injective is called **bijective** or **invertible**, as it allows for an inversion. In such a case we can introduce the inverse

$$f^{-1} : \mathscr{W} \to \mathscr{V}$$

such that the composition gives the identity mapping on the domain $\mathscr{V}$

$$f^{-1} f = I_{\mathscr{V}}$$

and vice versa

$$f f^{-1} = I_{\mathscr{W}}$$

the identity on the range $\mathscr{W}$.

Clearly, the composition of f and g is bijective, if and only if f and g are bijective. In such cases

$$(f g)^{-1} = g^{-1} f^{-1}$$

holds for the inverse.

Example. Let $\mathscr{M}$ be any set and

$$\varphi : \mathscr{M} \rightarrow \mathscr{R}^n$$

an injective function. Then we call φ a (global) chart or **coordinate system** (COOS) of $\mathscr{M}$. With the help of it we can uniquely identify each element $x \in \mathscr{M}$ by its n **coordinates** φ^i

$$\varphi : x \mapsto \{\varphi^1(x), \varphi^2(x), \ldots, \varphi^n(x)\} .$$

We will later on assume that such COOS are always both continuous and differentiable.

The following theorems will be often needed.

Theorem 1.1. *An inhomogeneous system of real linear equations*

$$\sum_{j=1}^{M} A_{ij} x_j = y_i \qquad i = 1, \ldots, M \geq 1$$

has a unique solution $\{x_j, j = 1, \ldots, M\}$ *if and only if the determinant* $\det[A_{ij}]$ *of its coefficients is non-zero.*

Theorem 1.2. *A homogeneous system of real linear equations*

$$\sum_{j=1}^{M} A_{ij} x_j = 0 \qquad i = 1, \ldots, M$$

has non-zero solutions, only if the determinant $\det[A_{ij}]$ *of its coefficients is zero. In this case, the solution is not unique, as any scalar multiple of it again is a solution.*

1.1 Repetitions from Vector Algebra

The following concepts should be familiar to the reader.

Definition 1.3. A real **vector space** or a **linear space** $\mathscr{V}$ is a set (of vectors) endowed with an addition

$$"+": \mathscr{V} \times \mathscr{V} \rightarrow \mathscr{V} \quad | \quad (\mathbf{a}, \mathbf{b}) \mapsto \mathbf{a} + \mathbf{b}$$

and a multiplication by a scalar

$$"\ ": \mathscr{R} \times \mathscr{V} \rightarrow \mathscr{V} \quad | \quad (\alpha, \mathbf{a}) \mapsto \alpha\,\mathbf{a}$$

with the following axioms

(commutative)	$\mathbf{a} + \mathbf{b} = \mathbf{b} + \mathbf{a}$	$\forall\, \mathbf{a}, \mathbf{b} \in \mathscr{V}$
(associative)	$(\mathbf{a} + \mathbf{b}) + \mathbf{c} = \mathbf{a} + (\mathbf{b} + \mathbf{c})$	$\forall\, \mathbf{a}, \mathbf{b}, \mathbf{c} \in \mathscr{V}$
(zero element)	$\mathbf{a} + \mathbf{o} = \mathbf{a}$	$\forall\, \mathbf{a} \in \mathscr{V}$
(negative element)	$\mathbf{a} + (-\mathbf{a}) = \mathbf{o}$	$\forall\, \mathbf{a} \in \mathscr{V}$
(associative)	$(\alpha\,\beta)\,\mathbf{a} = \alpha\,(\beta\,\mathbf{a})$	$\forall\, \alpha, \beta \in \mathscr{R}, \forall\, \mathbf{a} \in \mathscr{V}$
(unit element)	$1\,\mathbf{a} = \mathbf{a}$	$\forall\, \mathbf{a} \in \mathscr{V}$
(distributive)	$\alpha\,(\mathbf{a} + \mathbf{b}) = \alpha\,\mathbf{a} + \alpha\,\mathbf{b}$	$\forall\, \alpha \in \mathscr{R}, \forall\, \mathbf{a}, \mathbf{b} \in \mathscr{V}$
(distributive)	$(\alpha + \beta)\,\mathbf{a} = \alpha\,\mathbf{a} + \beta\,\mathbf{a}$	$\forall\, \alpha, \beta \in \mathscr{R}, \forall\, \mathbf{a} \in \mathscr{V}$

From this structure, concepts like linear (in-)dependence, vector basis, dimension of a vector space, etc. can be deduced in the usual way. In the sequel we will always consider real, three-dimensional vector spaces, if not otherwise stated.

Definition 1.4. Let $\mathscr{V}$ and $\mathscr{W}$ be vector spaces and f a mapping

$$f: \mathscr{V} \rightarrow \mathscr{W}.$$

f is called **linear** if it is compatible with the two linear structures, *i.e.*,

$$f(\mathbf{v} + \mathbf{w}) = f(\mathbf{v}) + f(\mathbf{w}) \qquad \forall\, \mathbf{v}, \mathbf{w} \in \mathscr{V}$$

$$f(\alpha\,\mathbf{v}) = \alpha f(\mathbf{v}) \qquad \forall\, \alpha \in \mathscr{R}, \forall\, \mathbf{v} \in \mathscr{V}$$

Note that by an abuse of notation, the $+$ on the left side stands for the addition on $\mathscr{V}$ and the $+$ on the right side for that on $\mathscr{W}$ without introducing a second symbol.

The composition of two linear functions is again linear. However, the reverse statement does not hold. If the composition of two functions is linear, then we cannot conclude that these two functions are necessarily linear.

Some vector spaces are additionally endowed with another operation.

Definition 1.5. An **inner product** or a **scalar product** is a mapping

$$" \cdot ": \mathscr{V} \times \mathscr{V} \to \mathscr{R} \quad | \quad (\mathbf{a}, \mathbf{b}) \mapsto \mathbf{a} \cdot \mathbf{b}$$

for which the following rules hold

(commutative)	$\mathbf{a} \cdot \mathbf{b} = \mathbf{b} \cdot \mathbf{a}$	$\forall\, \mathbf{a}, \mathbf{b} \in \mathscr{V}$
(associative)	$(\alpha\, \mathbf{a}) \cdot \mathbf{b} = \alpha(\mathbf{a} \cdot \mathbf{b})$	$\forall\, \alpha \in \mathscr{R}, \forall\, \mathbf{a}, \mathbf{b} \in \mathscr{V}$
(distributive)	$(\mathbf{a} + \mathbf{b}) \cdot \mathbf{c} = (\mathbf{a} \cdot \mathbf{c}) + (\mathbf{b} \cdot \mathbf{c})$	$\forall\, \mathbf{a}, \mathbf{b}, \mathbf{c} \in \mathscr{V}$
(positive-definite)	$\mathbf{a} \cdot \mathbf{a} > 0$	$\forall\, \mathbf{a} \neq \mathbf{o} \in \mathscr{V}$

A vector space with an inner product is called an **inner-product space**. Of course, there are examples for vector spaces with and without (canonical) inner products.

An inner product induces the **norm** or the **length** of a vector

$$|\mathbf{v}| := (\mathbf{v} \cdot \mathbf{v})^{1/2},$$

the **angle** between two non-zero vectors $\mathbf{u}$ and $\mathbf{v}$

$$\angle(\mathbf{u}, \mathbf{v}) := \arccos \frac{\mathbf{u} \cdot \mathbf{v}}{|\mathbf{u}||\mathbf{v}|}$$

and the **distance** between $\mathbf{u}$ and $\mathbf{v}$

$$d(\mathbf{u}, \mathbf{v}) := |\mathbf{u} - \mathbf{v}|.$$

Accordingly, an inner-product space is also a normed space, a metric space, and a EUCLIDean[1] space, in which we can perform EUCLIDean geometry.

For 3-dimensional inner product spaces, a **cross product** or a **vector product** is a mapping

$$" \times ": \mathscr{V} \times \mathscr{V} \to \mathscr{V} \quad | \quad (\mathbf{a}, \mathbf{b}) \mapsto \mathbf{a} \times \mathbf{b}$$

which is *bilinear, i.e.*, linear in both arguments

$$(\alpha\, \mathbf{a} + \mathbf{b}) \times \mathbf{c} = \alpha\,(\mathbf{a} \times \mathbf{c}) + \mathbf{b} \times \mathbf{c}$$

$$\mathbf{a} \times (\alpha\, \mathbf{b} + \mathbf{c}) = \alpha\,(\mathbf{a} \times \mathbf{b}) + \mathbf{a} \times \mathbf{c} \qquad \forall\, \alpha \in \mathscr{R}, \forall\, \mathbf{a}, \mathbf{b}, \mathbf{c} \in \mathscr{V}$$

and *anticommutative* or *skew*

$$\mathbf{a} \times \mathbf{b} = -\,\mathbf{b} \times \mathbf{a} \qquad \forall\, \mathbf{a}, \mathbf{b} \in \mathscr{V}.$$

[1] Euclid (approx. 320-260 a. c.)

These rules determine the cross product only up to its sign. If we, moreover, define an **orientation** for such a space, the vector product becomes unique. This is normally done by declaring the orientation of three vectors as positive, if the right-hand rule can be applied.

In the sequel $\mathscr{V}$ denotes a real three-dimensional inner-product space with orientation.

The cross product has the following properties

- $\mathbf{b} \times \mathbf{c} = \mathbf{o} \quad \Leftrightarrow \quad \mathbf{b}, \mathbf{c}$ are linearly dependent (parallel) $\qquad \mathbf{b}, \mathbf{c} \in \mathscr{V}$
- $(\mathbf{b} \times \mathbf{c}) \cdot (\mathbf{b} \times \mathbf{c}) = (\mathbf{b} \cdot \mathbf{b})(\mathbf{c} \cdot \mathbf{c}) - (\mathbf{b} \cdot \mathbf{c})^2 \qquad \forall\, \mathbf{a}, \mathbf{b}, \mathbf{c} \in \mathscr{V}.$

For the double cross product we have the rule

(1.1) $$\mathbf{a} \times (\mathbf{b} \times \mathbf{c}) = \mathbf{b}\,(\mathbf{a} \cdot \mathbf{c}) - \mathbf{c}\,(\mathbf{a} \cdot \mathbf{b}) \qquad \forall\, \mathbf{a}, \mathbf{b}, \mathbf{c} \in \mathscr{V}.$$

By means of the vector and the scalar product we define the **triple product**

$$\text{"}[\cdot, \cdot, \cdot]\text{"} : \mathscr{V} \times \mathscr{V} \times \mathscr{V} \to \mathscr{R} \quad | \quad (\mathbf{a}, \mathbf{b}, \mathbf{c}) \mapsto [\mathbf{a}, \mathbf{b}, \mathbf{c}]$$

by

$$[\mathbf{a}, \mathbf{b}, \mathbf{c}] := \mathbf{a} \cdot (\mathbf{b} \times \mathbf{c}) = (\mathbf{a} \times \mathbf{b}) \cdot \mathbf{c}\,.$$

It has the following properties for all $\mathbf{a}, \mathbf{b}, \mathbf{c} \in \mathscr{V}$

- trilinear: linear in the three arguments
- $[\mathbf{a}, \mathbf{b}, \mathbf{c}] = [\mathbf{b}, \mathbf{c}, \mathbf{a}] = [\mathbf{c}, \mathbf{a}, \mathbf{b}]$ anticommutative
  $= -[\mathbf{b}, \mathbf{a}, \mathbf{c}] = -[\mathbf{c}, \mathbf{b}, \mathbf{a}] = -[\mathbf{a}, \mathbf{c}, \mathbf{b}]$
- $[\mathbf{a}, \mathbf{b}, \mathbf{c}] = 0 \quad \Leftrightarrow \quad \mathbf{a}, \mathbf{b}, \mathbf{c}$ are linearly dependent.

A linear functional

$$\mathbf{v}^* : \mathscr{V} \to \mathscr{R}$$

on a vector space $\mathscr{V}$ is called a **covector** or a **dual vector**. If $\mathscr{V}$ is a finite dimensional vector space, then the set of all covectors forms the dual space $\mathscr{V}^*$ of $\mathscr{V}$ and has a natural linear structure of the same dimension as $\mathscr{V}$. If, additionally, $\mathscr{V}$ is endowed with an inner product, then there exists a natural identification of $\mathscr{V}$ with its dual (theorem of RIESZ[2]), and it is no longer necessary to distinguish between vectors and covectors. By such merging, however, the character of physical quantities as vectors or as covectors gets lost. And the same happens as a consequence on all levels of tensor spaces over $\mathscr{V}$. In BERTRAM (1989) this identification was not performed, so that the reader can compare the two approaches.

---

[2] Friedrich Riesz (1880-1956)

Representations of vectors by **components** $\mathbf{v}$, $\mathbf{w} \in \mathscr{V}$ with respect to a vector basis $\{\mathbf{g}_i\} = \{\mathbf{g}_1, \mathbf{g}_2, \mathbf{g}_3\} \subset \mathscr{V}$ are linear combinations

$$\mathbf{v} = v^i\, \mathbf{g}_i \qquad \text{and} \qquad \mathbf{w} = w^i\, \mathbf{g}_i$$

(sum convention[3]). With this we get

$$\mathbf{v} + \mathbf{w} = (v^i + w^i)\, \mathbf{g}_i$$

$$\alpha\, \mathbf{v} = (\alpha\, v^i)\, \mathbf{g}_i\,.$$

It is often advantageous to use **dual** or **reciprocal bases** $\{\mathbf{g}_i\}$, $\{\mathbf{g}^i\} \subset \mathscr{V}$ defined by the duality relation

$$\mathbf{g}_i \cdot \mathbf{g}^j = \delta_i^j \qquad i, j = 1, 2, 3$$

with the KRONECKER symbol

$$\delta_i^j := \begin{cases} 1 & \text{if } i = j \\ 0 & \text{if } i \neq j \end{cases}$$

For each vector basis there is a unique dual basis, which results from Theorem 1.1. The dual basis of the dual is again the original basis. Therefore, dual bases have equal rights. We can represent every vector with respect to both of them

$$\mathbf{v} = v^i\, \mathbf{g}_i = v_i\, \mathbf{g}^i$$

with

(1.2) $$v^i = \mathbf{v} \cdot \mathbf{g}^i = v_k\, \mathbf{g}^k \cdot \mathbf{g}^i = v_k\, G^{ki} \qquad i = 1, 2, 3$$

and

$$v_i = \mathbf{v} \cdot \mathbf{g}_i = v^k\, \mathbf{g}_k \cdot \mathbf{g}_i = v^k\, G_{ki} \qquad i = 1, 2, 3$$

with the **metric coefficients** of these bases

(1.3) $$G_{ik} := \mathbf{g}_i \cdot \mathbf{g}_k$$

$$G^{ik} := \mathbf{g}^i \cdot \mathbf{g}^k \qquad i, k = 1, 2, 3.$$

These are symmetric since the scalar products commute

$$G^{ik} = G^{ki} \qquad G_{ik} = G_{ki}\,.$$

We should note, however, that every vector has a natural representation, in which we should better leave it.

The advantage of the use of dual bases becomes clear when using scalar products

$$\mathbf{v} \cdot \mathbf{w} = v^i\, \mathbf{g}_i \cdot w_k\, \mathbf{g}^k = v^i\, w_k\, \delta_i^k = v^i\, w_i\,.$$

---

[3] If a product term contains an index twice, once superimposed and once as a subscript, then it has to be summed from 1 to N, where the latter is usually 3, the dimension of $\mathscr{V}$.

If we use the representation of both vectors with respect to the same basis, we obtain the longer result

$$\mathbf{v} \cdot \mathbf{w} = v^i\, \mathbf{g}_i \cdot w^k\, \mathbf{g}_k = v^i\, w^k\, G_{ik}$$

or

$$= v_i\, \mathbf{g}^i \cdot w_k\, \mathbf{g}^k = v_i\, w_k\, G^{ik}.$$

For the cross product we obtain the representations

$$\mathbf{a} \times \mathbf{b} = \pm\sqrt{det[G_{mn}]}\ \varepsilon_{ijk}\, a^i\, b^j\, \mathbf{g}^k = \pm\sqrt{det[G^{mn}]}\ \varepsilon^{ijk}\, a_i\, b_j\, \mathbf{g}_k$$

with $det[G_{mn}]$ denoting the determinant of the matrix $[G_{mn}]$ with elements G_{mn} and with the **permutation symbol** or the **LEVI-CIVITA**[4]**-symbol**

$$\varepsilon_{ijk} = \varepsilon^{ijk} = \begin{cases} 1 & \text{if } ijk \text{ is an even permutation of } 1,\ 2,\ 3 \\ -1 & \text{if } ijk \text{ is an odd permutation of } 1,\ 2,\ 3 \\ 0 & \text{if } ijk \text{ is no permutation of } 1,\ 2,\ 3 \end{cases}$$

The sign of this product depends on the orientation of the basis. If it is positively oriented, then the $+$ holds here.

For the triple product we obtain the component form

$$[\mathbf{a}\,,\, \mathbf{b}\,,\, \mathbf{c}] = \pm\sqrt{det[G_{mn}]}\ \varepsilon_{ijk}\, a^i\, b^j\, c^k = \pm\sqrt{det[G^{mn}]}\ \varepsilon^{ijk}\, a_i\, b_j\, c_k\,.$$

A positively oriented basis $\{\mathbf{e}_i\} \subset \mathscr{V}$ is called **orthonormal**, if

$$\mathbf{e}_i \cdot \mathbf{e}_k = \delta_{ik}$$

holds. Such an orthonormal basis (ONB) coincides with its dual. With respect to an ONB we obtain

$$\mathbf{v} \cdot \mathbf{w} = v^i\, w^i$$

$$\mathbf{v} \times \mathbf{w} = \varepsilon_{pqr}\, v^p\, w^q\, \mathbf{e}_r$$

$$[\mathbf{u}\,,\, \mathbf{v}\,,\, \mathbf{w}] = \varepsilon_{pqr}\, u^p\, v^q\, w^r.$$

By the rule of the double cross product (1.1) we conclude that

$$\varepsilon_{jkl}\, \varepsilon_{ilm} = \delta_{ik}\, \delta_{jm} - \delta_{ij}\, \delta_{km}$$

with respect to an ONB.

---

[4] Tullio Levi-Civita (1873-1941)

1.2 Tensor Algebra

A Comment on the Literature. The following books give introductions to tensor algebra: AKIVIS/ GOLDBERG (2003), BISHOP/ GOLDBERG (1968), DE BOER (1982), BOWEN/ WANG (1976), BRILLOUIN (1964), DIMITRIENKO (2002), GONZALEZ/ STUART (2008), HALMOS (1974), ITSKOV (2007), LEBEDEV/ CLOUD (2003), NOLL (1987), RUIZ-TOLOSA/ CASTILLO (2005), SCHADE (1997), SOKOLNIKOFF (1951, 1964), TROSTEL (1993, 1997). With respect to matrices we recommend LÜTKEPOHL (1996). In many textbooks on continuum mechanics one finds at the beginning a short introduction to tensors, like in those by BERTRAM (1989), BILLINGTON (1986), CHADWICK (1976), GURTIN (1981), IRGENS (2008), LEIGH (1968), NARASIMHAN (1993), OGDEN (1984a), SMITH (1993), TABER (2004).

Tensors

In the sequel we give a brief introduction to tensor calculus.

Definition 1.6. Let $\mathscr{V}$ be a vector space. A linear mapping

$$\mathbf{A} : \mathscr{V} \rightarrow \mathscr{V}$$

is a (*2*nd-order) **tensor** on $\mathscr{V}$.

Consequently, for each tensor $\mathbf{A}$ we have

$$\mathbf{A}(\mathbf{v} + \alpha\,\mathbf{w}) = \mathbf{A}(\mathbf{v}) + \alpha\,\mathbf{A}(\mathbf{w}) \qquad \forall\, \alpha \in \mathscr{R}, \mathbf{v}, \mathbf{w} \in \mathscr{V}.$$

Usually one drops the brackets of the argument and writes only

$$\mathbf{A}\,\mathbf{v} := \mathbf{A}(\mathbf{v}).$$

The set of all tensors on $\mathscr{V}$ will be denoted by $\mathscr{Lin}$.

The product of two tensors is introduced as the composition of two linear mappings

$$(\mathbf{A}\,\mathbf{B})\,\mathbf{v} := \mathbf{A}\,(\mathbf{B}\,\mathbf{v}) \qquad \forall\, \mathbf{A}, \mathbf{B} \in \mathscr{Lin}, \mathbf{v} \in \mathscr{V}.$$

Thus, $(\mathbf{A}\,\mathbf{B}) \in \mathscr{Lin}$. This product does not commute, but is associative

$$(\mathbf{A}\,\mathbf{B})\,\mathbf{C} = \mathbf{A}\,(\mathbf{B}\,\mathbf{C}) \qquad \forall\, \mathbf{A}, \mathbf{B}, \mathbf{C} \in \mathscr{Lin}.$$

Therefore, the brackets are not needed and further on dropped. We will write multiple products of the same tensor as powers, like $\mathbf{A}^2 := \mathbf{A}\,\mathbf{A}$.

We can push the linear structure of $\mathscr{V}$ forward to $\mathscr{Lin}$ by defining the linear operations on the values

$$(\mathbf{A}+\mathbf{B})\,\mathbf{v} \;:=\; \mathbf{A}\,\mathbf{v}+\mathbf{B}\,\mathbf{v} \qquad \forall\ \mathbf{A},\mathbf{B}\in \mathcal{Lin},\ \mathbf{v}\in\mathcal{V}$$

$$(\alpha\,\mathbf{A})\,\mathbf{v} \;:=\; \alpha\,(\mathbf{A}\,\mathbf{v}) \qquad \forall\ \mathbf{A}\in \mathcal{Lin},\ \mathbf{v}\in\mathcal{V},\ \alpha\in\mathcal{R}\,.$$

By these operations $\mathcal{Lin}$ also becomes a linear space.

A distinguished element of $\mathcal{Lin}$ is the **zero tensor** $\mathbf{0}$, which maps all vectors onto the zero vector

$$\mathbf{0}\,\mathbf{v} = \mathbf{o} \qquad \forall\,\mathbf{v}\in\mathcal{V},$$

and the *2*nd-order **identity tensor** $\mathbf{I}$, which maps each vector onto itself

$$\mathbf{I}\,\mathbf{v} = \mathbf{v} \qquad \forall\,\mathbf{v}\in\mathcal{V}.$$

A scalar multiple of the identity tensor $\alpha\,\mathbf{I}$ with $\alpha\in\mathcal{R}$ is called a **spherical tensor**.

The following rules hold $\forall\ \alpha\in\mathcal{R},\ \forall\ \mathbf{A},\mathbf{B},\mathbf{C}\in\mathcal{Lin}$:

$\mathbf{A}+\mathbf{B} = \mathbf{B}+\mathbf{A}$	(commutative)
$\alpha\,(\mathbf{A}\,\mathbf{B}) = (\alpha\,\mathbf{A})\,\mathbf{B} = \mathbf{A}\,(\alpha\,\mathbf{B})$	(associative)
$\mathbf{A}\,(\mathbf{B}+\mathbf{C}) = \mathbf{A}\,\mathbf{B}+\mathbf{A}\,\mathbf{C}$	(distributive)
$(\mathbf{A}+\mathbf{B})\,\mathbf{C} = \mathbf{A}\,\mathbf{C}+\mathbf{B}\,\mathbf{C}$	(distributive)
$\mathbf{A}\,\mathbf{0} = \mathbf{0}\,\mathbf{A} = \mathbf{0}$	(zero element)
$\mathbf{A}+\mathbf{0} = \mathbf{A}$	(neutral element of addition)
$\mathbf{A}\,\mathbf{I} = \mathbf{I}\,\mathbf{A} = \mathbf{A}$	(neutral element of composition)

If a tensor $\mathbf{A}$ is invertible (as a mapping), its **inverse tensor** $\mathbf{A}^{-1}$ gives

$$\mathbf{A}^{-1}\,(\mathbf{A}\,\mathbf{v}) = \mathbf{v} \qquad \forall\,\mathbf{v}\in\mathcal{V},$$

so that

$$\mathbf{A}^{-1}\,\mathbf{A} = \mathbf{A}\,\mathbf{A}^{-1} = \mathbf{I}$$

and

$$(\mathbf{A}^{-1})^{-1} = \mathbf{A}\,.$$

In particular the identity tensor is invertible and coincides with its inverse

$$\mathbf{I}^{-1} = \mathbf{I}\,.$$

A spherical tensor $\alpha\,\mathbf{I}$ is invertible if and only if $\alpha \neq 0$, as

$$(\alpha\,\mathbf{I})^{-1} = \alpha^{-1}\,\mathbf{I}\,.$$

The set of all invertible tensors on $\mathcal{V}$ is denoted by $\mathcal{Inv}$, the **general linear group**. We then have for the products of two tensors $\mathbf{A}$ and $\mathbf{B}$

$$\mathbf{A},\mathbf{B}\in\mathcal{Inv} \quad\Leftrightarrow\quad \mathbf{A}\,\mathbf{B}\in\mathcal{Inv}$$

and for this case

$$(\mathbf{A}\,\mathbf{B})^{-1} \;=\; \mathbf{B}^{-1}\,\mathbf{A}^{-1}.$$

We used the concept *group* in the following sense.

Definition 1.7. A set of invertible mappings $\mathcal{G}$ is called a **group** (under composition) if it fulfils the following axioms:

(*G1*) $\mathbf{A}_1\,, \mathbf{A}_2 \in \mathcal{G} \;\Rightarrow\; (\mathbf{A}_1\,\mathbf{A}_2) \in \mathcal{G}$ (closed under composition)

(*G2*) $\mathbf{A}_1\,, \mathbf{A}_2\,, \mathbf{A}_3 \in \mathcal{G} \;\Rightarrow\; (\mathbf{A}_1\,\mathbf{A}_2)\,\mathbf{A}_3 = \mathbf{A}_1\,(\mathbf{A}_2\,\mathbf{A}_3)$ (associative)

(*G3*) $\mathbf{I} \in \mathcal{G}$ (neutral element)

(*G4*) $\mathbf{A} \in \mathcal{G} \;\Rightarrow\; \mathbf{A}^{-1} \in \mathcal{G}$ (inverse element)

It is easy to verify that $\mathcal{Inv}$ forms a group.

Transposition of Tensors

In the following definition we make use of the inner product on $\mathcal{V}$. For every tensor $\mathbf{A} \in \mathcal{Lin}$ there exists uniquely a **transposed** tensor $\mathbf{A}^T \in \mathcal{Lin}$, such that

(1.4) $$\mathbf{v} \cdot (\mathbf{A}^T\,\mathbf{w}) \;=\; \mathbf{w} \cdot (\mathbf{A}\,\mathbf{v}) \qquad \forall\, \mathbf{v}\,, \mathbf{w} \in \mathcal{V}$$

holds. The following rules are valid.

$$(\mathbf{A}\,\mathbf{B})^T \;=\; \mathbf{B}^T\,\mathbf{A}^T \qquad \forall\, \mathbf{A}\,, \mathbf{B} \in \mathcal{Lin}$$

$$(\mathbf{A} + \alpha\,\mathbf{B})^T \;=\; \mathbf{A}^T + \alpha\,(\mathbf{B}^T) \qquad \forall\, \mathbf{A}\,, \mathbf{B} \in \mathcal{Lin}\,, \alpha \in \mathcal{R}$$

$$(\mathbf{A}^{-1})^T \;=\; (\mathbf{A}^T)^{-1} \;=:\; \mathbf{A}^{-T} \qquad \forall\, \mathbf{A} \in \mathcal{Inv}\,.$$

A tensor is called **symmetric** (or *self-adjoint*), if it coincides with its transposed tensor, and **anti(sym)metric** or **skew**, if it coincides with the negative of its transposed tensor. We write $\mathcal{Sym}$ for the set of all symmetric tensors, and $\mathcal{Skw}$ for the skew ones. Examples are $\mathbf{I}\,, \mathbf{0} \in \mathcal{Sym}$ and $\mathbf{0} \in \mathcal{Skw}$. The sets $\mathcal{Sym}$ and $\mathcal{Skw}$ form subspaces of $\mathcal{Lin}$, *i.e.*, they are closed under linear combinations

$$\mathbf{A}\,, \mathbf{B} \in \mathcal{Sym} \quad\Rightarrow\quad \mathbf{A} + \alpha\,\mathbf{B} \in \mathcal{Sym} \qquad \forall \alpha \in \mathcal{R}$$

$$\mathbf{A}\,, \mathbf{B} \in \mathcal{Skw} \quad\Rightarrow\quad \mathbf{A} + \alpha\,\mathbf{B} \in \mathcal{Skw} \qquad \forall \alpha \in \mathcal{R}.$$

The intersection of these subspaces consists only of the zero tensor

$$\mathcal{Sym} \cap \mathcal{Skw} \;=\; \{\mathbf{0}\}\,.$$

One can additively decompose each tensor $\mathbf{A} \in \mathcal{Lin}$ uniquely into its symmetric part $sym(\mathbf{A})$ and its skew part $skw(\mathbf{A})$

(1.5) $$\mathbf{A} \;=\; sym(\mathbf{A}) + skw(\mathbf{A})$$

with $\quad sym(\mathbf{A}) := \frac{1}{2}(\mathbf{A} + \mathbf{A}^T) \quad \in \mathcal{Sym}$

and $\quad skw(\mathbf{A}) := \frac{1}{2}(\mathbf{A} - \mathbf{A}^T) \quad \in \mathcal{Skw}$.

Principal Invariants of Tensors

Let $\{\mathbf{f}, \mathbf{g}, \mathbf{h}\}$ and $\{\mathbf{l}, \mathbf{m}, \mathbf{n}\}$ be two arbitrary bases in $\mathcal{V}$, and let $\mathbf{A} \in \mathcal{Lin}$. Due to the linearity of the triple product in all arguments, the following equalities hold[5]

$$\frac{[\mathbf{Af},\mathbf{g},\mathbf{h}] + [\mathbf{f},\mathbf{Ag},\mathbf{h}] + [\mathbf{f},\mathbf{g},\mathbf{Ah}]}{[\mathbf{f},\mathbf{g},\mathbf{h}]} = \frac{[\mathbf{Al},\mathbf{m},\mathbf{n}] + [\mathbf{l},\mathbf{Am},\mathbf{n}] + [\mathbf{l},\mathbf{m},\mathbf{An}]}{[\mathbf{l},\mathbf{m},\mathbf{n}]}$$

$$\frac{[\mathbf{Af},\mathbf{Ag},\mathbf{h}] + [\mathbf{f},\mathbf{Ag},\mathbf{Ah}] + [\mathbf{Af},\mathbf{g},\mathbf{Ah}]}{[\mathbf{f},\mathbf{g},\mathbf{h}]} = \frac{[\mathbf{Al},\mathbf{Am},\mathbf{n}] + [\mathbf{l},\mathbf{Am},\mathbf{An}] + [\mathbf{Al},\mathbf{m},\mathbf{An}]}{[\mathbf{l},\mathbf{m},\mathbf{n}]}$$

$$\frac{[\mathbf{Af},\mathbf{Ag},\mathbf{Ah}]}{[\mathbf{f},\mathbf{g},\mathbf{h}]} = \frac{[\mathbf{Al},\mathbf{Am},\mathbf{An}]}{[\mathbf{l},\mathbf{m},\mathbf{n}]} .$$

Therefore, for each tensor $\mathbf{A}$ there exist three real numbers $I_\mathbf{A}$, $II_\mathbf{A}$, $III_\mathbf{A}$ independent of the chosen basis $\{\mathbf{a}, \mathbf{b}, \mathbf{c}\}$, called the **principal invariants** of $\mathbf{A}$, such that

(1.6)
$$I_\mathbf{A} := \frac{[\mathbf{Aa},\mathbf{b},\mathbf{c}] + [\mathbf{a},\mathbf{Ab},\mathbf{c}] + [\mathbf{a},\mathbf{b},\mathbf{Ac}]}{[\mathbf{a},\mathbf{b},\mathbf{c}]}$$

$$II_\mathbf{A} := \frac{[\mathbf{a},\mathbf{Ab},\mathbf{Ac}] + [\mathbf{Aa},\mathbf{b},\mathbf{Ac}] + [\mathbf{Aa},\mathbf{Ab},\mathbf{c}]}{[\mathbf{a},\mathbf{b},\mathbf{c}]}$$

$$III_\mathbf{A} := \frac{[\mathbf{Aa},\mathbf{Ab},\mathbf{Ac}]}{[\mathbf{a},\mathbf{b},\mathbf{c}]} .$$

In other words, we have three mappings

[5] A detailed proof of this statement can be found in BILLINGTON (1986) p. 17 ff and in SMITH (1993) p. 23 ff.

$$I_{\mathbf{A}},\ II_{\mathbf{A}},\ III_{\mathbf{A}} : \mathcal{Lin} \to \mathcal{R}$$

which assign to each tensor its three principal invariants (where we did not distinguish between functions and values in the notation).

The first and the third invariant have names:

$I_{\mathbf{A}} = tr(\mathbf{A})$ **trace** of a tensor $\mathbf{A}$

$III_{\mathbf{A}} = det(\mathbf{A})$ **determinant** of a tensor $\mathbf{A}$.

The trace is a linear mapping

$$tr(\mathbf{A} + \alpha\,\mathbf{B}) = tr(\mathbf{A}) + \alpha\, tr(\mathbf{B}) \qquad \forall\, \mathbf{A}\,,\mathbf{B} \in \mathcal{Lin}\,,\ \alpha \in \mathcal{R}.$$

It obeys the following rules.

$$tr(\mathbf{A}) = tr(\mathbf{A}^T) = tr(sym(\mathbf{A})) \qquad \forall\, \mathbf{A} \in \mathcal{Lin}$$

$$tr(skw(\mathbf{A})) = 0 \qquad \forall\, \mathbf{A} \in \mathcal{Lin}$$

$$tr(\mathbf{A\,B}) = tr(\mathbf{B\,A}) \qquad \forall\, \mathbf{A}\,,\mathbf{B} \in \mathcal{Lin}$$

$$tr(\mathbf{0}) = 0$$

$$tr(\alpha\,\mathbf{I}) = 3\alpha \quad \text{in three dimensions} \qquad \forall\, \alpha \in \mathcal{R}.$$

Traceless tensors are called **deviators**. We can uniquely decompose every tensor

(1.7) $$\mathbf{A} = \mathbf{A}' + \mathbf{A}^{\circ}$$

into its deviatoric part $\mathbf{A}' := \mathbf{A} - {}^{1}/_{3}\, tr(\mathbf{A})\,\mathbf{I}$

and its spherical part $\mathbf{A}^{\circ} := {}^{1}/_{3}\, tr(\mathbf{A})\,\mathbf{I}$.

The trace induces a **scalar product** on $\mathcal{Lin}$ by

(1.8) $$\mathbf{A} \cdot \mathbf{B} := tr(\mathbf{A}^T\,\mathbf{B}) = tr(\mathbf{A}\,\mathbf{B}^T)$$

such that the tensor space becomes an inner product space. This inner product depends on the inner product of $\mathcal{V}$. We will denote all inner products by the same dot " · " regardless of the specific linear space under consideration. By this inner product of tensors, we can introduce the **value**, **length**, or **norm** of a tensor $\mathbf{A}$ by

$$|\,\mathbf{A}\,| := (\mathbf{A} \cdot \mathbf{A})^{1/2},$$

and the **distance** between two tensors by

$$d(\mathbf{A}\,,\mathbf{B}) := |\,\mathbf{A} - \mathbf{B}\,| = |\,\mathbf{B} - \mathbf{A}\,|$$

as well as the **angle** between two (non-zero) tensors $\mathbf{A}$ and $\mathbf{B}$ by

$$\angle(\mathbf{A}\,,\mathbf{B}) := arccos\frac{\mathbf{A}\cdot\mathbf{B}}{|\mathbf{A}|\,|\mathbf{B}|}\,.$$

For three tensors $\mathbf{A}\,,\mathbf{B}\,,\mathbf{C} \in \mathcal{Lin}$ we obtain by analogy to (1.4)

(1.9) $$\mathbf{A} \cdot (\mathbf{B\,C}) = (\mathbf{A\,C}^T) \cdot \mathbf{B} = (\mathbf{B}^T \mathbf{A}) \cdot \mathbf{C}\,.$$

Moreover, we get

$$\mathbf{A} \cdot \mathbf{B} = \mathbf{A}^T \cdot \mathbf{B}^T$$

and

$$\mathbf{A} \cdot \mathbf{I} = \mathbf{I} \cdot \mathbf{A} = tr(\mathbf{A})\,.$$

If the scalar product between two tensors is zero, then they are orthogonal (or perpendicular). In this sense, the symmetric and the skew tensors are mutually orthogonal

$$sym(\mathbf{A}) \cdot skw(\mathbf{B}) = 0 \qquad \forall\, \mathbf{A}, \mathbf{B} \in \mathcal{Lin}\,.$$

Also, the deviatoric tensors and the spherical tensors are orthogonal

$$\mathbf{A}' \cdot \mathbf{B}^\circ = 0 \qquad \forall\, \mathbf{A}, \mathbf{B} \in \mathcal{Lin}\,.$$

The determinant is not linear, but multiplicative and commutative

$$det(\mathbf{B\,A}) = det(\mathbf{A\,B}) = det(\mathbf{A})\, det(\mathbf{B})$$

(1.10) $$det(\mathbf{A}) = det(\mathbf{A}^T)$$

and in three dimensions

(1.11) $$det(\alpha\, \mathbf{A}) = \alpha^3\, det(\mathbf{A}) \qquad \forall\, \mathbf{A}, \mathbf{B} \in \mathcal{Lin}\,,\ \alpha \in \mathcal{R}\,.$$

The determinant of a tensor gives a simple criterion for the invertibility, which is a consequence of Theorem 1.1.

Theorem 1.8. *A tensor is invertible, if and only if its determinant is non-zero.*

This is closely related to the following statement.

Theorem 1.9. *Let* $\{\mathbf{g}_i\}$ *be a basis of* $\mathcal{V}$. *A tensor* $\mathbf{F}$ *is invertible, if and only if* $\{\mathbf{F\,g}_i\}$ *is also a basis of* $\mathcal{V}$.

The other principal invariants can be also expressed by traces

(1.12) $$II_\mathbf{A} = {}^1\!/_2\, \{tr^2(\mathbf{A}) - tr(\mathbf{A}^2)\}$$
$$= {}^1\!/_2\, \{(\mathbf{I} \cdot \mathbf{A})^2 - \mathbf{I} \cdot \mathbf{A}^2\}$$

(1.13) $$III_\mathbf{A} = {}^1\!/_6\, tr^3(\mathbf{A}) - {}^1\!/_2\, tr(\mathbf{A})\, tr(\mathbf{A}^2) + {}^1\!/_3\, tr(\mathbf{A}^3)$$
$$= {}^1\!/_6\, (\mathbf{I} \cdot \mathbf{A})^3 - {}^1\!/_2\, (\mathbf{I} \cdot \mathbf{A})\, (\mathbf{I} \cdot \mathbf{A}^2) + {}^1\!/_3\, \mathbf{I} \cdot \mathbf{A}^3.$$

For an invertible tensor $\mathbf{A}$ we obtain

$$I_{(\mathbf{A}^{-1})} = II_\mathbf{A} \,/\, III_\mathbf{A}$$

$$II_{(\mathbf{A}^{-1})} = I_\mathbf{A} \,/\, III_\mathbf{A}$$

$$III_{(\mathbf{A}^{-1})} = 1 \,/\, III_\mathbf{A}\,.$$

For the deviatoric part $\mathbf{A'}$ of a tensor $\mathbf{A}$ we obtain

$$I_{\mathbf{A'}} = 0$$

$$II_{\mathbf{A'}} = -\,{}^{1}\!/_{2}\, tr(\mathbf{A'}^{2})$$

$$III_{\mathbf{A'}} = {}^{1}\!/_{3}\, tr(\mathbf{A'}^{3})\,.$$

In particular we have for the unit tensor $\mathbf{I}$

$$I_{\mathbf{I}} = II_{\mathbf{I}} = 3 \qquad\qquad III_{\mathbf{I}} = 1$$

and for the zero tensor $\mathbf{0}$

$$I_{\mathbf{0}} = II_{\mathbf{0}} = III_{\mathbf{0}} = 0\,.$$

There are also non-zero tensors with all principal invariants zero.

The invertible tensors form the **general linear group** $\mathscr{Inv}$. The sign of the determinant indicates if the tensor changes the orientation of a basis or keeps it. Therefore, we call a tensor with positive determinant **orientation preserving**. The orientation preserving tensors form the **special linear group** $\mathscr{Inv}^{+}$. Tensors with determinant equal to ± 1 are called **unimodular**. Their set is denoted by $\mathscr{Unim}$, the **general unimodular group**. Tensors with determinant $+1$ form the **special unimodular group** $\mathscr{Unim}^{+}$.

Theorem 1.10. *Every invertible tensor* $\mathbf{F} \in \mathscr{Inv}$ *can be multiplicatively decomposed*

$$\mathbf{F} = \mathbf{F}^{\circ}\,\overline{\mathbf{F}} = \overline{\mathbf{F}}\,\mathbf{F}^{\circ}$$

into

- *its (unique) unimodular part* $\overline{\mathbf{F}} := III_{\mathbf{F}}^{\,-1/3}\,\mathbf{F}$
- *and its (unique) spherical part* $\mathbf{F}^{\circ} := III_{\mathbf{F}}^{\,1/3}\,\mathbf{I}$

1.2.1 Tensor Product and Tensor Components

The **tensor product** or **dyadic product** is a mapping

$$\otimes : \mathscr{V} \times \mathscr{V} \to \mathscr{Lin} \quad | \quad (\mathbf{u}\,,\mathbf{v}) \mapsto \mathbf{u} \otimes \mathbf{v}$$

defined by its action on an arbitrary vector $\mathbf{a} \in \mathscr{V}$

$$\mathbf{u} \otimes \mathbf{v}\,(\mathbf{a}) := (\mathbf{v} \cdot \mathbf{a})\,\mathbf{u}\,.$$

Obviously, this constitutes a linear mapping of vectors into vectors. The **(simple) dyad** $\mathbf{u} \otimes \mathbf{v}$ therefore is a (specific) tensor, which maps every vector in the direction of $\mathbf{u}$. The tensor product is linear in both factors

$$(\mathbf{u} + \alpha\,\mathbf{v}) \otimes \mathbf{w} = \mathbf{u} \otimes \mathbf{w} + \alpha\,(\mathbf{v} \otimes \mathbf{w})$$

$$\mathbf{u} \otimes (\mathbf{v} + \alpha\, \mathbf{w}) = \mathbf{u} \otimes \mathbf{v} + \alpha\, (\mathbf{u} \otimes \mathbf{w})$$

$\forall\, \mathbf{u}, \mathbf{v}, \mathbf{w} \in \mathscr{V}$, $\alpha \in \mathscr{R}$. Moreover, we have

$$(\mathbf{u} \otimes \mathbf{v})^T = \mathbf{v} \otimes \mathbf{u} \,.$$

Such a dyad is symmetric, if and only if $\mathbf{u}$ is parallel to $\mathbf{v}$

$$\mathbf{u} \otimes \mathbf{v} \in \mathscr{Sym} \quad \Leftrightarrow \quad \mathbf{u} \times \mathbf{v} = \mathbf{o} \,.$$

For the composition of a dyad and a tensor we obtain

$$\mathbf{A}\, (\mathbf{u} \otimes \mathbf{v}) = (\mathbf{A}\, \mathbf{u}) \otimes \mathbf{v}$$

$$(\mathbf{u} \otimes \mathbf{v})\, \mathbf{A} = \mathbf{u} \otimes (\mathbf{A}^T \mathbf{v}) \qquad \forall\, \mathbf{u}, \mathbf{v} \in \mathscr{V}, \mathbf{A} \in \mathscr{Lin} \,.$$

The composition of two dyads is related to the inner product of $\mathscr{V}$ by

$$(\mathbf{u} \otimes \mathbf{v})\, (\mathbf{w} \otimes \mathbf{x}) = (\mathbf{v} \cdot \mathbf{w})\, \mathbf{u} \otimes \mathbf{x} \,.$$

By (1.6) we immediately see that the second and third principal invariant of a (simple) dyad is zero. Only the trace

$$tr(\mathbf{u} \otimes \mathbf{v}) = \mathbf{I} \cdot \mathbf{u} \otimes \mathbf{v} = \mathbf{u} \cdot \mathbf{v}$$

can be non-zero. By this equation we see the relation of the inner product on $\mathscr{V}$ and on $\mathscr{Lin}$. For every tensor $\mathbf{A} \in \mathscr{Lin}$ we get

$$tr(\mathbf{A}\, \mathbf{u} \otimes \mathbf{v}) = (\mathbf{A}\, \mathbf{u}) \cdot \mathbf{v} \qquad \forall\, \mathbf{u}, \mathbf{v} \in \mathscr{V}.$$

The dyadic product is not only a useful representation for specific tensors, but also leads to a component representation for all tensors. For that purpose we choose an arbitrary basis $\{\mathbf{g}_i\} \subset \mathscr{V}$ and its dual $\{\mathbf{g}^i\} \subset \mathscr{V}$. Let $\mathbf{A}$ be a tensor. Then $(\mathbf{A}\, \mathbf{g}_j)$ is a vector for $j = 1, 2, 3$, the component representation of which with respect to the same basis is denoted by

$$\mathbf{A}\, \mathbf{g}_j = A^p{}_j\, \mathbf{g}_p \qquad j = 1, 2, 3$$

with components

$$A^p{}_j = \mathbf{g}^p \cdot (\mathbf{A}\, \mathbf{g}_j) = \mathbf{A} \cdot (\mathbf{g}^p \otimes \mathbf{g}_j) \qquad p, j = 1, 2, 3$$

after (1.2). If we know all of them, the tensor $\mathbf{A}$ is uniquely determined. In fact, if we choose an arbitrary vector $\mathbf{v} = v^i\, \mathbf{g}_i$, we obtain

$$\begin{aligned} &(\mathbf{A} - A^p{}_j\, \mathbf{g}_p \otimes \mathbf{g}^j)\, \mathbf{v} \\ &= (\mathbf{A} - A^p{}_j\, \mathbf{g}_p \otimes \mathbf{g}^j)\, v^i\, \mathbf{g}_i \\ &= (\mathbf{A}\, \mathbf{g}_i - A^p{}_j\, \mathbf{g}_p\, \delta_i^{\,j})\, v^i \\ &= (A^p{}_i\, \mathbf{g}_p - A^p{}_i\, \mathbf{g}_p)\, v^i = \mathbf{o} \,. \end{aligned}$$

Therefore, the **component representation of a tensor** is

$$\mathbf{A} = A^p{}_j\, \mathbf{g}_p \otimes \mathbf{g}^j$$

with respect to the **tensor basis** $\{\mathbf{g}_p \otimes \mathbf{g}^j\}$, and its application to a vector can be written as

$$\mathbf{A}\,\mathbf{v} = A^p{}_i\, v^i\, \mathbf{g}_p\,.$$

An analogous reasoning leads to the other component representations

$$\mathbf{A} = A_p{}^j\, \mathbf{g}^p \otimes \mathbf{g}_j = A^{pj}\, \mathbf{g}_p \otimes \mathbf{g}_j = A_{pj}\, \mathbf{g}^p \otimes \mathbf{g}^j$$

with

$$\begin{aligned} A_p{}^j &:= \mathbf{g}_p \cdot (\mathbf{A}\,\mathbf{g}^j) = \mathbf{A} \cdot (\mathbf{g}_p \otimes \mathbf{g}^j) \\ A^{pj} &:= \mathbf{g}^p \cdot (\mathbf{A}\,\mathbf{g}^j) = \mathbf{A} \cdot (\mathbf{g}^p \otimes \mathbf{g}^j) \qquad\qquad p\,,j = 1,\,2,\,3 \\ A_{pj} &:= \mathbf{g}_p \cdot (\mathbf{A}\,\mathbf{g}_j) = \mathbf{A} \cdot (\mathbf{g}_p \otimes \mathbf{g}_j) \end{aligned}$$

and their applications to a vector $\mathbf{v}$ can be calculated as

$$\mathbf{A}\,\mathbf{v} = A_p{}^i\, v_i\, \mathbf{g}^p = A^{pi}\, v_i\, \mathbf{g}_p = A_{pi}\, v^i\, \mathbf{g}^p.$$

Note that the four sets of nine tensor components $A^p{}_j$, $A_p{}^j$, A^{pj}, A_{pj} are generally all different, although they all represent the same tensor $\mathbf{A}$. Often these components are put into a 3×3-matrix, the **matrix of the components**, such as

$$[A_{pj}] := \begin{bmatrix} A_{11} & A_{12} & A_{13} \\ A_{21} & A_{22} & A_{23} \\ A_{31} & A_{32} & A_{33} \end{bmatrix}.$$

It is important to note that such a matrix only determines the tensor, if the used tensor basis is given.

If we want to determine the relation between the components $A^p{}_j$, $A_p{}^j$, A^{pj}, A_{pj}, we firstly verify the following representation of the identity tensor

$$\begin{aligned} \mathbf{I} &= \delta^j{}_q\, \mathbf{g}_j \otimes \mathbf{g}^q = \mathbf{g}_j \otimes \mathbf{g}^j \\ &= \delta_r{}^s\, \mathbf{g}^r \otimes \mathbf{g}_s = \mathbf{g}^r \otimes \mathbf{g}_r \qquad (1.14) \\ &= G_{kp}\, \mathbf{g}^k \otimes \mathbf{g}^p = G^{lm}\, \mathbf{g}_l \otimes \mathbf{g}_m \end{aligned}$$

with the metric coefficients G_{kj} and G^{lm} of these vector bases after (1.3). For an arbitrary vector $\mathbf{v} = v^i\, \mathbf{g}_i \in \mathscr{V}$ we get indeed

$$\begin{aligned} (\mathbf{g}_k \otimes \mathbf{g}^k)\, \mathbf{v} &= (\mathbf{g}_k \otimes \mathbf{g}^k)\,(v^i\, \mathbf{g}_i) = v^i\, \mathbf{g}_k\, (\mathbf{g}^k \cdot \mathbf{g}_i) = v^i\, \mathbf{g}_k\, \delta^k{}_i \\ &= v^i\, \mathbf{g}_i = \mathbf{v} \end{aligned}$$

which proves the first representation. The transpose of it gives the second form for $\mathbf{I}$ being symmetric. The third is shown by

$$\begin{aligned} (G_{kp}\, \mathbf{g}^k \otimes \mathbf{g}^p)\, \mathbf{v} &= (G_{kp}\, \mathbf{g}^k \otimes \mathbf{g}^p)\,(v^i\, \mathbf{g}_i) = G_{kp}\, v^i\, (\mathbf{g}^p \cdot \mathbf{g}_i)\, \mathbf{g}^k \\ &= G_{kp}\, v^i\, \delta^p{}_i\, \mathbf{g}^k = G_{ki}\, v^i\, \mathbf{g}^k = (\mathbf{g}_k \cdot \mathbf{g}_i)\, v^i\, \mathbf{g}^k = \mathbf{g}^k \otimes \mathbf{g}_k (v^i\, \mathbf{g}_i) \\ &= \mathbf{v} \end{aligned}$$

etc. The metric coefficients G_{kj} and G^{lm} are therefore the components of the identity tensor with respect to the tensor bases $\mathbf{g}^k \otimes \mathbf{g}^p$ and $\mathbf{g}_k \otimes \mathbf{g}_p$, respectively. These coefficients are mutually inverse, as

$$G_{ij}\, G^{jk} = (\mathbf{g}_i \cdot \mathbf{g}_j)\,(\mathbf{g}^j \cdot \mathbf{g}^k) = \mathbf{g}_i \cdot \big((\mathbf{g}_j \otimes \mathbf{g}^j)\,\mathbf{g}^k\big) = \mathbf{g}_i \cdot (\mathbf{I}\,\mathbf{g}^k)$$
$$= \mathbf{g}_i \cdot \mathbf{g}^k = \delta_i^{\,k} \qquad i\,,\, k = 1, 2, 3.$$

By substituting these component representations of $\mathbf{I}$ into $\mathbf{I}\,\mathbf{v} = \mathbf{v}$, we can verify the following transformations of the components of $\mathbf{v}$

$$v^r = G^{ri}\, v_i \qquad \text{and} \qquad v_r = G_{ri}\, v^i \qquad r = 1, 2, 3.$$

In a similar way we obtain from $\mathbf{I}\,\mathbf{A}\,\mathbf{I} = \mathbf{A}$ the transformations of the components of an arbitrary tensor $\mathbf{A}$

(1.15) $$A^{ik} = G^{im} A_m^{\ k} = G^{im} A_{mp}\, G^{pk} = A^i_{\ p}\, G^{pk} \qquad i\,,\, k = 1, 2, 3$$

etc. If we, however, use an ONB $\{\mathbf{e}_i\} \subset \mathscr{V}$, all these representations coincide with

$$\mathbf{A} = A^{ik}\, \mathbf{e}_i \otimes \mathbf{e}_k \qquad \text{and} \qquad \mathbf{I} = \mathbf{e}_i \otimes \mathbf{e}_i\,,$$

and it does not matter anymore whether an index is upper or lower.

For the transpose of a tensor we obtain

$$\mathbf{A}^T = A^p_{\ j}\, \mathbf{g}^j \otimes \mathbf{g}_p = A_p^{\ j}\, \mathbf{g}_j \otimes \mathbf{g}^p = A^{pj}\, \mathbf{g}_j \otimes \mathbf{g}_p = A_{pj}\, \mathbf{g}^j \otimes \mathbf{g}^p.$$

$\mathbf{A}$ is symmetric, if and only if

$$A^p_{\ j} = A_j^{\ p} \;\Leftrightarrow\; A^{pj} = A^{jp} \;\Leftrightarrow\; A_{pj} = A_{jp} \;\Leftrightarrow\; A_p^{\ j} = G^{jr} A_r^{\ s}\, G_{sp}$$
$$p\,,\, j = 1, 2, 3$$

etc. hold for all components. By the last condition we see that the symmetry of a tensor and of its matrix of components is not always equivalent.

The sum of two tensors is then

$$\mathbf{A} + \mathbf{B} = (A^r_{\ q} + B^r_{\ q})\, \mathbf{g}_r \otimes \mathbf{g}^q$$
$$= (A_q^{\ r} + B_q^{\ r})\, \mathbf{g}^q \otimes \mathbf{g}_r$$
$$= (A_{rq} + B_{rq})\, \mathbf{g}^r \otimes \mathbf{g}^q$$
$$= (A^{rq} + B^{rq})\, \mathbf{g}_r \otimes \mathbf{g}_q$$

and their composition

$$\mathbf{A}\,\mathbf{B} = (A^k_{\ q}\, \mathbf{g}_k \otimes \mathbf{g}^q)\,(B^r_{\ l}\, \mathbf{g}_r \otimes \mathbf{g}^l) = A^k_{\ q}\, B^r_{\ l}\, (\mathbf{g}^q \cdot \mathbf{g}_r)\, \mathbf{g}_k \otimes \mathbf{g}^l$$
$$= A^k_{\ q}\, B^q_{\ l}\, \mathbf{g}_k \otimes \mathbf{g}^l$$

and analogously

$$= A^k_{\ q}\, B^{ql}\, \mathbf{g}_k \otimes \mathbf{g}_l = A^k_{\ r}\, G^{rq}\, B_{ql}\, \mathbf{g}_k \otimes \mathbf{g}^l$$
$$= A_{kq}\, B^q_{\ l}\, \mathbf{g}^k \otimes \mathbf{g}^l = A_{kq}\, B^{ql}\, \mathbf{g}^k \otimes \mathbf{g}_l = A_k^{\ r}\, G_{rq}\, B^{ql}\, \mathbf{g}^k \otimes \mathbf{g}_l$$
$$= A^{kq}\, B_{ql}\, \mathbf{g}_k \otimes \mathbf{g}^l = A^{kq}\, B_q^{\ l}\, \mathbf{g}_k \otimes \mathbf{g}_l = A^{kr}\, G_{rq}\, B^{ql}\, \mathbf{g}_k \otimes \mathbf{g}_l$$

$$= A_k{}^q\, B_q{}^l\, \mathbf{g}^k \otimes \mathbf{g}_l = A_k{}^q\, B_{ql}\, \mathbf{g}^k \otimes \mathbf{g}^l = A_k{}^r\, G_{rq}\, B^q{}_l\, \mathbf{g}^k \otimes \mathbf{g}^l$$

etc. For the scalar product we obtain

$$\mathbf{A} \cdot \mathbf{B} \;=\; A^r{}_q\, B_r{}^q \;=\; A_r{}^q\, B^r{}_q \;=\; A^{rq}\, B_{rq} \;=\; A_{rq}\, B^{rq}$$

$$= A_r{}^s\, G_{sq}\, B^{rq} \;=\; A_{rq}\, G^{rs}\, B_s{}^q = A^{rs}\, G_{sq}\, B_r{}^q \;=\; A_{rs}\, G^{rj}\, G^{sq}\, B_{jq}$$

etc.

If we want to determine the principal invariants by the components, we identify $\mathbf{a} \equiv \mathbf{g}_1$, $\mathbf{b} \equiv \mathbf{g}_2$, $\mathbf{c} \equiv \mathbf{g}_3$ in (1.6) and use $\mathbf{A} = A^r{}_q\, \mathbf{g}_r \otimes \mathbf{g}^q$, so that we have $\mathbf{A}\, \mathbf{g}_j = A^p{}_j\, \mathbf{g}_p$. If we now apply the rules for the triple product, we obtain

$$[\mathbf{A}\, \mathbf{g}_1, \mathbf{g}_2, \mathbf{g}_3] \;=\; A^p{}_1\, [\mathbf{g}_p, \mathbf{g}_2, \mathbf{g}_3] = A^1{}_1\, [\mathbf{g}_1, \mathbf{g}_2, \mathbf{g}_3]$$

and for the other terms accordingly. This gives

$$I_\mathbf{A} \;=\; tr(\mathbf{A}) \;=\; A^1{}_1 + A^2{}_2 + A^3{}_3 \;=\; A^p{}_p \;=\; A_p{}^p\,,$$

whereas the representations A_{pp} and A^{pp} for the trace are only valid with respect to an ONB. For the other invariants we obtain

$$II_\mathbf{A} \;=\; ½\, \{tr^2(\mathbf{A}) - tr(\mathbf{A}^2)\} \;=\; ½\, \{(A_p{}^p)^2 - A_p{}^r\, A_r{}^p\}$$

$$= ½\, \{(A^p{}_p)^2 - A^p{}_r\, A^r{}_p\}$$

$$= A^1{}_1\, A^2{}_2 - A^1{}_2\, A^2{}_1 + A^2{}_2\, A^3{}_3 - A^2{}_3\, A^3{}_2 + A^3{}_3\, A^1{}_1 - A^3{}_1\, A^1{}_3$$

$$III_\mathbf{A} \;=\; det(\mathbf{A}) \;=\; \varepsilon_{pqr}\, A^p{}_1\, A^q{}_2\, A^r{}_3 = det[A^p{}_j]$$

where the latter indicates the determinant of the matrix of components with respect to a mixed tensor base.

1.2.2 Eigenvalue Problem

In mechanics, we are often confronted with the problem of finding directions or vectors, which remain invariant under the action of a given tensor.

To make this more precise, let $\mathbf{A} \in \mathcal{Lin}$. A vector $\mathbf{p} \neq \mathbf{o}$ is called **(right) eigenvector** or **proper vector** of $\mathbf{A}$, if there exists a scalar λ, the corresponding **eigenvalue**, such that

(1.16) $$\mathbf{A}\, \mathbf{p} \;=\; \lambda\, \mathbf{p}$$

holds. This equation is always trivially fulfilled for the zero vector and an arbitrary λ, so that we further on exclude this case. Equivalently we have

(1.17) $$(\mathbf{A} - \lambda\, \mathbf{I})\, \mathbf{p} \;=\; \mathbf{o}$$

or componentwise with respect to a basis $\{\mathbf{g}_i\} \subset \mathcal{V}$ with dual $\{\mathbf{g}^i\} \subset \mathcal{V}$

$$(A^i{}_k - \lambda\, \delta^i{}_k)\, p^k \;=\; 0 \qquad \text{for } i = 1, 2, 3.$$

Obviously, if $\mathbf{p}$ is an eigenvector, then so is $\alpha\,\mathbf{p}$ for all non-zero $\alpha \in \mathscr{R}$ (corresponding to the same eigenvalue). Therefore, a normalisation to unit length is reasonable.

The **left eigenvector** is defined as the right one of the transposed tensor. In the sequel we will mainly consider right eigenvectors.

By Theorem 1.1. we know that the above homogeneous system of linear equations has non-trivial solutions p^k if and only if its determinant is zero

$$det(A^i{}_k - \lambda\,\delta^i{}_k) = 0 = det(\mathbf{A} - \lambda\,\mathbf{I})\,.$$

By (1.6) we can express this condition by quotients like

$$\frac{[\mathbf{Aa} - \lambda\,\mathbf{a}, \mathbf{Ab} - \lambda\,\mathbf{b}, \mathbf{Ac} - \lambda\,\mathbf{c}]}{[\mathbf{a}, \mathbf{b}, \mathbf{c}]} = 0$$

for any linear independent vectors $\mathbf{a}$, $\mathbf{b}$, $\mathbf{c} \in \mathscr{V}$. By some ordering after powers of λ and considering the definitions of the other principal invariants (1.6), we obtain the **characteristic polynomial** of a tensor $\mathbf{A}$

(1.18) $$\lambda^3 - I_{\mathbf{A}}\,\lambda^2 + II_{\mathbf{A}}\,\lambda - III_{\mathbf{A}} = 0$$

From the theory of polynomials we have some knowledge about its roots. As the principal invariants are always real, we conclude the following.

Theorem 1.11. *A tensor has either*

- *three (not necessarily different) real eigenvalues, or*
- *one real and two conjugate complex eigenvalues.*

In the latter case, the corresponding eigenvectors are also conjugate complex, if the tensor is real. In fact, by taking the conjugate complex of (1.16), we obtain for real tensors $\mathbf{A}$

(1.19) $$\mathbf{A}\,\overline{\mathbf{p}} = \overline{\lambda}\,\overline{\mathbf{p}}\,.$$

After VIETA's[6] theorem on roots of polynomials, we obtain for the characteristic polynomial of a tensor

$$\lambda^3 - I_{\mathbf{A}}\,\lambda^2 + II_{\mathbf{A}}\,\lambda - III_{\mathbf{A}} = (\lambda - \lambda_1)\,(\lambda - \lambda_2)\,(\lambda - \lambda_3)$$

and by comparison the useful relations

(1.20) $$\begin{aligned} I_{\mathbf{A}} &= \lambda_1 + \lambda_2 + \lambda_3 \\ II_{\mathbf{A}} &= \lambda_1\,\lambda_2 + \lambda_2\,\lambda_3 + \lambda_3\,\lambda_1 \\ III_{\mathbf{A}} &= \lambda_1\,\lambda_2\,\lambda_3\,. \end{aligned}$$

From the latter we see by Theorem 1.8 that a tensor is invertible if and only if all eigenvalues are non-zero.

---

[6] Francois Viète (1540-1603)

Due to the determinant rule (1.10) the characteristic polynomial of the tensor and of its transpose are identical. Consequently, both have the same eigenvalues and principal invariants. The same holds for similar tensors. We call two tensors $\mathbf{A}$ and $\mathbf{C}$ **similar**, if an invertible tensor $\mathbf{B} \in \mathcal{Inv}$ exists, such that $\mathbf{C} = \mathbf{B\,A\,B}^{-1}$.

Theorem 1.12. *Let* $\mathbf{A} \in \mathcal{Lin}$ and $\mathbf{B} \in \mathcal{Inv}$. *Then* $\mathbf{A}$ *and* $\mathbf{B\,A\,B}^{-1}$ *have the same principal invariants and the same eigenvalues.*

If $\mathbf{p}$ *is an eigenvector for* $\mathbf{A}$, *then so is* $\mathbf{B\,p}$ *for* $\mathbf{B\,A\,B}^{-1}$.

Note that the symmetry of $\mathbf{A}$ does not imply the symmetry of $\mathbf{B\,A\,B}^{-1}$, but it does imply the existence of real eigenvalues of both tensors.

By (1.17), we can see some immediate consequences. The eigenvectors corresponding to different eigenvalues, are linearly independent. If two or more eigenvalues are identical, then any linear combination of the corresponding eigenvectors also give eigenvectors for the same eigenvalue. In other words, all eigenvectors corresponding to the same eigenvalue, span a subspace of $\mathcal{V}$, the **eigenspace** of this eigenvalue. Its dimension is equal or less the multiplicity of the eigenvalue. If it is equal, then we can find a basis $\{\mathbf{p}_k\} \subset \mathcal{V}$ consisting of eigenvectors, called the (right) **eigenbasis** of the tensor. In general, however, this eigenbasis is not unique.

Example. We consider a simple dyad $\mathbf{a} \otimes \mathbf{b}$. The characteristic polynomial degenerates to

$$\lambda^3 - (\mathbf{a} \cdot \mathbf{b})\,\lambda^2 = 0$$

and has two roots zero and one root $\mathbf{a} \cdot \mathbf{b}$. The corresponding eigenvectors are all those perpendicular to $\mathbf{b}$, which span a plane, and one in the direction of $\mathbf{a}$. Two special cases are of interest.

The dyad is symmetric if and only if $\mathbf{a}$ and $\mathbf{b}$ are parallel and, thus, of the form $\mathbf{a} \otimes \mathbf{a}$. Then the two eigenspaces are mutually orthogonal.

In contrast to this, let $\mathbf{a}$ and $\mathbf{b}$ be two non-zero vectors being mutually orthogonal ($\mathbf{a} \cdot \mathbf{b} = 0$). Then $\mathbf{a} \otimes \mathbf{b}$ is a non-symmetric tensor, all principal invariants of which are zero. This dyad has three eigenvalues zero, without being the zero tensor. The corresponding eigenspace is again perpendicular to $\mathbf{b}$ and, thus, two-dimensional.

If a tensor $\mathbf{A}$ has three real eigenvalues λ_r and a corresponding eigenbasis $\{\mathbf{p}_r\} \subset \mathcal{V}$, then for the dual basis $\{\mathbf{p}^r\} \subset \mathcal{V}$ we obtain

$$\begin{aligned} \mathbf{p}^q \cdot (\mathbf{A}\,\mathbf{p}_r) &= \lambda_r\,\mathbf{p}^q \cdot \mathbf{p}_r = \lambda_r\,\delta^q{}_r && \text{(no sum)} \\ &= (\mathbf{A}^T\,\mathbf{p}^q) \cdot \mathbf{p}_r && \text{for } r, q = 1, 2, 3. \end{aligned}$$

We conclude that the dual of the right eigenbasis forms the left eigenbasis belonging to the same eigenvalues. This allows for a **spectral form** of the tensor

(1.21) $$\mathbf{A} = \sum_{r=1}^{3} \lambda_r \, \mathbf{p}_r \otimes \mathbf{p}^r ,$$

and we call such a tensor **diagonalisable.** Note that we can only normalise one of these eigenbases, the left *or* the right one, because otherwise we would destroy the duality between them.

Not all tensors are diagonalisable. The dyad $\mathbf{a} \otimes \mathbf{b}$ of the above example with $\mathbf{a} \cdot \mathbf{b} = 0$ is not diagonalisable, as the eigenvectors corresponding to the three eigenvalues 0 span a plane, which is only two-dimensional.

Let $\mathbf{p}$ be an eigenvector of $\mathbf{A}$ corresponding to a real eigenvalue λ. By repeatedly applying $\mathbf{A}$ to equation (1.17), we obtain

$$\mathbf{A}^n \, \mathbf{p} = \lambda^n \, \mathbf{p}$$

for all natural numbers n, *i.e.*, $\mathbf{p}$ is eigenvector for all powers $\mathbf{A}^n$ of $\mathbf{A}$ corresponding to the eigenvalue λ^n. Furthermore, $\mathbf{p}$ is for all $\alpha \in \mathcal{R}$ eigenvector of $\alpha \mathbf{A}$ corresponding to the eigenvalue $\alpha \lambda$. If $\sum_{n=1}^{R} \alpha_n \mathbf{A}^n$ is a real tensor polynomial of $\mathbf{A}$, then $\sum_{n=1}^{R} \alpha_n \lambda^n$ is eigenvalue of this polynomial, again corresponding to the same eigenvector $\mathbf{p}$. If we apply this fact to the tensor polynomial

$$\mathbf{A}^3 - I_{\mathbf{A}} \mathbf{A}^2 + II_{\mathbf{A}} \mathbf{A} - III_{\mathbf{A}} \mathbf{I}$$

then we obtain its eigenvalues

$$\lambda^3 - I_{\mathbf{A}} \lambda^2 + II_{\mathbf{A}} \lambda - III_{\mathbf{A}}$$

which are all zero after the characteristic polynomial (1.18) of $\mathbf{A}$. For diagonalisable tensors this means that the tensor polynomial itself is zero. This conclusion holds for all tensors[7] because of the following

Theorem 1.13. of CAYLEY[8]-HAMILTON[9]
Each tensor $\mathbf{A} \in \mathcal{L}in$ *fulfils its characteristic equation*

$$\mathbf{A}^3 - I_{\mathbf{A}} \mathbf{A}^2 + II_{\mathbf{A}} \mathbf{A} - III_{\mathbf{A}} \mathbf{I} = \mathbf{0}$$

This equation can be used to substitute higher tensor polynomials by a combination of three lower ones like $\mathbf{A}^2$, $\mathbf{A}$ and $\mathbf{I}$.

For invertible $\mathbf{A}$, we can multiply this equation by $\mathbf{A}^{-1}$ and obtain a representation for its inverse

---

[7] For a general proof of this theorem, see, *e.g.*, SCHADE (1997) p. 184 ff.
[8] Arthur Cayley (1821-1895)
[9] William Rowan Hamilton (1805-1865)

$$\mathbf{A}^2 - I_\mathbf{A}\,\mathbf{A} + II_\mathbf{A}\,\mathbf{I} - III_\mathbf{A}\,\mathbf{A}^{-1} = \mathbf{0}\,.$$

The Eigenvalue Problem for Symmetric Tensors

Next we shall particularise these findings to symmetric tensors. For them, right and left eigenvectors coincide. Let λ and $\bar{\lambda}$ be two conjugate complex eigenvalues of a symmetric tensor $\mathbf{A}$ with corresponding conjugate complex eigenvectors $\mathbf{p}$ and $\bar{\mathbf{p}}$. Then by (1.16) and (1.19) we obtain

$$\mathbf{p} \cdot (\mathbf{A}\,\bar{\mathbf{p}}) = \bar{\lambda}\,\mathbf{p} \cdot \bar{\mathbf{p}}$$

$$= \bar{\mathbf{p}} \cdot (\mathbf{A}^T\,\mathbf{p}) = \bar{\mathbf{p}} \cdot (\mathbf{A}\,\mathbf{p}) = \lambda\,\mathbf{p} \cdot \bar{\mathbf{p}}$$

or

$$(\lambda - \bar{\lambda})\,\mathbf{p} \cdot \bar{\mathbf{p}} = 0\,.$$

If the two eigenvectors are conjugate complex, then their inner product is positive. Therefore, the difference of the eigenvalues must be zero, which means that their imaginary parts equal zero.

Theorem 1.14. *Symmetric tensors have only real eigenvalues.*

Let λ_1 and λ_2 be two different eigenvalues of a symmetric tensor $\mathbf{A}$ with corresponding eigenvectors $\mathbf{p}_1$ and $\mathbf{p}_2$, respectively. Then a similar calculation as the one above leads to

$$(\lambda_1 - \lambda_2)\,\mathbf{p}_1 \cdot \mathbf{p}_2 = 0\,.$$

Theorem 1.15. *Eigenvectors of symmetric tensors corresponding to different eigenvalues are mutually orthogonal.*

The resulting spectral form of a symmetric tensor is then

(1.22) $$\mathbf{A} = \sum_{r=1}^{3} \lambda_r\,\mathbf{p}_r \otimes \mathbf{p}_r \qquad \in \mathcal{Sym}$$

as a special case of (1.21) where the eigenbasis $\{\mathbf{p}_r\}$ forms an ONB. Accordingly, all symmetric tensors are also diagonalisable. This representation is also possible for multiple eigenvalues, but then loses its uniqueness.

The remedy for this non-uniqueness is the use of **eigenprojectors**. These are defined for an N_s-multiple eigenvalue λ_s corresponding to orthogonal and normalised eigenvectors $\mathbf{p}_r$ by

$$\mathbf{P}_s := \sum_{r=1}^{N_s} \mathbf{p}_r \otimes \mathbf{p}_r \qquad \in \mathcal{Sym}\,.$$

For equal eigenvalues like $\lambda_2 \equiv \lambda_3$, we obtain the eigenspace projectors

$$\mathbf{P}_1 = \mathbf{p}_1 \otimes \mathbf{p}_1 \quad \text{and} \quad \mathbf{P}_2 = \mathbf{I} - \mathbf{p}_1 \otimes \mathbf{p}_1\,.$$

For a triple eigenvalue, only one projector remains, namely

$$\mathbf{P}_1 = \mathbf{I} .$$

In general, the **eigenprojector representation** for a symmetric tensor is

(1.23)
$$\mathbf{A} = \sum_{r=1}^{K} \lambda_r \, \mathbf{P}_r \qquad \in \mathcal{Sym}$$

with K the number of distinct eigenvalues of $\mathbf{A}$, which is either 1, 2 or 3. This form is in all cases unique. If all eigenvalues are distinct, then this form is identical to the spectral form above.

The eigenprojectors $\mathbf{P}_r$ fulfil the projector rules

(1.24)

$\mathbf{P}_r \, \mathbf{P}_r = \mathbf{P}_r$	idempotent
$\mathbf{P}_r \, \mathbf{P}_s = \mathbf{0}$ for $r \neq s$	orthogonal
$\sum_{r=1}^{K} \mathbf{P}_r = \mathbf{I}$	complete

We can bring the eigenvalue equation (1.17) in the form

$$\mathbf{A} - \lambda_k \mathbf{I} = \sum_{r=1}^{K} \lambda_r \, \mathbf{P}_r - \lambda_k \sum_{r=1}^{K} \mathbf{P}_r = \sum_{r=1}^{K} (\lambda_r - \lambda_k) \, \mathbf{P}_r$$

for $k = 1 , ... , K$. By multiplication of two of these equations, we obtain

(1.25)
$$\mathbf{P}_r = \frac{\prod_{k \neq r}^{K} (\mathbf{A} - \lambda_k \mathbf{I})}{\prod_{k \neq r}^{K} (\lambda_r - \lambda_k)} .$$

This is **SYLVESTER's[10] formula**. It can be used for the determination of the eigenprojector, once we know its eigenvalues.

The same can be analogously achieved also for non-symmetric, but diagonalisable tensors, where the eigenprojectors are introduced as

$$\mathbf{P}_s := \sum_{r=1}^{N_s} \mathbf{p}_r \otimes \mathbf{p}^r .$$

Tensors with equal eigenvectors are called **coaxial**. The following tensors are examples for coaxiality: $\mathbf{A}$, $\mathbf{A}^2$, $\mathbf{A}^{-1}$ (if invertible), $\alpha \, \mathbf{A}$ with $\alpha \neq 0$, $\mathbf{A}'$ (deviator) for all $\mathbf{A} \in \mathcal{Sym}$.

The following result can easily be seen by using spectral representations for the involved tensors.

[10] James Joseph Sylvester (1814-1897)

Theorem 1.16. *Two symmetric tensors* $\mathbf{A}$, $\mathbf{B} \in \mathcal{Sym}$ *are coaxial, if and only if the composition commutes*

$$\mathbf{A}\,\mathbf{B} \;=\; \mathbf{B}\,\mathbf{A}$$

and, hence, is symmetric.

For invertible symmetric tensors $\mathbf{A}$ we get the spectral form

$$\mathbf{A}^{-1} \;=\; \sum_{r=1}^{3} \lambda_r^{\;-1}\, \mathbf{p}_r \otimes \mathbf{p}_r \;=\; \sum_{r=1}^{K} \lambda_r^{\;-1}\, \mathbf{P}_r$$

for the inverse. For the norm of a symmetric tensor holds

$$|\,\mathbf{A}\,|^{2} = \lambda_1^{\,2} + \lambda_2^{\,2} + \lambda_3^{\,2} \;=\; I_{\mathbf{A}}^{\;2} - 2\, II_{\mathbf{A}}\,.$$

For any power L (integer) of a symmetric tensor, we obtain for its invariants

$$I_{(\mathbf{A}^L)} \;=\; \lambda_1^{\;L} + \lambda_2^{\;L} + \lambda_3^{\;L}$$

$$II_{(\mathbf{A}^L)} \;=\; \lambda_1^{\;L}\,\lambda_2^{\;L} + \;\lambda_2^{\;L}\,\lambda_3^{\;L} + \lambda_3^{\;L}\,\lambda_1^{\;L}$$

$$III_{(\mathbf{A}^L)} \;=\; \lambda_1^{\;L}\,\lambda_2^{\;L}\,\lambda_3^{\;L}\,.$$

Spectral forms are used to define **tensor-functions** of symmetric tensors. Let

$$f : \mathcal{R} \rightarrow \mathcal{R}$$

be a function like

$$f(\lambda) \;\equiv\; e^{\lambda}\,,\, sin\,(\lambda)\,,\, tan(\lambda)\,,$$

then we define

$$F : \mathcal{Sym} \rightarrow \mathcal{Sym} \;\;\Big|\;\; F(\mathbf{A}) \;:=\; \sum_{r=1}^{3} f(\lambda_r)\; \mathbf{p}_r \otimes \mathbf{p}_r \;\in \mathcal{Sym}\,.$$

If we restrict this to tensors with positive eigenvalues, we can also define in the same way

$$F(\mathbf{A}) \equiv ln(\mathbf{A})\,,\, log(\mathbf{A})\,,\, \sqrt{\mathbf{A}}$$

etc. Consequently, the square root of such a tensor is unique, and

$$\sqrt{\mathbf{A}}\,\sqrt{\mathbf{A}} \;=\; \mathbf{A}\,.$$

An alternative choice of introducing tensor-functions of (not necessarily symmetric) tensors is by means of series. The tensorial exponential mapping, *e.g.*, is

$$exp(\mathbf{A}) \;:=\; \mathbf{I} + \frac{1}{1!}\,\mathbf{A} + \frac{1}{2!}\,\mathbf{A}^{2} + \frac{1}{3!}\,\mathbf{A}^{3} + \ldots \qquad \forall\; \mathbf{A} \in \mathcal{Lin}$$

and the tensorial logarithm is

$$ln(\mathbf{I} + \mathbf{A}) \;:=\; \mathbf{A} \;-\, {}^{1}\!/_{2}\,\mathbf{A}^{2} \;+\; {}^{1}\!/_{3}\,\mathbf{A}^{3} \;-\; \ldots \qquad \forall\; \mathbf{A} \in \mathcal{Lin}$$

if the argument lies within the range of convergence.

Bilinear Forms

A tensor $\mathbf{A} \in \mathcal{Lin}$ can not only be used as a linear mapping between vectors, but also as a **bilinear form**, *i.e.*, a mapping

$$\mathbf{A} : \mathcal{V} \times \mathcal{V} \to \mathcal{R} \qquad | \qquad (\mathbf{u}, \mathbf{v}) \mapsto \mathbf{u} \cdot (\mathbf{A}\, \mathbf{v}) .$$

In particular, we obtain a **quadratic form**, if $\mathbf{u}$ and $\mathbf{v}$ are identical

$$\mathbf{A} : \mathcal{V} \to \mathcal{R} \qquad | \qquad \mathbf{u} \mapsto \mathbf{u} \cdot (\mathbf{A}\, \mathbf{u}) .$$

In such a form only the symmetric part of $\mathbf{A}$ is significant. By the sign of such forms, tensors are categorised.

Definition 1.17. A tensor $\mathbf{A} \in \mathcal{Lin}$ is called

- **positive-definite**, if $\mathbf{v} \cdot (\mathbf{A}\, \mathbf{v}) > 0 \qquad \forall\, \mathbf{v} \neq \mathbf{o} \in \mathcal{V}$
- **positive semi-definite**, if $\mathbf{v} \cdot (\mathbf{A}\, \mathbf{v}) \geq 0 \qquad \forall\, \mathbf{v} \neq \mathbf{o} \in \mathcal{V}$
- **negative-definite**, if $\mathbf{v} \cdot (\mathbf{A}\, \mathbf{v}) < 0 \qquad \forall\, \mathbf{v} \neq \mathbf{o} \in \mathcal{V}$
- **negative semi-definite**, if $\mathbf{v} \cdot (\mathbf{A}\, \mathbf{v}) \leq 0 \qquad \forall\, \mathbf{v} \neq \mathbf{o} \in \mathcal{V}$
- **indefinite** otherwise

The following can easily be proved by the spectral form.

Theorem 1.18. *For symmetric tensors* $\mathbf{A}$ *the following holds:*

$\mathbf{A}$ *is positive-definite* $\Leftrightarrow$ *all eigenvalues are positive*

$\mathbf{A}$ *is positive semi-definite* $\Leftrightarrow$ *all eigenvalues are non-negative*

$\mathbf{A}$ *is negative-definite* $\Leftrightarrow$ *all eigenvalues are negative*

$\mathbf{A}$ *is negative semi-definite* $\Leftrightarrow$ *all eigenvalues are non-positive*

We will denote the set of symmetric and positive-definite tensors by $\mathcal{Psym}$. Therefore, $\mathcal{Psym} \subset \mathcal{Sym}$. The set $\mathcal{Psym}$ is important as it corresponds to the inner products on $\mathcal{V}$. In fact, let $\mathbf{v}, \mathbf{w} \in \mathcal{V}$ and $\mathbf{G} \in \mathcal{Psym}$, then the mapping

$$< \mathbf{v}, \mathbf{w} >_{\mathbf{G}} \; := \; \mathbf{v} \cdot (\mathbf{G}\, \mathbf{w})$$

defines an inner product on $\mathcal{V}$. Only for the identity $\mathbf{G} \equiv \mathbf{I} \in \mathcal{Psym}$, this inner product coincides with the standard one.

A geometrical interpretation of symmetric tensors is given by **tensor surfaces**. These are introduced in the following way. Let us consider all vectors $\mathbf{v}$ that solve the equation

$$\mathbf{v} \cdot (\mathbf{A}\, \mathbf{v}) = 1 .$$

If we interpret these vectors as position vectors in space, they describe a surface manifold. For example, let the tensor be the identity tensor or a spherical tensor, then this surface is a sphere. The normal of this surface has the same direction as $(\mathbf{A}\,\mathbf{v})$. This can easily be seen if we interpret $\mathbf{v} \cdot (\mathbf{A}\,\mathbf{v})$ for a given tensor $\mathbf{A}$ as a scalar valued vector function, and calculate its gradient (see Chap. 1.3).

By the form of such a surface, we can categorise the tensors. In the following table, we have ordered the eigenvalues such that $\lambda_1 \geq \lambda_2 \geq \lambda_3$.

sign of eigenvalue	quadratic form	$\mathbf{p}_1$ - $\mathbf{p}_2$ -plane	$\mathbf{p}_2$ - $\mathbf{p}_3$ - -plane	$\mathbf{p}_1$ - $\mathbf{p}_3$ - -plane	tensor surface
+ + +	positive def.	ellipse	ellipse	ellipse	ellipsoid
+ + 0	pos. semi-def.	ellipse	2 parallel straight lines	2 parallel straight lines	elliptical cylinder
+ 0 0	pos. semi-def.	2 par. str. lines	-	2 par. str. lines	2 parallel planes
+ + –	indefinite	ellipse	hyperbola	hyperbola	one-fold hyperboloid
+ – –	indefinite	hyperbola	-	hyperbola	two-fold hyperboloid

Let $\mathbf{A}$ be any tensor. Then $\mathbf{A}^T \mathbf{A}$ and $\mathbf{A}\,\mathbf{A}^T$ are symmetric tensors, having identical eigenvalues and principal invariants.

Theorem 1.19. *A tensor* $\mathbf{A}$ *is invertible, if and only if* $\mathbf{A}^T \mathbf{A}$ *(or* $\mathbf{A}\,\mathbf{A}^T$ *) is positive-definite.*

1.2.3 Special Tensors

Skew Tensors

Let $\mathbf{w} \in \mathcal{V}$. The mapping

$$\mathbf{W} : \mathcal{V} \to \mathcal{V} \quad | \quad \mathbf{x} \mapsto \mathbf{w} \times \mathbf{x}$$

induced by $\mathbf{w}$ is linear and, hence, a tensor $\mathbf{W} \in \mathcal{L}in$, in our notation

$$\mathbf{W}\,\mathbf{x} = \mathbf{w} \times \mathbf{x}\,. \tag{1.26}$$

After the definition of the transposition and the rules for the triple product, we obtain for all $\mathbf{x}, \mathbf{y} \in \mathcal{V}$

$$(\mathbf{W}^T \mathbf{y}) \cdot \mathbf{x} = \mathbf{y} \cdot (\mathbf{W}\,\mathbf{x})$$

$$= \mathbf{y} \cdot (\mathbf{w} \times \mathbf{x}) = -(\mathbf{w} \times \mathbf{y}) \cdot \mathbf{x} = -(\mathbf{W}\,\mathbf{y}) \cdot \mathbf{x}\,,$$

from which we conclude that **W** is skew. In fact, every skew tensor **W** uniquely corresponds to an **axial vector w**, which performs (1.26).

We now determine the components of a skew tensor **W** with respect to an ONB $\{\mathbf{e}_i\}$

$$\mathbf{W} = W^{ij}\, \mathbf{e}_i \otimes \mathbf{e}_j \,.$$

Because of

$$\mathbf{e}_i \cdot (\mathbf{W}\, \mathbf{e}_j) = W^{ij} = -\,\mathbf{e}_j \cdot (\mathbf{W}\, \mathbf{e}_i) = -\,W^{ji} \qquad \text{for } i, j = 1, 2, 3$$

we achieve

$$W^{ii} = 0 \qquad \text{for } i = 1, 2, 3 \text{ (no sum).}$$

Hence, skew tensors are traceless or deviatoric. The matrix of the coefficients of **W** has only three independent entries (the same as the vector **w**)

$$\begin{bmatrix} 0 & W^{12} & W^{13} \\ -W^{12} & 0 & W^{23} \\ -W^{13} & -W^{23} & 0 \end{bmatrix}.$$

With respect to any ONB we obtain

$$\mathbf{W}\, \mathbf{e}_j = W^{ij}\, \mathbf{e}_i = \mathbf{w} \times \mathbf{e}_j = w^k\, \varepsilon_{kji}\, \mathbf{e}_i = -\varepsilon_{ijk}\, w^k\, \mathbf{e}_i$$

and, thus, for the components

$$W^{ij} = -\,\varepsilon_{ijk}\, w^k \quad \Leftrightarrow \quad w^i = -\,{}^1\!/_2\, \varepsilon_{ijk}\, W^{jk}.$$

This gives

$$w^1 = W^{32} \qquad w^2 = W^{13} \qquad w^3 = W^{21}.$$

As any other tensor, **W** has at least one real eigenvalue, say λ_1. Let **p** be the corresponding (normalised) eigenvector. Because of the skewness

$$\lambda_1 = \mathbf{p} \cdot (\mathbf{W}\, \mathbf{p}) = \mathbf{p} \cdot (\mathbf{W}^T\, \mathbf{p}) = \mathbf{p} \cdot (-\,\mathbf{W}\, \mathbf{p}) = -\,\lambda_1$$

we conclude that every real eigenvalue of a skew tensor must be zero. If we choose the ONB $\{\mathbf{e}_i\}$ in such a way that $\mathbf{e}_1$ is the corresponding eigenvector, then

$$\mathbf{W}\, \mathbf{e}_1 = \mathbf{o} \qquad \Rightarrow \qquad W^{21} = W^{31} = -\,W^{12} = -\,W^{13} = 0\,.$$

The only non-zero components are possibly

$$W^{23} = \mathbf{e}_2 \cdot (\mathbf{W}\, \mathbf{e}_3) = -\,\mathbf{e}_3 \cdot (\mathbf{W}\, \mathbf{e}_2) = -\,W^{32} =: -\,w$$

and we obtain the representation by the tensor product

$$\mathbf{W} = w\, (\mathbf{e}_3 \otimes \mathbf{e}_2 - \mathbf{e}_2 \otimes \mathbf{e}_3)$$

or as a matrix of components

$$\begin{bmatrix} 0 & 0 & 0 \\ 0 & 0 & -w \\ 0 & w & 0 \end{bmatrix}.$$

We now calculate the axial vector by

$$\mathbf{W}\,\mathbf{x} = w\,(\mathbf{e}_3 \otimes \mathbf{e}_2 - \mathbf{e}_2 \otimes \mathbf{e}_3)\,\mathbf{x} = w\,(\mathbf{x}\cdot\mathbf{e}_2)\,\mathbf{e}_3 - w\,(\mathbf{x}\cdot\mathbf{e}_3)\,\mathbf{e}_2$$

$$= w\,\mathbf{e}_1 \times \big((\mathbf{x}\cdot\mathbf{e}_1)\,\mathbf{e}_1 + (\mathbf{x}\cdot\mathbf{e}_2)\,\mathbf{e}_2 + (\mathbf{x}\cdot\mathbf{e}_3)\,\mathbf{e}_3\big) = w\,\mathbf{e}_1 \times \mathbf{x}$$

$$= \mathbf{w} \times \mathbf{x} \qquad \forall\, \mathbf{x} \in \mathscr{V}.$$

Thus, $\mathbf{w} = w\,\mathbf{e}_1$ is the axial vector of the skew tensor $\mathbf{W}$. For $w \equiv 0$, $\mathbf{W}$ is the zero-tensor, which is also skew.

The principal invariants of a skew tensor $\mathbf{W}$ are by (1.20)

$$I_{\mathbf{W}} = 0 = \lambda_2 + \lambda_3$$

$$II_{\mathbf{W}} = w^2 = \lambda_2\,\lambda_3$$

$$III_{\mathbf{W}} = 0\,.$$

The first equation tells us that the real parts of the other eigenvalues are zero. From the second we obtain the solution

$$\lambda_{2,3} = \pm\, i\, w$$

so that the other two eigenvalues are either imaginary or zero.

For each skew tensor we have

$$\mathbf{W}^2 = \mathbf{w} \otimes \mathbf{w} - w^2\,\mathbf{I} \qquad \in \mathscr{Sym}$$

$$\mathbf{W}^3 = -\,w^2\,\mathbf{W} \qquad \in \mathscr{Skw}.$$

Let $\mathbf{b}\,,\mathbf{c} \in \mathscr{V}$ be arbitrary vectors. Then

$$\mathbf{b} \otimes \mathbf{c} - \mathbf{c} \otimes \mathbf{b}$$

is skew. From (1.1) we obtain for the double cross product

$$(\mathbf{c} \times \mathbf{b}) \times \mathbf{x} = \mathbf{x} \times (\mathbf{b} \times \mathbf{c})$$

$$= (\mathbf{c}\cdot\mathbf{x})\,\mathbf{b} - (\mathbf{b}\cdot\mathbf{x})\,\mathbf{c} = (\mathbf{b} \otimes \mathbf{c} - \mathbf{c} \otimes \mathbf{b})\,\mathbf{x} \qquad \forall\, \mathbf{x} \in \mathscr{V}.$$

Accordingly, $\mathbf{c} \times \mathbf{b}$ is the axial vector of the skew tensor $\mathbf{b} \otimes \mathbf{c} - \mathbf{c} \otimes \mathbf{b}$.

Orthogonal Tensors

While linear mappings (tensors) are compatible with the linear structure of domain and range, the following class of tensors is, in addition, compatible with the inner product structure of $\mathscr{V}$.

Definition 1.20. Let $\mathscr{V}$ be a vector space with inner product. A tensor $\mathbf{Q}$ is called **orthogonal** if

$$(\mathbf{Q\,a}) \cdot (\mathbf{Q\,b}) = \mathbf{a} \cdot \mathbf{b} \qquad \forall\, \mathbf{a}\,,\mathbf{b} \in \mathscr{V}.$$

Equivalently

$$\mathbf{a} \cdot \mathbf{b} = \mathbf{a} \cdot (\mathbf{Q}^T \mathbf{Q\,b})$$

or, because of the arbitrariness of the two vectors,

$$\mathbf{Q}^T \mathbf{Q} = \mathbf{I} \qquad \Leftrightarrow \qquad \mathbf{Q}^T = \mathbf{Q}^{-1}.$$

As inner products determine the length and the angle of vectors, orthogonal tensors leave them invariant. This can be achieved by rotations and reflections only.

Examples. The identity tensor $\mathbf{I}$ as well as its negative $-\mathbf{I}$ are orthogonal. The latter inverts all vectors. Also

$$\mathbf{e}_1 \otimes \mathbf{e}_1 + \mathbf{e}_2 \otimes \mathbf{e}_2 - \mathbf{e}_3 \otimes \mathbf{e}_3$$

is orthogonal, which reflects all vectors at the $\mathbf{e}_1$ - $\mathbf{e}_2$ -plane. And

$$\mathbf{e}_1 \otimes \mathbf{e}_1 - \mathbf{e}_2 \otimes \mathbf{e}_2 - \mathbf{e}_3 \otimes \mathbf{e}_3$$

is orthogonal, which rotates all vectors around $\mathbf{e}_1$ through an angle of $180°$.

If we want to see the restrictions of orthogonality upon the components of such a tensor, we represent it with respect to an ONB $\{\mathbf{e}_i\}$

$$\mathbf{Q} = Q^{rs}\, \mathbf{e}_r \otimes \mathbf{e}_s \qquad \Leftrightarrow \qquad \mathbf{Q}^T = Q^{sr}\, \mathbf{e}_r \otimes \mathbf{e}_s\,.$$

The orthogonality condition is then

$$\mathbf{Q}^T \mathbf{Q} = (Q^{sr}\, \mathbf{e}_r \otimes \mathbf{e}_s)\,(Q^{lm}\, \mathbf{e}_l \otimes \mathbf{e}_m) = Q^{sr} Q^{sm}\, \mathbf{e}_r \otimes \mathbf{e}_m$$

$$= \mathbf{I} = \delta^{rm}\, \mathbf{e}_r \otimes \mathbf{e}_m$$

$$\Leftrightarrow \qquad Q^{sr} Q^{sm} = \delta^{rm} \qquad \text{for } r, m = 1, 2, 3.$$

The three line vectors of the matrix of coefficients are normed to 1 and mutually orthogonal, and the same holds for the three column vectors.

The determinant of an orthogonal tensor is ± 1, because

$$det(\mathbf{I}) = 1 = det(\mathbf{Q}^T \mathbf{Q}) = det(\mathbf{Q}^T)\, det(\mathbf{Q}) = det(\mathbf{Q})^2.$$

For $det\, \mathbf{Q} = +1$ the orthogonal tensor is called **proper orthogonal** or a **versor** and describes a pure rotation (without reflection).

Let $\mathbf{p}$ be the eigenvector corresponding to a real eigenvalue λ of $\mathbf{Q}$

$$\mathbf{Q\,p} = \lambda\, \mathbf{p} \qquad \Rightarrow \qquad \mathbf{I\,p} = \mathbf{p} = \mathbf{Q}^T \mathbf{Q\,p} = \lambda\, \mathbf{Q}^T \mathbf{p}\,.$$

Thus, λ^{-1} is the eigenvalue of $\mathbf{Q}^T$ for the same eigenvector $\mathbf{p}$. As we already know that the eigenvalues of a tensors are also eigenvalues of its transpose, we conclude that all real eigenvalues of an orthogonal tensor must be ± 1.

If we now choose the ONB $\{\mathbf{e}_i\} \subset \mathscr{V}$ such that $\mathbf{e}_3$ is parallel to $\mathbf{p}$, then we have

$$\mathbf{e}_3 \cdot (\mathbf{Q}\,\mathbf{e}_1) = \mathbf{e}_1 \cdot (\mathbf{Q}^T\,\mathbf{e}_3) = \mathbf{e}_1 \cdot \mathbf{e}_3 = 0$$

$$\mathbf{e}_3 \cdot (\mathbf{Q}\,\mathbf{e}_2) = \mathbf{e}_2 \cdot (\mathbf{Q}^T\,\mathbf{e}_3) = \mathbf{e}_2 \cdot \mathbf{e}_3 = 0$$

i.e. all vectors perpendicular to $\mathbf{e}_3$ remain perpendicular under $\mathbf{Q}$. If we restrict our considerations to proper orthogonal tensors $\mathbf{Q}$, then it rotates the basis $\{\mathbf{e}_i\}$ rigidly around the $\mathbf{e}_3$-axis by an angular amount of some θ, so that

$$\mathbf{e}_1 \cdot (\mathbf{Q}\,\mathbf{e}_1) = \cos\theta = \mathbf{e}_2 \cdot (\mathbf{Q}\,\mathbf{e}_2)$$

hold. This can be achieved by the EULER[11]-RODRIGUES[12] representation of a versor

$$\mathbf{Q} = \mathbf{e}_3 \otimes \mathbf{e}_3 + (\mathbf{e}_1 \otimes \mathbf{e}_1 + \mathbf{e}_2 \otimes \mathbf{e}_2)\cos\theta + (\mathbf{e}_2 \otimes \mathbf{e}_1 - \mathbf{e}_1 \otimes \mathbf{e}_2)\sin\theta . \tag{1.27}$$

The matrix of coefficients with respect to the ONB $\{\mathbf{e}_i\}$ is

$$\begin{bmatrix} \cos\theta & -\sin\theta & 0 \\ \sin\theta & \cos\theta & 0 \\ 0 & 0 & 1 \end{bmatrix} .$$

This holds for all ONB for which the $\mathbf{e}_3$-direction coincides with the axis of rotation. By this representation we can easily calculate the principal invariants of a versor

$$I_{\mathbf{Q}} = II_{\mathbf{Q}} = 1 + 2\cos\theta \qquad III_{\mathbf{Q}} = 1.$$

The characteristic polynomial can be reduced to

$$\lambda^2 - (2\cos\theta)\,\lambda + 1 = 0$$

with two conjugate complex solutions

$$\cos\theta \pm i\sin\theta = e^{\pm i\theta}$$

for the two remaining eigenvalues.

If we rotate an arbitrary tensor $\mathbf{A}$ by an orthogonal tensor $\mathbf{Q}$

$$\mathbf{Q}\,\mathbf{A}\,\mathbf{Q}^T$$

then this is a special form of the similarity transformation and, according to Theorem 1.12, its eigenvalues and principal invariants remain the same

$$I_{\mathbf{A}} = I_{\mathbf{QAQ}^T}$$

[11] Leonhard Euler (1707-1783)
[12] Olinde Rodrigues (1794-1851)

$$II_{\mathbf{A}} = II_{\mathbf{QAQ}^T}$$

$$III_{\mathbf{A}} = III_{\mathbf{QAQ}^T}$$

and just the eigenvectors are rotated by $\mathbf{Q}$.

Polar Decomposition

Besides the additive decomposition of tensors into symmetric and skew parts after (1.5), the following multiplicative decomposition plays an important role in continuum mechanics.

> **Theorem 1.21. of polar decomposition** (FINGER 1892)
> *Every invertible tensor* $\mathbf{F}$ *can be uniquely decomposed in two ways*
>
> $$\mathbf{F} = \mathbf{R}\,\mathbf{U} = \mathbf{V}\,\mathbf{R}$$
>
> *into* $\mathbf{R}$: *orthogonal*
>
> $\mathbf{V}, \mathbf{U}$: *positive-definite and symmetric.*
>
> *If* $det\,\mathbf{F} > 0$ *, then* $\mathbf{R}$ *is proper orthogonal.*

Proof. We have already seen in Theorem 1.19 that the tensors

$$\mathbf{C} := \mathbf{F}^T\,\mathbf{F} \qquad \text{and} \qquad \mathbf{B} := \mathbf{F}\,\mathbf{F}^T$$

are symmetric and positive-definite if $\mathbf{F}$ is invertible. Such tensors have unique positive-definite and symmetric roots

$$\mathbf{U} = \sqrt{\mathbf{C}} \qquad \text{and} \qquad \mathbf{V} = \sqrt{\mathbf{B}}\,.$$

Let

$$\mathbf{R} := \mathbf{F}\,\mathbf{U}^{-1}$$

then

$$\mathbf{R}\,\mathbf{R}^T = \mathbf{F}\,\mathbf{U}^{-1}\,\mathbf{U}^{-T}\,\mathbf{F}^T = \mathbf{F}\,\mathbf{U}^{-2}\,\mathbf{F}^T$$

$$= \mathbf{F}\,\mathbf{C}^{-1}\,\mathbf{F}^T = \mathbf{F}\,\mathbf{F}^{-1}\,\mathbf{F}^{-T}\,\mathbf{F}^T = \mathbf{I}\,.$$

Thus, $\mathbf{R}$ is orthogonal. For

$$det(\mathbf{F}) = det(\mathbf{R})\,det(\mathbf{U}) > 0$$

follows $det(\mathbf{R}) > 0$ and $\mathbf{R}$ is proper orthogonal. For proving the uniqueness of this decomposition, we assume two decompositions

$$\mathbf{F} = \mathbf{R}_1\,\mathbf{U}_1 = \mathbf{R}_2\,\mathbf{U}_2$$

exist with $\mathbf{R}_i$ orthogonal and $\mathbf{U}_i$ positive-definite and symmetric for $i = 1\,,\,2$. Then

$$\mathbf{F}^T = \mathbf{U}_1^T\,\mathbf{R}_1^T = \mathbf{U}_2^T\,\mathbf{R}_2^T = \mathbf{U}_1\,\mathbf{R}_1^{-1} = \mathbf{U}_2\,\mathbf{R}_2^{-1}$$

and

$$\mathbf{C} = \mathbf{F}^T \mathbf{F} = \mathbf{U}_1 \mathbf{R}_1^{-1} \mathbf{R}_1 \mathbf{U}_1 = \mathbf{U}_2 \mathbf{R}_2^{-1} \mathbf{R}_2 \mathbf{U}_2 = \mathbf{U}_1^2 = \mathbf{U}_2^2.$$

As a symmetric positive-definite tensor has a unique square root

$$\mathbf{U}_1 = \mathbf{U}_2$$

the orthogonal part is also unique

$$\mathbf{R}_1 = \mathbf{R}_2 \,.$$

Similarly, one shows that with

$$\mathbf{V} = \sqrt{\mathbf{B}} = \sqrt{(\mathbf{F}\,\mathbf{F}^T)}$$

also

$$\underline{\mathbf{R}} := \mathbf{V}^{-1}\mathbf{F}$$

is orthogonal. Then we have

$$\mathbf{F} = \mathbf{R}\,\mathbf{U} = \mathbf{V}\,\underline{\mathbf{R}} = \underline{\mathbf{R}}\,\underline{\mathbf{R}}^T\mathbf{V}\,\underline{\mathbf{R}}\,.$$

As $\underline{\mathbf{R}}$ is orthogonal and $\underline{\mathbf{R}}^T\mathbf{V}\,\underline{\mathbf{R}}$ positive-definite and symmetric, we conclude from the uniqueness of the decomposition

$$\underline{\mathbf{R}} = \mathbf{R}$$

and

$$\mathbf{U} = \mathbf{R}^T\mathbf{V}\,\mathbf{R}\,; \quad q.\,e.\,d.$$

The following abbreviations for sets will be used further on.

$\mathscr{V}$	vector space
$\mathscr{L}in$	linear mappings from $\mathscr{V}$ to $\mathscr{V}$ (2nd-order tensors)
$\mathscr{I}nv \subset \mathscr{L}in$	invertible tensors
$\mathscr{U}nim \subset \mathscr{I}nv$	tensors with determinant ± 1 (unimodular)
$\mathscr{O}rth \subset \mathscr{U}nim$	orthogonal tensors
$\mathscr{S}ym \subset \mathscr{L}in$	symmetric tensors
$\mathscr{P}sym \subset \mathscr{S}ym$	symmetric and positive-definite tensors
$\mathscr{S}kw \subset \mathscr{L}in$	skew tensors

A superimposed $^+$ stands for the restriction to tensors with positive determinant. $\mathscr{R}^+$ stands for the positive reals.

We have already seen that the tensors with their addition and multiplication by a scalar form a 9-dimensional linear space, which contains $\mathscr{S}ym$ as a 6-dimensional subspace and $\mathscr{S}kw$ as a 3-dimensional one.

With respect to the composition, the following sets form groups.

$\mathscr{Inv}$	general linear group
$\mathscr{Inv}^+ = \mathscr{Lin}^+$	special linear group
$\mathscr{Unim}$	general unimodular group
$\mathscr{Unim}^+$	special unimodular group
$\mathscr{Orth}$	general orthogonal group
$\mathscr{Orth}^+$	special orthogonal group (rotational group)

1.2.4 Tensors of Higher Order

Up to now we only considered tensors of *2*nd-order as linear mappings on a vector space. If $\mathscr{V}$ is an inner product space, we can use a tensor **A** also as a bilinear mapping (*i.e.* linear in two input variables)

$$\mathbf{A}: \mathscr{V} \times \mathscr{V} \to \mathscr{R}$$

by the prescription

$$(\mathbf{u}, \mathbf{v}) \mapsto \mathbf{u} \cdot (\mathbf{A}\,\mathbf{v}) = \mathbf{A} \cdot \mathbf{u} \otimes \mathbf{v}\,.$$

By generalising this, we define *K*th-order tensors as multilinear forms

$$\overset{\langle K \rangle}{\mathbf{A}} : \mathscr{V}^K \to \mathscr{R}.$$

Scalars

For $K = 0$ we put

$$\overset{\langle 0 \rangle}{\mathbf{A}} : \mathscr{R} \to \mathscr{R}\,.$$

$\overset{\langle 0 \rangle}{\mathbf{A}}$ can be interpreted as a scalar by the multiplication of real numbers, which is obviously linear

$$x \mapsto \overset{\langle 0 \rangle}{\mathbf{A}}\, x\,.$$

Therefore, we can identify *0*th-order tensors by the reals $\mathscr{R}$.

Vectors

For $K = 1$ we have a linear mapping

$$\overset{\langle 1 \rangle}{\mathbf{A}} : \mathscr{V} \to \mathscr{R} .$$

$\overset{\langle 1 \rangle}{\mathbf{A}}$ can be interpreted by the inner product of $\mathscr{V}$, which is linear by definition, as a (co)vector by putting

$$\mathbf{v} \mapsto \overset{\langle 1 \rangle}{\mathbf{A}} \cdot \mathbf{v}$$

(simple contraction). Therefore, we can identify the *1*st-order tensors by the vector space $\mathscr{V}$ (RIESZ's representation theorem). After choosing a basis $\{\mathbf{g}_i\}$ and its dual $\{\mathbf{g}^i\}$ we have the component representations

$$\overset{\langle 1 \rangle}{\mathbf{A}} = a^i \, \mathbf{g}_i = a_i \, \mathbf{g}^i$$

with the components

$$a^i = \overset{\langle 1 \rangle}{\mathbf{A}} \cdot \mathbf{g}^i \quad \text{and} \quad a_i = \overset{\langle 1 \rangle}{\mathbf{A}} \cdot \mathbf{g}_i \qquad \text{for } i = 1, 2, 3.$$

Dyads

For $K = 2$ we have a bilinear mapping

$$\overset{\langle 2 \rangle}{\mathbf{A}} : \mathscr{V} \times \mathscr{V} \to \mathscr{R}.$$

As we have already seen, we can represent *2*nd-order tensors after choosing a basis as

$$\overset{\langle 2 \rangle}{\mathbf{A}} = A_{ik} \, \mathbf{g}^i \otimes \mathbf{g}^k$$

with generally $3^2 = 9$ independent components, and the application to two vectors $\mathbf{u}, \mathbf{v} \in \mathscr{V}$ is defined by

$$\overset{\langle 2 \rangle}{\mathbf{A}} (\mathbf{u}, \mathbf{v}) = A_{ik} \, (\mathbf{g}^i \cdot \mathbf{u}) \, (\mathbf{g}^k \cdot \mathbf{v}) = A_{ik} \, u^i \, v^k .$$

We have shown before in (1.8) that the trace induces an inner product between such tensors, which gives

$$\overset{\langle 2 \rangle}{\mathbf{A}} (\mathbf{u}, \mathbf{v}) = \overset{\langle 2 \rangle}{\mathbf{A}} \cdot (\mathbf{u} \otimes \mathbf{v}) .$$

For the components of a dyad we obtain after (1.15)

$$A_{ik} = \overset{\langle 2 \rangle}{\mathbf{A}} \cdot (\mathbf{g}_i \otimes \mathbf{g}_k) \qquad A_i^{\ k} = \overset{\langle 2 \rangle}{\mathbf{A}} \cdot (\mathbf{g}_i \otimes \mathbf{g}^k)$$

$$A^{ik} = \overset{\langle 2 \rangle}{\mathbf{A}} \cdot (\mathbf{g}^i \otimes \mathbf{g}^k) \qquad A^i_{\ k} = \overset{\langle 2 \rangle}{\mathbf{A}} \cdot (\mathbf{g}^i \otimes \mathbf{g}_k) .$$

Triads

For $K = 3$ we have a trilinear mapping

$$\overset{\langle 3\rangle}{\mathbf{A}} : \mathscr{V} \times \mathscr{V} \times \mathscr{V} \to \mathscr{R}.$$

Such a **triad** is the generalisation of the dyadic product to three factors

$$\overset{\langle 3\rangle}{\mathbf{A}} = A_{ikl}\, \mathbf{g}^i \otimes \mathbf{g}^k \otimes \mathbf{g}^l$$

and has $3^3 = 27$ independent components. The application of such triad to three vectors $\mathbf{u}, \mathbf{v}, \mathbf{w} \in \mathscr{V}$ is defined by

$$\overset{\langle 3\rangle}{\mathbf{A}} (\mathbf{u}, \mathbf{v}, \mathbf{w}) = A_{ikl}\, (\mathbf{g}^i \cdot \mathbf{u})\, (\mathbf{g}^k \cdot \mathbf{v})\, (\mathbf{g}^l \cdot \mathbf{w}) = A_{ikl}\, u^i\, v^k\, w^l,$$

which is automatically linear in the three inputs. This gives rise to introduce an inner product between triads by generalising

$$\overset{\langle 3\rangle}{\mathbf{A}} (\mathbf{g}_i, \mathbf{g}_k, \mathbf{g}_l) =: \overset{\langle 3\rangle}{\mathbf{A}} \cdot (\mathbf{g}_i \otimes \mathbf{g}_k \otimes \mathbf{g}_l) = A_{ikl}$$

or

$$\overset{\langle 3\rangle}{\mathbf{A}} (\mathbf{g}^i, \mathbf{g}^k, \mathbf{g}^l) =: \overset{\langle 3\rangle}{\mathbf{A}} \cdot (\mathbf{g}^i \otimes \mathbf{g}^k \otimes \mathbf{g}^l) = A^{ikl}$$

to arbitrary linear combinations of triads in the right factor. We obtain analogously

$$\overset{\langle 3\rangle}{\mathbf{A}} \cdot \overset{\langle 3\rangle}{\mathbf{B}} = A_{ikl}\, B^{ikl} = A^{ikl}\, B_{ikl}$$

with respect to dual bases.

One can also interpret such a triad as a linear mapping of a 2nd-order tensor onto a vector by assigning

$$\overset{\langle 3\rangle}{\mathbf{A}} (\mathbf{T}) = A_{ikl}\, T^{mn}\, (\mathbf{g}^k \cdot \mathbf{g}_m)\, (\mathbf{g}^l \cdot \mathbf{g}_n)\, \mathbf{g}^i = A_{ikl}\, T^{kl}\, \mathbf{g}^i \qquad \in \mathscr{V}$$

(double contraction), or as a linear mapping of a vector onto a 2nd-order tensor by assigning

$$\overset{\langle 3\rangle}{\mathbf{A}} (\mathbf{u}) = A_{ikl}\, (\mathbf{g}^l \cdot \mathbf{u})\, \mathbf{g}^i \otimes \mathbf{g}^k = A_{ikl}\, u^l\, \mathbf{g}^i \otimes \mathbf{g}^k \qquad \in \mathscr{Lin}$$

(simple contraction).

An **example** for a triad gives the $\overset{\langle 3\rangle}{\boldsymbol{\varepsilon}}$-tensor, which has the permutation symbols as components with respect to any ONB $\{\mathbf{e}_i\}$

$$\overset{\langle 3\rangle}{\boldsymbol{\varepsilon}} = \varepsilon^{ijk}\, \mathbf{e}_i \otimes \mathbf{e}_j \otimes \mathbf{e}_k .$$

It stands for a tensorial form of the triple product of $\mathbf{t}, \mathbf{u}, \mathbf{v} \in \mathscr{V}$

$$\overset{\langle 3\rangle}{\boldsymbol{\varepsilon}}\,(\mathbf{t}\,,\mathbf{u}\,,\mathbf{v}) \;=\; \varepsilon^{ijk}\,(\mathbf{e}_i\cdot\mathbf{t})(\mathbf{e}_j\cdot\mathbf{u})(\mathbf{e}_k\cdot\mathbf{v}) \;=\; \varepsilon^{ijk}\,t_i\,u_j\,v_k$$

$$=\;[\mathbf{t}\,,\mathbf{u}\,,\mathbf{v}] \;=\; \mathbf{t}\cdot(\mathbf{u}\times\mathbf{v})\,.$$

$\overset{\langle 3\rangle}{\boldsymbol{\varepsilon}}$ can also describe the cross product of two vectors

$$\overset{\langle 3\rangle}{\boldsymbol{\varepsilon}}\,(\mathbf{u}\,,\mathbf{v}) \;:=\; \varepsilon^{ijk}\,(\mathbf{e}_j\cdot\mathbf{u})(\mathbf{e}_k\cdot\mathbf{v})\,\mathbf{e}_i \;=\; \varepsilon^{jki}\,u_j\,v_k\,\mathbf{e}_i \;=\; \mathbf{u}\times\mathbf{v}\,.$$

Tetrads

For $K = 4$ we have a quadrilinear mapping

$$\overset{\langle 4\rangle}{\mathbf{A}} : \mathscr{V}\times\mathscr{V}\times\mathscr{V}\times\mathscr{V}\to\mathscr{R}.$$

Such tetrads are quite important in mechanics. Therefore, we introduce a distinct notation as $\boldsymbol{A} = \overset{\langle 4\rangle}{\mathbf{A}}$. A tetrad can be represented by tensor products of 4th-order

$$\boldsymbol{A} \;=\; A_{iklm}\,\mathbf{g}^i\otimes\mathbf{g}^k\otimes\mathbf{g}^l\otimes\mathbf{g}^m$$

with $3^4 = 81$ independent components. The application to four vectors $\mathbf{u}\,,\mathbf{v}\,,\mathbf{w}\,,\mathbf{x} \in \mathscr{V}$ is defined by

$$\boldsymbol{A}(\mathbf{u}\,,\mathbf{v}\,,\mathbf{w}\,,\mathbf{x}) \;=\; A_{iklm}\,(\mathbf{g}^i\cdot\mathbf{u})\,(\mathbf{g}^k\cdot\mathbf{v})\,(\mathbf{g}^l\cdot\mathbf{w})\,(\mathbf{g}^m\cdot\mathbf{x})$$

$$=\; A_{iklm}\,u^i\,v^k\,w^l\,x^m$$

(fourfold contraction). One can introduce an inner product on tetrads by generalising

$$\boldsymbol{A}\,(\mathbf{g}_i\,,\mathbf{g}_j\,,\mathbf{g}_k\,,\mathbf{g}_l) \;=:\; \boldsymbol{A}\cdot(\mathbf{g}_i\otimes\mathbf{g}_j\otimes\mathbf{g}_k\otimes\mathbf{g}_l) \;=\; A_{ijkl}$$

or

$$\boldsymbol{A}\,(\mathbf{g}^i,\mathbf{g}^j,\mathbf{g}^k,\mathbf{g}^l) \;=:\; \boldsymbol{A}\cdot(\mathbf{g}^i\otimes\mathbf{g}^j\otimes\mathbf{g}^k\otimes\mathbf{g}^l) \;=\; A^{ijkl}$$

to arbitrary linear combinations of tetrads in the right factor. This gives analogously the inner product of two arbitrary 4th-order tensors

$$\boldsymbol{A}\cdot\boldsymbol{B} \;=\; A_{ijkl}\,B^{ijkl} \;=\; A^{ijkl}\,B_{ijkl}$$

with respect to dual bases.

We can also use a tetrad for linearly mapping a 2nd-order tensor

$$\mathbf{T} \;=\; T^{op}\,\mathbf{g}_o\otimes\mathbf{g}_p \qquad \in \mathscr{Lin}$$

into another one

$$\boldsymbol{A}[\mathbf{T}] \;:=\; A_{iklm}\,\mathbf{g}^i\otimes\mathbf{g}^k(\mathbf{g}^l\cdot\mathbf{g}_o)\,(\mathbf{g}^m\cdot\mathbf{g}_p)\,T^{op}$$

$$=\; A_{iklm}\,T^{lm}\,\mathbf{g}^i\otimes\mathbf{g}^k \qquad \in \mathscr{Lin}\,.$$

(double contraction). In this way, the tetrads are the general linear mappings between *2*nd-order tensors

$$\boldsymbol{A} : \mathcal{Lin} \to \mathcal{Lin} \quad \Big| \quad \mathbf{T} \mapsto \boldsymbol{A}[\mathbf{T}] .$$

and in this role they are very important in mechanics. The composition of tetrads is defined as it is for functions and written as, *e.g.*, $\boldsymbol{A}\,\boldsymbol{B}$.

A tetrad $\boldsymbol{A}$ is called **invertible** if another tetrad $\boldsymbol{A}^{-1}$ exists such that the composition gives the identity:

$$\boldsymbol{A}^{-1}\,\boldsymbol{A}[\mathbf{T}] = \mathbf{T} \qquad \forall\, \mathbf{T} \in \mathcal{Lin} .$$

Otherwise it is called singular.

We can now generalise many concepts of *2*nd-order tensors to *4*th-order tensors. In doing so, we can pose the eigenvalue problem for tetrads, leading to eigentensors, to a characteristic polynomial, and, as its coefficients, to principal invariants.

Examples for Tetrads. The *4*th-order identity tensor maps each tensor into itself

$$\boldsymbol{I} : \mathbf{T} \mapsto \mathbf{T} \qquad \forall\, \mathbf{T} \in \mathcal{Lin} .$$

It has a simple representation with respect to an arbitrary basis $\{\mathbf{g}_i\}$ and its dual $\{\mathbf{g}^i\}$

$$\boldsymbol{I} = \mathbf{g}_i \otimes \mathbf{g}_k \otimes \mathbf{g}^i \otimes \mathbf{g}^k .$$

The *4*th-order **transposer** maps all tensors into their transposed ones

$$\boldsymbol{T} : \mathbf{T} \mapsto \mathbf{T}^T \qquad \forall\, \mathbf{T} \in \mathcal{Lin} .$$

It has the representation

(1.28) $$\boldsymbol{T} = \mathbf{g}_k \otimes \mathbf{g}_i \otimes \mathbf{g}^i \otimes \mathbf{g}^k .$$

The **symmetriser** maps all tensors into their symmetric part

$$\boldsymbol{I}^S : \mathcal{Lin} \to \mathcal{Sym} \quad \Big| \quad \mathbf{T} \mapsto sym(\mathbf{T}) .$$

It has the representations

(1.29) $$\begin{aligned} \boldsymbol{I}^S &= \tfrac{1}{2}\,(\boldsymbol{I} + \boldsymbol{T}) \\ &= \tfrac{1}{2}\,(\mathbf{g}_i \otimes \mathbf{g}_k + \mathbf{g}_k \otimes \mathbf{g}_i) \otimes \mathbf{g}^i \otimes \mathbf{g}^k \\ &= \tfrac{1}{2}\,(\delta^i{}_k\, \delta^j{}_l + \delta^j{}_k\, \delta^i{}_l)\, \mathbf{g}_i \otimes \mathbf{g}_j \otimes \mathbf{g}^k \otimes \mathbf{g}^l . \end{aligned}$$

The **anti(sym)metriser** maps all tensors into their skew part

$$\boldsymbol{I}^A : \mathcal{Lin} \to \mathcal{Skew} \quad \Big| \quad \mathbf{T} \mapsto skw(\mathbf{T}) .$$

It has the representations

(1.30) $\boldsymbol{I}^A = \tfrac{1}{2}(\boldsymbol{I} - \boldsymbol{T})$

$= \tfrac{1}{2}(\mathbf{g}_i \otimes \mathbf{g}_k - \mathbf{g}_k \otimes \mathbf{g}_i) \otimes \mathbf{g}^i \otimes \mathbf{g}^k$

$= \tfrac{1}{2}(\delta^i{}_k\, \delta^j{}_l - \delta^j{}_k\, \delta^i{}_l)\, \mathbf{g}_i \otimes \mathbf{g}_j \otimes \mathbf{g}^k \otimes \mathbf{g}^l .$

This leads to the following decomposition of the *4*th-order identity tensor

$$\boldsymbol{I} = \boldsymbol{I}^S + \boldsymbol{I}^A .$$

The **deviatoriser** maps all tensors into their deviatoric part

$$\boldsymbol{D} : \mathbf{T} \mapsto \mathbf{T}' \qquad \forall\, \mathbf{T} \in \mathcal{Lin} .$$

It has the representation

(1.31) $\boldsymbol{D} = \boldsymbol{I} - \tfrac{1}{3}\mathbf{I} \otimes \mathbf{I} .$

All these tetrads are subspace projectors on $\mathcal{Lin}$. We can decompose the identity tensor now into three orthogonal subspace projectors

$$\boldsymbol{I} = \tfrac{1}{3}\mathbf{I} \otimes \mathbf{I} + \boldsymbol{D}\,\boldsymbol{I}^S + \boldsymbol{I}^A$$

mapping every tensor into the sum of its spherical part, its symmetric deviator, and its skew deviator.

The **transposed** $\boldsymbol{A}^T$ of a tetrad $\boldsymbol{A}$ is defined in analogy to (1.4) by the bilinear form of tensors

(1.32) $\mathbf{S} \cdot \boldsymbol{A}^T[\mathbf{T}] = \mathbf{T} \cdot \boldsymbol{A}[\mathbf{S}] \qquad \forall\, \mathbf{S}, \mathbf{T} \in \mathcal{Lin} .$

A tetrad has the (major) **symmetry** if $\boldsymbol{A}^T = \boldsymbol{A}$. For its components this leads to

$$A_{iklm} = A_{lmik} .$$

The identity tensor is symmetric, just same as $\boldsymbol{I}^S$, $\boldsymbol{I}^A$, and $\boldsymbol{D}$.

Other symmetries are also often used.

The **right subsymmetry** of a tetrad $\boldsymbol{A}$ is given, if the following equivalent conditions are valid:

$\boldsymbol{A}[\mathbf{T}^T] = \boldsymbol{A}[\mathbf{T}] \qquad \forall\, \mathbf{T} \in \mathcal{Lin}$

$\Leftrightarrow \quad \boldsymbol{A} = \boldsymbol{A}\,\boldsymbol{T}$

$\Leftrightarrow \quad \boldsymbol{A} = \boldsymbol{A}\,\boldsymbol{I}^S$

$\Leftrightarrow \quad \boldsymbol{A}\,\boldsymbol{I}^A = \boldsymbol{0}$

$\Leftrightarrow \quad A_{ik}{}^{lm} = A_{ik}{}^{ml} .$

A tetrad $\boldsymbol{A}$ has the **left subsymmetry** if its transposed has the right subsymmetry. This is equivalent to all of the following conditions:

$(\boldsymbol{A}[\mathbf{T}])^T = \boldsymbol{A}[\mathbf{T}] \qquad \forall\, \mathbf{T} \in \mathcal{Lin}$

$\Leftrightarrow \qquad \boldsymbol{A} = \boldsymbol{T}\boldsymbol{A}$

$\Leftrightarrow \qquad \boldsymbol{A} = \boldsymbol{I}^S \boldsymbol{A}$

$\Leftrightarrow \qquad \boldsymbol{I}^A \boldsymbol{A} = \boldsymbol{0}$

$\Leftrightarrow \qquad A_{ik}{}^{lm} = A_{ki}{}^{lm}.$

The following theorem contains immediate results of these definitions.

Theorem 1.22.

1) A tetrad with the left subsymmetry maps all tensors into symmetric ones.

2) For a symmetric tetrad, the left and the right subsymmetry are equivalent.

3) The identity tensor is symmetric, but has no subsymmetry.

4) The symmetriser is symmetric and has both subsymmetries.

5) A tetrad with any subsymmetry is singular (non-invertible).

6) A tetrad has $3^4 = 81$ *independent components.*

7) A tetrad with the two subsymmetries has $6 \times 6 = 36$ *independent components.*

8) A symmetric tetrad has $9\,(9+1)\,/\,2 = 45$ *independent components.*

9) A symmetric tetrad with the two subsymmetries has $6\,(6+1)\,/\,2 = 21$ *independent components.*

For the linear mappings between symmetric tensors the two subsymmetries can be generally assumed. By doing so, we reduce the number of independent components from *81* to *36*. In this case, it is convenient to consider symmetric *2*nd-order tensors as 6 dimensional vectors according to

$$\begin{aligned}
\mathbf{T} &= T_{11}\, \mathbf{e}_1 \otimes \mathbf{e}_1 + T_{22}\, \mathbf{e}_2 \otimes \mathbf{e}_2 + T_{33}\, \mathbf{e}_3 \otimes \mathbf{e}_3 \\
&\quad + T_{23}\, (\mathbf{e}_2 \otimes \mathbf{e}_3 + \mathbf{e}_3 \otimes \mathbf{e}_2) \\
&\quad + T_{31}\, (\mathbf{e}_3 \otimes \mathbf{e}_1 + \mathbf{e}_1 \otimes \mathbf{e}_3) \\
&\quad + T_{12}\, (\mathbf{e}_1 \otimes \mathbf{e}_2 + \mathbf{e}_2 \otimes \mathbf{e}_1) \\
&= T_{11}\, \mathbf{e}_{V1} + T_{22}\, \mathbf{e}_{V2} + T_{33}\, \mathbf{e}_{V3} \\
&\quad + \sqrt{2}\, T_{23}\, \mathbf{e}_{V4} + \sqrt{2}\, T_{31}\, \mathbf{e}_{V5} + \sqrt{2}\, T_{12}\, \mathbf{e}_{V6} \qquad \in \mathcal{Sym}
\end{aligned}$$

with the coefficients

$$\{T_{11}\,,\, T_{22}\,,\, T_{33}\,,\, \sqrt{2}\, T_{23}\,,\, \sqrt{2}\, T_{31}\,,\, \sqrt{2}\, T_{12}\}$$

with respect to the tensorial ONB in $\mathcal{Sym}$

$$\begin{aligned}
&\mathbf{e}_{V1} = \mathbf{e}_1 \otimes \mathbf{e}_1 \qquad \mathbf{e}_{V2} = \mathbf{e}_2 \otimes \mathbf{e}_2 \qquad \mathbf{e}_{V3} = \mathbf{e}_3 \otimes \mathbf{e}_3 \\
(1.33) \qquad &\mathbf{e}_{V4} = {}^1/_{\sqrt{2}}\, (\mathbf{e}_2 \otimes \mathbf{e}_3 + \mathbf{e}_3 \otimes \mathbf{e}_2) \\
&\mathbf{e}_{V5} = {}^1/_{\sqrt{2}}\, (\mathbf{e}_1 \otimes \mathbf{e}_3 + \mathbf{e}_3 \otimes \mathbf{e}_1) \\
&\mathbf{e}_{V6} = {}^1/_{\sqrt{2}}\, (\mathbf{e}_2 \otimes \mathbf{e}_1 + \mathbf{e}_1 \otimes \mathbf{e}_2)
\end{aligned}$$

where the factor $1/\sqrt{2}$ results from the normalisation of the basis $\{\mathbf{e}_{V\alpha}\}$ like, *e.g.*

$$^1/_{\sqrt{2}}\,(\mathbf{e}_1 \otimes \mathbf{e}_2 + \mathbf{e}_2 \otimes \mathbf{e}_1) \cdot {}^1/_{\sqrt{2}}\,(\mathbf{e}_1 \otimes \mathbf{e}_2 + \mathbf{e}_2 \otimes \mathbf{e}_1) = 1\,.$$

Such an ONB allows for the **VOIGT**[13] **representation** of a tetrad with both subsymmetries as

(1.34)
$$\mathbf{C} = C_{ijkl}\,\mathbf{e}_i \otimes \mathbf{e}_j \otimes \mathbf{e}_k \otimes \mathbf{e}_l = \sum_{\alpha,\beta=1}^{6} c_V^{\alpha\beta}\,\mathbf{e}_{V\alpha} \otimes \mathbf{e}_{V\beta}$$

by a 6×6 matrix of components

$$\begin{bmatrix} C_{1111} & C_{1122} & C_{1133} & \sqrt{2}C_{1123} & \sqrt{2}C_{1131} & \sqrt{2}C_{1112} \\ C_{2211} & C_{2222} & C_{2233} & \sqrt{2}C_{2223} & \sqrt{2}C_{2231} & \sqrt{2}C_{2212} \\ C_{3311} & C_{3322} & C_{3333} & \sqrt{2}C_{3323} & \sqrt{2}C_{3331} & \sqrt{2}C_{3312} \\ \sqrt{2}C_{2311} & \sqrt{2}C_{2322} & \sqrt{2}C_{2333} & 2C_{2323} & 2C_{2331} & 2C_{2312} \\ \sqrt{2}C_{3111} & \sqrt{2}C_{3122} & \sqrt{2}C_{3133} & 2C_{3123} & 2C_{3131} & 2C_{3112} \\ \sqrt{2}C_{1211} & \sqrt{2}C_{1222} & \sqrt{2}C_{1233} & 2C_{1223} & 2C_{1231} & 2C_{1212} \end{bmatrix}$$

which is similar to that of a *2*nd-order tensor. Note that in the literature the normalisation[14] is often omitted, so that the coefficients are different (without *2* and $\sqrt{2}$).

Such a tetrad has the (major-) symmetry if and only if this matrix is symmetric. In this case it has only 21 independent values.

Tensors of fifth order are called **pentadics**, and those of sixth order **hexadics**.

Tensors of Higher Order

By generalising the above concepts in an analogous way, one can represent a tensor of arbitrary order $\overset{\langle K\rangle}{\mathbf{A}}$ $(K > 0)$ with respect to a basis $\{\mathbf{g}^i\}$ as linear combinations like

$$\overset{\langle K\rangle}{\mathbf{A}} = A_{ik\,...\,l}\,\mathbf{g}^i \otimes \mathbf{g}^k \otimes ... \otimes \mathbf{g}^l$$

with 3^K components. This representation is not the only possible one. By means of the metric coefficients one can exchange one or several of these basis vectors by its duals $\{\mathbf{g}_j\}$ similar to (1.15) like, *e.g.*,

$$\overset{\langle K\rangle}{\mathbf{A}} = A_{i\,...\,s\,...\,l}\,G^{sp}\,\mathbf{g}^i \otimes ... \otimes \mathbf{g}_p \otimes ... \otimes \mathbf{g}^l\,.$$

---

[13] Woldemar Voigt (1850-1919)

[14] The normalisation has already been suggested by KELVIN (THOMSON 1856).

For most physical tensors there is a natural representation with respect to either basis, so that such a change of basis is rarely necessary.

The inner product between two tensors of the same order is

$$\overset{\langle K\rangle}{\mathbf{A}} \cdot \overset{\langle K\rangle}{\mathbf{B}} = A_{ik\,\dots\,l}\, B^{ik\,\dots\,l} = A^{ik\,\dots\,l}\, B_{ik\,\dots\,l}\,.$$

The general linear mapping of a Kth-order tensor $\overset{\langle K\rangle}{\mathbf{B}}$ into an Lth-order tensor $\overset{\langle L\rangle}{\mathbf{C}}$ is represented by a $(K{+}L)$th-order tensor $\overset{\langle L+K\rangle}{\mathbf{A}}$ as

$$\overset{\langle L\rangle}{\mathbf{C}} = \overset{\langle L+K\rangle}{\mathbf{A}} \, [\, \overset{\langle K\rangle}{\mathbf{B}} \,]$$

$$= A_{\,i_1 \cdots\, i_L\; i_{L+1} \cdots\, i_{L+K}} \; B^{\,i_{L+1}\, \cdots\, i_{L+K}} \; \mathbf{g}^{\,i_1} \otimes \dots \otimes \mathbf{g}^{\,i_L}$$

(K-fold contraction).

Eigenvalue problems can only be posed for *endomorphisms*, *i.e.*, for mappings of some linear space into itself. This interpretation is always possible for tensors of even order $2K$ ($K > 0$), since we can use them as mappings between Kth-order tensors

$$\overset{\langle 2K\rangle}{\mathbf{T}} : \overset{\langle K\rangle}{\mathbf{A}} \mapsto \overset{\langle K\rangle}{\mathbf{B}}\,.$$

Their eigenvalue problem is then

$$(\, \overset{\langle 2K\rangle}{\mathbf{T}} - \lambda_r \overset{\langle 2K\rangle}{\mathbf{I}} \,)\,[\, \overset{\langle K\rangle}{\mathbf{P}}_r \,] = \overset{\langle K\rangle}{\mathbf{0}}$$

(K-fold contraction) with the $2K$th-order identity tensor $\overset{\langle 2K\rangle}{\mathbf{I}}$, the Kth-order zero tensor $\overset{\langle K\rangle}{\mathbf{0}}$, the Kth-order **eigentensor** $\overset{\langle K\rangle}{\mathbf{P}}_r$ and the corresponding eigenvalue λ_r. By analogy to 2nd-order tensors, this again leads to a characteristic polynomial, the coefficients of which are the principal invariants of $\overset{\langle 2K\rangle}{\mathbf{T}}$.

By analogy to (1.23), a projector decomposition of an arbitrary symmetric tensor of even order $\overset{\langle 2K\rangle}{\mathbf{T}}$ is of the form

$$\overset{\langle 2K\rangle}{\mathbf{T}} = \sum_{r=1}^{K} \lambda_r \overset{\langle 2K\rangle}{\mathbf{P}_r}$$

with K being the number of different eigenvalues λ_r of $\overset{\langle 2K\rangle}{\mathbf{T}}$. The projectors $\overset{\langle 2K\rangle}{\mathbf{P}_r}$ again fulfil the projector rules (1.24).

We should finally remark that one can generalise the above tensor concept to linear mappings between *different* inner-product spaces $\mathscr{V}$ and $\mathscr{W}$ of arbitrary dimensions

$$\mathbf{T}: \mathscr{V} \rightarrow \mathscr{W}.$$

A representation of such mapping can be obtained by introducing the dyadic product between vectors of the corresponding spaces

$$\mathbf{w} \otimes \mathbf{v} \qquad \text{with } \mathbf{v} \in \mathscr{V}, \mathbf{w} \in \mathscr{W}$$

which maps any vector $\mathbf{a} \in \mathscr{V}$ into $\mathscr{W}$ according to

$$\mathbf{w} \otimes \mathbf{v}\,(\mathbf{a}) \ := \ (\mathbf{v} \cdot \mathbf{a})\,\mathbf{w}\,.$$

By using the bases $\{\mathbf{f}^i\} \subset \mathscr{V}$ and $\{\mathbf{g}_i\} \subset \mathscr{W}$, we can represent such a tensor in the form

$$\mathbf{T} = T^i{}_j\,\mathbf{g}_i \otimes \mathbf{f}^j.$$

If the two spaces are different, however, an eigenvalue problem of such a tensor does not exist.

The **RAYLEIGH**[15] **product** maps all basis vectors of a tensor simultaneously without changing its components. To be more precise, let $\overset{\langle K\rangle}{\mathbf{T}}$ be a tensor of *K*th-order ($K \geq 0$) and $\mathbf{A}$ a 2nd-order tensor. Then the RAYLEIGH product of $\overset{\langle K\rangle}{\mathbf{T}}$ by $\mathbf{A}$ is defined as

$$\mathbf{A} * \overset{\langle K\rangle}{\mathbf{T}} = \mathbf{A} * (T^{ik\ldots l}\,\mathbf{g}_i \otimes \mathbf{g}_k \otimes \ldots \otimes \mathbf{g}_l)$$

$$(1.35) \qquad := T^{ik\ldots l}\,(\mathbf{A}\,\mathbf{g}_i) \otimes (\mathbf{A}\,\mathbf{g}_k) \otimes \ldots \otimes (\mathbf{A}\,\mathbf{g}_l)\,.$$

Of course, the result does not depend on the choice of the basis. If $\mathbf{A}$ is a versor, then the product is a rotation of $\overset{\langle K\rangle}{\mathbf{T}}$.

The RAYLEIGH product does not commute, but it is associative in the left factor

$$\mathbf{A} * (\mathbf{B} * \overset{\langle K\rangle}{\mathbf{T}}) = (\mathbf{A}\,\mathbf{B}) * \overset{\langle K\rangle}{\mathbf{T}} \qquad \forall\, \mathbf{A}, \mathbf{B} \in \mathscr{Lin}\,.$$

Thus we can omit the brackets. In this product, the identity tensor also gives the identity mapping

$$\mathbf{I} * \overset{\langle K\rangle}{\mathbf{T}} = \overset{\langle K\rangle}{\mathbf{T}}$$

[15] John William Strutt, since 1873 Lord Rayleigh (1842-1919)

and the inversion is done by

$$\mathbf{A} * (\mathbf{A}^{-1} * \overset{\langle K \rangle}{\mathbf{T}}) = \overset{\langle K \rangle}{\mathbf{T}} \qquad \forall\, \mathbf{A} \in \mathscr{Inv}\,.$$

Moreover, if $\mathbf{S}$ and $\mathbf{T}$ are tensors of arbitrary order, then we have

$$\mathbf{A} * (\mathbf{S} \otimes \mathbf{T}) = (\mathbf{A} * \mathbf{S}) \otimes (\mathbf{A} * \mathbf{T}) \qquad \forall\, \mathbf{A} \in \mathscr{Lin}\,.$$

This would not hold, if we replace the tensor product by the composition or an arbitrary contraction, unless $\mathbf{A}$ is orthogonal. For $K \equiv 0$ (scalars) the RAYLEIGH product coincides with the identity. For $K \equiv 1$ the RAYLEIGH product coincides with a linear mapping

$$\mathbf{A} * \mathbf{t} = \mathbf{A}\,\mathbf{t} \qquad \forall\, \mathbf{A} \in \mathscr{Lin}\,, \mathbf{t} \in \mathscr{V},$$

and for $K \equiv 2$ we have

$$\mathbf{A} * \mathbf{T} = \mathbf{A}\,\mathbf{T}\,\mathbf{A}^T \qquad \forall\, \mathbf{A}\,, \mathbf{T} \in \mathscr{Lin}\,.$$

For invertible *2*nd-order tensors we obtain

$$(\mathbf{A} * \mathbf{T})^{-1} = \mathbf{A}^{-T} * \mathbf{T}^{-1} \qquad \forall\, \mathbf{A}\,, \mathbf{T} \in \mathscr{Inv}\,.$$

A tensor of arbitrary order $\overset{\langle K \rangle}{\mathbf{T}}$ is called an **isotropic tensor** if it is invariant under the RAYLEIGH product with any orthogonal tensor:

$$\mathbf{Q} * \overset{\langle K \rangle}{\mathbf{T}} = \overset{\langle K \rangle}{\mathbf{T}} \qquad \forall\, \mathbf{Q} \in \mathscr{Orth}\,.$$

It this holds only for proper orthogonal tensors $\mathbf{Q} \in \mathscr{Orth}^{+}$, then the tensor $\overset{\langle K \rangle}{\mathbf{T}}$ is called hemitropic. For even-order tensors hemitropy and isotropy coincide, but not for odd-order ones.

Examples. All spherical tensors are isotropic *2*nd-order tensors, the same as all scalar multiples of even-order unit tensors, or as the following *4*th-order tensor with respect to an ONB $\{\mathbf{e}_i\}$

$$\alpha\, \mathbf{e}_i \otimes \mathbf{e}_i \otimes \mathbf{e}_j \otimes \mathbf{e}_j + \beta\, \mathbf{e}_i \otimes \mathbf{e}_j \otimes \mathbf{e}_i \otimes \mathbf{e}_j + \gamma\, \mathbf{e}_i \otimes \mathbf{e}_j \otimes \mathbf{e}_j \otimes \mathbf{e}_i$$

with $\alpha\,, \beta\,, \gamma \in \mathscr{R}$. The triad $\overset{\langle 3 \rangle}{\boldsymbol{\varepsilon}}$ is only hemitropic,

1.2.5 Isotropic Tensor-Functions

An application of such RAYLEIGH products is the specification of symmetry properties for functions between tensor spaces of arbitrary order

$$f : \overset{\langle K \rangle}{\mathbf{X}} \mapsto \overset{\langle M \rangle}{\mathbf{M}} \qquad K, M \geq 0\,.$$

We call a *2*nd-order tensor $\mathbf{A}$ a **symmetry transformation** of f, if for all $\overset{\langle K\rangle}{\mathbf{X}}$ in its domain the following holds

$$f(\mathbf{A} * \overset{\langle K\rangle}{\mathbf{X}}) = \mathbf{A} * f(\overset{\langle K\rangle}{\mathbf{X}}) .$$

However, this is only one possibility, and there are others depending on the nature of the argument and the image, as we will see later. If all orthogonal tensors (versors) are symmetry transformations, then we call f an **isotropic tensor-function**. If this invariance holds only for proper orthogonal tensors, it is called **hemitropic tensor function**. However, if K and M are even, then these two definitions coincide.

Isotropy of tensor-functions plays an important role in material theory, because it provides possibilities to find representations for such functions. We will give two examples.

Definition 1.23. A function

$$\varphi : \mathcal{Sym} \rightarrow \mathcal{R} \qquad \Big| \qquad \mathbf{C} \mapsto \varphi(\mathbf{C})$$

is called a **real isotropic tensor-function** if

$$\varphi(\mathbf{C}) = \varphi(\mathbf{Q} * \mathbf{C}) \qquad \forall\, \mathbf{C} \in \mathcal{Sym} ,\ \forall\, \mathbf{Q} \in \mathcal{Orth}^+ .$$

Theorem 1.24. Representation of real isotropic tensor-functions
A real isotropic tensor-function φ *can be represented*

- *either as a function of the principal invariants of the tensor*

$$\varphi(\mathbf{C}) = \phi(I_{\mathbf{C}}, II_{\mathbf{C}}, III_{\mathbf{C}})$$

- *or as a symmetric function of the eigenvalues of* $\mathbf{C}$

$$\varphi(\mathbf{C}) = \underline{\phi}(\lambda_1, \lambda_2, \lambda_3) = \underline{\phi}(\lambda_2, \lambda_1, \lambda_3) = \underline{\phi}(\lambda_3, \lambda_2, \lambda_1)$$

Proof. Let φ be a real isotropic tensor-function. As the characteristic equations of a symmetric tensor $\mathbf{C}$ and of the similar $\mathbf{Q}\,\mathbf{C}\,\mathbf{Q}^T$ coincide for all $\mathbf{Q} \in \mathcal{Orth}$, they have also the same eigenvalues and principal invariants

$$I_{\mathbf{C}} = I_{\mathbf{QCQ}^T} \qquad II_{\mathbf{C}} = II_{\mathbf{QCQ}^T} \qquad III_{\mathbf{C}} = III_{\mathbf{QCQ}^T} .$$

The inverse also holds. If two symmetric tensors $\mathbf{C}$ and $\underline{\mathbf{C}}$ have the same eigenvalues (or principal invariants), then we can find a $\mathbf{Q} \in \mathcal{Orth}$ such that

$$\mathbf{Q} * \mathbf{C} = \mathbf{Q}\,\mathbf{C}\,\mathbf{Q}^T = \underline{\mathbf{C}}$$

holds. In fact, let $\mathbf{u}_i$ be the eigenvectors of $\mathbf{C}$ and $\underline{\mathbf{u}}_i$ those of $\underline{\mathbf{C}}$, then by an appropriate numbering, this is performed by the orthogonal tensor

$$\mathbf{Q} \equiv \underline{\mathbf{u}}_i \otimes \mathbf{u}_i .$$

Accordingly, the dependence on $\mathbf{C}$ can be reduced on that of its principal invariants. This completes the proof for the first representation.

To prove the second, we use the spectral form

$$\mathbf{C} = \sum_{i=1}^{3} \lambda_i \, \mathbf{u}_i \otimes \mathbf{u}_i$$

with three (not necessarily different) real eigenvalues and the orthonormal eigenbasis $\{\mathbf{u}_i\}$. Any function of this tensor can be considered as a function of the eigenvalues and the eigenvectors

$$\varphi(\mathbf{C}) = \phi(\lambda_1 , \lambda_2 , \lambda_3 , \mathbf{u}_1 , \mathbf{u}_2 , \mathbf{u}_3) .$$

Because of the assumed isotropy of φ, we conclude

$$\varphi(\mathbf{C}) = \varphi(\mathbf{Q\,C\,Q}^T) = \phi(\lambda_1 , \lambda_2 , \lambda_3 , \mathbf{Q\,u}_1 , \mathbf{Q\,u}_2 , \mathbf{Q\,u}_3)$$

for any $\mathbf{Q} \in \mathcal{O}rth$, which can only be possible, if ϕ does not depend on the eigenvectors

$$\phi(\lambda_1 , \lambda_2 , \lambda_3 , \mathbf{u}_1 , \mathbf{u}_2 , \mathbf{u}_3) = \phi(\lambda_1 , \lambda_2 , \lambda_3)$$

where we used the same symbol for the reduced function. By the versor

$$\mathbf{Q} \equiv \mathbf{u}_2 \otimes \mathbf{u}_1 - \mathbf{u}_1 \otimes \mathbf{u}_2 + \mathbf{u}_3 \otimes \mathbf{u}_3 \in \mathcal{O}rth$$

we obtain

$$\mathbf{Q\,C\,Q}^T = \lambda_1 \, \mathbf{u}_2 \otimes \mathbf{u}_2 + \lambda_2 \, \mathbf{u}_1 \otimes \mathbf{u}_1 + \lambda_3 \, \mathbf{u}_3 \otimes \mathbf{u}_3$$

and, thus,

$$\varphi(\mathbf{C}) = \phi(\lambda_1 , \lambda_2 , \lambda_3) = \varphi(\mathbf{Q\,C\,Q}^T) = \phi(\lambda_2 , \lambda_1 , \lambda_3) .$$

Analogously, we find

$$\phi(\lambda_1 , \lambda_2 , \lambda_3) = \phi(\lambda_3 , \lambda_2 , \lambda_1) .$$

This renders the second representation. The other direction of the proof is trivial in both cases; *q. e. d.*

Definition 1.25. A function

$$k : \mathcal{S}ym \to \mathcal{S}ym \qquad | \qquad \mathbf{C} \mapsto k(\mathbf{C})$$

is called a **tensorial isotropic tensor-function**, if

$$\mathbf{Q} * k(\mathbf{C}) = k(\mathbf{Q} * \mathbf{C}) \qquad \forall\, \mathbf{C} \in \mathcal{S}ym , \mathbf{Q} \in \mathcal{O}rth .$$

Theorem 1.26. Representation of tensorial isotropic tensor-functions
A tensorial isotropic tensor-function k *can be represented as*

- $$k(\mathbf{C}) = \eta_0 \, \mathbf{I} + \eta_1 \, \mathbf{C} + \eta_2 \, \mathbf{C}^2$$

with three real functions η_i *of the principal invariants of* $\mathbf{C}$ *(RICHTER*[16] *representation); or*

- *by the spectral form*

$$k(\mathbf{C}) = \sigma(\lambda_1, \lambda_2, \lambda_3) \, \mathbf{u}_1 \otimes \mathbf{u}_1 + \sigma(\lambda_2, \lambda_3, \lambda_1) \, \mathbf{u}_2 \otimes \mathbf{u}_2 + \sigma(\lambda_3, \lambda_1, \lambda_2) \, \mathbf{u}_3 \otimes \mathbf{u}_3$$

with one real function σ *of three real arguments, being symmetric in the second and the third, if* $\mathbf{C}$ *has the eigenvalues* λ_i *and the unit eigenvectors* $\mathbf{u}_i$.

Proof of the RICHTER representation (RICHTER 1948, RIVLIN[17]/ ERICKSEN 1955, SERRIN 1959, see TRUESDELL/ NOLL 1965, p. 32 f).

At first, we prove the following property of tensorial isotropic tensor-functions k : *The eigenvectors of* $\mathbf{C}$ *are also eigenvectors of* $k(\mathbf{C})$.

Let $\{\mathbf{u}_1, \mathbf{u}_2, \mathbf{u}_3\}$ be the orthonormal eigenbasis of $\mathbf{C}$ and λ_i the corresponding eigenvalues. Then $\mathbf{C}$ has the spectral form

$$\mathbf{C} = \sum_{i=1}^{3} \lambda_i \, \mathbf{u}_i \otimes \mathbf{u}_i \, .$$

Let

$$\mathbf{Q} \equiv \mathbf{u}_1 \otimes \mathbf{u}_1 - \mathbf{u}_2 \otimes \mathbf{u}_2 - \mathbf{u}_3 \otimes \mathbf{u}_3 \qquad \in \mathcal{O}rth.$$

This is a *180*°-rotation around $\mathbf{u}_1$, which leaves this particular $\mathbf{C}$ invariant, as we will see now. We have

$$\mathbf{Q} \, \mathbf{u}_1 = \mathbf{u}_1 \qquad \text{and} \qquad \mathbf{Q} \, \mathbf{u}_i = -\mathbf{u}_i \quad \text{for } i = 2, 3.$$

Therefore

$$\mathbf{Q} \, \mathbf{C} \, \mathbf{Q}^T = \mathbf{C}$$

and for each tensorial isotropic tensor-function

$$k(\mathbf{C}) = k(\mathbf{Q} \, \mathbf{C} \, \mathbf{Q}^T) = \mathbf{Q} \, k(\mathbf{C}) \, \mathbf{Q}^T$$

or

$$k(\mathbf{C}) \, \mathbf{Q} = \mathbf{Q} \, k(\mathbf{C}) \, .$$

Applied to $\mathbf{u}_1$ we obtain

---

[16] Hans Richter (1912-1978)
[17] Ronald Ssamuel Rivlin (1915-2005)

$$k(\mathbf{C})\,\mathbf{Q}\,\mathbf{u}_1 = k(\mathbf{C})\,\mathbf{u}_1 = \mathbf{Q}\,k(\mathbf{C})\,\mathbf{u}_1 .$$

As those vectors parallel to $\mathbf{u}_1$ are the only ones which $\mathbf{Q}$ maps into their direction, we conclude

$$k(\mathbf{C})\,\mathbf{u}_1 \parallel \mathbf{u}_1$$

and $\mathbf{u}_1$ is also an eigenvector of $k(\mathbf{C})$. The same holds for $\mathbf{u}_2$ and $\mathbf{u}_3$.

In the same way, the values

$$\sigma_i = k(\mathbf{C}) \cdot \mathbf{u}_i \otimes \mathbf{u}_i \qquad \Leftrightarrow \qquad k(\mathbf{C})\,\mathbf{u}_i = \sigma_i\,\mathbf{u}_i \qquad \text{(no sum)}$$

(depending on $\mathbf{C}$), represent the eigenvalues of $k(\mathbf{C})$ corresponding to $\mathbf{u}_i$, such that its spectral decomposition is

$$k(\mathbf{C}) = \sum_{i=1}^{3} \sigma_i\,\mathbf{u}_i \otimes \mathbf{u}_i .$$

Let $m \leq 3$ be the number of different eigenvalues of $\mathbf{C}$.

Case 1: $m = 3$, *i.e.*, $\lambda_1 \neq \lambda_2 \neq \lambda_3 \neq \lambda_1$.

We make the ansatz

$$\sum_{i=1}^{3} \sigma_i\,\mathbf{u}_i \otimes \mathbf{u}_i = \eta_0 \sum_{i=1}^{3} \mathbf{u}_i \otimes \mathbf{u}_i + \eta_1 \sum_{i=1}^{3} \lambda_i\,\mathbf{u}_i \otimes \mathbf{u}_i + \eta_2 \sum_{i=1}^{3} \lambda_i^2\,\mathbf{u}_i \otimes \mathbf{u}_i$$

with three real functions $\eta_i(\mathbf{C})$, which will be specified later. By a comparison of the components, we obtain the inhomogeneous system of linear equations for η_i

$$\sigma_1 = \eta_0 + \eta_1\,\lambda_1 + \eta_2\,\lambda_1^2$$

$$\sigma_2 = \eta_0 + \eta_1\,\lambda_2 + \eta_2\,\lambda_2^2$$

$$\sigma_3 = \eta_0 + \eta_1\,\lambda_3 + \eta_2\,\lambda_3^2 .$$

The determinant of its coefficient matrix is

$$(\lambda_1 - \lambda_2)\,(\lambda_2 - \lambda_3)\,(\lambda_3 - \lambda_1),$$

which is non-zero, and the system has a unique solution $\{\eta_0, \eta_1, \eta_2\}$ after Theorem 1.1.

The ansatz can be directly notated as

$$k(\mathbf{C}) = \eta_0\,\mathbf{I} + \eta_1\,\mathbf{C} + \eta_2\,\mathbf{C}^2 .$$

Case 2: $m = 2$, *i.e.*, $\lambda_1 \neq \lambda_2 = \lambda_3$.

We use the ansatz

$$\sum_{i=1}^{3} \sigma_i\,\mathbf{u}_i \otimes \mathbf{u}_i = \eta_0 \sum_{i=1}^{3} \mathbf{u}_i \otimes \mathbf{u}_i + \eta_1 \sum_{i=1}^{3} \lambda_i\,\mathbf{u}_i \otimes \mathbf{u}_i ,$$

leading to the system of equations

$$\sigma_1 = \eta_0 + \eta_1\,\lambda_1$$

$$\sigma_2 = \eta_0 + \eta_1 \lambda_2$$

$$\sigma_3 = \eta_0 + \eta_1 \lambda_3$$

where the last two equations are identical. The determinant of the coefficients of the first two equations is

$$\lambda_2 - \lambda_1 \neq 0$$

and we again expect a unique solution. With $\eta_2 \equiv 0$ we again obtain the above representation.

Case 3: $m = 1$, *i.e.*, $\lambda_1 = \lambda_2 = \lambda_3$.

Here the analogous ansatz is simply

$$\sum_{i=1}^{3} \sigma_i \, \mathbf{u}_i \otimes \mathbf{u}_i = \eta_0 \sum_{i=1}^{3} \mathbf{u}_i \otimes \mathbf{u}_i$$

and thus all

$$\sigma_i = \eta_0 \, .$$

With $\eta_1 \equiv 0 \equiv \eta_2$ this also leads to the same representation.

By the isotropy condition we see that the three real functions η_i, $i = 0, 1, 2$ have to also be isotropic

$$\eta_i(\mathbf{C}) = \eta_i(\mathbf{Q}\,\mathbf{C}\,\mathbf{Q}^T) \qquad \forall\, \mathbf{C} \in \mathcal{Sym}, \ \forall\, \mathbf{Q} \in \mathcal{Orth}.$$

After the previous

Theorem **1.24.**, we can in this case reduce the dependence on $\mathbf{C}$ to the principal invariants

$$\eta_i(\mathbf{C}) = \eta_i(I_\mathbf{C}, II_\mathbf{C}, III_\mathbf{C}) \, .$$

Proof of the inverse direction. We already saw that real functions of the principal invariants are isotropic tensor-functions. Apart from that, for all integers m and all $\mathbf{C} \in \mathcal{Sym}$ and $\mathbf{Q} \in \mathcal{Orth}$ we have

$$(\mathbf{Q}\,\mathbf{C}\,\mathbf{Q}^T)^m = \mathbf{Q}\,\mathbf{C}^m\,\mathbf{Q}^T$$

and also for all tensor polynomials

$$\sum_m \alpha_m (\mathbf{Q}\,\mathbf{C}\,\mathbf{Q}^T)^m = \mathbf{Q}\,(\sum_m \alpha_m\, \mathbf{C}^m)\,\mathbf{Q}^T.$$

This completes the proof of the first isotropic representation.

Proof of the spectral form. We use the ansatz from above

$$k(\mathbf{C}) = \sum_{i=1}^{3} \sigma_i \, \mathbf{u}_i \otimes \mathbf{u}_i$$

with three functions of $\mathbf{C}$ in the form

$$\sigma_i = \underline{\sigma_i}(\lambda_i, \lambda_{i+1}, \lambda_{i+2}, \mathbf{u}_i, \mathbf{u}_{i+1}, \mathbf{u}_{i+2}) \qquad i = 1, 2, 3 \text{ modulo } 3.$$

By the isotropy of k we see

$$\underline{\sigma}_i(\lambda_i, \lambda_{i+1}, \lambda_{i+2}, \mathbf{u}_i, \mathbf{u}_{i+1}, \mathbf{u}_{i+2})$$
$$= \underline{\sigma}_i(\lambda_i, \lambda_{i+1}, \lambda_{i+2}, \mathbf{Q}\,\mathbf{u}_i, \mathbf{Q}\,\mathbf{u}_{i+1}, \mathbf{Q}\,\mathbf{u}_{i+2})$$

for each $\mathbf{Q} \in \mathcal{O}rth$, which can only hold if the $\underline{\sigma}_i$ do not depend on the eigenvectors

$$\underline{\sigma}_i(\lambda_i, \lambda_{i+1}, \lambda_{i+2}, \mathbf{u}_i, \mathbf{u}_{i+1}, \mathbf{u}_{i+2}) = \sigma_i(\lambda_i, \lambda_{i+1}, \lambda_{i+2}).$$

By the rotation around the diagonal

$$\mathbf{Q} \equiv \mathbf{u}_2 \otimes \mathbf{u}_1 + \mathbf{u}_3 \otimes \mathbf{u}_2 + \mathbf{u}_1 \otimes \mathbf{u}_3 \qquad \in \mathcal{O}rth$$

we get

$$\mathbf{Q}\,\mathbf{C}\,\mathbf{Q}^T = \lambda_1\,\mathbf{u}_2 \otimes \mathbf{u}_2 + \lambda_2\,\mathbf{u}_3 \otimes \mathbf{u}_3 + \lambda_3\,\mathbf{u}_1 \otimes \mathbf{u}_1$$

and

$$\begin{aligned} k(\mathbf{Q}\,\mathbf{C}\,\mathbf{Q}^T) &= \sigma_1(\lambda_3, \lambda_1, \lambda_2)\,\mathbf{u}_1 \otimes \mathbf{u}_1 \\ &+ \sigma_2(\lambda_1, \lambda_2, \lambda_3)\,\mathbf{u}_2 \otimes \mathbf{u}_2 \\ &+ \sigma_3(\lambda_2, \lambda_3, \lambda_1)\,\mathbf{u}_3 \otimes \mathbf{u}_3 \\ = \mathbf{Q}\,k(\mathbf{C})\,\mathbf{Q}^T &= \sigma_1(\lambda_1, \lambda_2, \lambda_3)\,\mathbf{u}_2 \otimes \mathbf{u}_2 \\ &+ \sigma_2(\lambda_2, \lambda_3, \lambda_1)\,\mathbf{u}_3 \otimes \mathbf{u}_3 \\ &+ \sigma_3(\lambda_3, \lambda_1, \lambda_2)\,\mathbf{u}_1 \otimes \mathbf{u}_1 . \end{aligned}$$

By comparison we conclude the identity of the functions

$$\sigma_1 = \sigma_2 = \sigma_3 =: \sigma .$$

Similarly, a $90°$-rotation around $\mathbf{u}_3$

$$\mathbf{Q} \equiv \mathbf{u}_2 \otimes \mathbf{u}_1 - \mathbf{u}_1 \otimes \mathbf{u}_2 + \mathbf{u}_3 \otimes \mathbf{u}_3 \qquad \in \mathcal{O}rth$$

transforms

$$\mathbf{Q}\,\mathbf{C}\,\mathbf{Q}^T = \lambda_1\,\mathbf{u}_2 \otimes \mathbf{u}_2 + \lambda_2\,\mathbf{u}_1 \otimes \mathbf{u}_1 + \lambda_3\,\mathbf{u}_3 \otimes \mathbf{u}_3$$

so that the isotropy condition gives

$$\begin{aligned} k(\mathbf{Q}\,\mathbf{C}\,\mathbf{Q}^T) &= \sigma(\lambda_2, \lambda_1, \lambda_3)\,\mathbf{u}_1 \otimes \mathbf{u}_1 \\ &+ \sigma(\lambda_1, \lambda_3, \lambda_2)\,\mathbf{u}_2 \otimes \mathbf{u}_2 \\ &+ \sigma(\lambda_3, \lambda_2, \lambda_1)\,\mathbf{u}_3 \otimes \mathbf{u}_3 \\ = \mathbf{Q}\,k(\mathbf{C})\,\mathbf{Q}^T &= \sigma(\lambda_1, \lambda_2, \lambda_3)\,\mathbf{u}_2 \otimes \mathbf{u}_2 \\ &+ \sigma(\lambda_2, \lambda_3, \lambda_1)\,\mathbf{u}_1 \otimes \mathbf{u}_1 \\ &+ \sigma(\lambda_3, \lambda_1, \lambda_2)\,\mathbf{u}_3 \otimes \mathbf{u}_3 . \end{aligned}$$

Therefore, σ must be symmetric in the last two arguments. This gives the second representation.

The proof of the reverse direction is obvious after what we have seen already; *q. e. d.*

If the argument $\mathbf{C}$ is invertible, we can eliminate $\mathbf{C}^2$ in the first representation by means of Theorem 1.13. of CAYLEY-HAMILTON and obtain an alternative RICHTER representation in the form

$$k(\mathbf{C}) = \underline{\eta}_0\,\mathbf{I} + \underline{\eta}_1\,\mathbf{C} + \underline{\eta}_{-1}\,\mathbf{C}^{-1} \qquad \text{with } \underline{\eta}_i(I_\mathbf{C}, II_\mathbf{C}, III_\mathbf{C}) .$$

given by

$$\underline{\eta}_0 = \eta_0 - \eta_2\, II_\mathbf{C}$$

$$\underline{\eta}_1 = \eta_1 + \eta_2\, I_\mathbf{C}$$

$$\underline{\eta}_{-1} = \eta_2\, III_\mathbf{C} .$$

1.3 Tensor Analysis

A Comment on the Literature. There are many books available on tensor analysis, like AKIVIS/ GOLDBERG (2003), BISHOP/ GOLDBERG (1968), BORISENKO/ TARAPOV (1968), BOWEN/ WANG (1976), BRAND (1947), BRILLOUIN (1964), ERICKSEN (1960), ITSKOV (2007), LEBEDEV/ CLOUD (2003), McCONNELL (1957), PACH/ FREY (1964), SCHADE (1997), SHOUTEN (1990), SOKOLNIKOFF (1951, 1964), TROSTEL (1997).

The aim of this chapter is to generalise the concept of a differential from a real function to tensor-functions of arbitrary dimension and order. First we need some concepts of topology. Let

$$\mathbf{f} : \mathscr{V} \to \mathscr{W}$$

be a (non-linear) function between two arbitrary normed linear spaces. Let $\mathbf{v}_o \in \mathscr{V}$, then we call $\mathbf{w}_o \in \mathscr{W}$ the **limit** of $\mathbf{f}$ in $\mathbf{v}_o$, or

$$\lim_{\mathbf{v}\to\mathbf{v}_o} \mathbf{f}(\mathbf{v}) = \mathbf{w}_o ,$$

if for each positive number ε a positive number δ exists such that

$$0 < |\mathbf{v} - \mathbf{v}_o| < \delta$$

implies

$$|\mathbf{f}(\mathbf{v}) - \mathbf{w}_o| < \varepsilon .$$

Such a function is called **continuous** in $\mathbf{v}_o$, if the limit exists and equals $\mathbf{f}(\mathbf{v}_o)$. The function is called continuous, if it is so in each point of its domain.

Linear functions between finite dimensional linear spaces are always continuous.

We will next introduce the **directional** or **GATEAUX**[18] **differential** for such a function. The aim of this is to approximate a (generally non-linear) function with a linear one at a certain point of its domain.

Definition 1.27. Let $\mathbf{v}$, $\mathbf{dv} \in \mathscr{V}$. The **differential** *of* $\mathbf{f}$ *in* $\mathbf{v}$ *in the direction of* $\mathbf{dv}$ is a mapping

$$d\mathbf{f}: \mathscr{V} \times \mathscr{V} \to \mathscr{W}$$

defined by the limit

$$d\mathbf{f}(\mathbf{v}, \mathbf{dv}) := \lim_{h \to 0} \frac{1}{h} [\mathbf{f}(\mathbf{v} + h\,\mathbf{dv}) - \mathbf{f}(\mathbf{v})] .$$

The function $\mathbf{f}$ is called **differentiable**, if the differential $d\mathbf{f}(\mathbf{v}, \mathbf{dv})$ exists for all $\mathbf{v}$, $\mathbf{dv} \in \mathscr{V}$.

The limit can also be written as

$$d\mathbf{f}(\mathbf{v}, \mathbf{dv}) = \frac{d}{dh} \mathbf{f}(\mathbf{v} + h\,\mathbf{dv}) \,|_{h=0} .$$

This form is perhaps more familiar from calculus.

If $\mathbf{f}$ is sufficiently smooth (this will be assumed further on), then the differential is linear in the second argument $\mathbf{dv}$. This fact shall be shown next. Firstly, we show the homogeneity. Let $\alpha \in \mathscr{R}$. As the case $\alpha = 0$ is trivial, we assume $\alpha \neq 0$. Then, by definition, we have

$$\begin{aligned} d\mathbf{f}(\mathbf{v}, \alpha\,\mathbf{dv}) &= \lim_{h \to 0} \frac{1}{h} (\mathbf{f}(\mathbf{v} + h\alpha\,\mathbf{dv}) - \mathbf{f}(\mathbf{v})) \\ &= \lim_{h \to 0} \frac{\alpha}{\alpha h} (\mathbf{f}(\mathbf{v} + h\alpha\,\mathbf{dv}) - \mathbf{f}(\mathbf{v})) \\ &= \alpha \lim_{k \to 0} \frac{1}{k} (\mathbf{f}(\mathbf{v} + k\,\mathbf{dv}) - \mathbf{f}(\mathbf{v})) \\ &= \alpha\, d\mathbf{f}(\mathbf{v}, \mathbf{dv}) . \end{aligned}$$

We next show the additivity in the second argument of the differential. Let $\mathbf{u}$, $\mathbf{v}$, $\mathbf{w} \in \mathscr{V}$, then

$$\begin{aligned} d\mathbf{f}(\mathbf{v}, \mathbf{u} + \mathbf{w}) &= \lim_{h \to 0} \frac{1}{h} (\mathbf{f}(\mathbf{v} + h(\mathbf{u} + \mathbf{w})) - \mathbf{f}(\mathbf{v})) \\ &= \lim_{h \to 0} \frac{1}{h} (\mathbf{f}(\mathbf{v} + h\,\mathbf{u} + h\,\mathbf{w}) - \mathbf{f}(\mathbf{v} + h\,\mathbf{u}) + \mathbf{f}(\mathbf{v} + h\,\mathbf{u}) - \mathbf{f}(\mathbf{v})) \end{aligned}$$

[18] René Eugène Gâteaux (1889-1914)

$$= \lim_{h \to 0} \frac{1}{h} \left(\mathbf{f}(\mathbf{v} + h\,\mathbf{u} + h\,\mathbf{w}) - \mathbf{f}(\mathbf{v} + h\,\mathbf{u}) \right)$$

$$+ \lim_{h \to 0} \frac{1}{h} \left(\mathbf{f}(\mathbf{v} + h\,\mathbf{u}) - \mathbf{f}(\mathbf{v}) \right)$$

$$= d\mathbf{f}(\mathbf{v}, \mathbf{w}) + d\mathbf{f}(\mathbf{v}, \mathbf{u}) .$$

By interpreting linear functions between vectors as tensors, we introduce the **gradient** or the **FRECHET**[19] **derivative** of a differentiable function $\mathbf{f(v)}$ with respect to $\mathbf{v}$ as a tensor, for which different notations are used like

$$\frac{\mathbf{df(v)}}{\mathbf{dv}} = grad\,\mathbf{f(v)} = \mathbf{f}'(\mathbf{v}) .$$

It performs

$$\mathbf{f}'(\mathbf{v})\,\mathbf{dv} := d\mathbf{f}(\mathbf{v}, \mathbf{dv}) \qquad \forall\, \mathbf{v}, \mathbf{dv} \in \mathscr{V} .$$

In general, the derivative has no unique representation as a tensor, as we will see in the sequel. The derivative obeys the *chain rule* for the composition of functions. Namely, let $\mathscr{U}$, $\mathscr{V}$, and $\mathscr{W}$ be normed vector spaces, and

$$\mathbf{g} : \mathscr{U} \to \mathscr{V}$$

$$\mathbf{f} : \mathscr{V} \to \mathscr{W}$$

differentiable functions. Then the derivative of the composition equals the composition of the derivatives

$$(\mathbf{f\,g})'(\mathbf{u}) = \mathbf{f}'(\mathbf{g}(\mathbf{u}))\,\mathbf{g}'(\mathbf{u}) .$$

Consequently, for an invertible function, the gradient of the inverse is equal to the inverse of the gradient (which therefore exists)

$$(\mathbf{f}^{-1})' = (\mathbf{f}')^{-1} .$$

The derivative is (pointwise) linear

$$[\mathbf{f}_1(\mathbf{v}) + \alpha\,\mathbf{f}_2(\mathbf{v})]' = \mathbf{f}_1'(\mathbf{v}) + \alpha\,\mathbf{f}_2'(\mathbf{v}) \qquad \forall \alpha \in \mathscr{R}, \forall\, \mathbf{v} \in \mathscr{V} .$$

For arbitrary products $\oplus$ between tensor-functions $\mathbf{S(v)}$ and $\mathbf{T(v)}$ of arbitrary order, the *product rule* or *LEIBNIZ*[20] *rule* holds

$$\mathbf{d}\,(\mathbf{S(v)} \oplus \mathbf{T(v)}) = \mathbf{S}'(\mathbf{v})\,\mathbf{dv} \oplus \mathbf{T(v)} + \mathbf{S(v)} \oplus \mathbf{T}'(\mathbf{v})\,\mathbf{dv} .$$

Herein, $\oplus$ can be a scalar, tensor, cross product or any reasonable contraction.

We will next specify the concept of the differential for mappings between linear spaces of different order.

---

[19] Maurice René Fréchet (1878-1973)

[20] Gottfried Wilhelm Leibniz (1646-1716)

- Let f be a <u>real-valued real function</u>

$$f : \mathscr{R} \rightarrow \mathscr{R} .$$

Then

$$df(x , dx) := \lim_{h \to 0} \frac{1}{h} (f(x + h\, dx) - f(x))$$

$$= \frac{d}{dh} f(x + h\, dx) \mid_{h=0}$$

and for $h\, dx \equiv \Delta x$ we obtain the usual form from real analysis

$$df(x , dx) := \lim_{\Delta x \to 0} \frac{dx}{\Delta x} (f(x + \Delta x) - f(x)) = \frac{df}{dx} dx .$$

As the real axis has only one direction, the differential is independent of the direction.

- Let f be a <u>real vector function</u>

$$\mathrm{f} : \mathscr{V} \rightarrow \mathscr{R} .$$

Then $d\mathrm{f}(\mathbf{v} , \mathbf{dv})$ is linear in the second vectorial argument

$$d\mathrm{f} : \mathscr{V} \times \mathscr{V} \rightarrow \mathscr{R} .$$

This can always be represented by the inner product with a (co-) vector $\frac{\mathbf{df(v)}}{\mathbf{dv}}$ called gradient so that

$$d\mathrm{f}(\mathbf{v} , \mathbf{dv}) = \frac{\mathbf{df(v)}}{\mathbf{dv}} \cdot \mathbf{dv} .$$

For achieving a component form of the gradient, we choose a vector basis $\{\mathbf{g}_i\} \subset \mathscr{V}$ and represent $\mathbf{v} = v^i\, \mathbf{g}_i$ and $\mathbf{dv} = dv^i\, \mathbf{g}_i$ and the function

$$\mathrm{f}(\mathbf{v}) = \mathrm{f}(v^i\, \mathbf{g}_i) =: f(v^1, v^2, \dots , v^N)$$

such that its differential is

$$d\mathrm{f}(\mathbf{v} , \mathbf{dv}) = d\mathrm{f}(v^i\, \mathbf{g}_i , dv^i\, \mathbf{g}_i)$$

$$= d f(v^1, v^2, \dots , v^N, dv^1)$$

$$+ d f(v^1, v^2, \dots , v^N, dv^2)$$

$$+ \dots + d f(v^1, v^2, \dots , v^N, dv^N)$$

$$= \frac{\partial f(v^1, v^2, \; \dots \; , v^N)}{\partial v^i} dv^i = \frac{\partial f(v^1, v^2, \; \dots \; , v^N)}{\partial v^i} \mathbf{g}^i \cdot dv^j\, \mathbf{g}_j$$

$$= \frac{\partial f(v^1, v^2, \; \dots \; , v^N)}{\partial v^i} \mathbf{g}^i \cdot \mathbf{dv} .$$

By comparison we find that the gradient can be expressed by the partial derivatives of the function with respect to the components of the vector

(1.36)
$$\frac{df(\mathbf{v})}{d\mathbf{v}} = \frac{\partial f(v^1, v^2, \dots, v^N)}{\partial v^i} \, \mathbf{g}^i \qquad \in \mathscr{V}$$

- Let $\mathbf{f}$ be a <u>*vectorial vector function*</u>

$$\mathbf{f}: \mathscr{V} \to \mathscr{V}.$$

Its derivative is a linear mapping

$$\frac{d\mathbf{f}(\mathbf{v})}{d\mathbf{v}} : \mathscr{V} \to \mathscr{V}$$

and, thus, a *2*nd-order tensor. We again choose a basis $\{\mathbf{g}_i\} \subset \mathscr{V}$ and represent its value as real functions f^k of N scalar components

$$\mathbf{f}(\mathbf{v}) = \mathbf{f}(v^j \mathbf{g}_j) = f^k(v^1, v^2, \dots, v^N) \, \mathbf{g}_k .$$

Its differential is

$$d\mathbf{f}(\mathbf{v}, d\mathbf{v}) = d\mathbf{f}(v^i \mathbf{g}_i, dv^j \mathbf{g}_j) = d\mathbf{f}(v^i \mathbf{g}_i, \mathbf{g}_j) \, dv^j$$

$$= df^k(v^1, v^2, \dots, v^N, dv^j) \, \mathbf{g}_k = \frac{\partial f^k(v^1, v^2, \dots, v^N)}{\partial v^j} \, dv^j \, \mathbf{g}_k$$

$$= \frac{\partial f^k(v^1, v^2, \dots, v^N)}{\partial v^j} \, \mathbf{g}_k \, (\mathbf{g}^j \cdot d\mathbf{v})$$

$$= \Big(\frac{\partial f^k(v^1, v^2, \dots, v^N)}{\partial v^j} \, \mathbf{g}_k \otimes \mathbf{g}^j \Big) \, d\mathbf{v} .$$

The derivative of $\mathbf{f}$ can therefore be represented as

(1.37)
$$grad\, \mathbf{f}(\mathbf{v}) = \frac{d\mathbf{f}(\mathbf{v})}{d\mathbf{v}} = \frac{\partial f^k(v^1, v^2, \dots, v^N)}{\partial v^j} \, \mathbf{g}_k \otimes \mathbf{g}^j \qquad \in \mathscr{Lin} .$$

We have expressed the differential of a vector function by partial derivatives of the real functions of the components. This simple form, however, can only be achieved when using the duality of the bases.

The **divergence** of $\mathbf{f}$ is the trace of the gradient

$$div\, \mathbf{f} := tr \frac{d\mathbf{f}(\mathbf{v})}{d\mathbf{v}} = \frac{\partial f^k(v^1, v^2, \dots, v^N)}{\partial v^k} \qquad \in \mathscr{R} .$$

In three dimensions, the **curl** of $\mathbf{f}$ is twice the axial vector of the gradient

$$curl\, \mathbf{f} := \frac{\partial f^k(v^1, v^2, \dots, v^N)}{\partial v^j} \, \mathbf{g}^j \times \mathbf{g}_k \qquad \in \mathscr{V} .$$

- Let $\mathbf{F}$ be a <u>*tensorial vector function*</u>

$$\mathbf{F} : \mathscr{V} \rightarrow \mathscr{Lin} .$$

Its differential is a function

$$d\mathbf{F} : \mathscr{V} \times \mathscr{V} \rightarrow \mathscr{Lin} .$$

For its determination we use the same method as before. We represent the value $d\mathbf{F}$ and the arguments by choosing a basis of $\mathscr{V}$

$$\begin{aligned} d\mathbf{F}(\mathbf{v}, \mathbf{dv}) &= dF^{kl}(v^i \mathbf{g}_i , dv^j \mathbf{g}_j)\, \mathbf{g}_k \otimes \mathbf{g}_l \\ &= \frac{\mathbf{dF}(\mathbf{v})}{\mathbf{dv}}\, \mathbf{dv} = dF^{kl}(v^i \mathbf{g}_i , \mathbf{g}_j)\, \mathbf{g}_k \otimes \mathbf{g}_l\, dv^j \\ &= \frac{\partial F^{kl}(v^1, v^2, \ldots, v^N)}{\partial v^j}\, \mathbf{g}_k \otimes \mathbf{g}_l\, (\mathbf{g}^j \cdot \mathbf{dv}) \\ &= \Big(\frac{\partial F^{kl}(v^1, v^2, \ldots, v^N)}{\partial v^j}\, \mathbf{g}_k \otimes \mathbf{g}_l \otimes \mathbf{g}^j \Big)\, \mathbf{dv} \end{aligned}$$

(simple contraction). The derivative of $\mathbf{F}$ is therefore the triad

$$grad\, \mathbf{F}(\mathbf{v}) = \frac{\mathbf{dF}(\mathbf{v})}{\mathbf{dv}} = \frac{\partial F^{kl}(v^1, v^2, \ldots, v^N)}{\partial v^j}\, \mathbf{g}_k \otimes \mathbf{g}_l \otimes \mathbf{g}^j .$$

The **divergence** of $\mathbf{F}(\mathbf{v})$ is defined as that particular vector function $div\, \mathbf{F}(\mathbf{v})$ that gives for all constant vectors $\mathbf{a}$

$$(div\, \mathbf{F}) \cdot \mathbf{a} := div(\mathbf{F}^T \mathbf{a}) .$$

This is achieved by

$$div\, \mathbf{F} = \frac{\partial F^{kj}(v^1, v^2, \ldots, v^N)}{\partial v^j}\, \mathbf{g}_k \qquad \in \mathscr{V}.$$

- We often need the derivative of a <u>real tensor-function</u>

$$f : \mathscr{Lin} \rightarrow \mathscr{R} .$$

Its differential is a function

$$df : \mathscr{Lin} \times \mathscr{Lin} \rightarrow \mathscr{R}$$

again being linear in the second argument. The derivative

$$\frac{df(\mathbf{T})}{d\mathbf{T}} : \mathscr{Lin} \rightarrow \mathscr{R}$$

can be expressed by means of the scalar product between tensors

$$df(\mathbf{T}, \mathbf{dT}) = \frac{df(\mathbf{T})}{d\mathbf{T}} \cdot \mathbf{dT} = tr\Big(\frac{df(\mathbf{T})}{d\mathbf{T}}\, \mathbf{dT}^T\Big) ,$$

wherein the tensor of the derivative can be expressed by its components

$$\frac{df(\mathbf{T})}{d\mathbf{T}} = \frac{\partial f(T^{11}, T^{12}, \ldots, T^{NN})}{\partial T^{ij}} \, \mathbf{g}^i \otimes \mathbf{g}^j \qquad \in \mathscr{Lin}$$

if

$$\mathbf{T} = T^{ij} \, \mathbf{g}_i \otimes \mathbf{g}_j \qquad \in \mathscr{Lin} .$$

$\mathbf{dT}$ has the component form

$$\mathbf{dT} = dT^{ij} \, \mathbf{g}_i \otimes \mathbf{g}_j \qquad \in \mathscr{Lin}$$

and, thus

$$df(\mathbf{T}, \mathbf{dT}) = \Big(\frac{\partial f(T^{11}, T^{12}, \ldots, T^{NN})}{\partial T^{ij}} \, \mathbf{g}^i \otimes \mathbf{g}^j\Big) \cdot (dT^{kl} \, \mathbf{g}_k \otimes \mathbf{g}_l)$$

$$= \frac{\partial f(T^{11}, T^{12}, \ldots, T^{NN})}{\partial T^{ij}} \, dT^{ij} \qquad \in \mathscr{R} .$$

If f is defined only on symmetric tensors, then also the derivative acts only on symmetric tensors $\mathbf{dT} \in \mathscr{Sym}$

$$df(\mathbf{T}, \mathbf{dT}) = \frac{df}{d\mathbf{T}} \cdot \mathbf{dT} = sym\Big(\frac{df}{d\mathbf{T}}\Big) \cdot \mathbf{dT} \qquad \in \mathscr{R} .$$

and the skew part remains indeterminate. In this case one can always symmetrise the derivative.

As an **example** of a real tensor-function, we consider the trace of a power of the tensor

$$f(\mathbf{T}) \equiv tr(\mathbf{T}^k) = \mathbf{T}^k \cdot \mathbf{I}$$

for $k > 0$. By applying the definition of the differential, we obtain for, *e.g.*, $k \equiv 3$ after the product rule

$$df(\mathbf{T}, \mathbf{dT})$$

$$= \frac{d}{dh} \{tr((\mathbf{T} + h\,\mathbf{dT})\,\mathbf{T}\,\mathbf{T}) + tr\,(\mathbf{T}\,(\mathbf{T} + h\,\mathbf{dT})\,\mathbf{T})$$

$$+ tr(\mathbf{T}\,\mathbf{T}\,(\mathbf{T} + h\,\mathbf{dT}))\} \,\big|_{h=0}$$

$$= 3\; tr(\mathbf{T}\,\mathbf{T}\,\mathbf{dT}) = 3\,(\mathbf{T}^2)^T \cdot \mathbf{dT} .$$

In general, we have

$$df(\mathbf{T}, \mathbf{dT}) = k\, tr(\mathbf{T}^{k-1}\, \mathbf{dT}) = k\, \mathbf{T}^{k-1} \cdot \mathbf{dT}^T = k\, (\mathbf{T}^{k-1})^T \cdot \mathbf{dT}$$

and therefore

(1.38) $$f\,'(\mathbf{T}) = k\,(\mathbf{T}^{k-1})^T.$$

In particular we get for $k \equiv 1$

$$dI_{\mathbf{T}} = \mathbf{I} \cdot \mathbf{dT} = tr(\mathbf{dT}) = I_{\mathbf{dT}} .$$

With the representation for the second principal invariant (1.12)

$$II_{\mathbf{T}} = \tfrac{1}{2} \{tr^2(\mathbf{T}) - tr(\mathbf{T}^2)\}$$

we obtain by the same formula (1.38)

$$dII_{\mathbf{T}} = \{tr(\mathbf{T})\,\mathbf{I} - \mathbf{T}^T\} \cdot \mathbf{dT} .$$

In order to compute the derivative of the third principal invariant, the determinant, we first take the trace of the CAYLEY-HAMILTON equation of Theorem 1.13.

$$III_{\mathbf{T}}\, tr(\mathbf{I}) - II_{\mathbf{T}}\, tr(\mathbf{T}) + I_{\mathbf{T}}\, tr(\mathbf{T}^2) - tr(\mathbf{T}^3) = 0 ,$$

which can be used to express the determinant in terms of traces (1.13). By again applying formula (1.38), the differential at $\mathbf{T}$ in the direction of $\mathbf{dT}$ turns out to be

$$dIII_{\mathbf{T}} = \{\mathbf{T}^{2T} + II_{\mathbf{T}}\,\mathbf{I} - tr(\mathbf{T})\,\mathbf{T}^T\} \cdot \mathbf{dT} .$$

For invertible tensors $\mathbf{T}$ this is simpler

$$dIII_{\mathbf{T}} = III_{\mathbf{T}}\,\mathbf{T}^{-T} \cdot \mathbf{dT} .$$

We thus obtained the following results

(1.39)
$$\frac{d\,I_{\mathbf{T}}}{d\,\mathbf{T}} = \mathbf{I}$$
$$\frac{d\,II_{\mathbf{T}}}{d\,\mathbf{T}} = I_{\mathbf{T}}\,\mathbf{I} - \mathbf{T}^T$$
$$\frac{d\,III_{\mathbf{T}}}{d\,\mathbf{T}} = \mathbf{T}^{2T} - I_{\mathbf{T}}\,\mathbf{T}^T + II_{\mathbf{T}}\,\mathbf{I} = III_{\mathbf{T}}\,\mathbf{T}^{-T}$$

Later, we will need the differential of a real isotropic tensor-function. By the representation

Theorem **1.24.** this can be given the form

$$f(\mathbf{T}) \equiv f_{iso}(I_{\mathbf{T}}, II_{\mathbf{T}}, III_{\mathbf{T}}) .$$

We determine the gradient by the chain rule

(1.40)
$$f\,'(\mathbf{T}) = \frac{df}{d\mathbf{T}} = \frac{\partial f_{iso}}{\partial I_{\mathbf{T}}} \frac{dI_{\mathbf{T}}}{d\mathbf{T}} + \frac{\partial f_{iso}}{\partial II_{\mathbf{T}}} \frac{dII_{\mathbf{T}}}{d\mathbf{T}} + \frac{\partial f_{iso}}{\partial III_{\mathbf{T}}} \frac{dIII_{\mathbf{T}}}{d\mathbf{T}}$$
$$= \frac{\partial f_{iso}}{\partial I_{\mathbf{T}}}\,\mathbf{I} + \frac{\partial f_{iso}}{\partial II_{\mathbf{T}}}\,(I_{\mathbf{T}}\,\mathbf{I} - \mathbf{T}^T) + \frac{\partial f_{iso}}{\partial III_{\mathbf{T}}}\,(\mathbf{T}^{T2} - I_{\mathbf{T}}\,\mathbf{T}^T + II_{\mathbf{T}}\,\mathbf{I})$$
$$= \alpha_0\,\mathbf{I} + \alpha_1\,\mathbf{T}^T + \alpha_2\,\mathbf{T}^{T2}$$

with the three real functions of principle invariants of $\mathbf{T}$

$$\alpha_0(I_\mathbf{T}, II_\mathbf{T}, III_\mathbf{T}) := \frac{\partial f_{iso}}{\partial I_\mathbf{T}} + \frac{\partial f_{iso}}{\partial II_\mathbf{T}} I_\mathbf{T} + \frac{\partial f_{iso}}{\partial III_\mathbf{T}} II_\mathbf{T}$$

$$\alpha_1(I_\mathbf{T}, II_\mathbf{T}, III_\mathbf{T}) := -\frac{\partial f_{iso}}{\partial II_\mathbf{T}} - \frac{\partial f_{iso}}{\partial III_\mathbf{T}} I_\mathbf{T}$$

$$\alpha_2(I_\mathbf{T}, II_\mathbf{T}, III_\mathbf{T}) := \frac{\partial f_{iso}}{\partial III_\mathbf{T}} .$$

If $\mathbf{T}$ is invertible, we obtain

$$f'(\mathbf{T}) = \frac{\partial f_{iso}}{\partial I_\mathbf{T}} \mathbf{I} + \frac{\partial f_{iso}}{\partial II_\mathbf{T}} (I_\mathbf{T}\,\mathbf{I} - \mathbf{T}^T) + \frac{\partial f_{iso}}{\partial III_\mathbf{T}} (III_\mathbf{T}\,\mathbf{T}^{-T})$$

$$= \beta_0\,\mathbf{I} + \beta_1\,\mathbf{T}^T + \beta_{-1}\,\mathbf{T}^{-T}$$

with the three real functions

$$\beta_0(I_\mathbf{T}, II_\mathbf{T}, III_\mathbf{T}) := \frac{\partial f_{iso}}{\partial I_\mathbf{T}} + \frac{\partial f_{iso}}{\partial II_\mathbf{T}} I_\mathbf{T}$$

$$\beta_1(I_\mathbf{T}, II_\mathbf{T}, III_\mathbf{T}) := -\frac{\partial f_{iso}}{\partial II_\mathbf{T}}$$

$$\beta_{-1}(I_\mathbf{T}, II_\mathbf{T}, III_\mathbf{T}) := \frac{\partial f_{iso}}{\partial III_\mathbf{T}} III_\mathbf{T} .$$

If a real function depends on several real, vectorial, and tensorial arguments of arbitrary order $\mathbf{T}_1, \dots, \mathbf{T}_K$

$$f(\mathbf{T}_1, \mathbf{T}_2, \dots, \mathbf{T}_K),$$

the differential of such function is because of its linearity

$$\begin{aligned} & df(\mathbf{T}_1, \mathbf{T}_2, \dots, \mathbf{T}_K, d\mathbf{T}_1, d\mathbf{T}_2, \dots, d\mathbf{T}_K) \\ & = df(\mathbf{T}_1, \mathbf{T}_2, \dots, \mathbf{T}_K, d\mathbf{T}_1, \mathbf{0}, \dots, \mathbf{0}) \\ & + df(\mathbf{T}_1, \mathbf{T}_2, \dots, \mathbf{T}_K, \mathbf{0}, d\mathbf{T}_2, \dots, \mathbf{0}) \\ & + \dots \\ & + df(\mathbf{T}_1, \mathbf{T}_2, \dots, \mathbf{T}_K, \mathbf{0}, \mathbf{0}, \dots, d\mathbf{T}_K) \\ & =: \frac{\partial f}{\partial \mathbf{T}_i} \cdot d\mathbf{T}_i \end{aligned}$$

with the symbol for the partial derivative $\frac{\partial f}{\partial \mathbf{T}_i}$ of f with respect to $\mathbf{T}_i$, which are defined by the differential

$$\frac{\partial f}{\partial \mathbf{T}_i} \cdot \mathbf{dT}_i := df(\mathbf{T}_1, \mathbf{T}_2, \dots, \mathbf{T}_K, \mathbf{0}, \dots, \mathbf{dT}_i, \dots, \mathbf{0}) \qquad i = 1, \dots, K$$

(no sum). The dot $\cdot$ stands for the scalar product of tensors of arbitrary, but equal, order. Analogous forms can be obtained for tensorial functions, which we will consider next.

- Let $\mathbf{F}$ be a <u>*tensorial tensor-function*</u>

$$\mathbf{F} : \mathcal{Lin} \rightarrow \mathcal{Lin} .$$

Its differential is a mapping

$$d\mathbf{F} : \mathcal{Lin} \times \mathcal{Lin} \rightarrow \mathcal{Lin} .$$

The derivative of $\mathbf{F}$ is the *4*th-order tensor

(1.41)
$$grad\, \mathbf{F}(\mathbf{T}) = \frac{\mathbf{dF}(\mathbf{T})}{\mathbf{dT}} = \frac{\partial F^{ij}(T^{11}, T^{12}, \dots, T^{NN})}{\partial T^{kl}}\, \mathbf{g}_i \otimes \mathbf{g}_j \otimes \mathbf{g}^k \otimes \mathbf{g}^l ,$$

which, when applied to a *2*nd-order tensor

$$\mathbf{dT} = dT^{pr}\, \mathbf{g}_p \otimes \mathbf{g}_r ,$$

gives the differential of $\mathbf{F}$ at $\mathbf{T}$ in the direction of $\mathbf{dT}$

$$d\mathbf{F}(\mathbf{T}, \mathbf{dT}) = \frac{\partial F^{ij}(T^{11}, T^{12}, \dots, T^{NN})}{\partial T^{kl}}\, \mathbf{g}_i \otimes \mathbf{g}_j \otimes \mathbf{g}^k \otimes \mathbf{g}^l [dT^{pr}\, \mathbf{g}_p \otimes \mathbf{g}_r]$$

$$= \frac{\partial F^{ij}(T^{11}, T^{12}, \dots, T^{NN})}{\partial T^{kl}}\, dT^{kl}\, \mathbf{g}_i \otimes \mathbf{g}_j$$

(double contraction), which again is a *2*nd-order tensor.

Example. It is not always easy to find a short representation of the derivative from the differential. For example, the differential of the square of a tensor is by the LEIBNIZ rule

(1.42)
$$d(\mathbf{AA}) = \mathbf{dA\,A} + \mathbf{A\,dA} ,$$

obviously a linear function of $\mathbf{dA}$, and, as such can be represented as a tetrad C through

$$d(\mathbf{AA}) = \mathsf{C}[\mathbf{dA}] .$$

We did not yet introduce a symbolic notation for this tetrad. It can be represented with respect to a vector basis $\{\mathbf{g}_i\}$ with dual $\{\mathbf{g}^i\}$ by

$$\mathsf{C} = (G_{ik}\, A_{lj} + G_{lj}\, A_{ik})\, \mathbf{g}^i \otimes \mathbf{g}^j \otimes \mathbf{g}^k \otimes \mathbf{g}^l$$

with the metric coefficients $G_{ik} := \mathbf{g}_i \cdot \mathbf{g}_k$.

Example. We determine the differential of the function

$$f(\mathbf{T}) \equiv \mathbf{T}^{-1},$$

which assigns to each invertible tensor its inverse. The differential of $\mathbf{T}\,\mathbf{T}^{-1} = \mathbf{I}$ is by the LEIBNIZ rule

$$d(\mathbf{T}\,\mathbf{T}^{-1}) = \mathbf{0} = \mathbf{dT}\,\mathbf{T}^{-1} + \mathbf{T}\,d(\mathbf{T}^{-1})$$

and, thus, we obtain

$$d(\mathbf{T}^{-1}) = -\mathbf{T}^{-1}\,\mathbf{dT}\,\mathbf{T}^{-1}.$$

Example. We consider the decomposition of a tensor into its deviatoric and its spherical parts after (1.7)

$$\mathbf{A} = \mathbf{A}' + \mathbf{A}^\circ = (\boldsymbol{I} - {}^1\!/_3\,\mathbf{I} \otimes \mathbf{I})[\mathbf{A}] + {}^1\!/_3\,\mathbf{I} \otimes \mathbf{I}[\mathbf{A}] \qquad \in \mathcal{Lin}.$$

Here we have

$$\mathbf{dA}' = \frac{\mathbf{dA}'}{\mathbf{dA}}[\mathbf{dA}] = (\boldsymbol{I} - {}^1\!/_3\,\mathbf{I} \otimes \mathbf{I})[\mathbf{dA}]$$

and

$$\mathbf{dA}^\circ = \frac{\mathbf{dA}^\circ}{\mathbf{dA}}[\mathbf{dA}] = {}^1\!/_3\,\mathbf{I} \otimes \mathbf{I}\,[\mathbf{dA}]$$

and, thus

$$\frac{\mathbf{dA}'}{\mathbf{dA}} = \boldsymbol{I} - {}^1\!/_3\,\mathbf{I} \otimes \mathbf{I} \qquad \text{and} \qquad \frac{\mathbf{dA}^\circ}{\mathbf{dA}} = {}^1\!/_3\,\mathbf{I} \otimes \mathbf{I}.$$

The derivative of the deviatoric/spherical part of a tensor with respect to the tensor projects every tensor into its deviatoric/spherical part. Both derivatives are projectors. The differential of the above decomposition is

$$\begin{aligned}
\mathbf{dA} &= \mathbf{dA}' + \mathbf{dA}^\circ \\
&= \frac{\partial \mathbf{A}}{\partial \mathbf{A}'}[\mathbf{dA}'] + \frac{\partial \mathbf{A}}{\partial \mathbf{A}^\circ}[\mathbf{dA}^\circ] \\
&= \frac{\partial \mathbf{A}}{\partial \mathbf{A}'}[(\boldsymbol{I} - {}^1\!/_3\,\mathbf{I} \otimes \mathbf{I})[\mathbf{dA}]] + \frac{\partial \mathbf{A}}{\partial \mathbf{A}^\circ}[{}^1\!/_3\,\mathbf{I} \otimes \mathbf{I}\,[\mathbf{dA}]]
\end{aligned}$$

which is also solved by

$$\frac{\partial \mathbf{A}}{\partial \mathbf{A}'} = \boldsymbol{I} - {}^1\!/_3\,\mathbf{I} \otimes \mathbf{I} \qquad \text{and} \qquad \frac{\partial \mathbf{A}}{\partial \mathbf{A}^\circ} = {}^1\!/_3\,\mathbf{I} \otimes \mathbf{I}.$$

Example. We consider the decomposition of an arbitrary tensor into its symmetric and skew parts (1.5)

$$\mathbf{A} = sym(\mathbf{A}) + skw(\mathbf{A}) = \boldsymbol{I}^S[\mathbf{A}] + \boldsymbol{I}^A[\mathbf{A}] \qquad \in \mathcal{Lin} .$$

By

$$d\,sym(\mathbf{A}) = \boldsymbol{I}^S[\mathbf{dA}] = \frac{\mathbf{d}\,sym(\mathbf{A})}{\mathbf{dA}}\,[\mathbf{dA}]$$

we conclude

$$\frac{\mathbf{d}\,sym(\mathbf{A})}{\mathbf{dA}} = \boldsymbol{I}^S$$

and by

$$d\,skw(\mathbf{A}) = \boldsymbol{I}^A[\mathbf{dA}] = \frac{\mathbf{d}\,skw(\mathbf{A})}{\mathbf{dA}}\,[\mathbf{dA}]$$

$$\frac{\mathbf{d}\,skw(\mathbf{A})}{\mathbf{dA}} = \boldsymbol{I}^A .$$

The derivative of the symmetric/ skew part of a tensor with respect to the tensor maps each tensor into its symmetric/ skew part.

The differential of the above decomposition is

$$\mathbf{dA} = d\,sym(\mathbf{A}) + d\,skw(\mathbf{A})$$

$$= \frac{\partial\mathbf{A}}{\partial sym(\mathbf{A})}\,[d\,sym(\mathbf{A})] + \frac{\partial\mathbf{A}}{\partial skw(\mathbf{A})}\,[d\,skw(\mathbf{A})]$$

$$= \frac{\partial\mathbf{A}}{\partial sym(\mathbf{A})}\,[\boldsymbol{I}^S[\mathbf{dA}]] + \frac{\partial\mathbf{A}}{\partial skw(\mathbf{A})}\,[\boldsymbol{I}^A[\mathbf{dA}]]$$

which is solved by

$$\frac{\partial\mathbf{A}}{\partial sym(\mathbf{A})} = \boldsymbol{I}^S \qquad \text{and} \qquad \frac{\partial\mathbf{A}}{\partial skw(\mathbf{A})} = \boldsymbol{I}^A ,$$

neither of which is unique.

Example. We consider a symmetric tensor $\mathbf{A}$ with three different eigenvalues λ_r in spectral representation

$$\mathbf{A} = \sum_{r=1}^{3} \lambda_r\, \mathbf{p}_r \otimes \mathbf{p}_r \qquad \in \mathcal{Sym}$$

where the eigenvectors $\mathbf{p}_r$ form an ONB. In the sequel, we will not make use of the sum convention, but instead explicitly notate all sums. We introduce the abbreviation of tensors of different orders by

$$\mathbf{P}_{ij\,...\,k} := \mathbf{p}_i \otimes \mathbf{p}_j \otimes ... \otimes \mathbf{p}_k \qquad \text{for } i, j, ...\, k = 1, 2, 3.$$

With this, we can also write

$$\mathbf{A} = \sum_{r=1}^{3} \lambda_r \, \mathbf{P}_{rr}$$

where we used a slightly different notation for the eigenprojectors as we did in Chap. 1.2.2. Each symmetric tensor is uniquely determined by its eigenvalues and the corresponding eigenvectors. Therefore, we take the tensor as a function of the eigenvalues (ordered after their magnitude) and the eigenprojectors $\mathbf{A}(\lambda_1, \lambda_2, \lambda_3, \mathbf{P}_{11}, \mathbf{P}_{22}, \mathbf{P}_{33})$. The differential of this function is

$$d\mathbf{A} = \sum_{r=1}^{3} \Big(\frac{\partial \mathbf{A}}{\partial \lambda_r}\, d\lambda_r + \frac{\partial \mathbf{A}}{\partial \mathbf{P}_{rr}} [\mathbf{dP}_{rr}]\Big)$$

$$= \sum_{r=1}^{3} (d\lambda_r \, \mathbf{P}_{rr} + \lambda_r \, \mathbf{dP}_{rr}) .$$

However, we must keep in mind that the three $\mathbf{dP}_{rr}$ are no independent tensors. The tripod of the unit eigenvectors $\{\mathbf{p}_1, \mathbf{p}_2, \mathbf{p}_3\}$ can only rotate in space. If $\mathbf{d\omega}$ is the axial vector, which describes the differential of this rotation as $\mathbf{dp}_r = \mathbf{d\omega} \times \mathbf{p}_r$, we obtain

(1.43) $$\mathbf{dP}_{rr} = (\mathbf{d\omega} \times \mathbf{p}_r) \otimes \mathbf{p}_r + \mathbf{p}_r \otimes (\mathbf{d\omega} \times \mathbf{p}_r) = \mathbf{d\omega} \times \mathbf{P}_{rr} - \mathbf{P}_{rr} \times \mathbf{d\omega}$$

where the cross product between vector and tensor is defined by

$$\mathbf{d\omega} \times \mathbf{P}_{rr} = \mathbf{d\omega} \times (\mathbf{p}_r \otimes \mathbf{p}_r) = (\mathbf{d\omega} \times \mathbf{p}_r) \otimes \mathbf{p}_r .$$

With the representation

$$\mathbf{d\omega} = \sum_{i=1}^{3} d\omega^i \, \mathbf{p}_i$$

we obtain, *e.g.*

$$\mathbf{dP}_{11} = d\omega^i \, \mathbf{p}_i \times \mathbf{p}_1 \otimes \mathbf{p}_1 - \mathbf{p}_1 \otimes \mathbf{p}_1 \times d\omega^i \, \mathbf{p}_i$$

$$= d\omega^3 (\mathbf{P}_{12} + \mathbf{P}_{21}) - d\omega^2 (\mathbf{P}_{13} + \mathbf{P}_{31}) .$$

After this equation, the differentials are perpendicular to all projectors

$$\mathbf{dP}_{rr} \cdot \mathbf{P}_{ss} = 0 \qquad \text{for } r, s = 1, 2, 3.$$

If we substitute into the above differential, we get

$$\begin{aligned} d\mathbf{A} &= d\lambda_1 \, \mathbf{P}_{11} + d\lambda_2 \, \mathbf{P}_{22} + d\lambda_3 \, \mathbf{P}_{33} \\ &+ \lambda_1 \, d\omega^3 (\mathbf{P}_{12} + \mathbf{P}_{21}) - \lambda_1 \, d\omega^2 (\mathbf{P}_{13} + \mathbf{P}_{31}) \\ &+ \lambda_2 \, d\omega^1 (\mathbf{P}_{23} + \mathbf{P}_{32}) - \lambda_2 \, d\omega^3 (\mathbf{P}_{21} + \mathbf{P}_{12}) \\ &+ \lambda_3 \, d\omega^2 (\mathbf{P}_{31} + \mathbf{P}_{13}) - \lambda_3 \, d\omega^1 (\mathbf{P}_{32} + \mathbf{P}_{23}) \end{aligned}$$

$$= \frac{\partial \mathbf{A}}{\partial \lambda_1} d\lambda_1 + \frac{\partial \mathbf{A}}{\partial \lambda_2} d\lambda_2 + \frac{\partial \mathbf{A}}{\partial \lambda_3} d\lambda_3$$

$$+ \frac{\partial \mathbf{A}}{\partial \mathbf{P}_{11}} [d\omega^3 (\mathbf{P}_{12} + \mathbf{P}_{21}) - d\omega^2 (\mathbf{P}_{13} + \mathbf{P}_{31})]$$

$$+ \frac{\partial \mathbf{A}}{\partial \mathbf{P}_{22}} [d\omega^1 (\mathbf{P}_{23} + \mathbf{P}_{32}) - d\omega^3 (\mathbf{P}_{21} + \mathbf{P}_{12})]$$

$$+ \frac{\partial \mathbf{A}}{\partial \mathbf{P}_{33}} [d\omega^2 (\mathbf{P}_{31} + \mathbf{P}_{13}) - d\omega^1 (\mathbf{P}_{32} + \mathbf{P}_{23})] .$$

Comparing the factors of the mutually independent $d\lambda^1$, $d\lambda^2$, $d\lambda^3$, $d\omega^1$, $d\omega^2$, $d\omega^3$, we find

$$\frac{\partial \mathbf{A}}{\partial \lambda_r} = \mathbf{P}_{rr} \qquad r = 1, 2, 3$$

and

$$\frac{\partial \mathbf{A}}{\partial \mathbf{P}_{rr}} = \lambda_r (\mathbf{P}_{r\,r+1\,r\,r+1} + \mathbf{P}_{r+1\,r\,r+1\,r} + \mathbf{P}_{r\,r+2\,r\,r+2} + \mathbf{P}_{r+2\,r\,r+2\,r})$$

for $r = 1, 2, 3$ *modulo 3*, like, *e.g.*

$$\frac{\partial \mathbf{A}}{\partial \mathbf{P}_{11}} = \lambda_1 (\mathbf{P}_{1212} + \mathbf{P}_{2121} + \mathbf{P}_{1313} + \mathbf{P}_{3131}) .$$

For these symmetric tetrads, $\mathbf{P}_{12}$, $\mathbf{P}_{21}$, $\mathbf{P}_{13}$, $\mathbf{P}_{31}$ are eigentensors corresponding to the eigenvalue λ_1, and all other $\mathbf{P}_{ij}$ to the eigenvalue 0.

Inversely, we immediately get

$$\frac{d\lambda_r}{d\mathbf{A}} = \mathbf{P}_{rr} \qquad r = 1, 2, 3.$$

$\frac{d\mathbf{P}_{rr}}{d\mathbf{A}}$ must solve the following equations

$$d\mathbf{P}_{rr} = \frac{d\mathbf{P}_{rr}}{d\mathbf{A}} [\mathbf{dA}] \qquad r = 1, 2, 3.$$

like, *e.g.*

$$d\mathbf{P}_{11} = \frac{d\mathbf{P}_{11}}{d\mathbf{A}} [\mathbf{dA}] = \frac{d\mathbf{P}_{11}}{d\mathbf{A}} \sum_{s=1}^{3} [d\lambda_s \mathbf{P}_{ss} + \lambda_s \mathbf{dP}_{ss}] .$$

By substituting the differentials and comparing the factors of the mutually independent $d\lambda^1$, $d\lambda^2$, $d\lambda^3$, $d\omega^1$, $d\omega^2$, $d\omega^3$, we see that these equations are solved by

$$\frac{d\mathbf{P}_{rr}}{d\mathbf{A}} = (\lambda_r - \lambda_{r+1})^{-1} (\mathbf{P}_{r\,r+1\,r\,r+1} + \mathbf{P}_{r+1\,r\,r+1\,r})$$

$$+ (\lambda_r - \lambda_{r+2})^{-1} (\mathbf{P}_{r\,r+2\,r\,r+2} + \mathbf{P}_{r+2\,r\,r+2\,r})$$

for $r = 1, 2, 3$ *modulo 3*, like, *e.g.*,

$$\frac{d\mathbf{P}_{11}}{d\mathbf{A}} = (\lambda_1 - \lambda_2)^{-1} (\mathbf{P}_{1212} + \mathbf{P}_{2121})$$

$$+ (\lambda_1 - \lambda_3)^{-1} (\mathbf{P}_{1313} + \mathbf{P}_{3131}) .$$

For this symmetric tetrad, $\mathbf{P}_{12}$ and $\mathbf{P}_{21}$ are eigentensors corresponding to the eigenvalue $(\lambda_1 - \lambda_2)^{-1}$, $\mathbf{P}_{13}$ and $\mathbf{P}_{31}$ to the eigenvalue $(\lambda_1 - \lambda_3)^{-1}$, and all other $\mathbf{P}_{ij}$ to the eigenvalue 0.

As a verification, we can use

$$\frac{d\mathbf{A}}{d\mathbf{A}} = \boldsymbol{I} = \sum_{r=1}^{3} \left(\frac{\partial \mathbf{A}}{\partial \lambda_r} \otimes \frac{d\lambda_r}{d\mathbf{A}} + \frac{\partial \mathbf{A}}{\partial \mathbf{P}_{rr}} \frac{d\mathbf{P}_{rr}}{d\mathbf{A}} \right)$$

with the *4*th-order identity tensor

$$\boldsymbol{I} = \sum_{i,j=1}^{3} \mathbf{P}_{ijij} .$$

In the literature, these tetrads are sometimes endowed with the left and right subsymmetries. This can be made, as $d\mathbf{A}$ and $d\mathbf{P}_{rr}$ are symmetric.

We list a possible choice of the derivatives for symmetric tensors with different eigenvalues.

(1.44)

$$\frac{\partial \mathbf{A}}{\partial \lambda_r} = \mathbf{P}_{rr}$$

$$\frac{d\lambda_r}{d\mathbf{A}} = \mathbf{P}_{rr}$$

$$\frac{\partial \mathbf{A}}{\partial \mathbf{P}_{rr}} = \lambda_r (\mathbf{P}_{r\,r+1\,r\,r+1} + \mathbf{P}_{r+1\,r\,r+1\,r} + \mathbf{P}_{r\,r+2\,r\,r+2} + \mathbf{P}_{r+2\,r\,r+2\,r})$$

$$\frac{d\mathbf{P}_{rr}}{d\mathbf{A}} = (\lambda_r - \lambda_{r+1})^{-1} (\mathbf{P}_{r\,r+1\,r\,r+1} + \mathbf{P}_{r+1\,r\,r+1\,r})$$

$$+ (\lambda_r - \lambda_{r+2})^{-1} (\mathbf{P}_{r\,r+2\,r\,r+2} + \mathbf{P}_{r+2\,r\,r+2\,r})$$

for $r = 1, 2, 3$ *modulo 3*

If the eigenvalues of $\mathbf{A}$ were not different (contrary to what we assumed), the eigenvectors $\mathbf{p}_i$ would no longer be uniquely defined, and some of the functions above would not be defined anymore.[21]

Example. For a *non-symmetric*, but still diagonalisable tensor $\mathbf{A}$ we have the spectral form (1.21)

$$\mathbf{A} = \sum_{r=1}^{3} \lambda_r \, \mathbf{p}_r \otimes \mathbf{p}^r =: \sum_{r=1}^{3} \lambda_r \, \mathbf{P}_r{}^r$$

with three right eigenvectors $\mathbf{p}_r$ and three left eigenvectors $\mathbf{p}^r$, *i.e.*,

$$\mathbf{A}\,\mathbf{p}_r = \lambda_r\,\mathbf{p}_r \qquad \text{and} \qquad \mathbf{A}^T\,\mathbf{p}^r = \lambda_r\,\mathbf{p}^r.$$

These eigenvector systems must be dual

$$\mathbf{p}_i \cdot \mathbf{p}^k = \delta_i{}^k.$$

Such a tensor is uniquely determined, if the eigenvalues and the corresponding left (or right) eigenvectors are known. For the differentials we conclude from the duality

$$d(\mathbf{p}_i \cdot \mathbf{p}^k) = 0 = \mathbf{dp}_i \cdot \mathbf{p}^k + \mathbf{p}_i \cdot \mathbf{dp}^k$$

and therefore

$$\mathbf{dp}_i \cdot \mathbf{p}^k = -\,\mathbf{p}_i \cdot \mathbf{dp}^k.$$

The differential of the first projector is

$$\begin{aligned}
d\mathbf{P}_1{}^1 &= \mathbf{dp}_1 \otimes \mathbf{p}^1 + \mathbf{p}_1 \otimes \mathbf{dp}^1 \\
&= \{\mathbf{p}^k \cdot (\mathbf{dp}_1 \otimes \mathbf{p}^1 + \mathbf{p}_1 \otimes \mathbf{dp}^1)\,\mathbf{p}_j\}\;\mathbf{p}_k \otimes \mathbf{p}^j \\
&= \{(\mathbf{p}^k \cdot \mathbf{dp}_1)(\mathbf{p}^1 \cdot \mathbf{p}_j) + (\mathbf{p}^k \cdot \mathbf{p}_1)(\mathbf{dp}^1 \cdot \mathbf{p}_j)\}\;\mathbf{p}_k \otimes \mathbf{p}^j \\
&= (\mathbf{p}^k \cdot \mathbf{dp}_1)\,\delta_j{}^1\,\mathbf{p}_k \otimes \mathbf{p}^j - \delta_1{}^k\,(\mathbf{p}^1 \cdot \mathbf{dp}_j)\,\mathbf{p}_k \otimes \mathbf{p}^j \\
&= (\mathbf{p}^k \cdot \mathbf{dp}_1)\,\mathbf{p}_k \otimes \mathbf{p}^1 - (\mathbf{p}^1 \cdot \mathbf{dp}_j)\,\mathbf{p}_1 \otimes \mathbf{p}^j \\
&= (\mathbf{p}^1 \cdot \mathbf{dp}_1)\,\mathbf{p}_1 \otimes \mathbf{p}^1 - (\mathbf{p}^1 \cdot \mathbf{dp}_1)\,\mathbf{p}_1 \otimes \mathbf{p}^1 \\
&\quad + (\mathbf{p}^2 \cdot \mathbf{dp}_1)\,\mathbf{p}_2 \otimes \mathbf{p}^1 - (\mathbf{p}^1 \cdot \mathbf{dp}_2)\,\mathbf{p}_1 \otimes \mathbf{p}^2 \\
&\quad + (\mathbf{p}^3 \cdot \mathbf{dp}_1)\,\mathbf{p}_3 \otimes \mathbf{p}^1 - (\mathbf{p}^1 \cdot \mathbf{dp}_3)\,\mathbf{p}_1 \otimes \mathbf{p}^3 \\
&= (\mathbf{P}_2{}^{12} + \mathbf{P}_3{}^{13})\,\mathbf{dp}_1 - \mathbf{P}_1{}^{21}\,\mathbf{dp}_2 - \mathbf{P}_1{}^{31}\,\mathbf{dp}_3
\end{aligned}$$

(simple contraction) with the triads $\mathbf{P}_i{}^{jk} := \mathbf{p}_i \otimes \mathbf{p}^j \otimes \mathbf{p}^k$. The differential of the entire tensor is

$$d\mathbf{A} = \frac{\partial \mathbf{A}}{\partial \lambda_1}\,d\lambda_1 + \frac{\partial \mathbf{A}}{\partial \lambda_2}\,d\lambda_2 + \frac{\partial \mathbf{A}}{\partial \lambda_3}\,d\lambda_3$$

---

[21] See VALLÉE/ HE/ LERINTIU (2006).

$$+ \frac{\partial \mathbf{A}}{\partial \mathbf{P}_1{}^1}[\mathbf{dP}_1{}^1] + \frac{\partial \mathbf{A}}{\partial \mathbf{P}_2{}^2}[\mathbf{dP}_2{}^2] + \frac{\partial \mathbf{A}}{\partial \mathbf{P}_3{}^3}[\mathbf{dP}_3{}^3]$$

$$= \frac{\partial \mathbf{A}}{\partial \lambda_1} d\lambda_1 + \frac{\partial \mathbf{A}}{\partial \lambda_2} d\lambda_2 + \frac{\partial \mathbf{A}}{\partial \lambda_3} d\lambda_3$$

$$+ \frac{\partial \mathbf{A}}{\partial \mathbf{P}_1{}^1}[(\mathbf{P}_2{}^{12} + \mathbf{P}_3{}^{13})\,\mathbf{dp}_1 - \mathbf{P}_1{}^{21}\,\mathbf{dp}_2 - \mathbf{P}_1{}^{31}\,\mathbf{dp}_3]$$

$$+ \frac{\partial \mathbf{A}}{\partial \mathbf{P}_2{}^2}[(\mathbf{P}_3{}^{23} + \mathbf{P}_1{}^{21})\,\mathbf{dp}_2 - \mathbf{P}_2{}^{32}\,\mathbf{dp}_3 - \mathbf{P}_2{}^{12}\,\mathbf{dp}_1]$$

$$+ \frac{\partial \mathbf{A}}{\partial \mathbf{P}_3{}^3}[(\mathbf{P}_1{}^{31} + \mathbf{P}_2{}^{32})\,\mathbf{dp}_3 - \mathbf{P}_3{}^{13}\,\mathbf{dp}_1 - \mathbf{P}_3{}^{23}\,\mathbf{dp}_2]$$

$$= d\lambda_1\,\mathbf{P}_1{}^1 + d\lambda_2\,\mathbf{P}_2{}^2 + d\lambda_3\,\mathbf{P}_3{}^3 + \lambda_1\,\mathbf{dP}_1{}^1 + \lambda_2\,\mathbf{dP}_2{}^2 + \lambda_3\,\mathbf{dP}_3{}^3$$

$$= d\lambda_1\,\mathbf{P}_1^1 + d\lambda_2\,\mathbf{P}_2^2 + d\lambda_3\,\mathbf{P}_3^3$$

$$+ \lambda_1\,\{(\mathbf{P}_2^{12} + \mathbf{P}_3^{13})\,\mathbf{dp}_1 - \mathbf{P}_1^{21}\,\mathbf{dp}_2 - \mathbf{P}_1^{31}\,\mathbf{dp}_3\}$$

$$+ \lambda_2\,\{(\mathbf{P}_3^{23} + \mathbf{P}_1^{21})\,\mathbf{dp}_2 - \mathbf{P}_2^{32}\,\mathbf{dp}_3 - \mathbf{P}_2^{12}\,\mathbf{dp}_1\}$$

$$+ \lambda_3\,\{(\mathbf{P}_1^{31} + \mathbf{P}_2^{32})\,\mathbf{dp}_3 - \mathbf{P}_3^{13}\,\mathbf{dp}_1 - \mathbf{P}_3^{23}\,\mathbf{dp}_2\}\,.$$

The above equation is solved for arbitrary $d\lambda_1$, $d\lambda_2$, $d\lambda_3$, $\mathbf{dp}_1$, $\mathbf{dp}_2$, $\mathbf{dp}_3$ by

$$\frac{\partial \mathbf{A}}{\partial \lambda_r} = \mathbf{P}_r{}^r \qquad r = 1, 2, 3$$

$$\frac{\partial \mathbf{A}}{\partial \mathbf{P}_1{}^1} = \lambda_1(\mathbf{P}_2{}^{12}{}_1 + \mathbf{P}_1{}^{21}{}_2 + \mathbf{P}_3{}^{13}{}_1 + \mathbf{P}_1{}^{31}{}_3)$$

$$\frac{\partial \mathbf{A}}{\partial \mathbf{P}_2{}^2} = \lambda_2(\mathbf{P}_3{}^{23}{}_2 + \mathbf{P}_2{}^{32}{}_3 + \mathbf{P}_1{}^{21}{}_2 + \mathbf{P}_2{}^{12}{}_1)$$

$$\frac{\partial \mathbf{A}}{\partial \mathbf{P}_3{}^3} = \lambda_3(\mathbf{P}_1{}^{31}{}_3 + \mathbf{P}_3{}^{13}{}_1 + \mathbf{P}_2{}^{32}{}_3 + \mathbf{P}_3{}^{23}{}_2)$$

with $\mathbf{P}_i{}^{jk}{}_l := \mathbf{p}_i \otimes \mathbf{p}^j \otimes \mathbf{p}^k \otimes \mathbf{p}_l$. The derivatives with respect to the right eigenvectors are

$$\frac{\partial \mathbf{A}}{\partial \mathbf{p}_1} = (\lambda_1 - \lambda_2)\,\mathbf{P}_2{}^{12} + (\lambda_1 - \lambda_3)\,\mathbf{P}_3{}^{13}$$

and the others analogously. The inverse differentials can be written as

$$\frac{d\lambda_r}{d\mathbf{A}} = \mathbf{P}^r{}_r \qquad r = 1, 2, 3$$

$$\frac{d\mathbf{P}_1{}^1}{d\mathbf{A}} = (\lambda_1 - \lambda_2)^{-1}(\mathbf{P}_1{}^{21}{}_2 + \mathbf{P}_2{}^{12}{}_1) + (\lambda_1 - \lambda_3)^{-1}(\mathbf{P}_1{}^{31}{}_3 + \mathbf{P}_3{}^{13}{}_1)$$

$$\frac{d\mathbf{P}_2{}^2}{d\mathbf{A}} = (\lambda_2 - \lambda_3)^{-1}(\mathbf{P}_2{}^{32}{}_3 + \mathbf{P}_3{}^{23}{}_2) + (\lambda_2 - \lambda_1)^{-1}(\mathbf{P}_2{}^{12}{}_1 + \mathbf{P}_1{}^{21}{}_2)$$

$$\frac{d\mathbf{P}_3{}^3}{d\mathbf{A}} = (\lambda_3 - \lambda_1)^{-1}(\mathbf{P}_3{}^{13}{}_1 + \mathbf{P}_1{}^{31}{}_3) + (\lambda_3 - \lambda_2)^{-1}(\mathbf{P}_3{}^{23}{}_2 + \mathbf{P}_2{}^{32}{}_3),$$

which can be verified by the condition

$$\frac{d\mathbf{A}}{d\mathbf{A}} = \mathbf{I} = \sum_{r=1}^{3}\left(\frac{\partial\mathbf{A}}{\partial\lambda_r}\otimes\frac{d\lambda_r}{d\mathbf{A}} + \frac{\partial\mathbf{A}}{\partial\mathbf{P}_r{}^r}\,\frac{d\mathbf{P}_r{}^r}{d\mathbf{A}}\right)$$

(double contraction between the last terms). Here $\mathbf{I}$ is the *4*th-order identity tensor with the representation

$$\mathbf{I} = \sum_{k,l=1}^{3} \mathbf{P}_k{}^{lk}{}_l\,.$$

We obtain the following rules.

(1.45)

$$\frac{\partial\mathbf{A}}{\partial\lambda_r} = \mathbf{P}_r{}^r$$

$$\frac{\partial\mathbf{A}}{\partial\mathbf{P}_r{}^r} = \lambda_r(\mathbf{P}_{r+1}{}^{r\,r+1}{}_r + \mathbf{P}_r{}^{r+1\,r}{}_{r+1} + \mathbf{P}_{r+2}{}^{r\,r+2}{}_r + \mathbf{P}_r{}^{r+2\,r}{}_{r+2})$$

$$\frac{d\lambda_r}{d\mathbf{A}} = \mathbf{P}^r{}_r$$

$$\frac{d\mathbf{P}_r{}^r}{d\mathbf{A}} = (\lambda_r - \lambda_{r+1})^{-1}(\mathbf{P}_r{}^{r+1\,r}{}_{r+1} + \mathbf{P}_{r+1}{}^{r\,r+1}{}_r)$$

$$+ (\lambda_r - \lambda_{r+2})^{-1}(\mathbf{P}_r{}^{r+2\,r}{}_{r+2} + \mathbf{P}_{r+2}{}^{r\,r+2}{}_r)$$

for $r = 1, 2, 3$ *modulo 3*

For a symmetric tensor $\mathbf{A}$ we again obtain the results of the foregoing example (1.44).

1.4 EUCLIDean Point Space

A Comment on the Literature. In some books on tensor calculus one finds introductions to the EUCLIDean space, like in ERICKSEN (1960), McCONNELL (1957), SOKOLNIKOFF (1951, 1964), TROSTEL (1997), STEINMANN (2015). An axiomatic introduction to the EUCLIDean space can be found in NOLL (1964) and BERTRAM (1989, p. 36-42). In the present context, however, we will proceed more pragmatically.

We describe the classical, *i.e.*, non-relativistic physical space as a point space, where we can perform EUCLIDean geometry like describing points, straight lines, planes, volumetric regions, etc., and where we can measure distances and angles.

First of all, the **EUCLIDean space** $\mathscr{E}$ consists of (spatial) **points** $x, y, z, \ldots \in \mathscr{E}$. For each pair of points we can measure their **distance**

$$d: \mathscr{E} \times \mathscr{E} \rightarrow \mathscr{R} \quad | \quad (x, y) \mapsto d(x, y) .$$

This distance or **EUCLIDean metric** makes a **metric space** out of $\mathscr{E}$.

One can associate with $\mathscr{E}$ a linear space by means of the following construction. We define an equivalence relation $\sim$ on all ordered point pairs like $(x, y) \in \mathscr{E} \times \mathscr{E}$

$(x, y) \sim (x', y')$

$\Leftrightarrow$

The directed straight line from x to y can be shifted by a parallel shift into that from x' to y'.

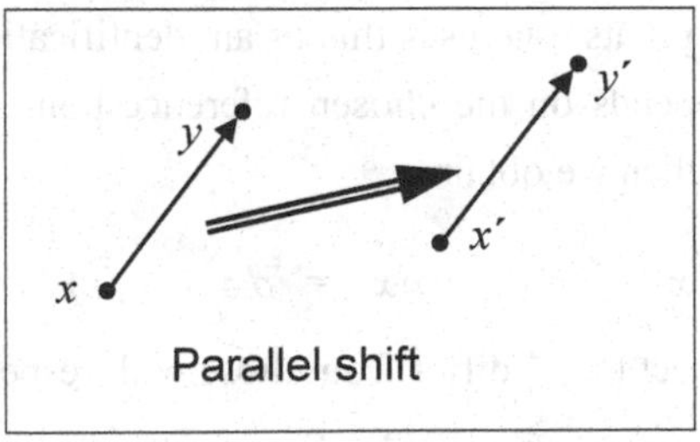

Parallel shift

The equivalence classes with respect to $\sim$ are called **EUCLIDean shifters** with the notation $\overrightarrow{xy}$. Accordingly, $(x, y), (x', y') \in \overrightarrow{xy} = \overrightarrow{x'y'}$.

The EUCLIDean shifters form a vector space in a natural way. The sum of two of them is defined by the parallelogram rule. We choose two appropriate point pairs for the two shifters, such that the endpoint of one coincides with the initial point of the other like (x, y) and (y, z). The sum is then

$$\overrightarrow{xy} + \overrightarrow{yz} = \overrightarrow{xz} \,.$$

The multiplication of a vector $\overrightarrow{xy}$ by a scalar $\alpha \in \mathscr{R}$ is executed by an elongation of the vector by the factor α. For negative α we additionally reverse the sense of the vector, like

$$(-1)\, \overrightarrow{xy} = \overrightarrow{yx} \,.$$

With these operations, the set $\mathscr{V}$ of all EUCLIDean shifters becomes a vector space, which is associated with the EUCLIDean point space $\mathscr{E}$. $(\mathscr{E}, \mathscr{V})$ form an **affine space**. In the sequel we will always assume that $\mathscr{V}$ is three-dimensional.

The EUCLIDean metric induces an inner product on $\mathscr{V}$ by

$$\overrightarrow{xy} \cdot \overrightarrow{yz} = \tfrac{1}{2} \{d(x, z)^2 - d(x, y)^2 - d(y, z)^2\}$$

(Cosine theorem). This is the only inner product on $\mathscr{V}$ that is compatible with the EUCLIDean metric since

$$\overrightarrow{xy} \cdot \overrightarrow{xy} = d(x, y)^2.$$

If we choose an arbitrary, but fixed point $o \in \mathscr{E}$ as a **reference point**, we obtain a bijection

$$\mathbf{r}_o : \mathscr{E} \to \mathscr{V} \quad | \quad x \mapsto \overrightarrow{ox} =: \mathbf{r}_o(x)$$

which uniquely assigns to each point x the **position vector** $\mathbf{r}_o(x)$. As it is often easier to handle vectors than points, one uses this as an identification between $\mathscr{E}$ and $\mathscr{V}$, which evidently depends on the chosen reference point. In fact, if we choose another one $o' \in \mathscr{E}$, then we obtain

$$\mathbf{r}_{o'}(x) = \overrightarrow{o'x} = \overrightarrow{o'o} + \overrightarrow{ox} = \overrightarrow{o'o} + \mathbf{r}_o(x) \,.$$

All position vectors with respect to o' differ from those with respect to o by the constant vector $\overrightarrow{o'o}$. However, in $\mathscr{E}$ no point is naturally distinguished, due to the *homogeneity of the EUCLIDean space*.

Coordinate Systems

(COOS) in $\mathcal{E}$ are pairs $\{\mathcal{U}, \varphi\}$ with $\mathcal{U} \subset \mathcal{E}$ being an open set and

$$\varphi : \mathcal{U} \to \mathcal{R}^3 \quad | \quad x \mapsto \{\varphi^1, \varphi^2, \varphi^3\}$$

being a continuous and injective mapping. If we compose φ with the i-th projection p^i of $\mathcal{R}^3$ into $\mathcal{R}$, then we obtain the ***i*-th coordinate**

$$\varphi^i := p^i \varphi : \mathcal{U} \to \mathcal{R} \quad | \quad x \mapsto \varphi^i \qquad i = 1, 2, 3.$$

Here, we did not distinguish between function and the value of the function, in order to not complicate the notation. In principle it does not matter whether we put the index of a coordinate in the upper or in the lower position. However, if we calculate the differentials, the upper indices match the sum convention better, as we will see. Therefore, we will prefer the upper position, even at the risk of it being confused with powers.

If for some COOS $\mathcal{U}$ coincides with the entire space $\mathcal{E}$, we call it a **global COOS**. For a cylinder COOS $\{r, \theta, z\}$, one usually leaves points with $r \equiv 0$ and $\theta \equiv 0$ out of $\mathcal{U}$ in order to avoid ambiguities. Such COOS are not global. In NEUTSCH (1995) one finds specific COOS.

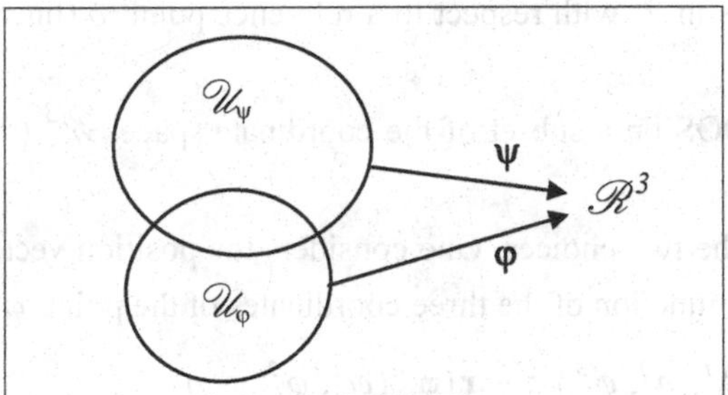

Change of COOS

We will generally use only such COOS, for which the change is differentiable. That means, if $\{\mathcal{U}_\varphi, \varphi\}$ and $\{\mathcal{U}_\psi, \psi\}$ are two COOS, then

$$\varphi\,\psi^{-1} : \psi(\mathcal{U}_\psi \cap \mathcal{U}_\varphi) \to \mathcal{R}^3$$

is assumed to be continuously differentiable. Then all the partial derivatives do exist

$$\frac{\partial \varphi^i}{\partial \psi^k} := \frac{\partial}{\partial \psi^k}(\varphi^i\,\psi^{-1}) \qquad i, k = 1, 2, 3$$

and

$$\frac{\partial \psi^k}{\partial \varphi^i} := \frac{\partial}{\partial \varphi^i}(\psi^k \boldsymbol{\varphi}^{-1}) \qquad i, k = 1, 2, 3.$$

The **JACOBI**[22] **matrix** of the COOS change

$$[J^i{}_k] = \left[\frac{\partial \varphi^i}{\partial \psi^k}\right] = \begin{bmatrix} \frac{\partial \varphi^1}{\partial \psi^1} & \frac{\partial \varphi^1}{\partial \psi^2} & \frac{\partial \varphi^1}{\partial \psi^3} \\ \frac{\partial \varphi^2}{\partial \psi^1} & \frac{\partial \varphi^2}{\partial \psi^2} & \frac{\partial \varphi^2}{\partial \psi^3} \\ \frac{\partial \varphi^3}{\partial \psi^1} & \frac{\partial \varphi^3}{\partial \psi^2} & \frac{\partial \varphi^3}{\partial \psi^3} \end{bmatrix}$$

exists everywhere in $\mathscr{U}_\psi \cap \mathscr{U}_\varphi$ and is non-singular. The two JACOBI matrices are mutually inverse

$$\frac{\partial \psi^i}{\partial \varphi^k}\,\frac{\partial \varphi^k}{\partial \psi^l} = \delta_l{}^i \qquad i, l = 1, 2, 3.$$

After the sum convention, we sum over all repeated indices in a counter position. The upper position in the denominator then corresponds to the lower position in the numerator, and vice versa.

We have the choice to identify the EUCLIDean space $\mathscr{E}$ globally or locally

- by position vectors in $\mathscr{V}$ with respect to a reference point o (linear structure),

or

- by means of a COOS on a subset of the coordinate space $\mathscr{R}^3$ (differentiable structure).

Often one combines the two choices. One considers the position vector $\mathbf{r}(x) \in \mathscr{V}$ of a point $x \in \mathscr{E}$ as a function of the three coordinates of the point $\varphi^i(x)$

$$\mathbf{r}_\varphi(\varphi^1, \varphi^2, \varphi^3) := \mathbf{r}(\boldsymbol{\varphi}^{-1}(\varphi^1, \varphi^2, \varphi^3)).$$

In each point we can determine the **tangent basis** or **covariant basis** consisting of the three vectors

(1.46)

$$\mathbf{r}_{\varphi i} := \frac{\partial \mathbf{r}}{\partial \varphi^i} \in \mathscr{V} \qquad i = 1, 2, 3$$

and the dual **gradient basis** or **contravariant basis**

$$\{\mathbf{r}_\varphi{}^i\} \subset \mathscr{V} \qquad i = 1, 2, 3$$

to the COOS. Duality means

[22] Carl Gustav Jacob Jacobi (1804-1851)

$$\mathbf{r}_{\varphi j} \cdot \mathbf{r}_{\varphi}{}^{i} = \delta_j{}^{i} \qquad i, j = 1, 2, 3,$$

which uniquely specifies the gradient basis. Under a change of the COOS, the tangent vectors transform according to the chain rule

$$\mathbf{r}_{\psi i} = \frac{\partial \mathbf{r}_{\psi}(\psi^1, \psi^2, \psi^3)}{\partial \psi^i} = \frac{\partial \mathbf{r}_{\varphi}(\varphi^1, \varphi^2, \varphi^3)}{\partial \varphi^k} \frac{\partial \varphi^k}{\partial \psi^i}$$

$$= \frac{\partial \varphi^k}{\partial \psi^i} \mathbf{r}_{\varphi k} \qquad i = 1, 2, 3,$$

i.e., by the transposed JACOBI matrix. Because of the duality

$$\delta_i{}^{r} = \mathbf{r}_{\psi i} \cdot \mathbf{r}_{\psi}{}^{r} = \frac{\partial \varphi^k}{\partial \psi^i} \mathbf{r}_{\varphi k} \cdot \frac{\partial \psi^r}{\partial \varphi^m} \mathbf{r}_{\varphi}{}^{m} \qquad i, r = 1, 2, 3,$$

the gradient vectors transform as

$$\mathbf{r}_{\psi}{}^{r} = \frac{\partial \psi^r}{\partial \varphi^m} \mathbf{r}_{\varphi}{}^{m} \qquad r = 1, 2, 3,$$

different from the tangent vectors.

The tangent and the gradient bases are called the **natural bases** of the COOS. They generally depend on the point (or on its coordinates), and are neither orthogonal nor normed. As long as we use only one COOS, we can omit the COOS index, as, *e.g.*, $\mathbf{r}^i$ instead of $\mathbf{r}_{\varphi}{}^{i}$.

Let $\mathscr{W}$ be a tensor space of arbitrary order. A function

$$\mathbf{f}: \mathscr{E} \to \mathscr{W}$$

is called a $\mathscr{W}$**-field**. By use of a COOS, we can substitute the point by its coordinates. We will generally assume the differentiability of all introduced fields, if not otherwise stated. The partial derivative of a field quantity with respect to the *i*-th coordinate is quite often notated by a comma such as

$$\mathbf{f}_{,i} := \frac{\partial \mathbf{f}}{\partial \varphi^i} \qquad i = 1, 2, 3.$$

If we represent a vector field with respect to the tangent basis

$$\mathbf{v} = v^i \, \mathbf{r}_i$$

we must keep in mind that the **contravariant component** v^i does not necessarily give us the magnitude of $\mathbf{v}$ in the direction of $\mathbf{r}_i$. Even for an orthogonal COOS, only after a normalisation

$$\mathbf{v} = v^{<i>} \frac{\mathbf{r}_i}{\sqrt{\mathbf{r}_i \cdot \mathbf{r}_i}} \qquad \text{(sum over } i\text{)}$$

we obtain the **physical component** of **v**

$$v^{<i>} := \sqrt{\mathbf{r}_i \cdot \mathbf{r}_i} \; v^i \qquad i = 1, 2, 3 \text{ (no sum)}.$$

We can also represent the vector with respect to the contravariant basis

$$\mathbf{v} = v_i \, \mathbf{r}^i$$

with the **covariant components** v_i , which can also be normalised to physical components, and are then identical to the ones above.

In general, we can represent tensor fields of arbitrary order with respect to the covariant, the contravariant, or a mixed tensor basis. The physical components are then defined analogously, like, *e.g.*

$$\mathbf{T} = T^i{}_j \; \mathbf{r}_i \otimes \mathbf{r}^j = T^{<ij>} \frac{\mathbf{r}_i}{\sqrt{\mathbf{r}_i \cdot \mathbf{r}_i}} \otimes \frac{\mathbf{r}^j}{\sqrt{\mathbf{r}^j \cdot \mathbf{r}^j}} \qquad \text{(sum over } i, j)$$

with

$$T^{<ij>} = T^i{}_j \; \sqrt{\mathbf{r}_i \cdot \mathbf{r}_i} \; \sqrt{\mathbf{r}^j \cdot \mathbf{r}^j} \qquad \text{(no sum over } i, j)$$

for an orthogonal CCOS.

1.4.1 Covariant Derivative

We have already shown how to calculate the gradient of a vectorial vector function by the partial derivatives of the component functions of the image vector with respect to those of the argument vector in (1.37). As an example, we consider a vector field **f** in the EUCLIDean space, where we describe the point by its position vector **r** , *i.e.*, $\mathbf{f}(\mathbf{r}) \in \mathscr{V}$.

We express the vector field with respect to a COOS $\boldsymbol{\varphi}$ and its natural basis $\{\mathbf{r}_k\}$

$$\mathbf{f}(\mathbf{r}) = f^k \, \mathbf{r}_k = f^k(\varphi^1, \varphi^2, \varphi^3) \; \mathbf{r}_k(\varphi^1, \varphi^2, \varphi^3)$$

wherein both the components and the basis vectors depend on the coordinates. With

$$d\mathbf{r}(\varphi^1, \varphi^2, \varphi^3, d\varphi^1, d\varphi^2, d\varphi^3) = \frac{\partial \mathbf{r}}{\partial \varphi^i} \, d\varphi^i = \mathbf{r}_i \, d\varphi^i$$

we obtain the differential of the vector field by the chain rule

$$d\mathbf{f}(\mathbf{r}, d\mathbf{r}) = d(f^k \, \mathbf{r}_k) = df^k \; \mathbf{r}_k + f^k \, d\mathbf{r}_k$$

$$= d\mathbf{f}(\mathbf{r}, \mathbf{r}_i \, d\varphi^i) = \frac{\partial f^k}{\partial \varphi^i} \, d\varphi^i \; \mathbf{r}_k + f^k \, \frac{\partial \mathbf{r}_k}{\partial \varphi^i} \, d\varphi^i$$

$$= d\mathbf{f}(\mathbf{r}\,,\mathbf{r}_i)\,d\varphi^i = (\frac{\partial f^k}{\partial\varphi^i}\,\mathbf{r}_k + f^k\,\frac{\partial\mathbf{r}_k}{\partial\varphi^i})\,(\mathbf{r}^i\cdot\mathbf{r}_j)\,d\varphi^j$$

$$= \mathbf{f}(\mathbf{r})'\,\mathbf{dr} = \{(\frac{\partial f^k}{\partial\varphi^i}\,\mathbf{r}_k + f^k\,\frac{\partial\mathbf{r}_k}{\partial\varphi^i})\otimes\mathbf{r}^i\}\,\mathbf{dr}$$

with the gradient of $\mathbf{f}$

$$\mathbf{f}(\mathbf{r})' = \frac{\partial}{\partial\varphi^i}(f^k\,\mathbf{r}_k)\otimes\mathbf{r}^i = (\frac{\partial f^k}{\partial\varphi^i}\,\mathbf{r}_k + f^k\,\frac{\partial\mathbf{r}_k}{\partial\varphi^i})\otimes\mathbf{r}^i.$$

The vector field

$$\frac{\partial\mathbf{r}_k}{\partial\varphi^i} = \frac{\partial^2\mathbf{r}}{\partial\varphi^k\partial\varphi^i} \qquad i\,,k = 1, 2, 3$$

can be put into a component representation with respect to its natural bases $\{\mathbf{r}_j\}$ or $\{\mathbf{r}^j\}$

$$\frac{\partial\mathbf{r}_k}{\partial\varphi^i} = \Gamma_{kji}\,\mathbf{r}^j = \Gamma_k{}^j{}_i\,\mathbf{r}_j \qquad i\,,k = 1, 2, 3$$

with the **CHRISTOFFEL**[23] **symbols**

$$\Gamma_{kji} := \frac{\partial\mathbf{r}_k}{\partial\varphi^i}\cdot\mathbf{r}_j = \frac{\partial^2\mathbf{r}}{\partial\varphi^i\partial\varphi^k}\cdot\mathbf{r}_j = \Gamma_k{}^m{}_i\,G_{mj} \qquad i\,,k\,,j = 1, 2, 3$$

$$\Gamma_k{}^j{}_i := \frac{\partial\mathbf{r}_k}{\partial\varphi^i}\cdot\mathbf{r}^j = \frac{\partial^2\mathbf{r}}{\partial\varphi^i\partial\varphi^k}\cdot\mathbf{r}^j = \Gamma_{kmi}\,G^{mj} \qquad i\,,k\,,j = 1, 2, 3.$$

They have three indices. The one in the middle transforms as all vector indices by metric coefficients $G_{ij} := \mathbf{r}_i\cdot\mathbf{r}_j$ and $G^{ij} := \mathbf{r}^i\cdot\mathbf{r}^j$ after (1.3).

The CHRISTOFFEL symbols are symmetric in the other two indices

$$\Gamma_k{}^j{}_i = \Gamma_i{}^j{}_k \qquad \Gamma_{kji} = \Gamma_{ijk} \qquad i\,,j\,,k = 1, 2, 3$$

according to SCHWARZ's[24] symmetry principle. Analogously, we introduce the derivatives of the gradient vectors as

$$\frac{\partial\mathbf{r}^k}{\partial\varphi^i} = \Gamma^{kj}{}_i\,\mathbf{r}_j = \Gamma^k{}_{ji}\,\mathbf{r}^j \qquad i\,,k = 1, 2, 3$$

with

---

[23] Elwin Bruno Christoffel (1829-1900)

[24] Hermann Amandus Schwarz (1843-1921)

$$\Gamma^{kj}{}_i := \frac{\partial \mathbf{r}^k}{\partial \varphi^i} \cdot \mathbf{r}^j = \Gamma^k{}_{mi}\, G^{mj} \qquad i, j, k = 1, 2, 3$$

and

$$\Gamma^k{}_{ji} := \frac{\partial \mathbf{r}^k}{\partial \varphi^i} \cdot \mathbf{r}_j = \Gamma^{km}{}_i\, G_{mj} \qquad i, j, k = 1, 2, 3.$$

Because of the duality $\mathbf{r}_i \cdot \mathbf{r}^j = \delta_i^j$ we have

$$\frac{\partial(\mathbf{r}_i \cdot \mathbf{r}^j)}{\partial \varphi^k} = 0 = \frac{\partial \mathbf{r}_i}{\partial \varphi^k} \cdot \mathbf{r}^j + \mathbf{r}_i \cdot \frac{\partial \mathbf{r}^j}{\partial \varphi^k}$$

$$= \Gamma_i{}^l{}_k\, \mathbf{r}_l \cdot \mathbf{r}^j + \mathbf{r}_i \cdot \Gamma^j{}_{lk}\, \mathbf{r}^l = \Gamma_i{}^j{}_k + \Gamma^j{}_{ik},$$

and therefore the CHRISTOFFEL symbols have the skewness

$$\Gamma^j{}_{ik} = -\Gamma_i{}^j{}_k \qquad i, j, k = 1, 2, 3.$$

By taking the derivatives of $G_{ij} := \mathbf{r}_i \cdot \mathbf{r}_j$ with respect to φ^k, substituting the CHRISTOFFEL symbols, and using the symmetry $\Gamma_{kji} = \Gamma_{ijk}$, one obtains the useful relation

$$\Gamma_{ijk} = \tfrac{1}{2}\left(\frac{\partial G_{ij}}{\partial \varphi^k} + \frac{\partial G_{jk}}{\partial \varphi^i} - \frac{\partial G_{ik}}{\partial \varphi^j}\right) \qquad i, j, k = 1, 2, 3.$$

For a Cartesian COOS, the natural bases are constant and all CHRISTOFFEL symbols are zero.

By differentiating the derivatives a second time, we obtain by again using SCHWARZ´s symmetry principle

$$\frac{\partial^2 \mathbf{r}_k}{\partial \varphi^i \partial \varphi^l} - \frac{\partial^2 \mathbf{r}_k}{\partial \varphi^l \partial \varphi^i} = \mathbf{o} \qquad i, k, l = 1, 2, 3$$

$$= \frac{\partial}{\partial \varphi^l}(\Gamma_k{}^j{}_i\, \mathbf{r}_j) - \frac{\partial}{\partial \varphi^i}(\Gamma_k{}^j{}_l\, \mathbf{r}_j)$$

$$= \frac{\partial}{\partial \varphi^l}(\Gamma_k{}^j{}_i)\, \mathbf{r}_j + \Gamma_k{}^j{}_i \frac{\partial \mathbf{r}_j}{\partial \varphi^l} - \frac{\partial}{\partial \varphi^i}(\Gamma_k{}^j{}_l)\, \mathbf{r}_j - \Gamma_k{}^j{}_l \frac{\partial \mathbf{r}_j}{\partial \varphi^i}$$

$$= \frac{\partial}{\partial \varphi^l}(\Gamma_k{}^m{}_i)\, \mathbf{r}_m + \Gamma_k{}^j{}_i\, \Gamma_j{}^m{}_l\, \mathbf{r}_m - \frac{\partial}{\partial \varphi^i}(\Gamma_k{}^m{}_l)\, \mathbf{r}_m - \Gamma_k{}^j{}_l\, \Gamma_j{}^m{}_i\, \mathbf{r}_m .$$

All components of these vectors must be zero

$$R^m{}_{kli} := \frac{\partial}{\partial \varphi^l}(\Gamma_k{}^m{}_i) - \frac{\partial}{\partial \varphi^i}(\Gamma_k{}^m{}_l) + \Gamma_k{}^j{}_i\, \Gamma_j{}^m{}_l - \Gamma_k{}^j{}_l\, \Gamma_j{}^m{}_i = 0$$

$$m, k, l, i = 1, 2, 3.$$

These are the components of **RIEMANN's**[25] **curvature tensor** of *4*th-order

(1.47) $$\boldsymbol{R} := R^m{}_{kli}\, \mathbf{r}_m \otimes \mathbf{r}^k \otimes \mathbf{r}^l \otimes \mathbf{r}^i$$

which is zero in all curvature-free spaces such as the EUCLIDean one. This is a necessary and sufficient condition for the integrability of the CHRISTOFFEL symbols to get tangent vectors again.

The gradient of a vector field can be put into the following forms

$$\mathbf{f(r)}' = \left(\frac{\partial f^k}{\partial \varphi^i}\, \mathbf{r}_k + f^k\, \frac{\partial \mathbf{r}_k}{\partial \varphi^i}\right) \otimes \mathbf{r}^i$$

$$= \left(\frac{\partial f^k}{\partial \varphi^i}\, \mathbf{r}_k + f^k\, \Gamma_k{}^j{}_i\, \mathbf{r}_j\right) \otimes \mathbf{r}^i$$

$$= \left(\frac{\partial f^j}{\partial \varphi^i} + f^k\, \Gamma_k{}^j{}_i\right) \mathbf{r}_j \otimes \mathbf{r}^i$$

$$= f^j\,|_i\, \mathbf{r}_j \otimes \mathbf{r}^i$$

with the coefficients

$$f^j\,|_i := \frac{\partial f^j}{\partial \varphi^i} + f^k\, \Gamma_k{}^j{}_i \qquad i, j = 1, 2, 3$$

called **covariant derivatives**. For a Cartesian COOS the covariant derivative coincides with the partial derivative of the components with respect to the coordinates.

We will now provide the corresponding formulae for fields of other orders.

- For <u>*scalar fields*</u>

$$f : \mathscr{E} \to \mathscr{R}$$

we have already considered the component functions. We obtain

$$df(\mathbf{r}, \mathbf{dr}) = \frac{\partial f}{\partial \varphi^i}\, d\varphi^i = \frac{\partial f}{\partial \varphi^i}\, \mathbf{r}^i \cdot \mathbf{r}_j\, d\varphi^j = \frac{\partial f}{\partial \varphi^i}\, \mathbf{r}^i \cdot \mathbf{dr}$$

$$= f(\mathbf{r})' \cdot \mathbf{dr}$$

with the gradient

$$f(\mathbf{r})' = grad\, f = \frac{\partial f}{\partial \varphi^i}\, \mathbf{r}^i .$$

[25] Bernhard Georg Friedrich Riemann (1826-1866)

- For *vector fields*

$$\mathbf{f} : \mathscr{E} \to \mathscr{V}$$

expressed in the form $\mathbf{f}(\mathbf{r}) = f_k\, \mathbf{r}^k$ we still need the representation with respect to the gradient basis

$$d\mathbf{f}(\mathbf{r}\,,\, d\mathbf{r}) \;=\; \frac{\partial f_k}{\partial \varphi^i}\, d\varphi^i\, \mathbf{r}^k + f_k\, \frac{\partial \mathbf{r}^k}{\partial \varphi^i}\, d\varphi^{\,i}$$

with the gradient

$$\mathbf{f}(\mathbf{r})' \;=\; grad\, f \;=\; \frac{\partial}{\partial \varphi^i}(f_j\, \mathbf{r}^j) \otimes \mathbf{r}^i \;=\; (\frac{\partial f_j}{\partial \varphi^i} \;+\; f_k\, \Gamma^k{}_{ji})\, \mathbf{r}^j \otimes \mathbf{r}^i$$

$$=:\; f_j \big|_i\; \mathbf{r}^j \otimes \mathbf{r}^i .$$

- For *tensor fields*

$$\mathbf{F} \;:\; \mathscr{E} \to \mathscr{Lin}$$

we obtain the representation of the gradient as a triad

$$\mathbf{F}(\mathbf{r})' \;=\; \frac{\partial}{\partial \varphi^i}\; (F^{jk}\, \mathbf{r}_j \otimes \mathbf{r}_k) \otimes \mathbf{r}^i$$

$$=\; (\frac{\partial F^{jk}}{\partial \varphi^i}\; \mathbf{r}_j \otimes \mathbf{r}_k + F^{jk}\, \frac{\partial \mathbf{r}_j}{\partial \varphi^i} \otimes \mathbf{r}_k + F^{jk}\, \mathbf{r}_j \otimes \frac{\partial \mathbf{r}_k}{\partial \varphi^i}) \otimes \mathbf{r}^i$$

$$=\; (\frac{\partial F^{jk}}{\partial \varphi^i} \;+\; F^{pk}\, \Gamma_p{}^j{}_i \;+\; F^{jl}\, \Gamma_l{}^k{}_i)\; \mathbf{r}_j \otimes \mathbf{r}_k \otimes \mathbf{r}^i$$

$$=:\; F^{jk}\big|_i\, \mathbf{r}_j \otimes \mathbf{r}_k \otimes \mathbf{r}^i$$

or

$$=\; \frac{\partial}{\partial \varphi^i}\; (F_{jk}\, \mathbf{r}^j \otimes \mathbf{r}^k) \otimes \mathbf{r}^i$$

$$=\; (\frac{\partial F_{jk}}{\partial \varphi^i} \;+\; F_{pk}\, \Gamma^p{}_{ji} \;+\; F_{jp}\, \Gamma^p{}_{ki})\; \mathbf{r}^j \otimes \mathbf{r}^k \otimes \mathbf{r}^i$$

$$=\; (\frac{\partial F_{jk}}{\partial \varphi^i} \;-\; F_{pk}\, \Gamma_j{}^p{}_i \;-\; F_{jp}\, \Gamma_k{}^p{}_i)\; \mathbf{r}^j \otimes \mathbf{r}^k \otimes \mathbf{r}^i$$

$$=:\; F_{jk}\big|_i\, \mathbf{r}^j \otimes \mathbf{r}^k \otimes \mathbf{r}^i$$

or

$$=\; \frac{\partial}{\partial \varphi^i}\; (F_j{}^k\, \mathbf{r}^j \otimes \mathbf{r}_k) \otimes \mathbf{r}^i$$

$$= (\frac{\partial F_j{}^k}{\partial \varphi^i} + F_p{}^k \Gamma^p{}_{ji} + F_j{}^p \Gamma_p{}^k{}_i)\, \mathbf{r}^j \otimes \mathbf{r}_k \otimes \mathbf{r}^i$$

$$= (\frac{\partial F_j{}^k}{\partial \varphi^i} - F_p{}^k \Gamma_j{}^p{}_i + F_j{}^p \Gamma_p{}^k{}_i)\, \mathbf{r}^j \otimes \mathbf{r}_k \otimes \mathbf{r}^i$$

$$=: F_j{}^k|_i\, \mathbf{r}^j \otimes \mathbf{r}_k \otimes \mathbf{r}^i$$

or

$$= \frac{\partial}{\partial \varphi^i}(F^j{}_k\, \mathbf{r}_j \otimes \mathbf{r}^k) \otimes \mathbf{r}^i$$

$$= (\frac{\partial F^j{}_k}{\partial \varphi^i} + F^p{}_k \Gamma_p{}^j{}_i + F^j{}_p \Gamma^p{}_{ki})\, \mathbf{r}_j \otimes \mathbf{r}^k \otimes \mathbf{r}^i$$

$$= (\frac{\partial F^j{}_k}{\partial \varphi^i} + F^p{}_k \Gamma_p{}^j{}_i - F^j{}_p \Gamma_k{}^p{}_i)\, \mathbf{r}_j \otimes \mathbf{r}^k \otimes \mathbf{r}^i$$

$$=: F^j{}_k|_i\, \mathbf{r}_j \otimes \mathbf{r}^k \otimes \mathbf{r}^i .$$

The procedure to determine the gradient of a tensor field of Kth-order is always the same:

1) represent the tensor field as a function of the coordinates, namely the components and the basis vectors,

2) differentiate the components and the basis vectors after the product rule partially with respect to all coordinates, and multiply the derivatives from the right tensorially by the corresponding gradient vector (this raises the order of the field by one to $K+1$).

As an example we determine the gradient of a tensor field of Kth-order

$$\mathbf{T} = T^{ij\ldots p}\, \underbrace{\mathbf{r}_i \otimes \mathbf{r}_j \otimes \ldots \otimes \mathbf{r}_p}_{K \text{ times}}$$

as

$$grad\, \mathbf{T} = \frac{\partial}{\partial \varphi^l}(T^{ij\ldots p}\, \mathbf{r}_i \otimes \mathbf{r}_j \otimes \ldots \otimes \mathbf{r}_p) \otimes \mathbf{r}^l$$

$$= (\frac{\partial}{\partial \varphi^l} T^{ij\ldots p})\, \mathbf{r}_i \otimes \mathbf{r}_j \otimes \ldots \otimes \mathbf{r}_p \otimes \mathbf{r}^l$$

$$+ T^{ij\ldots p} \frac{\partial \mathbf{r}_i}{\partial \varphi^l} \otimes \mathbf{r}_j \otimes \ldots \otimes \mathbf{r}_p \otimes \mathbf{r}^l$$

$$+ \; T^{ij\,...\,p}\,\mathbf{r}_i \otimes \frac{\partial \mathbf{r}_j}{\partial \varphi^l} \otimes ... \otimes \mathbf{r}_p \otimes \mathbf{r}^l$$

$$+ \; ...$$

$$+ \; T^{ij\,...\,p}\,\mathbf{r}_i \otimes \mathbf{r}_j \otimes ... \otimes \frac{\partial \mathbf{r}_p}{\partial \varphi^l} \otimes \mathbf{r}^l ,$$

where we can express the derivatives of the basis vectors by the CHRISTOFFEL symbols. We apply this gradient to an arbitrary tangent vector **dr** and obtain the differential of the field

$$d\mathbf{T}(\mathbf{r}\,,\mathbf{dr}) \;=\; \frac{\partial}{\partial \varphi^l}\,(T^{ij\,...\,p}\,\mathbf{r}_i \otimes \mathbf{r}_j \otimes ... \otimes \mathbf{r}_p) \otimes \mathbf{r}^l\,(d\varphi^r\,\mathbf{r}_r)$$

$$= \; \frac{\partial}{\partial \varphi^l}\,(T^{ij\,...\,p}\,\mathbf{r}_i \otimes \mathbf{r}_j \otimes ... \otimes \mathbf{r}_p)\,d\varphi^l ,$$

again a tensor field of *K*th-order.

By contraction of the two last basis vectors of the gradient, we obtain a generalisation of the **divergence** of a tensor field of *K*th-order

$$div\,\mathbf{T} \;=\; \frac{\partial}{\partial \varphi^l}\,[T^{ij\,...\,p}\,\mathbf{r}_i \otimes \mathbf{r}_j \otimes ... \otimes \mathbf{r}_{p-1}\,(\mathbf{r}_p] \cdot \mathbf{r}^l)$$

being a tensor field of $(K\!-\!1)$th-order. Here, the parenthesis [...] is the object for the differential operator, while (...) is it for the inner product. In particular, the divergence of a vector field **v** is

$$div\,\mathbf{v} \;=\; \frac{\partial}{\partial \varphi^l}\,(v^i\,\mathbf{r}_i) \cdot \mathbf{r}^l = (\frac{\partial v^i}{\partial \varphi^l} \;+\; v^m\,\Gamma_m{}^i{}_l)\,\mathbf{r}_i \cdot \mathbf{r}^l$$

$$= \; \frac{\partial v^i}{\partial \varphi^i} \;+\; v^m\,\Gamma_m{}^i{}_i \;=\; v^i\,|_i \, ,$$

a scalar field.

The divergence of a tensor field is

$$div\,\mathbf{T} \;=\; \frac{\partial}{\partial \varphi^l}\,(\,T^{ij}\,\mathbf{r}_i \otimes \mathbf{r}_j) \cdot \mathbf{r}^l$$

$$= \; (\frac{\partial T^{ij}}{\partial \varphi^l} \;+ T^{mj}\,\Gamma_m{}^i{}_l + T^{im}\,\Gamma_m{}^j{}_l)\,\mathbf{r}_i\,(\mathbf{r}_j \cdot \mathbf{r}^l)$$

$$= \; (\frac{\partial T^{ij}}{\partial \varphi^j} \;+ T^{mj}\,\Gamma_m{}^i{}_j + T^{im}\,\Gamma_m{}^j{}_j)\,\mathbf{r}_i$$

$$=: \; T^{ij}\,|_j\,\mathbf{r}_i \, ,$$

a vector field.

The **curl** of a vector field is

$$curl\,\mathbf{v} = \mathbf{r}^l \times \frac{\partial}{\partial \varphi^l}(v^i \mathbf{r}_i) = (\frac{\partial v^i}{\partial \varphi^l} + v^k \Gamma_k{}^i{}_l)\,\mathbf{r}^l \times \mathbf{r}_i$$

$$= (\frac{\partial v_i}{\partial \varphi^l} + v_k \Gamma^k{}_{il})\,\mathbf{r}^l \times \mathbf{r}^i .$$

All these operations can be notated by the **nabla**[26] **operator**

$$\boldsymbol{\nabla} := \frac{\partial}{\partial \varphi^l}\mathbf{r}^l .$$

The gradient of a scalar field $\phi(\mathbf{r})$ can be written as

$$grad\,\phi(\mathbf{r}) = \boldsymbol{\nabla}\phi = \phi\boldsymbol{\nabla}.$$

Here the algebraic connection between ϕ and $\boldsymbol{\nabla}$ is that of a multiplication of a (co-) vector by a scalar. We can put the nabla before or after the function.

For the gradient of a vector field $\mathbf{v(r)}$ this order is no more arbitrary, as we have to use the dyadic product between function and nabla

$$grad\,\mathbf{v(r)} = \mathbf{v} \otimes \boldsymbol{\nabla}.$$

For the divergence and curl of a vector field we obtain

$$div\,\mathbf{v(r)} = tr(\mathbf{v} \otimes \boldsymbol{\nabla}) = \mathbf{v} \cdot \boldsymbol{\nabla} = \boldsymbol{\nabla} \cdot \mathbf{v}$$

$$curl\,\mathbf{v(r)} = \boldsymbol{\nabla} \times \mathbf{v} = -\mathbf{v} \times \boldsymbol{\nabla}$$

and for the gradient of a tensor field $\mathbf{T(r)}$ of arbitrary order

$$grad\,\mathbf{T(r)} = \mathbf{T} \otimes \boldsymbol{\nabla}$$

and for its divergence

$$div\,\mathbf{T(r)} = \mathbf{T}\,\boldsymbol{\nabla}$$

(simple contraction).

The following rules hold for all differentiable scalar fields $\psi(\mathbf{r})$, $\varphi(\mathbf{r})$, vector fields $\mathbf{u(r)}$, $\mathbf{v(r)}$, and tensor fields $\mathbf{T(r)}$, $\mathbf{S(r)}$:

$$grad(\psi + \varphi) = grad(\psi) + grad(\varphi)$$

$$grad(\mathbf{u} + \mathbf{v}) = grad(\mathbf{u}) + grad(\mathbf{v})$$

$$div(\mathbf{u} + \mathbf{v}) = div(\mathbf{u}) + div(\mathbf{v})$$

$$curl(\mathbf{u} + \mathbf{v}) = curl(\mathbf{u}) + curl(\mathbf{v})$$

$$grad(\varphi\,\mathbf{v}) = (\varphi\,\mathbf{v}) \otimes \boldsymbol{\nabla} = \varphi\,(\mathbf{v} \otimes \boldsymbol{\nabla}) + \mathbf{v} \otimes \boldsymbol{\nabla}(\varphi)$$

[26] from the Greek word for *harp*

$$= \varphi\, grad(\mathbf{v}) + \mathbf{v} \otimes grad(\varphi)$$

$$div(\varphi\, \mathbf{v}) = \varphi\, div(\mathbf{v}) + \mathbf{v} \cdot grad(\varphi)$$

$$grad(\mathbf{u} \cdot \mathbf{v}) = grad(\mathbf{u})^T\, \mathbf{v} + grad(\mathbf{v})^T\, \mathbf{u}$$

$$div(\mathbf{u} \otimes \mathbf{v}) = \mathbf{u}\;\; div(\mathbf{v}) + grad(\mathbf{u})\, \mathbf{v}$$

$$div(\mathbf{T}\, \mathbf{v}) = \mathbf{T}^T \cdot grad(\mathbf{v}) \; + div(\mathbf{T}^T) \cdot \mathbf{v}$$

$$div(\varphi\, \mathbf{T}) = \varphi\, div(\mathbf{T}) + \mathbf{T}\, grad(\varphi)$$

$$div(\mathbf{T} + \mathbf{S}) = div(\mathbf{T}) + div(\mathbf{S})$$

Again, the chain rules hold for the gradients, divergences, and curls.

We obtain the following identities for all scalar fields $\varphi(\mathbf{r})$ and vector fields $\mathbf{v}(\mathbf{r})$:

$$curl\; grad\; \varphi = \mathbf{o}$$

$$div\; curl\; \mathbf{v} = 0$$

$$curl\; curl\; \mathbf{v} = grad\; div\; \mathbf{v} \; - \; div\; grad\; \mathbf{v}$$

$$div\; grad^T\, \mathbf{v} = grad\; div\; \mathbf{v}$$

$$div\; grad\; grad\; \varphi = grad\; div\; grad\; \varphi$$

$$div\; grad\; grad\; \mathbf{v} = grad\; div\; grad\; \mathbf{v}$$

$$div\; div\; grad\; \mathbf{v} = div\; grad\; div\; \mathbf{v}$$

$$div\; grad\; curl\; \mathbf{v} = curl\; div\; grad\; \mathbf{v}$$

Example. ***Cylindrical coordinates*** are introduced as $\{\varphi^1 \equiv r\,,\; \varphi^2 \equiv \theta\,,\; \varphi^3 \equiv z\}$ with two lengths r and z and one angle θ. Let $\{x\,, y\,, z\}$ be a Cartesian COOS and an associated ONB $\{\mathbf{e}_x\,, \mathbf{e}_y\,, \mathbf{e}_z\}$, then we obtain the coordinate transformations

$$x = r\cos\theta$$

$$y = r\sin\theta$$

$$z = z\,.$$

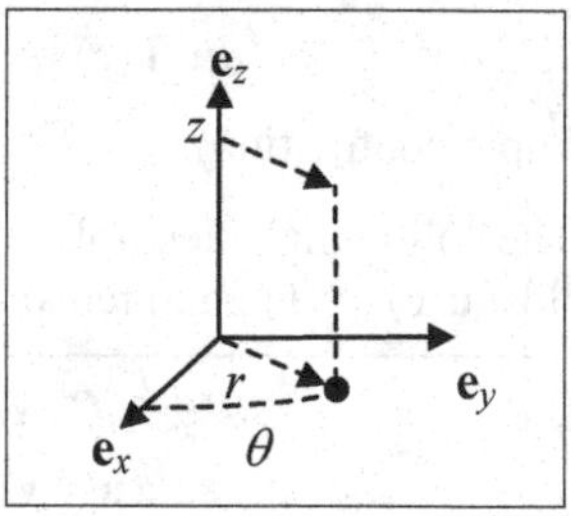

In each point with coordinates $\{r\,,\, \theta\,, z\}$ we can introduce a local ONB by

$$\mathbf{e}_r(\theta) \;:=\; \cos\theta\, \mathbf{e}_x \;+\; \sin\theta\, \mathbf{e}_y$$

$$\mathbf{e}_\theta(\theta) \;:=\; -\sin\theta\, \mathbf{e}_x \;+\; \cos\theta\, \mathbf{e}_y$$

$\mathbf{e}_z$.

The non-zero derivatives of them are

$$\frac{d\mathbf{e}_r}{d\theta} = \mathbf{e}_\theta \qquad \frac{d\mathbf{e}_\theta}{d\theta} = -\,\mathbf{e}_r\,.$$

The position vector is

$$\begin{aligned}\mathbf{r} &= x\,\mathbf{e}_x + y\,\mathbf{e}_y + z\,\mathbf{e}_z \\ &= r\,\mathbf{e}_r(\theta) + z\,\mathbf{e}_z\,.\end{aligned}$$

The tangent vectors to the cylindrical coordinates are

$$\mathbf{r}_{\varphi 1} := \frac{\partial \mathbf{r}}{\partial r} = \mathbf{e}_r(\theta) \qquad \mathbf{r}_{\varphi 2} := \frac{\partial \mathbf{r}}{\partial \theta} = r\,\mathbf{e}_\theta(\theta) \qquad \mathbf{r}_{\varphi 3} := \frac{\partial \mathbf{r}}{\partial z} = \mathbf{e}_z$$

and, dually, the gradient vectors

$$\mathbf{r}_\varphi^{\,1} = \mathbf{e}_r(\theta) \qquad \mathbf{r}_\varphi^{\,2} = r^{-1}\,\mathbf{e}_\theta(\theta) \qquad \mathbf{r}_\varphi^{\,3} = \mathbf{e}_z\,.$$

The matrices of the metric coefficients are

$$[G_{ij}] = \begin{bmatrix} 1 & 0 & 0 \\ 0 & r^2 & 0 \\ 0 & 0 & 1 \end{bmatrix} \qquad [G^{ij}] = \begin{bmatrix} 1 & 0 & 0 \\ 0 & r^{-2} & 0 \\ 0 & 0 & 1 \end{bmatrix}.$$

The only non-zero CHRISTOFFEL symbols are

$$\Gamma_{122} = \Gamma_{221} = r \qquad \Gamma_{212} = -r$$

$$\Gamma_{1\;2}^{\;2} = \Gamma_{2\;1}^{\;2} = r^{-1} \qquad \Gamma_{2\;2}^{\;1} = -\,r\,.$$

The gradient of a scalar field $\phi(r\,,\theta\,,z)$ is

$$grad\;\phi = \phi\,,_r\,\mathbf{e}_r + \phi\,,_\theta\,\frac{1}{r}\,\mathbf{e}_\theta + \phi\,,_z\,\mathbf{e}_z\,.$$

The gradient of a vector field

$$\mathbf{v}(r,\theta,z) = v^r\,\mathbf{e}_r + v^\theta\,\mathbf{e}_\theta + v^z\,\mathbf{e}_z$$

is

$$\begin{aligned} grad\,\mathbf{v} = {} & v^r,_r\,\mathbf{e}_r\otimes\mathbf{e}_r + r^{-1}\,(v^r,_\theta - v^\theta)\,\mathbf{e}_r\otimes\mathbf{e}_\theta + v^r,_z\,\mathbf{e}_r\otimes\mathbf{e}_z \\ & + v^\theta,_r\,\mathbf{e}_\theta\otimes\mathbf{e}_r + r^{-1}\,(v^\theta,_\theta + v^r)\,\mathbf{e}_\theta\otimes\mathbf{e}_\theta + v^\theta,_z\,\mathbf{e}_\theta\otimes\mathbf{e}_z \\ & + v^z,_r\,\mathbf{e}_z\otimes\mathbf{e}_r + r^{-1}\,v^z,_\theta\,\mathbf{e}_z\otimes\mathbf{e}_\theta + v^z,_z\,\mathbf{e}_z\otimes\mathbf{e}_z\,. \end{aligned}$$

Its divergence is

$$div\,\mathbf{v} = v^r,_r + r^{-1}\,(v^\theta,_\theta + v^r) + v^z,_z$$

and its curl

$$curl\,\mathbf{v} = (r^{-1} v^z,_\theta - v^\theta,_z)\,\mathbf{e}_r + (v^r,_z - v^z,_r)\,\mathbf{e}_\theta + [v^\theta,_r + r^{-1}(v^\theta - v^r,_\theta)]\,\mathbf{e}_z\,.$$

For the divergence of a tensor field $\mathbf{T} = T^{ij}\,\mathbf{e}_i \otimes \mathbf{e}_j$ we obtain

(1.48)
$$\begin{aligned} div\,\mathbf{T} = {} & (T^{rr},_r + r^{-1}\,T^{r\theta},_\theta + \frac{T^{rr} - T^{\theta\theta}}{r} + T^{rz},_z)\,\mathbf{e}_r \\ & + (T^{\theta r},_r + r^{-1}\,T^{\theta\theta},_\theta + \frac{T^{r\theta} + T^{\theta r}}{r} + T^{\theta z},_z)\,\mathbf{e}_\theta \\ & + (T^{zr},_r + r^{-1}\,T^{z\theta},_\theta + \frac{T^{zr}}{r} + T^{zz},_z)\,\mathbf{e}_z\,. \end{aligned}$$

Note that all of these representations make use of physical components with respect to normed bases $\{\mathbf{e}_i\}$.

Example. ***Spherical coordinates*** are introduced as $\{\varphi^1 \equiv r\,, \varphi^2 \equiv \theta\,, \varphi^3 \equiv \varphi\}$ with a length $0 < r < \infty$ and two angles $-\pi < \varphi < \pi$ and $0 < \theta < \pi$. With respect to a Cartesian COOS $\{x^1, x^2, x^3\}$ the transformations are

$$x^1 = r \sin\theta \cos\varphi$$
$$x^2 = r \sin\theta \sin\varphi$$
$$x^3 = r \cos\theta.$$

The tangent vectors have the following components with respect to the Cartesian basis

$$\mathbf{r}_1 = \mathbf{r}_r = \{\sin\theta \cos\varphi\,, \sin\theta \sin\varphi\,, \cos\theta\}$$
$$\mathbf{r}_2 = \mathbf{r}_\theta = r\,\{\cos\theta \cos\varphi\,, \cos\theta \sin\varphi\,, -\sin\theta\}$$
$$\mathbf{r}_3 = \mathbf{r}_\varphi = r\,\{-\sin\theta \sin\varphi\,, \cos\varphi \sin\theta\,, 0\}$$

and the gradient vectors

$$\mathbf{r}^1 = \mathbf{r}_1$$
$$\mathbf{r}^2 = r^{-2}\,\mathbf{r}_2$$
$$\mathbf{r}^3 = (r^2 \sin^2\theta)^{-1}\,\mathbf{r}_3\,.$$

The only non-zero metric coefficients are

$$G_{11} = 1 \qquad G_{22} = r^2 \qquad G_{33} = r^2 \sin^2\theta$$
$$G^{11} = 1 \qquad G^{22} = r^{-2} \qquad G^{33} = r^{-2} \sin^{-2}\theta.$$

The only non-zero CHRISTOFFEL symbols are

$$\Gamma_{122} = \Gamma_{221} = r \qquad \Gamma_{323} = -r^2 \sin\theta \cos\theta$$
$$\Gamma_{212} = -r \qquad \Gamma_{313} = -r \sin^2\theta$$

$$\Gamma_{133} = \Gamma_{331} = r\sin^2\theta \qquad \Gamma_{233} = \Gamma_{332} = r^2\sin\theta\cos\theta$$

$$\Gamma^{1}_{2\,2} = -r \qquad \Gamma^{1}_{3\,3} = -r\sin^2\theta$$

$$\Gamma^{2}_{1\,2} = \Gamma^{2}_{2\,1} = r^{-1} \qquad \Gamma^{2}_{3\,3} = -\sin\theta\cos\theta$$

$$\Gamma^{3}_{1\,3} = \Gamma^{3}_{3\,1} = r^{-1} \qquad \Gamma^{3}_{2\,3} = \Gamma^{3}_{3\,2} = \cot\theta .$$

1.4.2 Integral Theorems

are often needed in continuum mechanics for changing surface integrals into volumetric ones, and vice versa. This can be generally done by the following **GAUSS**[27] **transformation.**

Theorem of GAUSS 1.28. *Let* $\mathscr{B}$ *be a regular domain*[28] *in the EUCLIDean space* $\mathscr{E}$ *with surface* $\partial\mathscr{B}$, $\mathbf{n}$ *the outer surface normal on* $\partial\mathscr{B}$, $\mathbf{v}$ *a vector field being continuous on* $\mathscr{B}$ *and continuously differentiable in the interior of* $\mathscr{B}$ *. Then*

$$\int_{\partial\mathscr{B}} \mathbf{v}\cdot\mathbf{n}\,da = \int_{\mathscr{B}} div\,\mathbf{v}\,dv$$

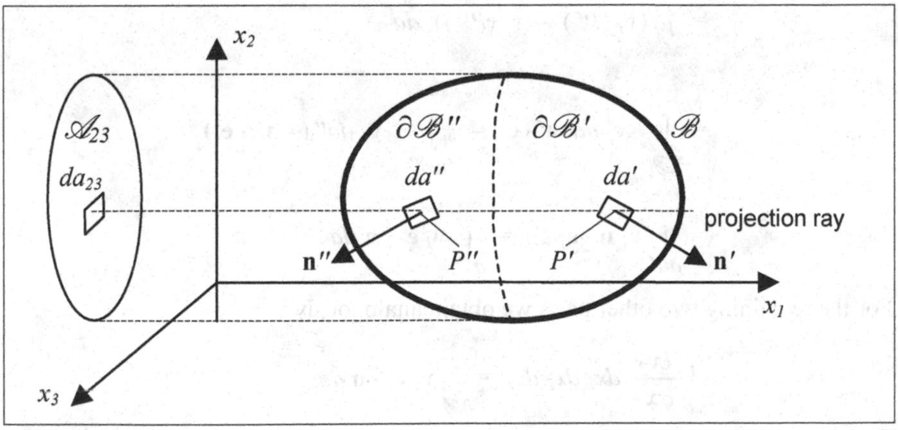

[27] Carl Friedrich Gauß (1777-1855)
[28] in the sense of KELLOGG (1929)

Proof. With respect to a constant ONB, the right-hand side is

$$\int_{\mathscr{B}} div\, \mathbf{v}\, dv = \int_{\mathscr{B}} \left(\frac{\partial v_1}{\partial x_1} + \frac{\partial v_2}{\partial x_2} + \frac{\partial v_3}{\partial x_3} \right) dx_1\, dx_2\, dx_3 \,.$$ Firstly, we assume that the domain is convex. We project it into the x_2-x_3-coordinate plane, which gives $\mathscr{A}_{23}$. We divide the surface $\partial\mathscr{B}$ of $\mathscr{B}$ into its positive part $\partial\mathscr{B}'$ with respect to the x_1-direction, and its negative part $\partial\mathscr{B}''$. A projective ray, which is parallel to the x_1-axis, penetrates $\partial\mathscr{B}$ in the points P' and P''. In these points we call the elements of area da' and da'', respectively, and the corresponding normal vectors by $\mathbf{n}'$ and $\mathbf{n}''$. The projections of these elements of area into the x_1-x_2-plane are

$$da_{23} = da' \cos(\mathbf{n}', \mathbf{e}_1) = da'' \cos(\mathbf{n}'', -\mathbf{e}_1) \,.$$

We can express the cosines of the angle between $\mathbf{n}'$ and $\mathbf{e}_1$ by the scalar product as

$$\cos(\mathbf{n}', \mathbf{e}_1) = \mathbf{n}' \cdot \mathbf{e}_1$$

and analogously

$$\cos(\mathbf{n}'', -\mathbf{e}_1) = -\mathbf{n}'' \cdot \mathbf{e}_1 \,.$$

We partially integrate the first part of the integral

$$\int_{\mathscr{B}} \frac{\partial v_1}{\partial x_1} dx_1\, dx_2\, dx_3 = \int_{\mathscr{A}_{23}} \int_{P''}^{P'} \frac{\partial v_1}{\partial x_1} dx_1\, da_{23}$$

$$= \int_{\mathscr{A}_{23}} (v_1(P') - v_1(P''))\, da_{23}$$

$$= \int_{\partial\mathscr{B}'} v_1\, da'\, \mathbf{n}' \cdot \mathbf{e}_1 - \int_{\partial\mathscr{B}''} v_1\, da''\, (-\mathbf{n}'' \cdot \mathbf{e}_1)$$

$$= \int_{\partial\mathscr{B}} v_1\, \mathbf{n} \cdot \mathbf{e}_1\, da = \int_{\partial\mathscr{B}} v_1\, \mathbf{e}_1 \cdot \mathbf{n}\, da \,.$$

For the remaining two other parts we obtain analogously

$$\int_{\mathscr{B}} \frac{\partial v_2}{\partial x_2} dx_1\, dx_2\, dx_3 = \int_{\partial\mathscr{B}} v_2\, \mathbf{e}_2 \cdot \mathbf{n}\, da$$

and

$$\int_{\mathscr{B}} \frac{\partial v_3}{\partial x_3}\, dx_1\, dx_2\, dx_3 = \int_{\partial\mathscr{B}} v_3\, \mathbf{e}_3 \cdot \mathbf{n}\; da$$

and for the sum

$$\int_{\mathscr{B}} div\, \mathbf{v}\; dv = \int_{\partial\mathscr{B}} \mathbf{v} \cdot \mathbf{n}\; da\,.$$

If the domain is not convex, then the projection ray would possibly penetrate the surface more than twice. Then we would have to also project these parts of $\partial\mathscr{B}$ into the coordinate plane, with the same result. If the domain contains internal surfaces like bubbles, we must include them also into the integration.

Example. If we take the volume integral over the divergence of the position vector, we obtain three times the volume of the domain of integration

$$\int_{\mathscr{B}} div\, \mathbf{r}\; dv = 3\, V.$$

In order to obtain other versions of this important integral theorem, we substitute $\mathbf{v}$ by the vector field $\mathbf{v} \equiv \alpha\, \mathbf{a}$ with a scalar field α and a constant vector field $\mathbf{a}$. Then we get

$$\int_{\partial\mathscr{B}} \alpha\, \mathbf{a} \cdot \mathbf{n}\; da = \mathbf{a} \cdot \int_{\partial\mathscr{B}} \alpha\, \mathbf{n}\; da = \int_{\mathscr{B}} div(\alpha\, \mathbf{a})\; dv = \mathbf{a} \cdot \int_{\mathscr{B}} grad(\alpha)\; dv$$

and because of the arbitrariness of $\mathbf{a}$

(1.49)

$$\boxed{\int_{\partial\mathscr{B}} \alpha\, \mathbf{n}\; da = \int_{\mathscr{B}} grad(\alpha)\; dv}$$

If we instead put $\mathbf{v} \equiv \mathbf{T}^T\, \mathbf{u}$ with a tensor field $\mathbf{T}$ and a vector field $\mathbf{u}$, then we obtain

$$\int_{\partial\mathscr{B}} (\mathbf{T}^T\, \mathbf{u}) \cdot \mathbf{n}\; da = \int_{\mathscr{B}} div(\mathbf{T}^T\, \mathbf{u})\; dv =$$

(1.50)

$$\boxed{\int_{\partial\mathscr{B}} \mathbf{u} \cdot (\mathbf{T}\, \mathbf{n})\; da = \int_{\mathscr{B}} div(\mathbf{T}) \cdot \mathbf{u}\; dv + \int_{\mathscr{B}} \mathbf{T} \cdot grad(\mathbf{u})\; dv}$$

If $\mathbf{u}$ happens to be a constant field, then we get

(1.51)

$$\int_{\partial\mathscr{B}} \mathbf{T}\,\mathbf{n}\,da = \int_{\mathscr{B}} div\,\mathbf{T}\,dv$$

Next we substitute **v** by the vector field

$$\mathbf{v} \equiv (\mathbf{a}\cdot\mathbf{u})\,\mathbf{b}$$

with **u** continuously differentiable, and **a** and **b** constant vector fields. Then

$$\int_{\partial\mathscr{B}} (\mathbf{a}\cdot\mathbf{u})\,(\mathbf{b}\cdot\mathbf{n})\,da = \int_{\mathscr{B}} div\{(\mathbf{a}\cdot\mathbf{u})\,\mathbf{b}\}\,dv$$

$$= \mathbf{a}\cdot\Big(\int_{\partial\mathscr{B}} \mathbf{u}\otimes\mathbf{n}\,da\Big)\,\mathbf{b} = \mathbf{a}\cdot\int_{\mathscr{B}} grad\,\mathbf{u}\,dv\,\mathbf{b}\,,$$

and, because of the arbitrariness of **a** and **b**

(1.52)

$$\int_{\partial\mathscr{B}} \mathbf{u}\otimes\mathbf{n}\,da = \int_{\mathscr{B}} grad\,\mathbf{u}\,dv$$

Generally, one can state the following form of the GAUSS transformation. If ϕ is a tensor field of arbitrary order, and $\oplus$ an arbitrary product for which the ollowing combinations make sense, then

(1.53)

$$\int_{\partial\mathscr{B}} \phi\oplus\mathbf{n}\;da = \int_{\mathscr{B}} \phi\oplus\boldsymbol{\nabla}\,dv$$

This includes all previous versions of the integral theorem.

2 Kinematics

A Comment on the Literature. The following books contain introductions to non-linear continuum mechanics: ALTENBACH (2012), BAŞAR/ WEICHERT (2000), BERTRAM (1989), BETTEN (1993, 2001), BILLINGTON (1986), CHADWICK (1976), CHAVES (2013), DIMITRIENKO (2011), DOGHRI (2000), GONZALEZ/ STUART (2008), GREVE (2003), GURTIN (1981), HOLZAPFEL (2000), IRGENS (2008), LEIGH (1968), LIU (2002), MURNAGHAN (1937), NARASIMHAN (1993), REDDY (2008), RUBIN (2021), SALENÇON (2001), SMITH (1993), STEINMANN (2015), TABER (2004). The following books provide extensive sources for further research: TRUESDELL/ NOLL (1965), WANG/ TRUESDELL (1973).

2.1 Placements of Bodies

In continuum mechanics we consider material bodies, which move smoothly through the EUCLIDean space and at each instant continuously occupy a certain spatial domain. The subject of kinematics is the description of such movements in their temporal and spatial aspects. For that purpose we apply *geometry* and *chronometry*. The causes of such movements are not under consideration in kinematics, but in dynamics, which will be dealt with in the following chapter.

In classical, *i.e.*, non-relativistic kinematics, space and time are considered as independent concepts. The **time**, which we parameterise by real numbers, is assumed to pass at the same rate in all places, *i.e.*, simultaneously. The question of where to put the reference point of time "zero" is rather irrelevant as we will mainly deal with time differences. We will identify the time axis by the real axis $\mathscr{R}$, both of them having a natural order which corresponds to the notion of earlier and later instants.

The classical physical **space** will be identified by the three-dimensional uncurved EUCLIDean space of the foregoing Chap. 1.4.

The abstract **material body** will be considered as a three-dimensional manifold with boundary, consisting of points, which we call *material* (in contrast to *spatial* points). The body becomes observable by us when it moves through the space. Mathematically, such a motion is a time-dependent embedding into the EUCLIDean space. These are specified by certain regularities, which make it possible to map the body at each instant into $\mathscr{R}^3$, such that all coordinate changes are

A. Bertram, *Elasticity and Plasticity of Large Deformations*,
https://doi.org/10.1007/978-3-030-72328-6_2

differentiable. At each instant, this embedding is called a **placement**[29] of the body $\mathscr{B}$ at a time $t \in \mathscr{R}$, and it is given by a mapping

$$\kappa(\cdot, t) : \mathscr{B} \to \mathscr{E} .$$

If restricted to the currently occupied spatial region

$$\mathscr{B}_t := \kappa(\mathscr{B}, t) \subset \mathscr{E}$$

$\kappa(\cdot, t)\,|_{\mathscr{B}_t}$ is a bijection.

A **motion** of the body can then be considered as a time-dependent sequence of placements, defined on a certain closed time interval $\mathscr{I} \subset \mathscr{R}$, thus

$$\kappa : \mathscr{B} \times \mathscr{I} \to \mathscr{E}$$

with the following properties.

- For all $t \in \mathscr{I}$, $\kappa(\cdot, t)$ is a placement.
- For all $X \in \mathscr{B}$, $\kappa(X, \cdot)$ is twice piecewise continuously differentiable.

The curve $\kappa(X, \cdot)$ is the **path** of the material point X.

It is rather customary (although not necessary) to introduce a **reference placement**

$$\kappa_0 : \mathscr{B} \to \mathscr{E} ,$$

i.e., to choose an arbitrary but fixed placement of the body. The region of space occupied by the reference placement is denoted by

$$\mathscr{B}_0 := \kappa_0(\mathscr{B}) \subset \mathscr{E} .$$

The composition of its inverse with a motion

$$\chi(\cdot, t) := \kappa(\kappa_0^{-1}(\cdot), t) : \mathscr{E} \supset \mathscr{B}_0 \to \mathscr{B}_t \subset \mathscr{E}$$

is the EUCLIDean description of a motion of the body, being for all times t a bijection between regions of the EUCLIDean space. This seems to be more practical, as it does not involve the abstract body manifold anymore. Hence, further on we will mean such by a motion a mapping and we will always assume its continuity and differentiability. Its inverse in the spatial argument is also a bijection denoted by

$$\chi^{-1}(\cdot, t) : \mathscr{B}_t \to \mathscr{B}_0 .$$

We will also assume that χ (and consequently also χ^{-1}) is orientation preserving.

---

[29] In the older literature the placement is often called *configuration.* This use, however, clashes with the common meaning of this word. We will therefore not use it in this sense (see the Preface of the 2nd edition of TRUESDELL/ NOLL, 1992).

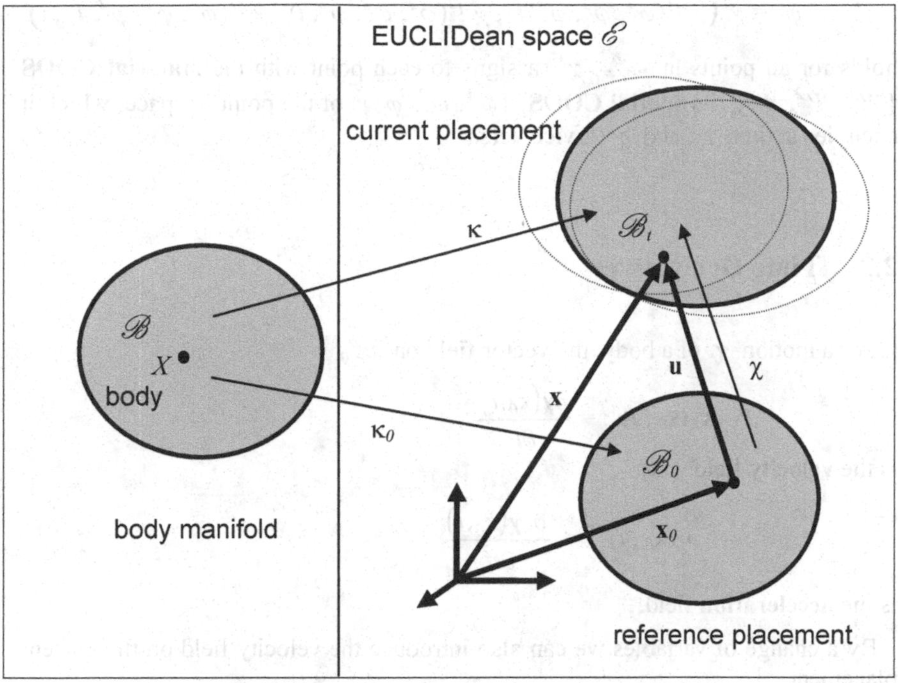

For describing the EUCLIDean space by means of position vectors with respect to an arbitrary fixed reference point $o \in \mathscr{E}$, the following notations will be used:

$$\mathbf{x} = \boldsymbol{\chi}(\mathbf{x}_0\,, t) \in \mathscr{V} \qquad \text{with } \mathbf{x}_0 := \overrightarrow{o\kappa_0(X)}$$

$$\mathbf{x}_0 = \boldsymbol{\chi}^{-1}(\mathbf{x}\,, t) \in \mathscr{V} \qquad \text{with } \mathbf{x} := \overrightarrow{o\kappa(X,t)}$$

for all $X \in \mathscr{B}$, *i.e.*, we substituted the spatial arguments of χ by position vectors.

The field of **displacement vectors** is the difference of the position vectors of a material point in the current and in the reference placement

$$\mathbf{u}(\mathbf{x}_0\,, t) := \boldsymbol{\chi}(\mathbf{x}_0\,, t) - \mathbf{x}_0\,.$$

Alternatively, we can describe the motion by a time-dependent change of coordinates. For this purpose, we choose two COOS, like $\{\varphi^i\}$ in $\mathscr{B}_t$ and $\{\Psi^i\}$ in $\mathscr{B}_0$. The motion can then be considered as a coordinate transformation

$$\varphi^i = \chi^i(\Psi^1, \Psi^2, \Psi^3, t) \qquad i = 1, 2, 3$$

or inversely

$$\Psi^i = \chi^{-1\,i}(\varphi^1, \varphi^2, \varphi^3, t) \qquad i = 1, 2, 3$$

such that

$$\varphi^i = \chi^i\left(\chi^{-1^1}(\varphi^1, \varphi^2, \varphi^3, t), \chi^{-1^2}(\varphi^1, \varphi^2, \varphi^3, t), \chi^{-1^3}(\varphi^1, \varphi^2, \varphi^3, t), t\right)$$

holds for all points in $\mathscr{B}_t$. χ^i assigns to each point with the **material COOS** $\{\Psi^1, \Psi^2, \Psi^3\}$ its **spatial COOS** $\{\varphi^1, \varphi^2, \varphi^3\}$ of the point in space, which it occupies at time t , and χ^{-1^i} vice versa.

2.2 Time Derivatives

For a motion $\boldsymbol{\chi}$ of a body, the vector field on $\mathscr{B}_0$

$$\mathbf{v}_L(\mathbf{x}_0, t) := \frac{\partial \boldsymbol{\chi}(\mathbf{x}_0, t)}{\partial t}$$

is the **velocity** field, and

(2.1)
$$\mathbf{a}_L(\mathbf{x}_0, t) := \frac{\partial^2 \boldsymbol{\chi}(\mathbf{x}_0, t)}{\partial t^2}$$

is the **acceleration** field.

By a change of variables we can also introduce the velocity field on the current placement

$$\mathbf{v}_L(\mathbf{x}_0, t) = \mathbf{v}_L(\boldsymbol{\chi}^{-1}(\mathbf{x}, t), t) =: \mathbf{v}_E(\mathbf{x}, t).$$

Such a transformation can be applied to any scalar, vector or tensor field ϕ of $\mathscr{B}_0$ to $\mathscr{B}_t$ or vice versa, namely by

$$\phi_L(\mathbf{x}_0, t) = \phi_L(\boldsymbol{\chi}^{-1}(\mathbf{x}, t), t) =: \phi_E(\mathbf{x}, t)$$

$$\phi_E(\mathbf{x}, t) = \phi_E(\boldsymbol{\chi}(\mathbf{x}_0, t), t) =: \phi_L(\mathbf{x}_0, t).$$

The fields on $\mathscr{B}_0$ are in the **LAGRANGEan**[30] or **material description**, those on $\mathscr{B}_t$ in the **EULERean** or **spatial description**. Which description is more advantageous depends on the specific problem. Mathematically, both descriptions are equivalent. We will indicate these descriptions by suffixes E and L only if it is important. Otherwise we will drop them for brevity.

If ϕ is a field of arbitrary order, being differentiable with respect to time, then we obtain by the chain rule

(2.2)
$$\phi^\bullet := \frac{\partial \phi_L}{\partial t} = \frac{\partial \phi_E}{\partial t} + \frac{\partial \phi_E}{\partial \mathbf{x}} \boldsymbol{\chi}^\bullet = \frac{\partial \phi_E}{\partial t} + grad(\phi_E)\, \mathbf{v}.$$

---

[30] Joseph Louis Lagrange (1736-1813)

We call the time derivatives

$\phi^\bullet$	substantial
$\dfrac{\partial \phi_L}{\partial t}$	material (as $\mathbf{x}_0 \in \mathscr{B}_0$ is kept constant)
$\dfrac{\partial \phi_E}{\partial t}$	local (as $\mathbf{x} \in \mathscr{B}_t$ is kept constant)
$grad(\phi_E)\, \mathbf{v}$	convective.

The algebraic product between $grad\ \phi_E$ and $\mathbf{v}$ is a simple contraction. If ϕ_E is a

- scalar field, then $grad\ \phi_E$ is a vector field, and $grad(\phi_E)\,\mathbf{v}$ a scalar product,
- vector field, then $grad\ \phi_E$ is a tensor field, and $grad(\phi_E)\,\mathbf{v}$ the application of a 2nd-order tensor to a vector,
- tensor field of Kth-order, then $grad(\phi_E)$ is a tensor field of $(K+1)$th-order, applied to the vector field $\mathbf{v}$ (simple contraction), which again gives a tensor field of Kth-order. If we choose a basis fixed in space, then we obtain the component form

$$(\phi_E^\bullet)^{i_1 \dots i_K} = \frac{\partial \phi_E^{i_1 \dots i_K}}{\partial t} + \frac{\partial \phi_E^{i_1 \dots i_K}}{\partial x^i}\, v^i .$$

2.3 Spatial Derivatives

Apart from time derivatives, we need spatial derivatives to compute gradients, divergences, curls, etc. of field variables. For this purpose it is necessary to distinguish between the derivatives of variables in the LAGRANGEan and in the EULERean description. The starting point of the analysis is the differential of a field, for which we keep the time variable constant

$$d\phi_L(\mathbf{x}_0, t, d\mathbf{x}_0) = Grad\ \phi_L(\mathbf{x}_0, t)\, d\mathbf{x}_0 = \phi_L(\mathbf{x}_0, t) \otimes \nabla_L\, d\mathbf{x}_0$$

$$d\phi_E(\mathbf{x}, t, d\mathbf{x}) = grad\ \phi_E(\mathbf{x}, t)\, d\mathbf{x} = \phi_E(\mathbf{x}, t) \otimes \nabla_E\, d\mathbf{x}$$

with the LAGRANGEan and the EULERean nabla operator

$$\nabla_L := \frac{\partial}{\partial \Psi^i}\, \mathbf{r}_\Psi{}^i \qquad \nabla_E := \frac{\partial}{\partial \varphi^i}\, \mathbf{r}_\varphi{}^i$$

with respect to a material COOS $\{\Psi^i\}$ in $\mathscr{B}_0$ with gradient basis $\{\mathbf{r}_\Psi{}^i\}$, and a spatial COOS $\{\varphi^i\}$ in $\mathscr{B}_t$ with $\{\mathbf{r}_\varphi{}^i\}$. We use the differential operators with lower case letters like *grad* and *div* etc. with respect to the EULERean variables and with upper case letters like *Grad* and *Div* etc. for the LAGRANGEan variables.

We obtain the relations

$$Grad\,\mathbf{x}_0 = \mathbf{x}_0 \otimes \nabla_L = \mathbf{I} \quad \Rightarrow \quad Div\,\mathbf{x}_0 = 3 \qquad Curl\,\mathbf{x}_0 = \mathbf{o}$$

$$grad\,\mathbf{x} = \mathbf{x} \otimes \nabla_E = \mathbf{I} \quad \Rightarrow \quad div\,\mathbf{x} = 3 \qquad curl\,\mathbf{x} = \mathbf{o}\,.$$

The following tensor field will be of elementary importance in the local deformation analysis

$$\mathbf{F}(\mathbf{x}_0\,, t) \;:=\; Grad\,\boldsymbol{\chi}(\mathbf{x}_0\,, t) \;=\; \boldsymbol{\chi}(\mathbf{x}_0\,, t) \otimes \nabla_L$$

or inversely

$$grad\,\boldsymbol{\chi}^{-1}(\mathbf{x}\,, t) \;=\; \boldsymbol{\chi}^{-1}(\mathbf{x}\,, t) \otimes \nabla_E \;=\; \mathbf{F}^{-1}(\mathbf{x}\,, t)\,.$$

$\mathbf{F}$ is usually called **deformation gradient**, although this name is rather misleading. It is not the gradient of a deformation, not even a gradient, which describes deformations, as it also contains rigid rotations. $\mathbf{F}$ is dimensionless and assumed to be invertible at each point and each time[31], which assures local invertibility after the inverse function theorem of advanced calculus. Note that the invertibility of $\mathbf{F}$ in each point does not guarantee the global invertibility of $\boldsymbol{\chi}$, but has to be additionally assured.

If the reference placement is a possible placement, then $\boldsymbol{\chi}$ is orientation preserving and so is the gradient, *i.e.*, $det(\mathbf{F}) > 0$. Therefore, we have $\mathbf{F}\,, \mathbf{F}^{-1} \in \mathscr{Inv}^+$.

By the chain rule we get for any field ϕ

$$\begin{aligned}
Grad\,\phi_L(\mathbf{x}_0\,, t) &= \phi_L(\mathbf{x}_0\,, t) \otimes \nabla_L \\
&= grad\,\phi_E\,(\boldsymbol{\chi}(\mathbf{x}_0\,, t), t)\; Grad\,\boldsymbol{\chi}(\mathbf{x}_0\,, t) \\
&= grad\,\phi_E\,(\boldsymbol{\chi}(\mathbf{x}_0\,, t)\,, t)\; \mathbf{F}(\mathbf{x}_0\,, t) \\
&= (\phi_E\,(\mathbf{x}\,, t) \otimes \nabla_E)\; \mathbf{F} \\
grad\,\phi_E(\mathbf{x}\,, t) &= \phi_E\,(\mathbf{x}\,, t) \otimes \nabla_E \\
&= Grad\,\phi_L(\boldsymbol{\chi}^{-1}(\mathbf{x}\,, t)\,, t)\; grad\,\boldsymbol{\chi}^{-1}(\mathbf{x}\,, t) \\
&= Grad\,\phi_L(\boldsymbol{\chi}^{-1}(\mathbf{x}\,, t)\,, t)\; \mathbf{F}^{-1}(\mathbf{x}\,, t) \\
&= (\phi_L(\mathbf{x}_0\,, t) \otimes \nabla_L)\; \mathbf{F}^{-1}.
\end{aligned}$$

For the transformation of the nabla operators, we obtain by a simple contraction

(2.3) $$\nabla_L = \nabla_E\,\mathbf{F} = \mathbf{F}^T\,\nabla_E \qquad \text{and} \qquad \nabla_E = \nabla_L\,\mathbf{F}^{-1} = \mathbf{F}^{-T}\,\nabla_L\,.$$

The deformation gradient transforms the material line elements between the reference and the current placement (EULER 1762)

(2.4) $$\begin{aligned} d\mathbf{x} &= d\boldsymbol{\chi}(\mathbf{x}_0\,, t\,, d\mathbf{x}_0) = \mathbf{F}\,d\mathbf{x}_0 \\ d\mathbf{x}_0 &= d\boldsymbol{\chi}^{-1}(\mathbf{x}\,, t\,, d\mathbf{x}) = \mathbf{F}^{-1}\,d\mathbf{x}\,. \end{aligned}$$

---

[31] In some context, like cavitation analysis, this restriction must be relaxed.

If we choose a material basis of tangent vectors $\{\mathbf{dx}_{0i}\}$ with dual $\{\mathbf{dx}_0^{\,i}\}$, then $\{\mathbf{dx}_i := \mathbf{F}\,\mathbf{dx}_{0i}\}$ also forms a basis after Theorem 1.9., and the deformation gradient has the form

$$\mathbf{F} = \mathbf{dx}_i \otimes \mathbf{dx}_0^{\,i}.$$

Therefore, the deformation gradient is sometimes called a two-point tensor. With the right leg it stands in the reference placement, with the left leg in the current placement.

The component form of $\mathbf{F}$ is achieved by the choice of a material COOS $\{\Psi^i\}$ in $\mathscr{B}_0$ with natural bases $\{\mathbf{r}_{\Psi i}\}$ and $\{\mathbf{r}_\Psi^{\,i}\}$, and a spatial COOS $\{\varphi^i\}$ in $\mathscr{B}_t$ with natural bases $\{\mathbf{r}_{\varphi i}\}$ and $\{\mathbf{r}_\varphi^{\,i}\}$. In the COOS domains we have

$$\varphi^i = \chi^i(\Psi^1, \Psi^2, \Psi^3, t) \qquad i = 1, 2, 3$$

and

$$\Psi^i = \chi^{-1\,i}(\varphi^1, \varphi^2, \varphi^3, t) \qquad i = 1, 2, 3.$$

The differentials of these coordinate transformations are

$$d\varphi^i = d\chi^i(\Psi^1, \Psi^2, \Psi^3, t, d\Psi^1, d\Psi^2, d\Psi^3)$$

$$= \frac{\partial \chi^i}{\partial \Psi^k}\, d\Psi^k = \frac{\partial \varphi^i}{\partial \Psi^k}\, d\Psi^k$$

(again keeping the time fixed), and inversely

$$d\Psi^i = d\chi^{-1\,i}(\varphi^1, \varphi^2, \varphi^3, t, d\varphi^1, d\varphi^2, d\varphi^3)$$

$$= \frac{\partial \chi^{-1\,i}}{\partial \varphi^k}\, d\varphi^k = \frac{\partial \Psi^i}{\partial \varphi^k}\, d\varphi^k.$$

The two JACOBI matrices are mutually inverse because of

$$\frac{\partial \varphi^i}{\partial \varphi^j} = \delta^i_j = \frac{\partial \varphi^i}{\partial \Psi^k}\,\frac{\partial \Psi^k}{\partial \varphi^j}\,.$$

The tangent vectors in the reference placement are the differentials

$$\mathbf{dx}_0 = \frac{\partial \mathbf{x}_0}{\partial \Psi^i}\, d\Psi^i = \mathbf{r}_{\Psi i}\, d\Psi^i$$

and for the spatial ones

$$\mathbf{dx} = \frac{\partial \mathbf{x}}{\partial \varphi^i}\, d\varphi^i = \mathbf{r}_{\varphi i}\, d\varphi^i = \mathbf{r}_{\varphi i}\, \frac{\partial \chi^i}{\partial \Psi^k}\, d\Psi^k$$

$$= \Big(\frac{\partial \chi^i}{\partial \Psi^k}\, \mathbf{r}_{\varphi i} \otimes \mathbf{r}_\Psi^{\,k}\Big)\, \mathbf{r}_{\Psi m}\, d\Psi^m = \mathbf{F}\,\mathbf{dx}_0$$

and thus

(2.5)
$$\mathbf{F} = \frac{\partial \varphi^k}{\partial \Psi^i}\, \mathbf{r}_{\varphi k} \otimes \mathbf{r}_\Psi{}^i$$

with inverse

$$\mathbf{F}^{-1} = \frac{\partial \Psi^i}{\partial \varphi^k}\, \mathbf{r}_{\Psi i} \otimes \mathbf{r}_\varphi{}^k$$

Both COOS can be chosen arbitrarily, and two special choices are always possible.

1) A COOS is so large that both $\mathscr{B}_0$ *and* $\mathscr{B}_t$ are covered by it. In this case one COOS and one set of natural bases are sufficient. However, even then it is necessary to distinguish between the EULERean and LAGRANGEan coordinates of a point.

2) We introduce an arbitrary COOS in the reference placement, which induces a COOS on the body. When the body moves through the space, we choose those particular coordinates which momentarily coincide with this fixed one. Then a material point has at all times the same material and spatial coordinate values with respect to a continuously changing COOS, and χ^i as well as χ^{-1i} become the identities in the *i*-th spatial argument

$$\Psi^i = \chi^i(\Psi^1, \Psi^2, \Psi^3, t) \qquad i = 1, 2, 3$$

and

$$\varphi_t{}^i = \chi^{-1\,i}(\varphi_t{}^1, \varphi_t{}^2, \varphi_t{}^3, t) \qquad i = 1, 2, 3.$$

The suffix t to the EULERean coordinates $\varphi_t{}^i$ indicates its time dependence. The components of $\mathbf{F}$ and $\mathbf{F}^{-1}$ become trivial

$$\frac{\partial \varphi_t{}^k}{\partial \Psi^i} = \delta^k{}_i \qquad\qquad \frac{\partial \Psi^i}{\partial \varphi_t{}^k} = \delta^i{}_k\,.$$

$\{\varphi_t{}^j\}$ are called **convected coordinates**. However, the time-dependent bases to which they correspond, are non-trivial. This special case, which renders the component forms much simpler, can always be chosen.

The basic concept for all local deformation measures is the deformation gradient. After Theorem 1.21. it allows for the following polar decomposition

(2.6)
$$\mathbf{F} = \mathbf{R}\,\mathbf{U} = \mathbf{V}\,\mathbf{R}$$

with

$\mathbf{F}$	$\in \mathscr{Inv}^+$	deformation gradient
$\mathbf{R}$	$\in \mathscr{Orth}^+$	**rotation tensor**
$\mathbf{U}\,,\mathbf{V}$	$\in \mathscr{Psym}$	**right** and **left stretch tensor**

and the **right CAUCHY[32]-GREEN[33] tensor** (1841), also known as **local configuration** (tensor),

$$\mathbf{C} := \mathbf{U}^2 = \mathbf{F}^T \mathbf{F} \qquad \in \mathcal{P}sym$$

and the **left CAUCHY-GREEN tensor**, also named after FINGER[34] (1894)

$$\mathbf{B} := \mathbf{V}^2 = \mathbf{F}\,\mathbf{F}^T = \mathbf{R}\,\mathbf{C}\,\mathbf{R}^T = \mathbf{R} * \mathbf{C} \qquad \in \mathcal{P}sym\,.$$

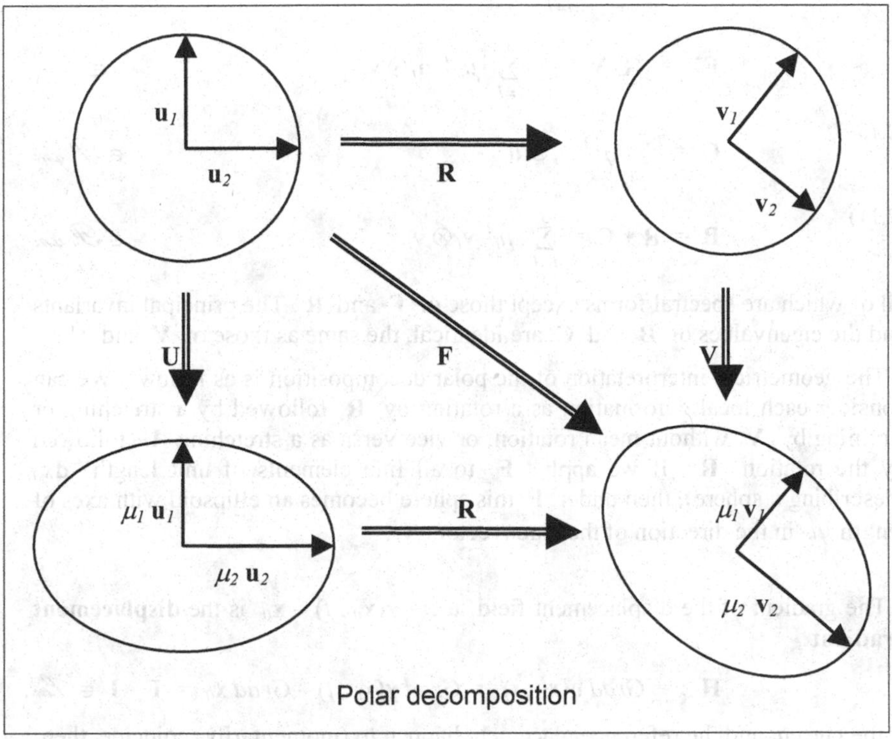

Polar decomposition

If we use the spectral form of the right stretch tensor with respect to its unit eigen-basis $\{\mathbf{u}_i\}$

(2.7)
$$\mathbf{U} = \sum_{i=1}^{3} \mu_i \; \mathbf{u}_i \otimes \mathbf{u}_i \qquad \in \mathcal{P}sym\,,$$

then $\mathbf{R}$ rotates it into the eigenbasis of $\mathbf{V}$

$$\mathbf{v}_i := \mathbf{R}\,\mathbf{u}_i$$

---

32 Augustin Louis Cauchy (1789-1857)
33 George Green (1793-1841)
34 Joseph Finger (1841-1925)

and the following representations are valid

$$\mathbf{R} = \sum_{i=1}^{3} \mathbf{v}_i \otimes \mathbf{u}_i \qquad \in \mathcal{O}rth^+$$

(2.8)
$$\mathbf{V} = \mathbf{R} * \mathbf{U} = \sum_{i=1}^{3} \mu_i \, \mathbf{v}_i \otimes \mathbf{v}_i \qquad \in \mathcal{P}sym$$

(2.9)
$$\mathbf{F} = \mathbf{V}\,\mathbf{R} = \sum_{i=1}^{3} \mu_i \, \mathbf{v}_i \otimes \mathbf{u}_i \qquad \in \mathcal{I}nv^+$$

$$\mathbf{F}^{-1} = \mathbf{R}^T \mathbf{V}^{-1} = \sum_{i=1}^{3} \mu_i^{-1} \, \mathbf{u}_i \otimes \mathbf{v}_i \qquad \in \mathcal{I}nv^+$$

(2.10)
$$\mathbf{C} = \sum_{i=1}^{3} \mu_i^2 \, \mathbf{u}_i \otimes \mathbf{u}_i \qquad \in \mathcal{P}sym$$

(2.11)
$$\mathbf{B} = \mathbf{R} * \mathbf{C} = \sum_{i=1}^{3} \mu_i^2 \, \mathbf{v}_i \otimes \mathbf{v}_i \qquad \in \mathcal{P}sym$$

all of which are spectral forms except those of $\mathbf{F}$ and $\mathbf{R}$. The principal invariants and the eigenvalues of $\mathbf{B}$ and $\mathbf{C}$ are identical, the same as those of $\mathbf{V}$ and $\mathbf{U}$.

The geometrical interpretation of the polar decomposition is as follows. We can consider each local deformation as a rotation by $\mathbf{R}$ followed by a stretching or straining by $\mathbf{V}$ without mean rotation, or vice versa as a stretching $\mathbf{U}$ followed by the rotation $\mathbf{R}$. If we apply $\mathbf{F}$ to all line elements of unit length $d\mathbf{x}_0$ (describing a sphere), then under $\mathbf{F}$ this sphere becomes an ellipsoid with axes of length μ_i in the direction of the eigenvectors $\mathbf{v}_i$.

The gradient of the displacement field $\mathbf{u} := \boldsymbol{\chi}(\mathbf{x}_0, t) - \mathbf{x}_0$ is the **displacement gradient**

$$\mathbf{H} := Grad\,\mathbf{u}(\mathbf{x}_0, t) = Grad\,\boldsymbol{\chi}(\mathbf{x}_0, t) - Grad\,\mathbf{x}_0 = \mathbf{F} - \mathbf{I} \in \mathcal{L}in .$$

If the current and the reference placement happen to (momentarily) coincide, then

$$\mathbf{F} \equiv \mathbf{R} \equiv \mathbf{V} \equiv \mathbf{U} \equiv \mathbf{B} \equiv \mathbf{C} \equiv \mathbf{I} \quad \text{and} \quad \mathbf{H} \equiv \mathbf{0} .$$

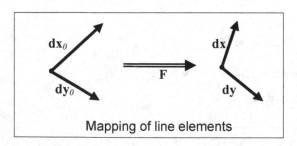

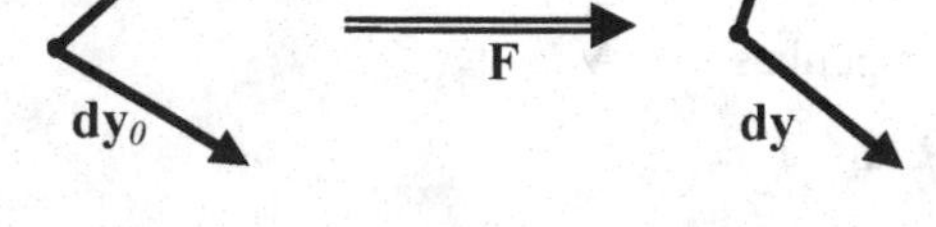

Mapping of line elements

Let $\mathbf{dx}_0$ and $\mathbf{dy}_0$ be two line elements in the reference placement, and $\mathbf{dx}$ and $\mathbf{dy}$ the same material line elements in the current placement, then their lengths and angles are determined by the scalar products after (2.4)

$$\mathbf{dx} \cdot \mathbf{dy} = (\mathbf{F}\,\mathbf{dx}_0) \cdot (\mathbf{F}\,\mathbf{dy}_0) = \mathbf{dx}_0 \cdot (\mathbf{F}^T\,\mathbf{F}\,\mathbf{dy}_0) = \mathbf{dx}_0 \cdot (\mathbf{C}\,\mathbf{dy}_0)$$

$$\mathbf{dx}_0 \cdot \mathbf{dy}_0 = (\mathbf{F}^{-1}\,\mathbf{dx}) \cdot (\mathbf{F}^{-1}\,\mathbf{dy}) = \mathbf{dx} \cdot (\mathbf{F}^{-T}\,\mathbf{F}^{-1}\,\mathbf{dy}) = \mathbf{dx} \cdot (\mathbf{B}^{-1}\,\mathbf{dy})$$

and their change by the difference

$$\begin{aligned} &\mathbf{dx} \cdot \mathbf{dy} - \mathbf{dx}_0 \cdot \mathbf{dy}_0 \\ &= \mathbf{dx}_0 \cdot \{(\mathbf{C} - \mathbf{I})\,\mathbf{dy}_0\} = \mathbf{dx}_0 \cdot (2\,\mathbf{E}^G\,\mathbf{dy}_0) \\ &= \mathbf{dx} \cdot \{(\mathbf{I} - \mathbf{B}^{-1})\,\mathbf{dy}\} = \mathbf{dx} \cdot (2\,\mathbf{E}^a\,\mathbf{dy}) \end{aligned}$$

with **GREEN´s strain tensor** (1841), also named after **de St. VENANT** (1844)

$$\mathbf{E}^G := \tfrac{1}{2}(\mathbf{C} - \mathbf{I}) \qquad \in \mathcal{Sym}$$

and **ALMANSI´s**[35] (1911) or **HAMEL´s**[36] **strain tensor** (1912)

$$\mathbf{E}^a := \tfrac{1}{2}(\mathbf{I} - \mathbf{B}^{-1}) \qquad \in \mathcal{Sym} .$$

These strain tensors are related by

$$\mathbf{E}^G = \mathbf{F}^T\,\mathbf{E}^a\,\mathbf{F} = \mathbf{F}^T * \mathbf{E}^a .$$

Other common strain tensors are the **spatial BIOT**[37] **strain tensor**, also named after **SWAINGER**

$$\mathbf{E}^b := \mathbf{I} - \mathbf{V}^{-1} \qquad \in \mathcal{Sym} ,$$

or the **material BIOT strain tensor**, also named after **CAUCHY** or **SWAINGER**

$$\mathbf{E}^B := \mathbf{U} - \mathbf{I} \qquad \in \mathcal{Sym} ,$$

the **spatial logarithmic** or **HENCKY**[38] **strain tensor** (1928)

$$\mathbf{E}^h := \ln \mathbf{V} = \tfrac{1}{2} \ln \mathbf{B} \qquad \in \mathcal{Sym} ,$$

and the **material logarithmic** or **HENCKY strain tensor**[39]

$$\mathbf{E}^H := \ln \mathbf{U} = \tfrac{1}{2} \ln \mathbf{C} \qquad \in \mathcal{Sym} .$$

All these tensors are dimensionless and can be determined by $\mathbf{F}$ (or, likewise by $\mathbf{H}$). For example,

$$\mathbf{E}^G = \tfrac{1}{2}(\mathbf{F}^T\,\mathbf{F} - \mathbf{I}) = \tfrac{1}{2}(\mathbf{H} + \mathbf{H}^T + \mathbf{H}^T\,\mathbf{H}) . \tag{2.12}$$

[35] Emilio Almansi (1869-1948)
[36] Georg Hamel (1877-1954)
[37] Maurice Anthony Biot (1905-1985)
[38] Heinrich Hencky (1885-1951), see LATORRE/ MONTÁNS (2014)
[39] See FITZGERALD (1980)

Because of the last term, $\mathbf{E}^G$ is non-linear in $\mathbf{H}$ and, thus, also non-linear in the displacements $\mathbf{u}$. Such non-linearity is common for all of the above strain tensors (*geometrical nonlinearity*).

We obtain the following spectral forms for these strain tensors

$$\mathbf{E}^G = \tfrac{1}{2} \sum_{i=1}^{3} (\mu_i^2 - 1)\, \mathbf{u}_i \otimes \mathbf{u}_i \qquad \in \mathcal{Sym}$$

$$\mathbf{E}^B = \sum_{i=1}^{3} (\mu_i - 1)\, \mathbf{u}_i \otimes \mathbf{u}_i \qquad \in \mathcal{Sym}$$

$$\mathbf{E}^H = \sum_{i=1}^{3} \ln(\mu_i)\, \mathbf{u}_i \otimes \mathbf{u}_i \qquad \in \mathcal{Sym}$$

$$\mathbf{E}^h = \sum_{i=1}^{3} \ln(\mu_i)\, \mathbf{v}_i \otimes \mathbf{v}_i \qquad \in \mathcal{Sym}$$

$$\mathbf{E}^b = \sum_{i=1}^{3} (1 - \mu_i^{-1})\, \mathbf{v}_i \otimes \mathbf{v}_i \qquad \in \mathcal{Sym}$$

$$\mathbf{E}^a = \tfrac{1}{2} \sum_{i=1}^{3} (1 - \mu_i^{-2})\, \mathbf{v}_i \otimes \mathbf{v}_i \qquad \in \mathcal{Sym}\,.$$

Between the principal invariants we have the following relations[40]

$$
\begin{aligned}
2\, I_{\mathbf{E}^G} &= I_{\mathbf{C}} - 3 & I_{\mathbf{C}} &= 2\, I_{\mathbf{E}^G} + 3 \\
4\, II_{\mathbf{E}^G} &= II_{\mathbf{C}} - 2\, I_{\mathbf{C}} + 3 & II_{\mathbf{C}} &= 4\, I_{\mathbf{E}^G} + 4\, II_{\mathbf{E}^G} + 3 \qquad (2.13) \\
8\, III_{\mathbf{E}^G} &= III_{\mathbf{C}} - II_{\mathbf{C}} + I_{\mathbf{C}} - 1 & III_{\mathbf{C}} &= 2\, I_{\mathbf{E}^G} + 4\, II_{\mathbf{E}^G} + 8\, III_{\mathbf{E}^G} + 1\,.
\end{aligned}
$$

We can conclude that in the geometrically non-linear deformation theory we have many different strain measures, which have the following properties in common.

- All strain tensors are symmetric.
- All material strain tensors are coaxial (with eigenvectors $\mathbf{u}_i$).
- All spatial strain tensors are also coaxial (with eigenvectors $\mathbf{v}_i$).
- The *i*-th eigenvalue of one of them can be expressed by the *i*-th eigenvalue of any other. Both are either positive (for straining), negative (for pressing), or zero.
- In the reference placement all strain tensors are zero, the same as for rigid body motions.
- If the state of deformation is uniaxial (because two eigenvalues are zero), then the same holds for all other strain measures.
- If the state of deformation is plane (because one eigenvalue is zero), then the same holds for all other strain measures.
- If the state of deformation is triaxial (because all eigenvalues are non-zero), then the same holds for all other strain measures.

---

[40] See DOYLE/ ERICKSEN (1956) p. 97.

- If the deformation is a dilatation (because all eigenvalues are equal), then the same holds for all other strain measures.
- The derivatives of the strain tensors with respect to μ_i at $\mu_i \equiv 1$ are 1.

It was suggested by HILL[41] (1968) to generalise the concept of a strain tensor such that these properties remain valid[42]. For this purpose we introduce a real function

$$f : \mathscr{R}^+ \to \mathscr{R}$$

with the following properties

- twice differentiable
- $f\,' > 0$ everywhere
- $f(1) = 0$
- $f(1)' = 1$.

This function maps the eigenvalues of the stretch tensors and, thus, defines a special isotropic tensor-function

$$\boldsymbol{F} : \mathscr{P}sym \to \mathscr{S}ym$$

which assigns to a symmetric tensor with spectral form

$$\mathbf{A} = \sum_{r=1}^{3} \mu_r \, \mathbf{p}_r \otimes \mathbf{p}_r \qquad \in \mathscr{P}sym$$

the coaxial tensor

$$\boldsymbol{F}(\mathbf{A}) := \sum_{r=1}^{3} f(\mu_r) \, \mathbf{p}_r \otimes \mathbf{p}_r \qquad \in \mathscr{S}ym \, .$$

For such an f we call

(2.14)
$$\mathbf{E}^{\mathrm{Gen}} := \boldsymbol{F}(\mathbf{U}) = \sum_{r=1}^{3} f(\mu_r) \, \mathbf{u}_r \otimes \mathbf{u}_r \qquad \in \mathscr{S}ym$$

a **generalised material strain tensor**, and

(2.15)
$$\mathbf{E}^{\mathrm{gen}} := \boldsymbol{F}(\mathbf{V}) = \sum_{r=1}^{3} f(\mu_r) \, \mathbf{v}_r \otimes \mathbf{v}_r \qquad \in \mathscr{S}ym$$

a **generalised spatial strain tensor**. As the only difference between these two is the rotation of the eigenvectors by $\mathbf{R}$, we can conclude, that each material strain tensor corresponds to a spatial one through

$$\mathbf{E}^{\mathrm{gen}} = \mathbf{R} * \mathbf{E}^{\mathrm{Gen}}$$

or inversely

[41] Rodney Hill (1921-2011)
[42] See also KARNI/ REINER (1964).

$$\mathbf{E}^{Gen} = \mathbf{R}^T * \mathbf{E}^{gen}.$$

Evidently, $\mathbf{E}^G$, $\mathbf{E}^B$, and $\mathbf{E}^H$ are generalised material strain tensors, and $\mathbf{E}^b$, $\mathbf{E}^a$, and $\mathbf{E}^h$ are spatial ones. A specific class of strain tensors is obtained by power functions of the form

(2.16) $$f(\mu_r) \equiv f_m(\mu_r) := m^{-1}\,(\mu_r^{\,m} - 1)$$

for all non-zero reals m (SETH[43] 1964). As a limit for $m \to 0$ we obtain the logarithmic function

$$f_0(\mu_r) = ln(\mu_r)\,.$$

For the derivatives of all of these functions we obtain

$$f_m(\mu_r)' = \mu_r^{\,m-1}.$$

SETH´s ansatz leads to strain tensors

$$\mathbf{E}^{Gen} \equiv m^{-1}\,(\mathbf{U}^m - \mathbf{I}) \qquad \text{for } m \neq 0$$

or

$$\mathbf{E}^{Gen} \equiv ln\,\mathbf{U} \qquad \text{for } m = 0$$

and

$$\mathbf{E}^{gen} \equiv m^{-1}\,(\mathbf{V}^m - \mathbf{I}) \qquad \text{for } m \neq 0$$

or

$$\mathbf{E}^{gen} \equiv ln\,\mathbf{V} \qquad \text{for } m = 0\,.$$

All of the above strain tensors are within this special class, either as material strain tensors

- for $m \equiv -2$ the material PIOLA[44] strain tensor
- for $m \equiv 0$ the material HENCKY strain tensor
- for $m \equiv 1$ the material BIOT strain tensor
- for $m \equiv 2$ GREEN´s strain tensor

or as spatial strain tensors

- for $m \equiv -2$ ALMANSI´s strain tensor
- for $m \equiv -1$ the spatial BIOT strain tensor
- for $m \equiv 0$ the spatial HENCKY strain tensor
- for $m \equiv 2$ FINGER´s strain tensor.

---

[43] Bhoj Raj Seth (1907-1979)
[44] Gabbrio Piola (1791-1850)

Thus, in the geometrical non-linear deformation theory, we have infinitely many strain tensors, all of which can be determined by the deformation gradient.

We now determine the component representation with respect to an arbitrary COOS $\{\Psi^i\}$ in the reference placement and $\{\varphi^i\}$ in the current placement and their natural bases. First of all, the identity tensor has the forms after (1.14)

$$\begin{aligned}\mathbf{I} &= \mathbf{r}_{\Psi i} \otimes \mathbf{r}_{\Psi}{}^{i} = \mathbf{r}_{\Psi}{}^{i} \otimes \mathbf{r}_{\Psi i} = G_{\Psi}{}^{ik}\, \mathbf{r}_{\Psi i} \otimes \mathbf{r}_{\Psi k} = G_{\Psi\, ik}\, \mathbf{r}_{\Psi}{}^{i} \otimes \mathbf{r}_{\Psi}{}^{k} \\ &= \mathbf{r}_{\varphi i} \otimes \mathbf{r}_{\varphi}{}^{i} = \mathbf{r}_{\varphi}{}^{i} \otimes \mathbf{r}_{\varphi i} = G_{\varphi}{}^{ik}\, \mathbf{r}_{\varphi i} \otimes \mathbf{r}_{\varphi k} = G_{\varphi\, ik}\, \mathbf{r}_{\varphi}{}^{i} \otimes \mathbf{r}_{\varphi}{}^{k} \\ &= (\mathbf{r}_{\varphi}{}^{i} \cdot \mathbf{r}_{\Psi k})\, \mathbf{r}_{\varphi i} \otimes \mathbf{r}_{\Psi}{}^{k} = (\mathbf{r}_{\varphi i} \cdot \mathbf{r}_{\Psi}{}^{k})\, \mathbf{r}_{\varphi}{}^{i} \otimes \mathbf{r}_{\Psi k} \qquad \in \mathcal{P}sym\,.\end{aligned}$$

For the deformation gradient we obtain the natural representations

$$\mathbf{F} = \frac{\partial \varphi^i}{\partial \Psi^k}\, \mathbf{r}_{\varphi i} \otimes \mathbf{r}_{\Psi}{}^{k} \in \mathcal{I}nv^{+} \qquad \mathbf{F}^{T} = \frac{\partial \varphi^i}{\partial \Psi^k}\, \mathbf{r}_{\Psi}{}^{k} \otimes \mathbf{r}_{\varphi i} \in \mathcal{I}nv^{+}$$

$$\mathbf{F}^{-1} = \frac{\partial \Psi^i}{\partial \varphi^k}\, \mathbf{r}_{\Psi i} \otimes \mathbf{r}_{\varphi}{}^{k} \in \mathcal{I}nv^{+} \qquad \mathbf{F}^{-T} = \frac{\partial \Psi^i}{\partial \varphi^k}\, \mathbf{r}_{\varphi}{}^{k} \otimes \mathbf{r}_{\Psi i} \in \mathcal{I}nv^{+}$$

and further on

$$\mathbf{H} = \Big(\frac{\partial \varphi^i}{\partial \Psi^k} - \mathbf{r}_{\varphi}{}^{i} \cdot \mathbf{r}_{\Psi k}\Big)\, \mathbf{r}_{\varphi i} \otimes \mathbf{r}_{\Psi}{}^{k} \qquad \in \mathcal{L}in$$

$$\mathbf{B} = \frac{\partial \varphi^i}{\partial \Psi^k}\, G_{\Psi}{}^{kl}\, \frac{\partial \varphi^m}{\partial \Psi^l}\, \mathbf{r}_{\varphi i} \otimes \mathbf{r}_{\varphi m} \qquad \in \mathcal{P}sym$$

$$\mathbf{C} = \frac{\partial \varphi^i}{\partial \Psi^k}\, G_{\varphi\, im}\, \frac{\partial \varphi^m}{\partial \Psi^l}\, \mathbf{r}_{\Psi}{}^{k} \otimes \mathbf{r}_{\Psi}{}^{l} \qquad \in \mathcal{P}sym$$

$$\mathbf{E}^{\mathrm{G}} = \tfrac{1}{2}\Big(\frac{\partial \varphi^i}{\partial \Psi^k}\, G_{\varphi\, im}\, \frac{\partial \varphi^m}{\partial \Psi^l} - G_{\Psi\, kl}\Big)\, \mathbf{r}_{\Psi}{}^{k} \otimes \mathbf{r}_{\Psi}{}^{l} \qquad \in \mathcal{S}ym$$

$$\mathbf{E}^{\mathrm{a}} = \tfrac{1}{2}\Big(G_{\varphi\, kl} - \frac{\partial \Psi^i}{\partial \varphi^k}\, G_{\Psi\, im}\, \frac{\partial \Psi^m}{\partial \varphi^l}\Big)\, \mathbf{r}_{\varphi}{}^{k} \otimes \mathbf{r}_{\varphi}{}^{l} \qquad \in \mathcal{S}ym\,.$$

For a convected COOS this is simpler

$$\mathbf{F} = \mathbf{r}_{\varphi i} \otimes \mathbf{r}_{\Psi}{}^{i} \qquad \in \mathcal{I}nv^{+}$$

$$\mathbf{H} = \mathbf{r}_{\varphi i} \otimes \mathbf{r}_{\Psi}{}^{i} - \mathbf{r}_{\varphi i} \otimes \mathbf{r}_{\varphi}{}^{i} = \mathbf{r}_{\varphi i} \otimes \mathbf{r}_{\Psi}{}^{i} - \mathbf{r}_{\Psi i} \otimes \mathbf{r}_{\Psi}{}^{i} \qquad \in \mathcal{L}in$$

$$\mathbf{B} = G_{\Psi}{}^{kl}\, \mathbf{r}_{\varphi k} \otimes \mathbf{r}_{\varphi l} \qquad \in \mathcal{P}sym$$

$$\mathbf{C} = G_{\varphi\, kl}\, \mathbf{r}_{\Psi}{}^{k} \otimes \mathbf{r}_{\Psi}{}^{l} \qquad \in \mathcal{P}sym$$

$$\mathbf{E}^G = \tfrac{1}{2}\,(G_{\varphi\, im} - G_{\Psi\, im})\; \mathbf{r}_\Psi{}^i \otimes \mathbf{r}_\Psi{}^m \qquad \in \mathcal{S}ym$$

$$\mathbf{E}^a = \tfrac{1}{2}\,(G_{\varphi\, im} - G_{\Psi\, im})\; \mathbf{r}_\varphi{}^i \otimes \mathbf{r}_\varphi{}^m \qquad \in \mathcal{S}ym\,.$$

These representations show clearly that

- **F** and **H** are two-point tensors,
- **B** and **C** are metrics of material line elements,
- $\mathbf{E}^G$ and $\mathbf{E}^a$ are differences of such metrics.

The volume elements dv in $\mathscr{B}_t$ and dv_0 in $\mathscr{B}_0$ can be represented as triple products by three linear independent tangent vectors (line elements)

$$dv = [\mathbf{dx}\,,\, \mathbf{dy}\,,\, \mathbf{dz}] \qquad dv_0 = [\mathbf{dx}_0\,,\, \mathbf{dy}_0\,,\, \mathbf{dz}_0]\,.$$

By the definition of the determinant of a tensor (1.6) we obtain the transformation rule (EULER 1762)

(2.17)
$$\frac{dv}{dv_0} = det(\mathbf{F}) =: J$$

$$= det(\mathbf{U}) = det(\mathbf{V}) = \sqrt{det(\mathbf{C})} = \sqrt{det(\mathbf{B})} = \mu_1\,\mu_2\,\mu_3\,.$$

If the motion of a body is such that $det(\mathbf{F})$ is always and everywhere constant, then its occupied volume is also constant and we call the motion *isochoric*.

The vectorial elements of area in the reference placement $\mathbf{da}_0$ and in the current placement $\mathbf{da}$ can be obtained by the cross product of two linear independent vectors (line elements)

$$\mathbf{da}_0 = \mathbf{dx}_0 \times \mathbf{dy}_0$$

$$\mathbf{da} = \mathbf{dx} \times \mathbf{dy}\,.$$

By taking the scalar product by an arbitrary vector $\mathbf{v}$, we obtain the triple product

$$\begin{aligned}
\mathbf{da} \cdot \mathbf{v} &= (\mathbf{dx} \times \mathbf{dy}) \cdot \mathbf{v} = [\mathbf{dx}\,,\, \mathbf{dy}\,,\, \mathbf{v}] \\
&= [\mathbf{F}\,\mathbf{dx}_0\,,\, \mathbf{F}\,\mathbf{dy}_0\,,\, \mathbf{F}\,\mathbf{F}^{-1}\,\mathbf{v}] \\
&= det(\mathbf{F})\,[\mathbf{dx}_0\,,\, \mathbf{dy}_0\,,\, \mathbf{F}^{-1}\,\mathbf{v}] \\
&= J\,(\mathbf{dx}_0 \times \mathbf{dy}_0) \cdot (\mathbf{F}^{-1}\,\mathbf{v}) = \{J\,\mathbf{F}^{-T}(\mathbf{dx}_0 \times \mathbf{dy}_0)\} \cdot \mathbf{v} \\
&= (J\,\mathbf{F}^{-T}\,\mathbf{da}_0) \cdot \mathbf{v}\,.
\end{aligned}$$

Because of the arbitrariness of $\mathbf{v}$, we conclude **NANSON´s**[45] **formula** (1878)

(2.18)
$$\mathbf{da} = det(\mathbf{F})\,\mathbf{F}^{-T}\,\mathbf{da}_0 = \mathbf{E}^F\,\mathbf{da}_0$$

with the **area placement tensor** (sometimes called the *cofactor* of **F**)

$$\mathbf{E}^F := det(\mathbf{F})\,\mathbf{F}^{-T} \in \mathcal{I}nv^+\,.$$

As a summary, these are the transformations for all non-trivial differential forms:

[45] Edward John Nanson (1850-1936)

for line elements	$\mathbf{dx} = \mathbf{F}\,\mathbf{dx}_0$
for area elements	$\mathbf{da} = det(\mathbf{F})\,\mathbf{F}^{-T}\,\mathbf{da}_0 = \mathbf{E}^{\mathrm{F}}\,\mathbf{da}_0$
for volume elements	$dv = det(\mathbf{F})\,dv_0 = J\,dv_0$

Often one also needs the time rates of the deformations. The spatial **velocity gradient** is

$$\mathbf{L}(\mathbf{x}\,,t) := grad\,\mathbf{v}_E(\mathbf{x}\,,t) = \mathbf{v}_E(\mathbf{x}\,,t)\otimes\nabla_E$$

$$= Grad\,\mathbf{v}_L(\mathbf{x}_0\,,t)\,\mathbf{F}^{-1} = Grad\,\frac{\partial\chi(\mathbf{x}_0,t)}{\partial t}\,\mathbf{F}^{-1}$$

$$= \frac{\partial^2\chi(\mathbf{x}_0,t)}{\partial t\,\partial\mathbf{x}_0}\,\mathbf{F}^{-1} = \frac{\partial\,Grad\,\chi(\mathbf{x}_0,t)}{\partial t}\,\mathbf{F}^{-1} =$$

(2.19) $$\mathbf{L} = \mathbf{F}^{\bullet}\,\mathbf{F}^{-1} \qquad \in \mathcal{Lin}\,.$$

The additive decomposition (1.5) of the velocity gradient gives the **CAUCHY-STOKES[46] decomposition**

$$\mathbf{L} = \mathbf{D} + \mathbf{W} \qquad \in \mathcal{Lin}$$

into the symmetric **stretching** or **rate of deformation tensor**

$$\mathbf{D} := \tfrac{1}{2}\,(\mathbf{L} + \mathbf{L}^T) \qquad \in \mathcal{Sym}$$

and the skew **spin tensor**

$$\mathbf{W} := \tfrac{1}{2}\,(\mathbf{L} - \mathbf{L}^T) \qquad \in \mathcal{Skw}\,.$$

The latter is related to the curl of the velocity field by

$$\mathbf{W}\,\mathbf{dx} = \mathbf{w} \times \mathbf{dx}$$

with the **spin vector**

$$\mathbf{w} := \tfrac{1}{2}\,curl\,\mathbf{v} \qquad \in \mathcal{V}$$

being the axial vector of $\mathbf{W}$. This vector indicates the angular velocity by which those material line elements rotate that currently coincide with the eigendirections of $\mathbf{D}$. If $\mathbf{w}$ is zero, we call the motion *irrotational*. For the interpretation of these parts, we consider the rate of a material line element

$$\mathbf{dx}^{\bullet} = (\mathbf{F}\,\mathbf{dx}_0)^{\bullet} = \mathbf{F}^{\bullet}\,\mathbf{dx}_0 = \mathbf{F}^{\bullet}\,\mathbf{F}^{-1}\,\mathbf{dx} = \mathbf{L}\,\mathbf{dx}\,.$$

For the rate of the scalar product between two material line elements we get

$$(\mathbf{dx}\cdot\mathbf{dy})^{\bullet} = \mathbf{dx}^{\bullet}\cdot\mathbf{dy} + \mathbf{dx}\cdot\mathbf{dy}^{\bullet}$$

$$= (\mathbf{L}\,\mathbf{dx})\cdot\mathbf{dy} + \mathbf{dx}\cdot(\mathbf{L}\,\mathbf{dy}) = \mathbf{dx}\cdot(\mathbf{L} + \mathbf{L}^T)\,\mathbf{dy}$$

[46] Georg Gabriel Stokes (1819-1903)

$$= \mathbf{dx} \cdot (2\, \mathbf{D}\, \mathbf{dy}) \,.$$

Thus, the rate of deformation tensor determines the rate at which lengths of and angles between two material line elements change. The relation to the rate of the material tensors is

$$\mathbf{E}^{G\bullet} = \tfrac{1}{2}\, \mathbf{C}^{\bullet} = \tfrac{1}{2}\, (\mathbf{F}^T \mathbf{F})^{\bullet} = \tfrac{1}{2}\, (\mathbf{F}^T \mathbf{F}^{\bullet} + \mathbf{F}^{T\bullet} \mathbf{F})$$

$$= \mathbf{F}^T\, \tfrac{1}{2}\, (\mathbf{F}^{\bullet} \mathbf{F}^{-1} + \mathbf{F}^{-T} \mathbf{F}^{T\bullet})\, \mathbf{F} = \mathbf{F}^T * \mathbf{D} \,.$$

The rates of the differential forms are

$$\mathbf{dx}^{\bullet} = \mathbf{L}\, \mathbf{dx}$$

(2.20) $$\mathbf{da}^{\bullet} = \{tr(\mathbf{L})\, \mathbf{I} - \mathbf{L}^T\}\, \mathbf{da} = \{div(\mathbf{v})\, \mathbf{I} - \mathbf{L}^T\}\, \mathbf{da}$$

$$dv^{\bullet} = tr(\mathbf{L})\, dv = tr(\mathbf{D})\, dv = div(\mathbf{v})\, dv = J^{\bullet}\, dv_0$$

$$\Rightarrow \qquad J^{\bullet} = J\, div(\mathbf{v}) = J\, tr(\mathbf{L}) = J\, \mathbf{F}^{\bullet} \cdot \mathbf{F}^{-T}$$

$$= (III_{\mathbf{F}})^{\bullet} = J\, \mathbf{F}^{-T} \cdot \mathbf{F}^{\bullet} = J\, \mathbf{I} \cdot \mathbf{F}^{\bullet}\, \mathbf{F}^{-1}$$

$$= (III_{\mathbf{C}}^{1/2})^{\bullet} = \tfrac{1}{2}\, J\, \mathbf{C}^{-1} \cdot \mathbf{C}^{\bullet} = J\, \mathbf{C}^{-1} \cdot \mathbf{E}^{G\bullet} .$$

These equations give some simple criteria for a motion to be isochoric.

Theorem 2.1. *A motion is isochoric, if and only if for all times everywhere in the body the following equivalent conditions hold:*

- $\mathbf{F} \in \mathcal{U}nim^+$
- $J = 1$
- $J^{\bullet} = 0$
- $tr(\mathbf{L}) = tr(\mathbf{D}) = div(\mathbf{v}) = 0$

Rigid Body Motions

can be represented by

(2.21) $$\boldsymbol{\chi}(\mathbf{x}_0\,, t) = \mathbf{Q}(t)\, \mathbf{x}_0 + \mathbf{c}(t)$$

with $\mathbf{c}(t) \in \mathcal{V}$ and $\mathbf{Q}(t) \in \mathcal{O}rth^+$, both differentiable functions of time, but spatially constant. $\mathbf{c}(t)$ presents a translation, and $\mathbf{Q}(t)$ a rotation. The inverse is then

$$\boldsymbol{\chi}^{-1}(\mathbf{x}\,, t) = \mathbf{Q}(t)^T (\mathbf{x} - \mathbf{c}(t)) \,.$$

The deformation gradient is now orthogonal and spatially constant

(2.22) $$\mathbf{F}(t) \equiv \mathbf{Q}(t) \equiv \mathbf{R}(t)$$

and there is no stretching

$$\mathbf{U} \equiv \mathbf{V} \equiv \mathbf{C} \equiv \mathbf{B} \equiv \mathbf{I}$$

and all strain tensors are zero

$$\mathbf{E}^{\mathrm{Gen}} \equiv \mathbf{E}^{\mathrm{gen}} \equiv \mathbf{0}$$

except the linear strain tensor which will be demonstrated below.

The velocity field of such a motion is

$$\mathbf{v} = \boldsymbol{\chi}(\mathbf{x}_0 , t)^{\bullet} = \mathbf{Q}(t)^{\bullet}\, \mathbf{x}_0 + \mathbf{c}(t)^{\bullet}$$

$$= \mathbf{Q}(t)^{\bullet}\, \mathbf{Q}(t)^T (\mathbf{x} - \mathbf{c}(t)) + \mathbf{c}(t)^{\bullet}.$$

The velocity gradient

$$\mathbf{L} \equiv \mathbf{Q}^{\bullet}\, \mathbf{Q}^T \equiv \mathbf{W} \qquad \in \mathscr{Skew}$$

is here skew, and therefore

$$\mathbf{D} \equiv \mathbf{0}\,.$$

The axial vector of the spin tensor is the spin vector $\mathbf{w}$, which gives

$$\mathbf{v} = \mathbf{w} \times (\mathbf{x} - \mathbf{c}) + \mathbf{c}^{\bullet}.$$

This is **EULER's velocity formula** for rigid body motions, after which the velocity field of a rigid body consists of a translatoric and a rotational part. The acceleration field of a rigid body motion is then

$$\mathbf{a} = \mathbf{v}^{\bullet} = \boldsymbol{\chi}(\mathbf{x}_0 , t)^{\bullet\bullet} = \{\mathbf{Q}(t)\, \mathbf{x}_0 + \mathbf{c}(t)\}^{\bullet\bullet}$$

$$= \mathbf{w}^{\bullet} \times (\mathbf{x} - \mathbf{c}) + \mathbf{w} \times (\mathbf{x}^{\bullet} - \mathbf{c}^{\bullet}) + \mathbf{c}^{\bullet\bullet}$$

$$= \mathbf{w}^{\bullet} \times (\mathbf{x} - \mathbf{c}) + \mathbf{w} \times \{\mathbf{w} \times (\mathbf{x} - \mathbf{c})\} + \mathbf{c}^{\bullet\bullet}.$$

Superimposed Rigid Body Motions

upon an arbitrary motion $\boldsymbol{\chi}(\mathbf{x}_0 , t)$ are introduced by an orthogonal tensor $\mathbf{Q}(t)$ and a vector $\mathbf{c}(t)$, both time dependent in general, as

(2.23) $$\boldsymbol{\chi}_{mod}(\mathbf{x}_0 , t) = \mathbf{Q}(t)\, \boldsymbol{\chi}(\mathbf{x}_0 , t) + \mathbf{c}(t)\,.$$

We obtain the modified deformation gradient as a function of the one of the original deformation by

$$\mathbf{F}_{mod} = \mathbf{Q}\,\mathbf{F}$$

and similarly

$$\mathbf{V}_{mod} = \mathbf{Q}\,\mathbf{V}\,\mathbf{Q}^T = \mathbf{Q} * \mathbf{V}$$

$$\mathbf{B}_{mod} = \mathbf{Q}\,\mathbf{B}\,\mathbf{Q}^T = \mathbf{Q} * \mathbf{B}$$

$$\mathbf{E}^{\mathrm{gen}}{}_{mod} = \mathbf{Q}\,\mathbf{E}^{\mathrm{gen}}\,\mathbf{Q}^T = \mathbf{Q} * \mathbf{E}^{\mathrm{gen}}$$

$$\mathbf{L}_{mod} = \mathbf{F}_{mod}{}^{\bullet}\, \mathbf{F}_{mod}{}^{-1} = \mathbf{Q}\,\mathbf{L}\,\mathbf{Q}^T + \mathbf{Q}^{\bullet}\, \mathbf{Q}^T$$

$$\mathbf{D}_{mod} = \mathbf{Q}\,\mathbf{D}\,\mathbf{Q}^T$$

$$\mathbf{W}_{mod} = \mathbf{Q}\,\mathbf{W}\,\mathbf{Q}^T + \mathbf{Q}^{\bullet}\,\mathbf{Q}^T.$$

The material tensors are invariant under such superimposed rigid body motions

$$\mathbf{U}_{mod} = \mathbf{U}$$

$$\mathbf{C}_{mod} = \mathbf{C}$$

$$\mathbf{E}^{\mathrm{Gen}}{}_{mod} = \mathbf{E}^{\mathrm{Gen}}.$$

Small Deformations

All of the introduced strain tensors are non-linear in the displacement gradient and, consequently, in the displacements. We refer to this non-linearity as a geometrical one. However, under certain conditions a linearisation is reasonable and advantageous.

A measure for the magnitude of deformations (strains and rotations) of material line elements is the norm of the displacement gradient

$$\varepsilon := |\mathbf{H}| = \sqrt{(\mathbf{H} \cdot \mathbf{H})}.$$

Of course, this measure depends on the choice of the reference placement. We achieve the **theory of small deformations** or the **geometrically linear theory**, if we neglect all terms of the order of ε^2 in comparison to linear ones. This shall be made more precise in the sequel.

Let $\mathcal{U}$ and $\mathcal{V}$ be normed linear spaces and

$$f : \mathcal{U} \rightarrow \mathcal{V}$$

a function. We call f **of the order** n if there is a constant $A \neq 0$ such that

$$\lim_{\varepsilon \to 0} \frac{|f(\mathbf{H})|}{\varepsilon^n} = A$$

holds for all zero sequences in $\mathcal{U}$. We notate this by the LANDAU[47] symbol $f = O(\varepsilon^n)$. Two functions

$$f, g : \mathcal{U} \rightarrow \mathcal{V}$$

are **equal of the order** n, if the difference $f - g$ is of the order n. If f and g are equal of order 2, we call them **almost equal** and write $f \approx g$.

The **linear strain tensor**

(2.24) $$\mathbf{E} := ½\,(\mathbf{H} + \mathbf{H}^T) = ½\,(Grad\,\mathbf{u} + Grad^T\,\mathbf{u}) = O(\varepsilon)$$

renders

[47] Edmund Landau (1877-1938)

$$\mathbf{E} \approx \mathbf{E}^{G}$$

in view of (2.12). It has the components

$$\varepsilon_{ij} = ½ \left(\frac{\partial u_i}{\partial X^j} + \frac{\partial u_j}{\partial X^i}\right)$$

with respect to a Cartesian COOS.

We next investigate the other deformation and strain tensors under linearisation. The right CAUCHY-GREEN tensor becomes

(2.25) $$\mathbf{C} = \mathbf{U}^2 = 2\,\mathbf{E}^G + \mathbf{I} \approx 2\,\mathbf{E} + \mathbf{I} \approx (\mathbf{I} + \mathbf{E})^2 .$$

The square root of this is $\mathbf{U}$. If we develop the root into a series, we obtain the linear approximation

$$\mathbf{U} = \sqrt{\mathbf{C}} = \sqrt{(\mathbf{I} + 2\,\mathbf{E}^G)} = \mathbf{I} + \mathbf{E}^G + O(\varepsilon^2) \approx \mathbf{I} + \mathbf{E} .$$

Its inverse is

$$\mathbf{U}^{-1} \approx \mathbf{I} - \mathbf{E}$$

because of

$$(\mathbf{I} + \mathbf{E})(\mathbf{I} - \mathbf{E}) \approx \mathbf{U}\,\mathbf{U}^{-1} = \mathbf{I} .$$

The rotatoric part of the deformation gradient is

$$\mathbf{R} = \mathbf{F}\,\mathbf{U}^{-1} \approx (\mathbf{I} + \mathbf{H})(\mathbf{I} - \mathbf{E}) \approx \mathbf{I} + ½\,(\mathbf{H} - \mathbf{H}^T) .$$

Thus, *small deformations* means small strains *and* small rotations, but arbitrary translations. The left stretch tensor becomes

$$\mathbf{V} = \mathbf{R}\,\mathbf{U}\,\mathbf{R}^T \approx \{\mathbf{I} + ½\,(\mathbf{H} - \mathbf{H}^T)\}\,(\mathbf{I} + \mathbf{E})\,\{\mathbf{I} - ½\,(\mathbf{H} - \mathbf{H}^T)\}$$
$$\approx \mathbf{I} + \mathbf{E}$$

and therefore

$$\mathbf{B} = \mathbf{V}^2 \approx \mathbf{C} \approx \mathbf{I} + 2\,\mathbf{E} .$$

Its inverse is

$$\mathbf{B}^{-1} \approx \mathbf{C}^{-1} \approx \mathbf{I} - 2\,\mathbf{E} .$$

With this we get also

$$\mathbf{E}^{a} \approx \mathbf{E} .$$

The deformation gradient is

$$\mathbf{F} = \mathbf{I} + \mathbf{H} = \mathbf{I} + \mathbf{E} + ½\,(\mathbf{H} - \mathbf{H}^T) \approx \mathbf{E} + \mathbf{R} .$$

The inverse of it is

$$\mathbf{F}^{-1} \approx \mathbf{I} - \mathbf{H} .$$

If we want to linearise the generalised strain tensors, we develop the function f of (2.14) around 1 into a TAYLOR series

$$f(\mu_r) = f(1) + f'(1)\,(\mu_r - 1) + \tfrac{1}{2} f''(1)\,(\mu_r - 1)^2 + \ldots$$
$$= \mu_r - 1 + \tfrac{1}{2} f''(1)\,(\mu_r - 1)^2 + \ldots \approx \mu_r - 1$$

$(\mu_r - 1)$ is of the same order as the eigenvalues of $\mathbf{E}$, and thus

$$\mathbf{E}^{\mathrm{Gen}} := \sum_{r=1}^{3} f(\mu_r)\, \mathbf{u}_r \otimes \mathbf{u}_r \approx \mathbf{U} - \mathbf{I} \approx \mathbf{E}\,.$$

$\mathbf{U}$ and $\mathbf{V}$ coincide in the linear theory, and so do the material and the spatial strain tensors

$$\mathbf{E}^{\mathrm{gen}} = \sum_{r=1}^{3} f(\mu_r)\, \mathbf{v}_r \otimes \mathbf{v}_r \approx \mathbf{V} - \mathbf{I} \approx \mathbf{E}\,.$$

For small deformations, all strain tensors can be substituted by the linear one.

The determinant is related to the eigenvalues ε_i of $\mathbf{E}$ by

$$J = det(\mathbf{F}) = det(\mathbf{U}) = det(\mathbf{V}) = \mu_1\, \mu_2\, \mu_3$$
$$\approx (1 + \varepsilon_1)(1 + \varepsilon_2)(1 + \varepsilon_3)$$
$$\approx 1 + \varepsilon_1 + \varepsilon_2 + \varepsilon_3 = 1 + tr(\mathbf{E})$$

and

$$J^2 = det(\mathbf{F}^2) = det(\mathbf{C}) = det(\mathbf{B}) \approx 1 + 2\, tr(\mathbf{E})\,.$$

The area placement tensor becomes in the linear theory

$$\mathbf{E}^{\mathrm{F}} \approx (1 + tr(\mathbf{E}))\,(\mathbf{I} - \mathbf{H}^T) \approx (1 + tr(\mathbf{E}))\,\mathbf{I} - \mathbf{H}^T.$$

We now assume that apart from $|\mathbf{H}|$ also $|\mathbf{H}^\bullet|$ is small, which involves a time scale. This renders

$$\mathbf{L} = grad\, \mathbf{v}_E = \mathbf{F}^\bullet\, \mathbf{F}^{-1} \approx \mathbf{H}^\bullet (\mathbf{I} - \mathbf{H}) \approx \mathbf{H}^\bullet$$
$$= (Grad\, \mathbf{u})^\bullet = Grad\, \mathbf{u}^\bullet = Grad\, \mathbf{v}_L\,,$$

i.e. the difference between the spatial and the material velocity gradient is negligible in the linear theory. The same holds for all fields being of the same order as the displacement gradient, as

$$\nabla_L = \nabla_E\, \mathbf{F} = \nabla_E\, (\mathbf{I} + \mathbf{H}) \approx \nabla_E\,.$$

Further on

$$\mathbf{D} \approx \tfrac{1}{2}(\mathbf{H} + \mathbf{H}^T)^\bullet = \mathbf{E}^\bullet$$

and

$$\mathbf{W} \approx \tfrac{1}{2}(\mathbf{H} - \mathbf{H}^T)^\bullet.$$

If a deformation results from a *rigid body motion*, then after (2.22) $\mathbf{F} \equiv \mathbf{R}$ is orthogonal and the linear strain tensor becomes

$$\mathbf{E} := \tfrac{1}{2}(\mathbf{H} + \mathbf{H}^T) \equiv \tfrac{1}{2}(\mathbf{R} + \mathbf{R}^T) - \mathbf{I}\,.$$

In fact, let $\mathbf{e}_3 \in \mathscr{V}$ be the axial vector and $\varphi \neq 0$ the angle of rotation of the versor $\mathbf{R}$, then with (1.27) we obtain a non-vanishing

$$\mathbf{E} \equiv (\cos\varphi - 1)\,(\mathbf{I} - \mathbf{e}_3 \otimes \mathbf{e}_3) \qquad \text{with } tr(\mathbf{E}) \equiv 2\,(\cos\varphi - 1)$$

and

$$\mathbf{W} \equiv \sin\varphi\,(\mathbf{e}_2 \otimes \mathbf{e}_1 - \mathbf{e}_1 \otimes \mathbf{e}_2)\,.$$

Only after a linearisation in φ we get

$$\mathbf{E} \approx \mathbf{0} \qquad \text{and} \qquad \mathbf{W} \approx \varphi\,(\mathbf{e}_2 \otimes \mathbf{e}_1 - \mathbf{e}_1 \otimes \mathbf{e}_2)\,.$$

Special Deformations

In the sequel we will consider two important examples of motions, namely the simple shear and the COUETTE-flow.

Example: ***Simple Shear***

An interesting and important motion is the simple shear of a body. We use a Cartesian COOS $\{X, Y, Z\}$ in the reference placement equal to the initial placement, and the same COOS $\{x, y, z\}$ for the current placement. Simple shear in the *x*-*y*-plane in the *x*-direction is defined by

$$x = \chi^1(X, Y, Z, t) = X + \gamma(t)\,Y$$

$$y = \chi^2(X, Y, Z, t) = Y$$

$$z = \chi^3(X, Y, Z, t) = Z\,.$$

The *shear number* $\gamma(t)$ is assumed to be a continuous and differentiable function of time. The inverse functions are then

$$X = \chi^{-1\,1}(x, y, z, t) = x - \gamma(t)\,y$$

$$Y = \chi^{-1\,2}(x, y, z, t) = y$$

$$Z = \chi^{-1\,3}(x, y, z, t) = z\,.$$

The displacement field has only one non-zero component

$$u_x = x - X = \gamma(t)\,Y\,.$$

The displacement gradient is then time-dependent, but constant in space

$$\mathbf{H} = \gamma(t)\,\mathbf{e}_x \otimes \mathbf{e}_y \qquad \in \mathscr{Lin}$$

such as the deformation gradient

(2.26) $$\mathbf{F} = \mathbf{I} + \gamma(t)\,\mathbf{e}_x \otimes \mathbf{e}_y \qquad \in \mathscr{Inv}^+$$

and its inverse

$$\mathbf{F}^{-1} = \mathbf{I} - \gamma(t)\, \mathbf{e}_x \otimes \mathbf{e}_y \qquad \in \mathcal{Inv}^+ .$$

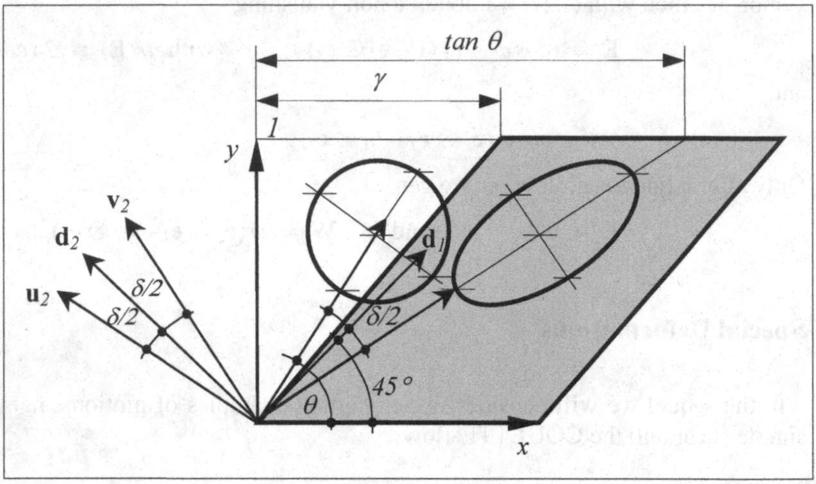

This deformation is homogeneous. As the determinant is $J = 1$, simple shear is also an isochoric motion. We obtain further

$$\mathbf{C} := \mathbf{F}^T \mathbf{F} = \mathbf{I} + \gamma(t)\,(\mathbf{e}_x \otimes \mathbf{e}_y + \mathbf{e}_y \otimes \mathbf{e}_x) + \gamma(t)^2\, \mathbf{e}_y \otimes \mathbf{e}_y \qquad \in \mathcal{Psym}$$

$$\mathbf{B} := \mathbf{F}\,\mathbf{F}^T = \mathbf{I} + \gamma(t)\,(\mathbf{e}_x \otimes \mathbf{e}_y + \mathbf{e}_y \otimes \mathbf{e}_x) + \gamma(t)^2\, \mathbf{e}_x \otimes \mathbf{e}_x \qquad \in \mathcal{Psym}$$

$$\mathbf{E}^G := \tfrac{1}{2}(\mathbf{C} - \mathbf{I}) = \tfrac{1}{2}\gamma(t)\,(\mathbf{e}_x \otimes \mathbf{e}_y + \mathbf{e}_y \otimes \mathbf{e}_x) + \tfrac{1}{2}\gamma(t)^2\, \mathbf{e}_y \otimes \mathbf{e}_y \quad \in \mathcal{Sym}$$

$$\mathbf{E}^a := \tfrac{1}{2}(\mathbf{I} - \mathbf{B}^{-1}) = \tfrac{1}{2}\gamma(t)\,(\mathbf{e}_x \otimes \mathbf{e}_y + \mathbf{e}_y \otimes \mathbf{e}_x) - \tfrac{1}{2}\gamma(t)^2\, \mathbf{e}_y \otimes \mathbf{e}_y \in \mathcal{Sym} .$$

The area placement tensor is

$$\mathbf{E}^F := det(\mathbf{F})\, \mathbf{F}^{-T} = \mathbf{I} - \gamma(t)\, \mathbf{e}_y \otimes \mathbf{e}_x \qquad \in \mathcal{Inv}^+ .$$

For small $|\gamma(t)|$, the current placement is close to the reference placement, and all strain tensors turn into the linear strain tensor

$$\mathbf{E} := \tfrac{1}{2}(\mathbf{H} + \mathbf{H}^T) = \tfrac{1}{2}\gamma(t)\,(\mathbf{e}_x \otimes \mathbf{e}_y + \mathbf{e}_y \otimes \mathbf{e}_x) \qquad \in \mathcal{Sym} .$$

This tensor is traceless or deviatoric, as we expect it for an isochoric deformation.

The eigenvalue problem of $\mathbf{C}$ (and $\mathbf{U}$) can be solved by finding unit direction vectors $\mathbf{u}_i$, for which

$$(\mathbf{F}\,\mathbf{u}_i) \cdot (\mathbf{F}\,\mathbf{u}_j) = \mathbf{u}_i \cdot (\mathbf{C}\,\mathbf{u}_j) = 0 \qquad \text{for } i \neq j .$$

Obviously $\mathbf{u}_3 = \mathbf{e}_z$ is an eigenvector. The others must be mutually orthogonal and, thus, lie in the x-y-plane. We can obtain them by rotating the ONB around $\mathbf{e}_z$ for an angle θ

$$\mathbf{u}_1 = \cos\theta\ \mathbf{e}_x + \sin\theta\ \mathbf{e}_y$$

$$\mathbf{u}_2 = -\sin\theta\ \mathbf{e}_x + \cos\theta\ \mathbf{e}_y\,.$$

The above eigenvector equation then becomes

$$\gamma\,(-\sin^2\theta + \cos^2\theta) + \gamma^2 \sin\theta\ \cos\theta = 0$$

or

$$\gamma = \tan\theta - \cot\theta = -2\cot(2\theta)$$

for the determination of θ. This transcendental equation can be uniquely solved for any given positive γ by one θ between $\pi/4$ and $\pi/2$. The corresponding eigenvalues are the RAYLEIGH quotients

$$\begin{aligned}\lambda_1 &= \mathbf{u}_1 \cdot (\mathbf{C}\,\mathbf{u}_1) / (\mathbf{u}_1 \cdot \mathbf{u}_1)\\ &= \cos^2\theta + 2\gamma\sin\theta\ \cos\theta + (1+\gamma^2)\sin^2\theta\\ &= \cos^2\theta + 2\sin^2\theta - 2\cos^2\theta + \sin^2\theta + (\tan^2\theta - 2 + \cot^2\theta)\sin^2\theta\\ &= (1+\tan^2\theta)\sin^2\theta = (\sin^2\theta + \cos^2\theta)\sin^2\theta / \cos^2\theta = \tan^2\theta\,.\end{aligned}$$

For $\mathbf{u}_3$ we can easily guess the eigenvalue 1. By $J = 1 = \lambda_1\,\lambda_2\,\lambda_3$ we conclude

$$\lambda_2 = \cot^2\theta\,.$$

These are the three eigenvalues of $\mathbf{C}$ and $\mathbf{B}$. Those of $\mathbf{U}$ and $\mathbf{V}$ are their square roots

$$\mu_1 = \sqrt{\lambda_1} = \tan\theta \qquad \mu_2 = \cot\theta \qquad \mu_3 = 1\,.$$

Simple shear is generally characterised by the eigenvalues of $\mathbf{U}$ and $\mathbf{V}$ $\{\mu\,,\,1\,,\,\mu^{-1}\}$, where $\mu > 0$. By the representation (2.9) we obtain the eigenvectors of $\mathbf{V}$ and $\mathbf{B}$ as

$$\mathbf{v}_i = \mu_i^{-1}\,\mathbf{F}\,\mathbf{u}_i$$

such as

$$\begin{aligned}\mathbf{v}_1 &= \mu_1^{-1}\,\mathbf{F}\,\mathbf{u}_1 = \tan^{-1}\theta\,\{(\cos\theta + \gamma\sin\theta)\,\mathbf{e}_x + \sin\theta\ \mathbf{e}_y\}\\ &= (\cos^2\theta / \sin\theta)\,\mathbf{e}_x + \gamma\,\cos\theta\,\mathbf{e}_x + \cos\theta\,\mathbf{e}_y\\ &= \sin\theta\ \mathbf{e}_x + \cos\theta\,\mathbf{e}_y\end{aligned}$$

and perpendicularly

$$\mathbf{v}_2 = -\cos\theta\ \mathbf{e}_x + \sin\theta\,\mathbf{e}_y$$

and

$$\mathbf{v}_3 = \mathbf{e}_z\,.$$

For the geometrical interpretation we introduce the angle

$$\delta := 2\theta - \pi/2\,.$$

By means of trigonometric relations we obtain

$$\gamma = \tan\theta - \cot\theta = -2\cot 2\theta = -2\tan(\pi/2 - 2\theta)$$
$$= 2\tan(2\theta - \pi/2) = 2\tan\delta .$$

The angle of the rotation **R** is determined by the scalar product

$$\mathbf{v}_1 \cdot \mathbf{u}_1 = 2\sin\theta\cos\theta = \sin 2\theta = \sin(\pi/2 + \delta) = \cos\delta ,$$

as well as δ. For small deformations this angle also becomes small. If, on the contrary, the shear number becomes very large, θ and δ converge to $\pi/2$, $\mathbf{u}_1$ to $\mathbf{e}_y$, and $\mathbf{v}_1$ to $\mathbf{e}_x$.

If we draw a unit circle in the reference placement on the body in the *x-y*-plane, then under the shear it becomes an ellipse with main axes μ_1 $\mathbf{v}_1$ and $\mu_2\,\mathbf{v}_2$.

The velocity field of a simple shear is in the LAGRANGEan description

$$\mathbf{v}_L(\mathbf{x}_0, t) = \gamma(t)^\bullet\, Y\, \mathbf{e}_x$$

and in the EULERean

$$\mathbf{v}_E(\mathbf{x}, t) = \gamma(t)^\bullet\, y\, \mathbf{e}_x .$$

The velocity gradient is then constant in space

(2.27) $$\mathbf{L}(t) = \gamma(t)^\bullet\, \mathbf{e}_x \otimes \mathbf{e}_y$$

with the stretching tensor

$$\mathbf{D}(t) = \tfrac{1}{2}\,\gamma(t)^\bullet\,(\mathbf{e}_x \otimes \mathbf{e}_y + \mathbf{e}_y \otimes \mathbf{e}_x)$$

and the spin tensor

$$\mathbf{W}(t) = \tfrac{1}{2}\,\gamma(t)^\bullet\,(\mathbf{e}_x \otimes \mathbf{e}_y - \mathbf{e}_y \otimes \mathbf{e}_x)$$

and the corresponding spin vector

$$\mathbf{w}(t) = -\tfrac{1}{2}\,\gamma(t)^\bullet\, \mathbf{e}_z .$$

The eigenvectors of the stretching tensors coincide with those material line elements which are currently strained at extreme rates

$$\mathbf{d}_1 := \frac{1}{\sqrt{2}}(\mathbf{e}_x + \mathbf{e}_y) \qquad \text{and} \qquad \mathbf{d}_2 := \frac{1}{\sqrt{2}}(\mathbf{e}_x - \mathbf{e}_y) .$$

The third eigenvector coincides with the *z*-axis.

Independent of the COOS used, for simple shear the following fact is characteristic. At each instant there are two orthogonal vectors **a** and **b** with

$$\mathbf{L} = \mathbf{a} \otimes \mathbf{b} \qquad \in \mathcal{Lin}$$

and

$$\mathbf{D} = sym(\mathbf{a} \otimes \mathbf{b}) \qquad \in \mathcal{Sym}$$

and

$$\mathbf{w} = -\tfrac{1}{2}\,\mathbf{a} \times \mathbf{b} \qquad \in \mathcal{V}.$$

Example. The ***COUETTE*** [48]***-flow*** is given by the coordinate transformation

$$r = \chi^1(R\,, \Theta, Z\,, t) = R$$

$$\vartheta = \chi^2(R\,, \Theta, Z\,, t) = \Theta + \alpha(R\,, t)$$

$$z = \chi^3(R\,, \Theta, Z\,, t) = Z$$

with a differentiable function of radius and time $\alpha(R\,,t)$ with initial value $\alpha(R\,,0) = 0$ with respect to a cylindrical COOS $\{R\,, \Theta, Z\}$ for the initial or reference placement, and another cylindrical COOS $\{r\,, \vartheta\,, z\}$ for the current placement. The position vector in the current placement is

$$\mathbf{r} = r\,\mathbf{e}_r(\vartheta) + z\,\mathbf{e}_z$$

and in the reference placement

$$\mathbf{r}_0 = R\,\mathbf{e}_r(\Theta) + Z\,\mathbf{e}_z$$

with respect to a local ONB $\{\mathbf{e}_r(\vartheta)\,, \mathbf{e}_\theta(\vartheta)\,, \mathbf{e}_z\}$. The tangent bases to these coordinates are

$$\mathbf{r}_{\varphi 1} = \mathbf{e}_r(\vartheta) \qquad \mathbf{r}_{\varphi 2} = r\,\mathbf{e}_\theta(\vartheta) \qquad \mathbf{r}_{\varphi 3} = \mathbf{e}_z$$

$$\mathbf{r}_{\Psi 1} = \mathbf{e}_r(\Theta) \qquad \mathbf{r}_{\Psi 2} = R\,\mathbf{e}_\theta(\Theta) \qquad \mathbf{r}_{\Psi 3} = \mathbf{e}_z$$

and the gradient bases

$$\mathbf{r}_\varphi^{\ 1} = \mathbf{e}_r(\vartheta) \qquad \mathbf{r}_\varphi^{\ 2} = r^{-1}\,\mathbf{e}_\theta(\vartheta) \qquad \mathbf{r}_{\varphi 3} = \mathbf{e}_z$$

$$\mathbf{r}_\Psi^{\ 1} = \mathbf{e}_r(\Theta) \qquad \mathbf{r}_\Psi^{\ 2} = R^{-1}\,\mathbf{e}_\theta(\Theta) \qquad \mathbf{r}_\Psi^{\ 3} = \mathbf{e}_z\,.$$

One can express one basis by the other

$$\mathbf{r}_{\Psi i} = (\mathbf{r}_{\Psi i} \cdot \mathbf{r}_\varphi^{\ k})\,\mathbf{r}_{\varphi k}\,,$$

i.e.,

$$\mathbf{r}_{\Psi 1} = \cos\alpha\ \mathbf{r}_{\varphi 1} - \sin\alpha\ \mathbf{r}_{\varphi 2}$$

$$\mathbf{r}_{\Psi 2} = \sin\alpha\ \mathbf{r}_{\varphi 1} + \cos\alpha\ \mathbf{r}_{\varphi 2}$$

$$\mathbf{r}_{\Psi 3} = \mathbf{r}_{\varphi 3}\,.$$

---

[48] Maurice Couette (1858-1943)

The velocity field in EULERean representation is

$$\mathbf{v}(r, \vartheta) = \mathbf{r}^\bullet = r\,\alpha^\bullet\,\mathbf{e}_\theta(\vartheta)$$

and the acceleration field

$$\mathbf{a}(r, \vartheta) = -r\,\alpha^{\bullet 2}\,\mathbf{e}_r(\vartheta) + r\,\alpha^{\bullet\bullet}\,\mathbf{e}_\theta(\vartheta)\,.$$

The deformation gradient is after (2.5)

$$\mathbf{F} = \frac{\partial \varphi^k}{\partial \Psi^i}\,\mathbf{r}_{\varphi\,k} \otimes \mathbf{r}_\Psi{}^i = \mathbf{r}_{\varphi\,k} \otimes \mathbf{r}_\Psi{}^k + \alpha'\,\mathbf{r}_{\varphi\,2} \otimes \mathbf{r}_\Psi{}^1$$

with $\alpha(R\,,\,t)' := \dfrac{\partial \alpha}{\partial R}$. This displacement is homogeneous with respect to z. The surfaces of a cylinder revolve around the z-axis. The determinant is $J=1$, hence this motion is isochoric. The displacement gradient is

$$\begin{aligned}\mathbf{H} = \mathbf{F} - \mathbf{I} &= \mathbf{r}_{\varphi\,k} \otimes \mathbf{r}_\Psi{}^k + \alpha'\,\mathbf{r}_{\varphi\,2} \otimes \mathbf{r}_\Psi{}^1 - \mathbf{r}_{\Psi\,k} \otimes \mathbf{r}_\Psi{}^k \\ &= (\mathbf{r}_{\varphi\,k} - \mathbf{r}_{\Psi\,k}) \otimes \mathbf{r}_\Psi{}^k + \alpha'\,\mathbf{r}_{\varphi\,2} \otimes \mathbf{r}_\Psi{}^1\end{aligned}$$

with the components

$$\begin{bmatrix} \cos\alpha - 1 - \alpha' R \sin\alpha & -\sin\alpha & 0 \\ \sin\alpha + \alpha' R \cos\alpha & \cos\alpha - 1 & 0 \\ 0 & 0 & 0 \end{bmatrix}$$

with respect to the tensorial basis $\{\mathbf{e}_i\,(\Theta) \otimes \mathbf{e}_j\,(\Theta)\}$. With respect to the same basis, $\mathbf{C}$ has the components

$$\begin{bmatrix} 1 + \alpha'^2 R^2 & \alpha' R & 0 \\ \alpha' R & 1 & 0 \\ 0 & 0 & 1 \end{bmatrix}$$

and $\mathbf{E}^G$

$$\tfrac{1}{2}\begin{bmatrix} \alpha'^2 R^2 & \alpha' R & 0 \\ \alpha' R & 0 & 0 \\ 0 & 0 & 0 \end{bmatrix}.$$

By the identification $\gamma(t) \equiv R\,\alpha'$, this is the same form as for simple shear. Thus, the COUETTE-flow is locally like a simple shear.

With respect to $\{\mathbf{e}_i\,(\vartheta) \otimes \mathbf{e}_j\,(\vartheta)\}$, $\mathbf{B}$ has the components

$$\begin{bmatrix} 1 & \alpha' r & 0 \\ \alpha' r & 1 + \alpha'^2 r^2 & 0 \\ 0 & 0 & 1 \end{bmatrix}$$

and $\mathbf{E}^a$

$$^1/_2 \begin{bmatrix} -\alpha'^2 r^2 & \alpha' r & 0 \\ \alpha' r & 0 & 0 \\ 0 & 0 & 0 \end{bmatrix} .$$

For arbitrary deformation rates the following decomposition theorem holds:

Theorem 2.2. *Each velocity gradient* $\mathbf{L} \in \mathcal{L}in$ *can be additively decomposed as*

$$\mathbf{L} = \mathbf{W} + \alpha\,\mathbf{I} + sym(\mathbf{a}_1 \otimes \mathbf{b}_1) + sym(\mathbf{a}_2 \otimes \mathbf{b}_2)$$

into

- *a spin* $\mathbf{W} \in \mathcal{S}kew$
- *a volumetric rate* $\alpha\,\mathbf{I}$ with $\alpha \in \mathcal{R}$
- *and two shear rates* $\mathbf{a}_i \otimes \mathbf{b}_i$ *with* $\mathbf{a}_i \cdot \mathbf{b}_i = 0$, $\mathbf{a}_i, \mathbf{b}_i \in \mathcal{V}$, $i = 1, 2$.

Proof. First of all, we can decompose each tensor into its symmetric part (**D**) and its skew part (**W**). The symmetric part can be further decomposed into its spherical part $\alpha\,\mathbf{I}$ and its deviatoric one $\mathbf{D}'$. Let the spectral form of the latter be

$$\mathbf{D}' = \sum_{r=1}^{3} d_r\, \mathbf{d}_r \otimes \mathbf{d}_r$$

with an arbitrary numbering of the eigenvalues. Because of its tracelessness we get

$$\begin{aligned} \mathbf{D}' &= d_1\, \mathbf{d}_1 \otimes \mathbf{d}_1 + d_2\, \mathbf{d}_2 \otimes \mathbf{d}_2 - (d_1 + d_2)\, \mathbf{d}_3 \otimes \mathbf{d}_3 \\ &= d_1\, (\mathbf{d}_1 \otimes \mathbf{d}_1 - \mathbf{d}_3 \otimes \mathbf{d}_3) + d_2\, (\mathbf{d}_2 \otimes \mathbf{d}_2 - \mathbf{d}_3 \otimes \mathbf{d}_3) \\ &= d_1\, sym\{(\mathbf{d}_1 + \mathbf{d}_3) \otimes (\mathbf{d}_1 - \mathbf{d}_3)\} \\ &\quad + d_2\, sym\{(\mathbf{d}_2 + \mathbf{d}_3) \otimes (\mathbf{d}_2 - \mathbf{d}_3)\} . \end{aligned}$$

These are the velocity gradients of two shears; *q.e.d.*

The same decomposition can also be applied to the linear strain tensor. Thus, in linear theory, the strains can be always decomposed additively into a dilatation and two shears.

For large deformations this does not hold in this form. By further specifying the polar decomposition from Theorem 1.21., however, we find the shears also here.

Theorem 2.3 Decomposition of the deformation gradient
Each invertible tensor $\mathbf{F} \in \mathscr{Inv}^+$ *can be decomposed in two ways*

$$\mathbf{F} = \mathbf{R}\,\mathbf{U}_2\,\mathbf{U}_1\,\mathbf{U}_0 = \mathbf{V}_2\,\mathbf{V}_1\,\mathbf{V}_0\,\mathbf{R}$$

with $\mathbf{R} \in \mathscr{Orth}^+$ *: rotation (orthogonal)*

$\mathbf{U}_0 = \mathbf{V}_0 \in \mathscr{Psym}$ *: dilatation (spherical tensor)*

$\mathbf{U}_1, \mathbf{U}_2, \mathbf{V}_1, \mathbf{V}_2 \in \mathscr{Psym}$ *: stretches of simple shears*

Proof. We already saw by the polar decomposition that the tensors $\mathbf{V}$, $\mathbf{U}$ are positive-definite and symmetric and possess the same eigenvalues $\mu_r > 0$. The volumetric parts are

$$\mathbf{U}_0 = \mathbf{V}_0 := J^{1/3}\,\mathbf{I} \qquad \text{with } J = \mu_1\,\mu_2\,\mu_3$$

and the unimodular ones

$$\overline{\mathbf{U}} := J^{-1/3}\,\mathbf{U} \in \mathscr{Unim}^+ \qquad \overline{\mathbf{V}} := J^{-1/3}\,\mathbf{V} \in \mathscr{Unim}^+$$

describe the *distortional* parts of

$$\mathbf{U} = \overline{\mathbf{U}}\,\mathbf{U}_0 = \mathbf{U}_0\,\overline{\mathbf{U}} \qquad \text{and of} \qquad \mathbf{V} = \overline{\mathbf{V}}\,\mathbf{V}_0 = \mathbf{V}_0\,\overline{\mathbf{V}}\,.$$

$\overline{\mathbf{U}}$ and $\overline{\mathbf{V}}$ have the eigenvalues $\underline{\mu}_r := J^{-1/3}\,\mu_r > 0$. Let the eigenvectors of $\mathbf{U}$ and $\overline{\mathbf{U}}$ be $\mathbf{u}_r$ and those of $\mathbf{V}$ and $\overline{\mathbf{V}}$ be $\mathbf{v}_r$. We pose

$$\mathbf{U}_1 = \underline{\mu}_1\,\mathbf{u}_1 \otimes \mathbf{u}_1 + \mathbf{u}_2 \otimes \mathbf{u}_2 + \underline{\mu}_1^{-1}\,\mathbf{u}_3 \otimes \mathbf{u}_3$$

$$\mathbf{U}_2 = \mathbf{u}_1 \otimes \mathbf{u}_1 + \underline{\mu}_2\,\mathbf{u}_2 \otimes \mathbf{u}_2 + \underline{\mu}_2^{-1}\,\mathbf{u}_3 \otimes \mathbf{u}_3$$

$$\mathbf{V}_1 = \underline{\mu}_1\,\mathbf{v}_1 \otimes \mathbf{v}_1 + \mathbf{v}_2 \otimes \mathbf{v}_2 + \underline{\mu}_1^{-1}\,\mathbf{v}_3 \otimes \mathbf{v}_3$$

$$\mathbf{V}_2 = \mathbf{v}_1 \otimes \mathbf{v}_1 + \underline{\mu}_2\,\mathbf{v}_2 \otimes \mathbf{v}_2 + \underline{\mu}_2^{-1}\,\mathbf{v}_3 \otimes \mathbf{v}_3\,.$$

These are the stretches of simple shears. One can easily calculate the decompositions of the theorem by them; *q.e.d.*

As the three parts of $\mathbf{U}$ and of $\mathbf{V}$ are coaxial, they all commute. Thus, the order of these deformations is arbitrary.

The relation between the decompositions of the last theorems is obtained by the following theorem, which can be easily proved.

Theorem 2.4. *If all parts of the previous decomposition are differentiable functions in time, then*

- $\mathbf{R}^\bullet\,\mathbf{R}^T$ *is skew*
- $\mathbf{U}_0^\bullet\,\mathbf{U}_0^{-1} = \mathbf{V}_0^\bullet\,\mathbf{V}_0^{-1} = J^\bullet/(3J)\,\mathbf{I}$ *is spherical*
- $\mathbf{U}_i^\bullet\,\mathbf{U}_i^{-1}$ and $\mathbf{V}_i^\bullet\,\mathbf{V}_i^{-1}$, $i = 1, 2$, *are deviatoric (traceless)*

Note that the latter need not be symmetric. **Problem.** (decomposition of $\mathbf{C}$ in distortional and dilatorischen Anteil)

3 Balance Laws

3.1 Mass

We assign to each body (as well as to its subbodies) a measure $m(\mathscr{B})$, *i.e.*, a non-negative real number, the **mass** of $\mathscr{B}$. In classical, non-relativistic mechanics we assume the **axiom of conservation of mass**: *The mass of the body remains constant under all deformations.*

The mass is assumed to be additive. If $\mathscr{B}_1$ and $\mathscr{B}_2$ are two disjoint bodies, then

$$m(\mathscr{B}_1) + m(\mathscr{B}_2) = m(\mathscr{B}_1 \cup \mathscr{B}_2) .$$

If the mass is sufficiently smooth, then we conclude the existence of a mass density by the theorem of RADON-NIKODYM from measure theory.

Definition 3.1. Let κ be a placement of the body, then its **mass density** in this placement κ is a scalar field $\rho = \rho_E(\mathbf{x}, t)$ on $\mathscr{B}_t$ such that

$$m(\mathscr{B}) = \int_{\mathscr{B}_t} \rho_E(\mathbf{x}, t)\, dv$$

After the axiom of conservation of mass, the mass does not depend on the placement. In the reference placement κ_0 we thus have also

$$m(\mathscr{B}) = \int_{\mathscr{B}_0} \rho_{0L}(\mathbf{x}_0)\, dv_0$$

with the **mass density** ρ_0 in the reference placement, being constant in time. Note that both fields ρ and ρ_0 can be considered in the LAGRANGEan or in the EULERean description. The values of ρ_E and ρ_{0E} at the same material point are, in general, different, the same as those of ρ_L and ρ_{0L} .

The **element of mass** in the current placement is

$$dm := \rho\, dv$$

and in the reference placement

$$dm_0 := \rho_0\, dv_0 .$$

After the axiom of conservation of mass, they are equal

$$dm_0 = dm$$

so that

A. Bertram, *Elasticity and Plasticity of Large Deformations*,
https://doi.org/10.1007/978-3-030-72328-6_3

$$\rho\, dv = \rho_0\, dv_0$$

holds. Thus we can suppress the suffix zero at m and dm .

For a sequence of volumes converging to zero, we obtain the limit in the current placement

$$\rho = \frac{dm}{dv} = \lim_{\Delta v \to 0} \frac{\Delta m}{\Delta v}$$

and in the reference placement

$$\rho_0 = \frac{dm}{dv_0} = \lim_{\Delta v_0 \to 0} \frac{\Delta m}{\Delta v_0} .$$

By the transformation of the volume elements (2.17) we obtain

(3.1)
$$\frac{\rho_0}{\rho} = J = det(\mathbf{F}) .$$

The derivative of the ratio of mass densities with respect to time is after (2.20)

$$(\frac{\rho_0}{\rho})^{\bullet} = -\frac{\rho_0}{\rho^2}\rho^{\bullet} = -\frac{\rho^{\bullet}}{\rho} J$$

$$= J^{\bullet} = J\, tr(\mathbf{L}) = J\, tr(\mathbf{D}) = J\, div\, \mathbf{v}_E$$

and the incremental form of the local balance of mass becomes

$$\rho_E^{\bullet} + \rho_E\, div\, \mathbf{v}_E = 0$$

or with the substantial derivative of the mass density

$$\rho_E^{\bullet} = \frac{\partial \rho_E}{\partial t} + (grad\, \rho_E) \cdot \mathbf{v}_E$$

this can be turned into

(3.2)
$$\frac{\partial \rho_E}{\partial t} + div(\rho_E\, \mathbf{v}_E) = 0$$

by use of the product rule.

3.2 General Balance Equation

For the balance equations, surface and volume integrals are needed, which can be transformed between the current and the reference placement. These describe quantities like the mass, which have global values, but also local densities. While the mass is a scalar quantity, we will now consider quantities, which are tensors of arbitrary order.

Let $\Gamma(\mathscr{B}, t)$ be a global physical quantity as a tensor of Kth-order, having a density (per unit volume) ϕ as a tensor field of the same order in the current placement, and a density ϕ_0 in the reference placement

$$\Gamma(\mathscr{B}, t) = \int_{\mathscr{B}_t} \phi_E \, dv = \int_{\mathscr{B}_0} \phi_L \, J \, dv_0 = \int_{\mathscr{B}_0} \phi_{0L} \, dv_0 \, .$$

If not otherwise stated, we assume that both densities are smooth functions of time and space. A comparison gives the transformation of these densities

$$\phi_0 = J \phi$$

equal to that of the mass density (3.1). Often we will need the time derivatives of such integral quantities. By transforming them onto the (time-independent) reference placement, one can differentiate under the integral, and afterwards transform it back again, if the result is preferred in the EULERean description. We obtain

$$\Gamma(\mathscr{B}, t)^\bullet := \frac{d}{dt} \Gamma(\mathscr{B}, t)$$

$$= \frac{d}{dt} \int_{\mathscr{B}_0} \phi_{0L} \, dv_0 = \int_{\mathscr{B}_0} \phi_{0L}{}^\bullet \, dv_0 = \int_{\mathscr{B}_0} (\phi_L{}^\bullet J + \phi_L J^\bullet) \, dv_0$$

$$= \frac{d}{dt} \int_{\mathscr{B}_t} \phi_E \, dv = \int_{\mathscr{B}_t} (\phi_E{}^\bullet \, dv + \phi_E \, dv^\bullet)$$

and by (2.2)

$$= \int_{\mathscr{B}_t} \left(\frac{\partial \phi_E}{\partial t} + grad(\phi_E) \, \mathbf{v}_E + \phi_E \, div \mathbf{v}_E\right) dv$$

$$= \int_{\mathscr{B}_t} \left(\frac{\partial \phi_E}{\partial t} + div(\phi_E \otimes \mathbf{v}_E)\right) dv$$

and by means of the GAUSS transformation

$$= \int_{\mathscr{B}_t} \frac{\partial \phi_E}{\partial t} \, dv + \int_{\partial \mathscr{B}_t} \phi_E \, (\mathbf{v}_E \cdot \mathbf{n}) \, da$$

and thus **REYNOLDS'**[49] **transport equation** in integral form

(3.3)
$$\frac{d}{dt} \int_{\mathscr{B}_t} \phi_E \, dv = \int_{\mathscr{B}_t} \frac{\partial \phi_E}{\partial t} \, dv + \int_{\partial \mathscr{B}_t} \phi_E \, (\mathbf{v}_E \cdot \mathbf{n}) \, da$$

[49] Osborne Reynolds (1842-1912)

If the domain of the integral is not fixed to the body, *i.e.*, material, but instead an arbitrarily moving control volume $\mathscr{D}_t$, then we obtain for the rate of the same quantity

$$\frac{d}{dt}\int_{\mathscr{D}_t} \phi_E \, dv = \int_{\mathscr{D}_t} \frac{\partial \phi_E}{\partial t} \, dv + \int_{\partial \mathscr{D}_t} \phi_E \, (\underline{\mathbf{v}}_E \cdot \mathbf{n}) \, da$$

where $\underline{\mathbf{v}}_E$ is the velocity of the control volume in space[50].

After dividing ϕ by the corresponding mass density, we obtain the **specific intensity** of this quantity in the current placement

$$\gamma := \frac{\phi}{\rho} = \frac{d\Gamma}{dm}$$

and analogously in the reference placement

$$\gamma_0 := \frac{\phi_0}{\rho_0} = \frac{J\phi}{J\rho} = \gamma$$

which turn out to be equal as a result of conservation of mass. With this, the global form

$$\Gamma(\mathscr{B}, t) = \int_{\mathscr{B}_t} \gamma_E \, dm = \int_{\mathscr{B}_0} \gamma_L \, dm$$

also holds. For the time derivative the specific quantities are advantageous, since

$$\Gamma(\mathscr{B}, t)^\bullet := \frac{d}{dt}\Gamma(\mathscr{B}, t)$$

$$= \Big(\int_{\mathscr{B}_0} \gamma_L \, dm\Big)^\bullet = \int_{\mathscr{B}_0} \gamma_L^\bullet \, dm$$

$$= \Big(\int_{\mathscr{B}_t} \gamma_E \, dm\Big)^\bullet = \int_{\mathscr{B}_t} \gamma_E^\bullet \, dm = \int_{\mathscr{B}_t} \Big(\frac{\partial \gamma_E}{\partial t} + grad(\gamma_E) \cdot \mathbf{v}_E\Big) \, dm \, .$$

We see that in spite of the time-dependence of the domain of the integral $\mathscr{B}_t$ in the EULERean description, we can differentiate under the integral, if we use specific intensities.

Since the above equations hold not only for the body itself, but also for all subbodies, the local form of the transport equation is again (2.2)

(3.4)
$$\gamma_E^\bullet = \frac{\partial \gamma_E}{\partial t} + grad(\gamma_E) \cdot \mathbf{v}_E \, .$$

50 For a proof of this formula see LIU (2002), p. 32.

Example: Continuity

Conservation of mass can be expressed by (2.20) as

$$m(\mathcal{B})^{\bullet} = 0 = \int_{\mathcal{B}_t} (\rho_E^{\bullet}\, dv + \rho_E\, dv^{\bullet}) = \int_{\mathcal{B}_t} (\rho_E^{\bullet} + \rho_E\, div\, \mathbf{v})\, dv$$

$$= \int_{\mathcal{B}_t} [\frac{\partial \rho_E}{\partial t} + grad(\rho_E) \cdot \mathbf{v}_E + \rho_E\, div\, \mathbf{v}_E]\, dv$$

$$= \int_{\mathcal{B}_t} [\frac{\partial \rho_E}{\partial t} + div\, (\rho_E\, \mathbf{v}_E)]\, dv$$

$$= \int_{\mathcal{B}_t} \frac{\partial \rho_E}{\partial t}\, dv + \int_{\partial \mathcal{B}_t} \rho_E\, \mathbf{v}_E \cdot \mathbf{n}\, da$$

or locally by the continuity equation

$$\rho_E^{\bullet} + \rho_E\, div\, \mathbf{v} = 0\,.$$

As **examples,** we consider the **linear momentum** of the body by the identification $\gamma \equiv \mathbf{v}$

(3.5)
$$\mathbf{p}(\mathcal{B}, t) := \int_{\mathcal{B}_t} \mathbf{v}\, dm = \int_{\mathcal{B}_t} \mathbf{r}^{\bullet}\, dm$$

and the **moment of momentum** of the body with respect to a point of reference $O \in \mathcal{E}$ by $\gamma \equiv \mathbf{r} \times \mathbf{v}$

(3.6)
$$\mathbf{d}_O(\mathcal{B}, t) := \int_{\mathcal{B}_t} \mathbf{r} \times \mathbf{r}^{\bullet}\, dm = \int_{\mathcal{B}_t} \mathbf{r} \times \mathbf{v}\, dm$$

with the position vector $\mathbf{r}$ with respect to O. The rates of these global quantities are

(3.7)
$$\mathbf{p}(\mathcal{B}, t)^{\bullet} = (\int_{\mathcal{B}_t} \mathbf{v}\, dm)^{\bullet} = \int_{\mathcal{B}_t} \mathbf{v}^{\bullet}\, dm = \int_{\mathcal{B}_t} \mathbf{a}\, dm$$

and

(3.8)
$$\mathbf{d}_O(\mathcal{B}, t)^{\bullet} = (\int_{\mathcal{B}_t} \mathbf{r} \times \mathbf{r}^{\bullet}\, dm)^{\bullet} = \int_{\mathcal{B}_t} (\mathbf{r}^{\bullet} \times \mathbf{r}^{\bullet} + \mathbf{r} \times \mathbf{r}^{\bullet\bullet})\, dm$$

$$= \int_{\mathcal{B}_t} \mathbf{r} \times \mathbf{a}\, dm$$

with the acceleration field $\mathbf{a} := \mathbf{r}^{\bullet\bullet}$. The **centre of mass** of the body is defined by the position vector

$$\mathbf{r}_M := \frac{1}{m} \int_{\mathscr{B}_t} \mathbf{r}\, dm$$

being a fixed material point only for rigid bodies. Its velocity and acceleration are in general

$$\mathbf{v}_M := \mathbf{r}_M^{\bullet} = \frac{1}{m} \int_{\mathscr{B}_t} \mathbf{v}\, dm \qquad \mathbf{a}_M := \mathbf{r}_M^{\bullet\bullet} = \frac{1}{m} \int_{\mathscr{B}_t} \mathbf{a}\, dm$$

which are neither those of material points. With this we obtain for the linear momentum

$$\mathbf{p}(\mathscr{B}, t) = \int_{\mathscr{B}_t} \mathbf{r}^{\bullet}\, dm = m\, \mathbf{r}_M^{\bullet} = m\, \mathbf{v}_M$$

and for its time derivative

$$\mathbf{p}(\mathscr{B}, t)^{\bullet} = m\, \mathbf{v}_M(t)^{\bullet} = m\, \mathbf{a}_M(t)\,.$$

Besides such volumetric quantities, we will often need quantities distributed over the surface of the body, which are fluxes into the body or out of the body through its boundary.

Let $\mathbf{q}$ be such a flux as a tensor field of arbitrary order (here denoted as a vector field). Its projection $\mathbf{q} \cdot \mathbf{n}$ in the normal direction of the surface of the body is the border crossing part of it. The net export of this quantity of the body is then

$$\int_{\partial\mathscr{B}_t} \mathbf{q} \cdot \mathbf{da} = \int_{\partial\mathscr{B}_t} \mathbf{q} \cdot \mathbf{n}\, da\,.$$

This can be pulled back to the reference placement by NANSON's formula (2.18)

$$\int_{\partial\mathscr{B}_t} \mathbf{q} \cdot \mathbf{n}\, da = \int_{\partial\mathscr{B}_0} \mathbf{q} \cdot \mathbf{E}^{F}\, \mathbf{da}_0$$

$$=: \int_{\partial\mathscr{B}_0} \mathbf{q}_0 \cdot \mathbf{da}_0 = \int_{\partial\mathscr{B}_0} \mathbf{q}_0 \cdot \mathbf{n}_0\, da_0\,.$$

Thus, the transformation of surface densities is generally

$$\mathbf{q}_0 = \mathbf{q}\, \mathbf{E}^{F}$$

(simple contraction) or for a vectorial flux in particular

(3.9) $$\mathbf{q}_0 = \mathbf{E}^{F\,T} \mathbf{q} = det(\mathbf{F})\, \mathbf{F}^{-1} \mathbf{q}\,.$$

Using the GAUSS transformation we can transform the surface densities into volumetric ones

$$\int_{\partial\mathcal{B}_t} \mathbf{q}\cdot\mathbf{n}\,da = \int_{\mathcal{B}_t} div\,\mathbf{q}\,dv = \int_{\mathcal{B}_0} div(\mathbf{q})\,J\,dv_0$$

$$= \int_{\partial\mathcal{B}_0} \mathbf{q}_0\cdot\mathbf{n}_0\,da_0 = \int_{\mathcal{B}_0} Div\,\mathbf{q}_0\,dv_0$$

and obtain the local transformation

$$J\,div\,\mathbf{q} = Div\,\mathbf{q}_0$$

called the **PIOLA identity**.

The general form of the **global balance equation** is assumed as

(3.10)
$$\Gamma(\mathcal{B},t)^\bullet = \int_{\mathcal{B}_t} r\,dm - \int_{\partial\mathcal{B}_t} \mathbf{q}\cdot\mathbf{n}\,da$$

(simple contraction in the last integral) after which the rate of a global quantity Γ is fed by

- a specific source or production part per unit of mass r, a tensor field of *K*th order, and
- a flux per unit area $\mathbf{q}$, a tensor field of $(K+1)$th-order (here notated as a vector field), the projection of which, in the direction of the outer normal to the surface $\mathbf{n}$ is the outflow of Γ through the surface.

The latter part can be transformed into a volumetric integral by GAUSS′ theorem, which gives the balance equation in the volumetric form

$$\int_{\mathcal{B}_t} \gamma^\bullet\,dm = \int_{\mathcal{B}_t} \left(r - \frac{1}{\rho}\,div\,\mathbf{q}\right) dm\,.$$

If it holds for the body and all its subbodies, then we obtain for sufficiently smooth fields the **local balance equation**

(3.11)
$$\rho\,\gamma^\bullet = \rho r - div\,\mathbf{q}$$

wherein the fields have to be in the EULERean description. By using the PIOLA identity, we analogously obtain in the reference placement

$$\rho_0\,\gamma^\bullet = \rho_0\,r - Div\,\mathbf{q}_0$$

with the fields in the LAGRANGEan description.

Jump Balances

In some problems the considered fields are not smooth, but only piecewise continuous. This happens if singular fronts travel through the body. These can be considered as jumps of the field variables at moving singular interfaces. We now assume that such a front $\mathscr{S}_t$ is a smooth interface at each instant, and that the fields in the body have two-sided limits at this interface. Let ϕ be such a field (we will write it as if it were a scalar field, although all results also apply for tensor fields of arbitrary order). We divide the domain $\mathscr{B}_t$ into two subdomains $\mathscr{B}_t^+$ and $\mathscr{B}_t^-$ on both sides of $\mathscr{S}_t$ such that

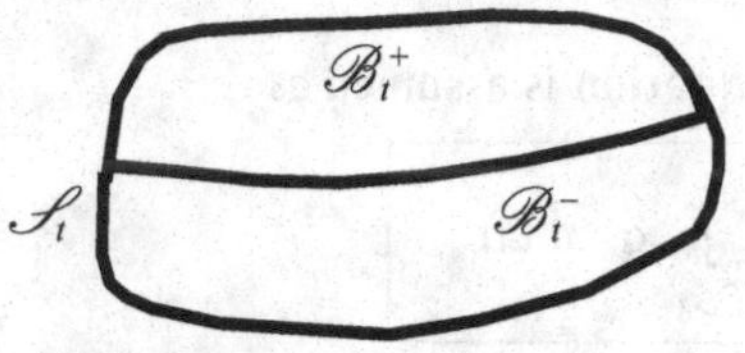

Singular interface in a body

$$\mathscr{B}_t^+ \cup \mathscr{B}_t^- = \mathscr{B}_t$$

$$\mathscr{B}_t^+ \cap \mathscr{B}_t^- = \mathscr{S}_t$$

and indicate the values of ϕ on the positive side of $\mathscr{S}_t$ by $+$ and on the negative side by $-$. On $\mathscr{S}_t$ itself, ϕ obtains the limits ϕ^+ and ϕ^-, and the ϕ-jump is defined as

$$[\phi] := \phi^+ - \phi^- .$$

For convenience, we describe $\mathscr{S}_t$ by a function

$$\varphi : \mathscr{E} \times \mathscr{I} \to \mathscr{R} \qquad | \quad (\mathbf{x}, t) \mapsto \varphi(\mathbf{x}, t)$$

the roots in $\mathscr{B}_t$ of which form at all times $\mathscr{S}_t$

$$\varphi(\mathbf{x}, t) = 0 \quad \Leftrightarrow \quad \mathbf{x} \in \mathscr{S}_t \qquad \forall\, \mathbf{x} \in \mathscr{B}_t$$

and with φ being positive on the positive side of $\mathscr{S}_t$

$$\varphi(\mathbf{x}, t) \geq 0 \quad \Leftrightarrow \quad \mathbf{x} \in \mathscr{B}_t^+$$

and negative on the other side. The current normal to the interface in a point of it is

$$\mathbf{n} := \frac{grad\,\varphi}{|grad\,\varphi|}.$$

We next choose a parameterised path in $\mathscr{E}$

$$\mathbf{r} : \mathscr{I} \rightarrow \mathscr{E} \quad \big| \quad t \mapsto \mathbf{r}(t)$$

which is at each instant on $\mathscr{S}_t$

$$\varphi(\mathbf{r}(t)\,, t) = 0 \qquad \forall\, t \in \mathscr{I}.$$

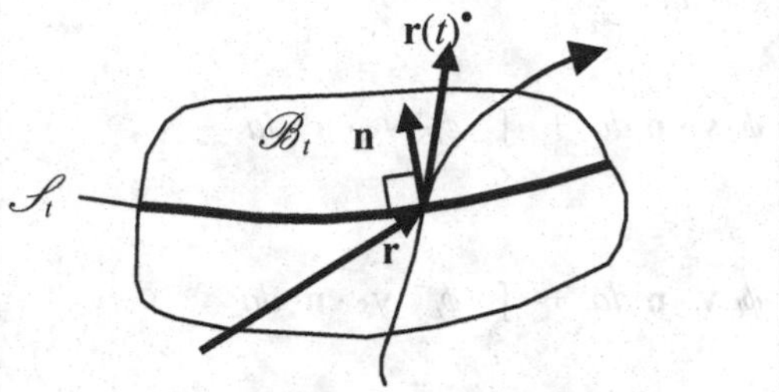

Path through a singular interface

The velocity of this path is the tangent vector $\mathbf{v}_S := \mathbf{r}(t)^\bullet$. By

$$\varphi(\mathbf{r}(t)\,, t)^\bullet = grad(\varphi) \cdot \mathbf{v}_S + \frac{\partial \varphi}{\partial t}$$

$$= |grad(\varphi)|\, v_n + \frac{\partial \varphi}{\partial t} = 0$$

we obtain the normal velocity of the interface

$$v_n := \mathbf{n} \cdot \mathbf{v}_S = -\frac{\partial \varphi}{\partial t}\, |grad(\varphi)|^{-1}.$$

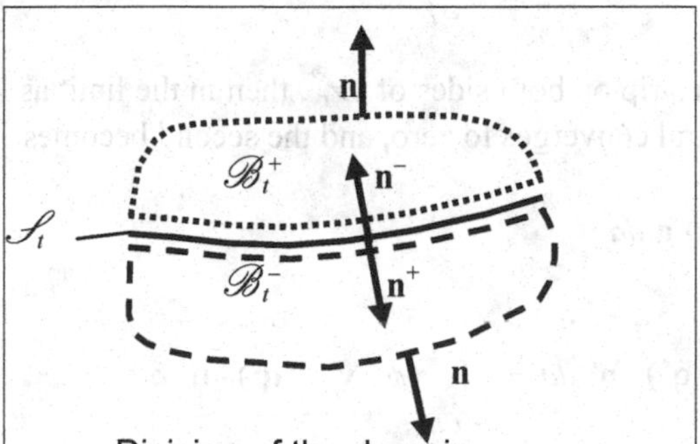

Division of the domain

While all other parts of $\mathbf{v}_S$ depend on the particular choice of the path, this normal velocity is independent of that choice.

The GAUSS transformation in the form (1.53) holds only in domains of continuous and differentiable fields. However, we can generalise it easily to domains with singular interfaces.

On $\mathscr{S}_t$, $\mathbf{n}^+ = -\mathbf{n}$ is the outer normal for $\mathscr{B}_t^+$, and $\mathbf{n}^- = +\mathbf{n}$ the one for $\mathscr{B}_t^-$.

Both subdomains meet the requirements of a regular domain, and for the time derivative of a global physical quantity we have again

$$\Gamma(\mathscr{B},t)^\bullet = \frac{d}{dt}\int_{\mathscr{B}_t} \phi_E \, dv = \frac{d}{dt}\Big(\int_{\mathscr{B}_t^+} \phi_E \, dv + \int_{\mathscr{B}_t^-} \phi_E \, dv\Big)$$

$$= \int_{\mathscr{B}_t^+} \frac{\partial \phi_E}{\partial t} dv + \int_{\partial\mathscr{B}_t^+ \setminus \mathscr{S}_t} \phi_E \, \mathbf{v} \cdot \mathbf{n} \, da + \int_{\mathscr{S}_t} \phi_E^+ \mathbf{v}_S \cdot \mathbf{n}^+ \, da$$

$$+ \int_{\mathscr{B}_t^-} \frac{\partial \phi_E}{\partial t} dv + \int_{\partial\mathscr{B}_t^- \setminus \mathscr{S}_t} \phi_E \, \mathbf{v} \cdot \mathbf{n} \, da + \int_{\mathscr{S}_t} \phi_E^- \, \mathbf{v}_S \cdot \mathbf{n}^- \, da$$

$$= \int_{\mathscr{B}_t} \frac{\partial \phi_E}{\partial t} dv + \int_{\partial\mathscr{B}_t} \phi_E \, \mathbf{v} \cdot \mathbf{n} \, da - \int_{\mathscr{S}_t} [\phi_E] \, \mathbf{v}_S \cdot \mathbf{n} \, da \, .$$

We further assume that this rate is fed not only by a flux $\mathbf{q}$ and a volumetric source r, but also by an area-distributed source r_a on $\mathscr{S}_t$

$$\Gamma(\mathscr{B},t)^\bullet = \int_{\mathscr{B}_t} r \, dm - \int_{\partial\mathscr{B}_t} \mathbf{q} \cdot \mathbf{n} \, da + \int_{\mathscr{S}_t} r_a \, da \, .$$

Thus

$$\int_{\mathscr{B}_t} \Big(\frac{\partial \phi_E}{\partial t} - r\,\rho\Big) dv + \int_{\partial\mathscr{B}_t} (\phi_E \, \mathbf{v} + \mathbf{q}) \cdot \mathbf{n} \, da - \int_{\mathscr{S}_t} ([\phi_E] \, \mathbf{n} \cdot \mathbf{v}_S + r_a) \, da = 0 \, .$$

If we now contract $\mathscr{B}_t$ to a thin strip on both sides of $\mathscr{S}_t$, then in the limit as its width goes to zero, the first integral converges to zero, and the second becomes

$$\int_{\partial\mathscr{B}_t} (\phi_E \, \mathbf{v} + \mathbf{q}) \cdot \mathbf{n} \, da$$

$$= \int_{\mathscr{S}_t} (\phi_E^+ \, \mathbf{v}^+ + \mathbf{q}^+) \cdot \mathbf{n}^- \, da + \int_{\mathscr{S}_t} (\phi_E^- \, \mathbf{v}^- + \mathbf{q}^-) \cdot \mathbf{n}^+ \, da$$

$$= \int_{\mathscr{S}_t} (\phi_E^+ \mathbf{v}^+ - \phi_E^- \mathbf{v}^- + \mathbf{q}^+ - \mathbf{q}^-) \cdot \mathbf{n} \, da$$

$$= \int_{\mathscr{S}_t} [\phi_E \mathbf{v}] \cdot \mathbf{n} \, da + \int_{\mathscr{S}_t} [q_n] \, da \qquad \text{with } q_n := \mathbf{q} \cdot \mathbf{n} \, .$$

If this holds for $\mathscr{S}_t$ and all of its subdomains, the **local jump balance equation**[51]

(3.12) $$[\phi_E (v - v_n)] + [q_n] - r_a = 0$$

holds with the normal part $v := \mathbf{n} \cdot \mathbf{v}$ of the (material) velocity $\mathbf{v}$, which can also be non-smooth at the front.

3.3 Observer-Dependent Laws of Motion

Each axiomatic theory is based on primitive concepts and assumptions, called axioms, the validity of which cannot be proven within the theory itself, in principle. These axioms are used to derive further statements. For continuum mechanics we choose conservation of mass and two more balance laws as fundamental axioms, namely **EULER´s laws of motion** (1744).

Assumption 3.2. Balance of linear momentum
The rate of change of the linear momentum $\mathbf{p}$ *equals the resultant force* $\mathbf{f}$ *acting on the body*

(3.13) $$\mathbf{p}(\mathscr{B}, t)^{\bullet} = \mathbf{f}(\mathscr{B}, t)$$

Assumption 3.3. Balance of moment of momentum
The rate of change of the moment of momentum $\mathbf{d}_O$ *with respect to a fixed point of reference* $O \in \mathscr{E}$ *equals the resultant moment* $\mathbf{m}_O$ *with respect to* O *acting on the body*

(3.14) $$\mathbf{d}_O(\mathscr{B}, t)^{\bullet} = \mathbf{m}_O(\mathscr{B}, t)$$

These two laws are assumed to be valid for all possible motions of all bodies (including their subbodies) at all times.

In the case of equilibrium or rest, linear momentum and moment of momentum are zero, and the laws give the **equilibrium conditions**

- of the forces $\mathbf{f}(\mathscr{B}, t) = \mathbf{o}$
- and of the moments $\mathbf{m}_O(\mathscr{B}, t) = \mathbf{o}$

51 See TRUESDELL/ TOUPIN (1960) Sect. 193, and DIMITRIENKO (2011).

which are already valid, if the linear momentum and moment of momentum are only constant in time.

The two laws of motion shall not be considered as definitions of forces and moments. Instead we will introduce these dynamical quantities as independent primitive concepts. Before we do so, however, we should explain the dependence of all these quantities upon the frame of reference or on the observer.

The above formulation of EULER´s laws contain statements, which need further explanation. One may ask, *e.g.*, what a *fixed point* of the EUCLIDean space is. We will see in the sequel, that this is a quite subjective notion. The EUCLIDean space becomes observable for an **observer** only if there are material bodies in this space. An *absolute space* cannot be identified by us. And each observer monitors the space in a different way. A body may appear at rest to one observer, while it is appearing to move to another, since each observer watches his or her individual EUCLIDean space.

All observers can probably agree to a unique metric of both space and time, but not to a common space-time-metric. Each observer will measure the spatial distance between non-simultaneous events differently. At a fixed instant of time, however, the EUCLIDean spaces of two observers will differ only by a distance preserving bijection (an isometry). If all observers describe their spaces by position vectors with respect to their individual points of reference, then

- the 1st observer sees the position vector of a spatial point $P \in \mathscr{E}$ at time t with respect to his point of reference $O \in \mathscr{E}$ as

$$\mathbf{r}(P, t) = \overrightarrow{OP}$$

and

- the 2nd observer (with *) with respect to his point of reference $O' \in \mathscr{E}^*$ as

$$\mathbf{r}^*(P^*, t) = \overrightarrow{O'P^*} = \overrightarrow{O'O^*} + \overrightarrow{O^*P^*}$$

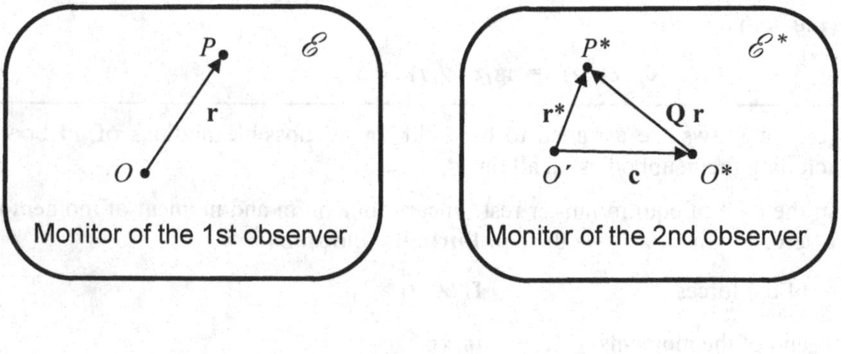

The distance between O and P and between O^* and P^* is the same for both observers. Therefore, the vector $\overrightarrow{O^* P^*}$ can be obtained by a (time-dependent) rotation from $\mathbf{r}$, which we can describe by an orthogonal tensor $\mathbf{Q}$

$$\overrightarrow{O^* P^*} = \mathbf{Q}(t)\, \mathbf{r}(P, t) \qquad \text{with } \mathbf{Q} \in \mathcal{O}rth$$

With $\mathbf{c} := \overrightarrow{O'O^*} \in \mathscr{V}$ this gives the

EUCLIDean transformation of the position vectors under change of observer

(3.15)
$$\mathbf{r}^*(P^*, t) = \mathbf{Q}(t)\, \mathbf{r}(P, t) + \mathbf{c}(t)$$

The 2nd observer records his space as being shifted by $\mathbf{c}(t)$ and rotated by $\mathbf{Q}(t)$ with respect to the space of the 1st observer.

In principle, one can also allow for a change of orientation between the observers. A natural or canonical orientation of the space does not exist. Every observer can freely choose his or her orientation of space. Accordingly, $\mathbf{Q}(t)$ can also be an improper orthogonal tensor.[52] Later we will see, however, that this choice has hardly any consequences in the context of simple materials. For gradient materials, however, this choice does have consequences, as we will demonstrate at the end of this book.

The important fact that the EUCLIDean transformations conserve the spatial distance between simultaneous events can be easily proven by considering the distance of two points P and R expressed by the norm of the difference vector

1st observer: $\quad |\mathbf{r}(P) - \mathbf{r}(R)| = \{[\mathbf{r}(P) - \mathbf{r}(R)] \cdot [\mathbf{r}(P) - \mathbf{r}(R)]\}^{1/2}$

2nd observer:
$$\begin{aligned} |\mathbf{r}^*(P^*) - \mathbf{r}^*(R^*)| &= |\mathbf{Q}(t)\, \mathbf{r}(P) + \mathbf{c}(t) - \mathbf{Q}(t)\, \mathbf{r}(R) - \mathbf{c}(t)| \\ &= |\mathbf{Q}(t)\, [\mathbf{r}(P) - \mathbf{r}(R)]| \\ &= \{(\mathbf{Q}(t)\, [\mathbf{r}(P) - \mathbf{r}(R)]) \cdot (\mathbf{Q}(t)\, [\mathbf{r}(P) - \mathbf{r}(R)])\}^{1/2} \\ &= \{[\mathbf{r}(P) - \mathbf{r}(R)] \cdot [\mathbf{r}(P) - \mathbf{r}(R)]\}^{1/2} \\ &= |\mathbf{r}(P) - \mathbf{r}(R)|\,. \end{aligned}$$

However, this holds only for simultaneous events, as $\mathbf{c}(t)$ and $\mathbf{Q}(t)$ are in general time-dependent, if the observers can move independently. We will further on assume that the used EUCLIDean transformations are differentiable with respect to time, *i.e.*, the derivatives $\mathbf{c}(t)^{\bullet}$, $\mathbf{c}(t)^{\bullet\bullet}$ and $\mathbf{Q}(t)^{\bullet}$, $\mathbf{Q}(t)^{\bullet\bullet}$ exist as continuous functions of time.

The EUCLIDean transformations form a group under composition. If a physical quantity ϕ is depending on the observer, then the change of observer or the

---

[52] In the previous editions of this book such orientation changes have not been considered.

EUCLIDean transformation induces a transformation of ϕ into ϕ^*, what is called an *action of the EUCLIDean group*. If we assume that the reference placement is observer-independent[53], then the action of this group on the motion is

(3.16) $$\mathbf{x} = \boldsymbol{\chi}(\mathbf{x}_0, t) \quad \Leftrightarrow \quad \mathbf{x}^* = \boldsymbol{\chi}^*(\mathbf{x}_0, t) = \mathbf{Q}(t)\,\boldsymbol{\chi}(\mathbf{x}_0, t) + \mathbf{c}(t)\,.$$

Note the formal similarity with the superimposed rigid body motion introduced in (2.23). However, there we considered *two* motions recorded by *one* observer, while we here consider *one* motion recorded by *two* observers. Moreover, the orthogonal $\mathbf{Q}(t)$ resulting from a change of observer can have a negative detereminant, while it must be proper orthogonal when resulting from a superimposed rigid body motion.

We can now determine the actions of the EUCLIDean group on the other kinematical quantities like the deformation gradient

(3.17) $$\mathbf{F}^* = Grad\,\boldsymbol{\chi}^* = \frac{\partial \chi^*}{\partial \mathbf{x}_0} = \frac{\partial \chi^*}{\partial \mathbf{x}}\,Grad\,\boldsymbol{\chi} = \mathbf{Q}\,\mathbf{F}$$

and, as immediate consequences,

$\mathbf{U}^* = \mathbf{U}$	right stretch tensor
$\mathbf{C}^* = \mathbf{C}$	right CAUCHY-GREEN tensor
$\mathbf{R}^* = \mathbf{Q}\,\mathbf{R}$	rotation tensor
$\mathbf{V}^* = \mathbf{Q}\,\mathbf{V}\,\mathbf{Q}^T$	left stretch tensor
$\mathbf{B}^* = \mathbf{Q}\,\mathbf{B}\,\mathbf{Q}^T$	left CAUCHY-GREEN tensor
$\mathbf{E}^{G*} = \mathbf{E}^G$	GREEN´s strain tensor
$\mathbf{E}^{Gen*} = \mathbf{E}^{Gen}$	generalised material strain tensor
$\mathbf{E}^{a*} = \mathbf{Q}\,\mathbf{E}^a\,\mathbf{Q}^T$	ALMANSI´s strain tensor
$\mathbf{E}^{gen*} = \mathbf{Q}\,\mathbf{E}^{gen}\,\mathbf{Q}^T$	generalised spatial strain tensor

etc. As all observers measure the same volume

$$J^* = det(\mathbf{F}^*) = det(\mathbf{Q}\,\mathbf{F}) = det(\mathbf{Q})\,det(\mathbf{F}) = det(\mathbf{F}) = J\,,$$

also the mass density is invariant

$$\rho^* = \rho\,.$$

Another example for the action of this group is the velocity of a material point moving through the space. It is

- for the 1st observer:

$$\mathbf{v} = \mathbf{x}^\bullet = \boldsymbol{\chi}(\mathbf{x}_0, t)^\bullet$$

---

[53] This is an assumption, which is part of the definition of the reference placement. One is of course free to introduce its observer-dependence differently like that of the current placement (RIVLIN 2002, see LIU 2004).

- for the 2nd observer by (3.16):

$$\begin{aligned}\mathbf{v}^* &= \mathbf{x}^{*\bullet} = \boldsymbol{\chi}^*(\mathbf{x}_0, t)^\bullet \\ &= \mathbf{Q}^\bullet\,\mathbf{x} + \mathbf{Q}\,\mathbf{x}^\bullet + \mathbf{c}^\bullet \\ &= \mathbf{Q}^\bullet\,\mathbf{Q}^T(\mathbf{x}^* - \mathbf{c}) + \mathbf{Q}\,\mathbf{v} + \mathbf{c}^\bullet \\ &= \boldsymbol{\omega} \times (\mathbf{x}^* - \mathbf{c}) + \mathbf{Q}\,\mathbf{v} + \mathbf{c}^\bullet .\end{aligned}$$

Besides the rotated part $\mathbf{Q}(t)\,\mathbf{v}$, also the relative translational velocity $\mathbf{c}(t)^\bullet$ and the relative angular velocity $\boldsymbol{\omega}(t)$ appear, which is the axial vector of the skew tensor $\mathbf{Q}^\bullet\,\mathbf{Q}^T$. By a second differentiation with respect to time we obtain the action of the EUCLIDean group on the acceleration. The two observers record the following accelerations

- the 1st observer:

$$\mathbf{a} = \mathbf{v}^\bullet = \boldsymbol{\chi}(\mathbf{x}_0, t)^{\bullet\bullet}$$

- the 2nd observer:

$$\begin{aligned}\mathbf{a}^* &= \mathbf{v}^{*\bullet} = \boldsymbol{\chi}^*(\mathbf{x}_0, t)^{\bullet\bullet} \\ &= \boldsymbol{\omega}^\bullet \times (\mathbf{x}^* - \mathbf{c}) + \boldsymbol{\omega} \times (\mathbf{x}^{*\bullet} - \mathbf{c}^\bullet) + \mathbf{Q}^\bullet\,\mathbf{v} + \mathbf{Q}\,\mathbf{v}^\bullet + \mathbf{c}^{\bullet\bullet} \\ &= \boldsymbol{\omega}^\bullet \times (\mathbf{x}^* - \mathbf{c}) + \boldsymbol{\omega} \times [\boldsymbol{\omega} \times (\mathbf{x}^* - \mathbf{c}) + \mathbf{Q}\,\mathbf{v}] \\ &\quad + \mathbf{Q}^\bullet\,\mathbf{Q}^T\,\mathbf{Q}\,\mathbf{v} + \mathbf{Q}\,\mathbf{a} + \mathbf{c}^{\bullet\bullet} \\ &= \boldsymbol{\omega}^\bullet \times (\mathbf{x}^* - \mathbf{c}) + \boldsymbol{\omega} \times [\boldsymbol{\omega} \times (\mathbf{x}^* - \mathbf{c})] + 2\,\boldsymbol{\omega} \times \mathbf{Q}\,\mathbf{v} + \mathbf{Q}\,\mathbf{a} + \mathbf{c}^{\bullet\bullet} \\ &= \boldsymbol{\omega}^\bullet \times (\mathbf{x}^* - \mathbf{c}) + \boldsymbol{\omega} \times [(\mathbf{x}^* - \mathbf{c}) \times \boldsymbol{\omega}] + 2\,\boldsymbol{\omega} \times (\mathbf{v}^* - \mathbf{c}^\bullet) \\ &\quad + \mathbf{Q}\,\mathbf{a} + \mathbf{c}^{\bullet\bullet}\end{aligned}$$

with the

relative acceleration	$\mathbf{Q}\,\mathbf{a}$
translational acceleration	$\mathbf{c}^{\bullet\bullet}$
angular acceleration	$\boldsymbol{\omega}^\bullet \times (\mathbf{x}^* - \mathbf{c})$
CORIOLIS[54] acceleration	$2\,\boldsymbol{\omega} \times (\mathbf{v}^* - \mathbf{c}^\bullet)$
centripetal acceleration	$\boldsymbol{\omega} \times [(\mathbf{x}^* - \mathbf{c}) \times \boldsymbol{\omega}]$.

Obviously, all observers record the acceleration, which they need for the laws of motion, in completely different amounts and directions.

The action on the velocity gradient is

(3.18)
$$\begin{aligned}\mathbf{L}^* &= \mathit{grad}^*\,\mathbf{v}^* = \mathbf{F}^{*\bullet}\,\mathbf{F}^{*-1} = (\mathbf{Q}\,\mathbf{F})^\bullet(\mathbf{Q}\,\mathbf{F})^{-1} \\ &= \mathbf{Q}\,\mathbf{F}^\bullet\,\mathbf{F}^{-1}\,\mathbf{Q}^T + \mathbf{Q}^\bullet\,\mathbf{F}\,\mathbf{F}^{-1}\,\mathbf{Q}^T = \mathbf{Q}\,\mathbf{L}\,\mathbf{Q}^T + \mathbf{Q}^\bullet\,\mathbf{Q}^T.\end{aligned}$$

[54] Gaspard Gustave de Coriolis (1792-1843)

If we decompose it into its symmetric and skew parts, we obtain

$$\mathbf{D}^* = \mathbf{Q\,D\,Q}^T$$

and

$$\mathbf{W}^* = \mathbf{Q\,W\,Q}^T + \mathbf{Q}^\bullet \mathbf{Q}^T.$$

A tensor $\mathbf{T}$ of arbitrary order is called **objective** (under change of observer), if it is just rotated by $\mathbf{Q}$ under the action of the EUCLIDean group

$$\mathbf{T}^* = \mathbf{Q} * \mathbf{T}.$$

Note that a negative determinant of $\mathbf{Q}$ would rule out in this expression, if $\mathbf{T}$ is of even order. If, on the other hand

$$\mathbf{T}^* = \mathbf{T}$$

holds, then we call it **invariant** (under change of observer). Note that in the literature the label *objective* is often used in a different way[55].

Accordingly, a scalar α, a vector $\mathbf{w}$, or a *2*nd-order tensor $\mathbf{T}$ are objective, if their transformations are

$$\alpha^* = \alpha \qquad \mathbf{w}^* = \mathbf{Q\,w} \qquad \mathbf{T}^* = \mathbf{Q\,T\,Q}^T.$$

For scalars objectivity and invariance coincide.

The following quantities are

- objective: $\mathbf{V}$, $\mathbf{B}$, $\mathbf{E}^a$, $\mathbf{E}^{gen}$, $\mathbf{D}$
- invariant: $\mathbf{U}$, $\mathbf{C}$, $\mathbf{E}^G$, $\mathbf{E}^{Gen}$.

Objective time-dependent quantities α, $\mathbf{w}$, and $\mathbf{T}$ in the EULERean description give objective derivatives $\alpha^\bullet$, *grad* α, *grad* $\mathbf{w}$, *div* $\mathbf{w}$, *grad* $\mathbf{T}$, and *div* $\mathbf{T}$.

However, $\mathbf{w}^\bullet$ and $\mathbf{T}^\bullet$ are not objective. Let, *e.g.*, $\mathbf{T}$ be an objective tensor of *2*nd-order, then if $\mathbf{T}^\bullet$ is its time-derivative for one observer, for the other it is

$$\begin{aligned}\mathbf{T}^{*\bullet} &= (\mathbf{Q\,T\,Q}^T)^\bullet = \mathbf{Q\,T}^\bullet\mathbf{Q}^T + \mathbf{Q}^\bullet\mathbf{T\,Q}^T + \mathbf{Q\,T\,Q}^{T\bullet}\\ &= \mathbf{Q\,T}^\bullet\mathbf{Q}^T + \mathbf{Q}^\bullet\mathbf{Q}^T\mathbf{Q\,T\,Q}^T + \mathbf{Q\,T\,Q}^T\mathbf{Q\,Q}^{T\bullet}\\ &= \mathbf{Q\,T}^\bullet\mathbf{Q}^T + \mathbf{Q}^\bullet\mathbf{Q}^T\mathbf{T}^* - \mathbf{T}^*\mathbf{Q}^\bullet\mathbf{Q}^T\end{aligned}$$

which renders the time rate as non-objective.

Examples for objective vectors are the forces and the moments

$$\mathbf{f}^* = \mathbf{Q\,f}$$

$$\mathbf{m}_O^* = \mathbf{Q\,m}_O$$

(by assumption and by an appropriate choice of the points of reference for the latter).

55 like, *e.g.*, HILL (1978), OGDEN (1984b), KOROBEYNIKOV (2008)

The velocities and accelerations are neither objective nor invariant vectors, and the same holds for the rates of linear momentum and moment of momentum. Thus, on the right-hand sides of EULER's laws of motion (3.13) and (3.14) we have objective vectors, while the left-hand sides are not objective. Therefore, these laws cannot be valid for all observers at the same time.

In order to find out for which class of observer they are valid, we will first assume, that there is at least one such observer. We will refer to her or him as an **inertial observer**. However, then they must also hold under all those changes of observers for which the accelerations transform as objective vectors (by an appropriate choice of the reference point in (3.14)). Such special EUCLIDean transformations are called **GALILEIan**[56] **transformations**. They are characterised by $\mathbf{c}(t)^{\bullet\bullet} \equiv \mathbf{o}$ and $\mathbf{Q}(t)^{\bullet} \equiv \mathbf{0}$ (and consequently also $\boldsymbol{\omega}(t) \equiv \mathbf{o}$). The GALILEIan transformations constitute a subgroup of the EUCLIDean group. EULER's laws of motion (3.13) and (3.14) are therefore also valid for all those observers who are generated from inertial ones by GALILEIan transformations.

We conclude that EULER's laws of motion in mechanics are only GALILEIan invariant, but not EUCLIDean invariant.

One can alternatively develop a mechanical theory, for which the laws of motion remain also valid under EUCLIDean transformation, but then one must skip the objectivity of forces and moments. This way was chosen in, *e.g.*, BERTRAM (1989). In the present context, however, we continue as outlined before, which is the usual axiomatic procedure.

3.4 Stress Analysis

The fundamental method of most science is the *principle of sections*, *i.e.*, to separate objects by a (thought) cut from their environment, and to substitute the influence of it on them by introducing appropriate concepts. In mechanics, we cut the body out of the universe, and substitute its mechanical influence on the body by the dynamical quantities **forces** and **moments** (EULER's cut principle). It is further assumed that these quantities consist of two parts, namely one which acts on the surface, and one which acts on the interior of the body. For forces, this gives the distribution into **surface** or **contact forces** $\mathbf{f}_c$ and **body forces** $\mathbf{f}_b$

$$\mathbf{f}(\mathscr{B}, t) = \mathbf{f}_c(\mathscr{B}, t) + \mathbf{f}_b(\mathscr{B}, t)$$

both resulting from densities

$$\mathbf{f}_c = \int_{\partial\mathscr{B}_t} \mathbf{t}\, da \qquad \text{and} \qquad \mathbf{f}_b = \int_{\mathscr{B}_t} \mathbf{b}\, dm\,.$$

[56] Galileo Galilei (1564-1642)

All these vector fields are assumed to be objective. For the moments we will introduce an analogous distribution.

The most important example of the body force density **b** is gravitation. On the surface of the earth, every body is attracted by the mass of the earth, and normally one uses a *constant* gravitational field as a first approximation. The gravitational forces between the different parts of the body itself are in most cases negligible, as their influence on the kinetics of the body is rather small compared to other forces.

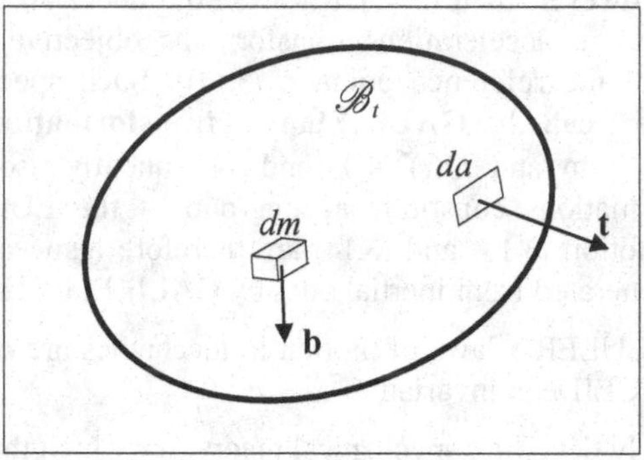

The density of the contact forces **t** is called the **traction** or **stress** (vector) field. It results from the contact of the body with its neighbouring bodies. Therefore, it depends on the section or, more precisely, on the locus and on the orientation of it, which we describe by the position vector and the outer normal to the surface **n**, respectively. This assumption is part of the **stress principle** of EULER and CAUCHY. After it we do not take into account other parameters like, *e.g.*, the curvature of the cut surface.

If an inertial observer substitutes these forces into the balance of linear momentum (3.13), he or she gets

$$\int_{\mathscr{B}_t} \mathbf{a} \, dm = \int_{\mathscr{B}_t} \mathbf{b} \, dm + \int_{\partial\mathscr{B}_t} \mathbf{t} \, da .$$

This corresponds to the form of the general global balance equation (3.10).

In order to specify the dependence of the stress vector on the normal **n**, we follow a rationale by CAUCHY and consider a tetrahedron with surfaces A_1, A_2, A_3, and A_4 and their piecewise constant normals $\mathbf{n}_1$ to $\mathbf{n}_4$ being oriented with respect to a Cartesian COOS as shown in the figure. As we further on want to shrink this tetrahedron by a limit process to the origin point of the COOS, other loci of the stress fields on these surfaces do not enter our analysis. According to the stress principle, the tractions are a function of the outer normals

$$\mathbf{t}_i = \mathbf{t}(\mathbf{n}_i) \qquad i = 1, 2, 3, 4.$$

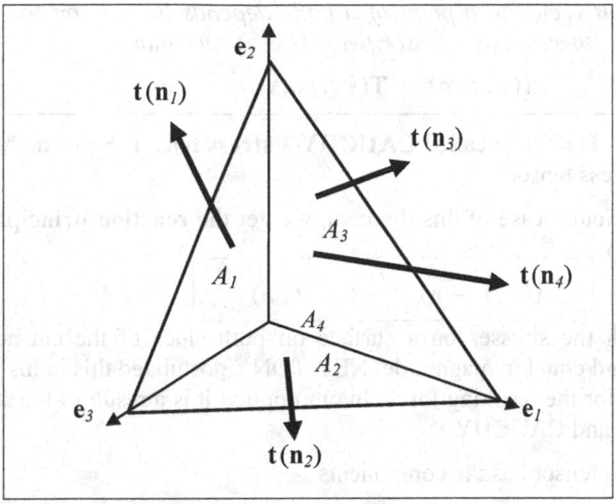

If we now use the balance of linear momentum and consider the limit, then the volume integrals are small of higher order than the surface integrals, and only the stresses remain important

$$\sum_{i=1}^{4} \mathbf{t}(\mathbf{n}_i)\, A_i = \mathbf{o}\,.$$

The first three normals are $\mathbf{n}_i \equiv -\,\mathbf{e}_i$. By the GAUSS transformation (1.49) we obtain with $\alpha \equiv 1$ for the normal of closed surfaces

$$\int_{\partial\mathcal{B}_t} \mathbf{n}\, da = \mathbf{o} = \mathbf{n}_1 A_1 + \mathbf{n}_2 A_2 + \mathbf{n}_3 A_3 + \mathbf{n}_4 A_4$$

so that the fourth normal can be expressed by the first three

$$\mathbf{n}_4 = -\frac{1}{A_4}\sum_{i=1}^{3} \mathbf{n}_i\ A_i = \frac{1}{A_4}\sum_{i=1}^{3} \mathbf{e}_i\ A_i\,.$$

Thus

$$-\,\mathbf{t}\Big(\frac{1}{A_4}\sum_{i=1}^{3} \mathbf{e}_i\, A_i\Big) = \frac{1}{A_4}\sum_{i=1}^{3} \mathbf{t}(-\,\mathbf{e}_i)\, A_i\,.$$

As this holds for all ratios between the areas, we have proven the linear dependence.

[58] Isaac Newton (1643-1727)

Theorem 3.4. of CAUCHY (1823)

The traction vector in a point of a body depends linearly on the normal of the surface, i.e., there exists a tensor field $\mathbf{T}(\mathbf{x}, t)$ *such that*

(3.19) $$\mathbf{t}(\mathbf{x}, t, \mathbf{n}) = \mathbf{T}(\mathbf{x}, t)\, \mathbf{n}(\mathbf{x}, t)$$

The tensor $\mathbf{T}(\mathbf{x}, t)$ is called **CAUCHY´s stress tensor**. Some authors also call it the **true** stress tensor.

As a particular case of this theorem, we get the **reaction principle (Lemma of CAUCHY)**

$$\mathbf{t}(\mathbf{x}, t, -\mathbf{n}) = -\mathbf{t}(\mathbf{x}, t, \mathbf{n})$$

after which the stresses on a surface on both sides of the cut are opposed in direction and equal in magnitude. NEWTON[58] postulated this in his 3rd law (*actio = reactio*) for the resulting force. In our context it is a result of the stress principle of EULER and CAUCHY.

The stress tensor has the components

$$\sigma_{ik} = \mathbf{e}_i \cdot (\mathbf{T}\, \mathbf{e}_k) \qquad i, k = 1, 2, 3$$

with respect to an ONB, which is the i-th stress component on the k-th surface (area with normal $\mathbf{e}_k$).

If we substitute CAUCHY´s equation from Theorem 3.4. into the balance of linear momentum (3.13), then we can transform all integrals into volumetric ones by means of the GAUSS transformation

$$\int_{\mathcal{B}_t} \mathbf{a}\; dm = \int_{\mathcal{B}_t} \mathbf{b}\; dm + \int_{\partial\mathcal{B}_t} \mathbf{T}\,\mathbf{n}\; da$$

$$= \int_{\mathcal{B}_t} \mathbf{b}\; dm + \int_{\mathcal{B}_t} div\, \mathbf{T}\; dv\,.$$

If the arguments are continuous and the domain of the integration is arbitrary, then we get the local version of the balance of linear momentum.

Theorem 3.5. CAUCHY´s first law of motion (1827)

At any point of the body, the following balance of linear momentum holds

(3.20) $$div\, \mathbf{T} + \rho\, \mathbf{b} = \rho\, \mathbf{a}$$

Next we consider the balance of moment of momentum. Again we assume that the moments are composed of a mass-distributed torque $\mathbf{m}_b$ and a surface-distributed or contact torque $\mathbf{m}_c$, apart from the moments induced by the forces

$$\mathbf{m}_O = \int_{\partial\mathcal{B}_t} \mathbf{m}_c\; da + \int_{\mathcal{B}_t} \mathbf{m}_b\; dm + \int_{\partial\mathcal{B}_t} \mathbf{r}_O \times \mathbf{t}\; da + \int_{\mathcal{B}_t} \mathbf{r}_O \times \mathbf{b}\; dm$$

(all in the EULERean description). One can now show by analogy to CAUCHY´s theorem that there exists a tensor $\mathbf{M}$ for the contact torques so that

$$\mathbf{m}_c(\mathbf{x}\,, t\,, \mathbf{n}) \;=\; \mathbf{M}(\mathbf{x}\,, t)\,\mathbf{n}(\mathbf{x}\,, t)$$

holds. We substitute this into the balance of moment of momentum (3.14)

$$\int_{\mathscr{B}_t} \mathbf{r}_O \times \mathbf{a}\, dm \;=\; \int_{\partial\mathscr{B}_t} (\mathbf{M} + \mathbf{r}_O \times \mathbf{T})\,\mathbf{n}\, da \;+\; \int_{\mathscr{B}_t} (\mathbf{m}_b + \mathbf{r}_O \times \mathbf{b})\, dm$$

where $\mathbf{r}_O \times \mathbf{T}$ is that tensor which gives for an arbitrary vector $\mathbf{n}$

$$(\mathbf{r}_O \times \mathbf{T})\,\mathbf{n} \;=\; \mathbf{r}_O \times (\mathbf{T}\,\mathbf{n})\,.$$

By the GAUSS transformation we obtain for sufficiently smooth arguments

$$\rho\,\mathbf{r}_O \times \mathbf{a} \;=\; div(\mathbf{M} + \mathbf{r}_O \times \mathbf{T}) + \rho\,\mathbf{m}_b + \rho\,\mathbf{r}_O \times \mathbf{b}\,.$$

By multiplying Eq. (3.20) by $\mathbf{r}_O$ we get

$$\rho\,\mathbf{r}_O \times \mathbf{a} \;=\; \mathbf{r}_O \times div\,\mathbf{T} + \rho\,\mathbf{r}_O \times \mathbf{b}\,.$$

The product rule gives

$$\begin{aligned} div(\mathbf{r}_O \times \mathbf{T}) &= (\mathbf{r}_O \times \mathbf{T})\,\nabla_E \\ &= \overset{\downarrow}{\mathbf{r}_O} \times \mathbf{T}\,\nabla_E \;+\; \mathbf{r}_O \times \overset{\downarrow}{\mathbf{T}}\,\nabla_E \\ &= 2\,\mathbf{t}^A \;+\; \mathbf{r}_O \times div\,\mathbf{T}\,, \end{aligned}$$

wherein $\mathbf{t}^A$ is the axial vector of $\mathbf{T}$. This can be demonstrated by using an ONB

$$\begin{aligned} \overset{\downarrow}{\mathbf{r}_O} \times \mathbf{T}\,\nabla_E &= \overset{\downarrow}{x_i}\;\mathbf{e}_i \times \sigma_{kl}\,\mathbf{e}_k \otimes \mathbf{e}_l \,(\frac{\partial}{\partial x_n}\,\mathbf{e}_n) \\ &= \frac{\partial x_i}{\partial x_n}\;\mathbf{e}_i \times \mathbf{e}_k\;\sigma_{kl}\,(\mathbf{e}_l \cdot \mathbf{e}_n) \\ &= \delta_{in}\;\varepsilon_{ikm}\;\mathbf{e}_m\;\sigma_{kl}\;\delta_{ln} \\ &= \varepsilon_{ikm}\;\mathbf{e}_m\;\sigma_{ki} \\ &= \mathbf{e}_1\,(\sigma_{32} - \sigma_{23}) \;+\; \mathbf{e}_2\,(\sigma_{13} - \sigma_{31}) + \mathbf{e}_3\,(\sigma_{21} - \sigma_{12}) \\ &= 2\,\mathbf{t}^A. \end{aligned}$$

With this, EULER´s second law reduces locally to

$$div\,\mathbf{M} + \rho\,\mathbf{m}_b \;+ 2\,\mathbf{t}^A \;=\; \mathbf{o}\,.$$

Material models for which such distributed moments are taken into account, are called *KOITER*[59] *continua* (1964). In the theory of polar media (called

---

[59] Warner Tjardus Koiter (1914-1997)

COSSERAT[60] *theories*) one introduces additional local degrees of freedom, called *micro spins*.

However, we will not follow this track, but instead focus our attention to **non-polar media**, for which

$$\mathbf{m}_c \equiv \mathbf{o} \qquad \text{and} \qquad \mathbf{m}_b \equiv \mathbf{o}$$

$$\Rightarrow \qquad \mathbf{M} = \mathbf{0}$$

holds by assumption. From the balance of moment of momentum we conclude in this case $\mathbf{t}^{A} = \mathbf{o}$.

Theorem 3.6. CAUCHY's second law of motion or **BOLTZMANN**[61]**'s axiom**
For every non-polar medium CAUCHY's stress tensor is symmetric

(3.21) $$\mathbf{T} = \mathbf{T}^{T}$$

The symmetry allows for a spectral form of CAUCHY's stress tensor

(3.22) $$\mathbf{T} = \sum_{r=1}^{3} \sigma_r\, \mathbf{s}_r \otimes \mathbf{s}_r$$

with the **principal stresses** σ_r and the eigenvectors $\mathbf{s}_r$.

We now generate integral equations, which are equivalent to the laws of motion. By multiplying (3.20) by an arbitrary differentiable vector field $\delta\mathbf{v}$ as a test function, we obtain

$$div(\mathbf{T}) \cdot \delta\mathbf{v} + \rho\, \mathbf{b} \cdot \delta\mathbf{v} = \rho\, \mathbf{a} \cdot \delta\mathbf{v}\,.$$

Integration over $\mathscr{B}_t$ gives

$$\int_{\mathscr{B}_t} div(\mathbf{T}) \cdot \delta\mathbf{v}\; dv + \int_{\mathscr{B}_t} \mathbf{b} \cdot \delta\mathbf{v}\; dm = \int_{\mathscr{B}_t} \mathbf{a} \cdot \delta\mathbf{v}\; dm\,.$$

By applying the product rule, we obtain

$$div(\mathbf{T}) \cdot \delta\mathbf{v} = \delta\mathbf{v} \cdot div\, \mathbf{T} = div(\mathbf{T}^{T}\, \delta\mathbf{v}) - \mathbf{T} \cdot grad\, \delta\mathbf{v}\,,$$

and by (3.21)

$$= div(\mathbf{T}\, \delta\mathbf{v}) - \mathbf{T} \cdot \delta\mathbf{D}$$

with

$$\delta\mathbf{D} := \tfrac{1}{2}\,(grad\, \delta\mathbf{v} + grad^{T}\, \delta\mathbf{v})\,.$$

By means of the GAUSS transformation and (3.21) we get

$$\int_{\mathscr{B}_t} div(\mathbf{T}\, \delta\mathbf{v})\; dv = \int_{\partial\mathscr{B}_t} (\mathbf{T}\, \delta\mathbf{v}) \cdot \mathbf{n}\; da = \int_{\partial\mathscr{B}_t} \delta\mathbf{v} \cdot (\mathbf{T}\, \mathbf{n})\; da = \int_{\partial\mathscr{B}_t} \mathbf{t} \cdot \delta\mathbf{v}\; da\,.$$

[60] Eugène Maurice Pierre Cosserat (1866-1931), Francois Cosserat (1852-1914)
[61] Ludwig Boltzmann (1844-1904)

We finally deduced the

Theorem 3.7. Principle of virtual power

The laws of motion (3.20) *and* (3.21) *are fulfilled, if and only if for all vector fields* $\delta\mathbf{v}$ *on the body and on all subbodies*

(3.23)
$$\int_{\partial\mathscr{B}_t} \mathbf{t}\cdot\delta\mathbf{v}\; da + \int_{\mathscr{B}_t} \mathbf{b}\cdot\delta\mathbf{v}\; dm = \int_{\mathscr{B}_t} \mathbf{a}\cdot\delta\mathbf{v}\; dm + \int_{\mathscr{B}_t} \mathbf{T}\cdot\delta\mathbf{D}\; dv$$

holds with $\delta\mathbf{D} := \frac{1}{2}(grad\,\delta\mathbf{v} + grad^T\,\delta\mathbf{v})$.

The other direction can be seen by identifying $\delta\mathbf{v}$ with a translational field (for (3.20) or a rotational field (for (3.21)). The equivalence gets lost, however, if we reduce the regularity of the fields. In such cases the principle of virtual power turns out to be weaker than EULER's laws of motion.

The equivalence of this principle to the laws of motion also gets lost if we only take the current velocity field $\mathbf{v}$ as a test function $\delta\mathbf{v}$

$$\int_{\partial\mathscr{B}_t} \mathbf{t}\cdot\mathbf{v}\; da + \int_{\mathscr{B}_t} \mathbf{b}\cdot\mathbf{v}\; dm = \int_{\mathscr{B}_t} \mathbf{a}\cdot\mathbf{v}\; dm + \int_{\mathscr{B}_t} \mathbf{T}\cdot\mathbf{D}\; dv\,.$$

Then

$$\mathbf{a}\cdot\mathbf{v} = \mathbf{x}^{\bullet\bullet}\cdot\mathbf{x}^{\bullet} = (\tfrac{1}{2}\,\mathbf{x}^{\bullet 2})^{\bullet}$$

leads to the rate of the kinetic energy.

Theorem 3.8. Balance of power

(3.24)
$$L_e = L + K^{\bullet}$$

In every admissible motion, the **power of the forces**

$$L_e := \int_{\partial\mathscr{B}_t} \mathbf{t}\cdot\mathbf{v}\; da + \int_{\mathscr{B}_t} \mathbf{b}\cdot\mathbf{v}\; dm$$

equals the **stress power**

$$L := \int_{\mathscr{B}_t} l\; dm \qquad \text{with} \qquad l := \frac{1}{\rho}\,\mathbf{T}\cdot\mathbf{L}$$

plus the rate of the **kinetic energy**

$$K := \tfrac{1}{2}\int_{\mathscr{B}_t} \mathbf{v}^2\; dm\,.$$

By integrating this balance along a motion between two instants, we obtain the **balance of work**

$$A_e = \Delta K + A$$

after which the **work done by the forces**

$$A_e := \int_{t_1}^{t_2} L_e \, dt$$

equals the **work done by the stresses**

$$A := \int_{t_1}^{t_2} L \, dt$$

plus the change of the kinetic energy

$$\Delta K = \int_{t_1}^{t_2} K^\bullet \, dt = K(t_2) - K(t_1) .$$

Because of the arbitrariness of the time interval, the balance of work is equivalent to the balance of power to hold at each instant. Both are consequences of (3.24) or CAUCHY's equations (3.20) and (3.21), but not vice versa.

We must keep in mind that all these equations of motion principally hold only for inertial observers and, thus, are only GALILEIan invariant.

Besides CAUCHY's stresses, many other stress tensors are in use, some of which we will now introduce. By transforming the vectorial element of area **da** back to the reference placement by means of NANSON's formula (2.18), we get for the incremental force vector

$$\mathbf{dt} = \mathbf{T}\,\mathbf{da} = \mathbf{T}\,\mathbf{E}^{\mathrm{F}}\,\mathbf{da}_0 = J\,\mathbf{T}\,\mathbf{F}^{-T}\,\mathbf{da}_0 = \mathbf{T}^{\mathrm{1PK}}\,\mathbf{da}_0$$

with

- the **1st PIOLA-KIRCHHOFF**[62] **stress tensor** (or LAGRANGEan stress tensor, sometimes also called engineering or nominal stress)

$$\mathbf{T}^{\mathrm{1PK}} := \mathbf{T}\,\mathbf{E}^{\mathrm{F}} = J\,\mathbf{T}\,\mathbf{F}^{-T} \qquad \in \mathcal{L}in .$$

If we also pull the tension vector back into the reference placement

$$\mathbf{F}^{-1}\,\mathbf{dt} = \mathbf{F}^{-1}\,\mathbf{T}\,\mathbf{da} = \mathbf{F}^{-1}\,\mathbf{T}^{\mathrm{1PK}}\,\mathbf{da}_0 = J\,\mathbf{F}^{-1}\,\mathbf{T}\,\mathbf{F}^{-T}\,\mathbf{da}_0 ,$$

we can make use of

- the **2nd PIOLA-KIRCHHOFF tensor** (or KAPPUS'[63] stress tensor)

$$\mathbf{T}^{\mathrm{2PK}} := J\,\mathbf{F}^{-1}\,\mathbf{T}\,\mathbf{F}^{-T} = \mathbf{F}^{-1}\,\mathbf{T}^{\mathrm{1PK}} \qquad \in \mathcal{L}in .$$

[62] Gustav Robert Kirchhoff (1824-1887)
[63] Robert Kappus (1904-1973)

Other stress tensors are

- the **KIRCHHOFF stress tensor**

$$\mathbf{T}^{\mathrm{K}} := J\,\mathbf{T} \qquad \in \mathcal{Lin}$$

- the **MANDEL**[64] **stress tensor**

$$\mathbf{T}^{\mathrm{M}} := \mathbf{C}\,\mathbf{T}^{2\mathrm{PK}} \qquad \in \mathcal{Lin}$$

- the **BIOT stress tensor**

$$\mathbf{T}^{\mathrm{B}} := \tfrac{1}{2}(\mathbf{T}^{2\mathrm{PK}}\,\mathbf{U} + \mathbf{U}\,\mathbf{T}^{2\mathrm{PK}}) = \tfrac{1}{2}(\mathbf{T}^{1\mathrm{PK}\,T}\,\mathbf{R} + \mathbf{R}^{T}\,\mathbf{T}^{1\mathrm{PK}}) \qquad \in \mathcal{Lin}$$

- the **convected stress tensor**

$$\mathbf{T}_{\mathrm{k}} := \mathbf{F}^{T}\,\mathbf{T}\,\mathbf{F} \qquad \in \mathcal{Lin}$$

- the **relative** (or back rotated) **stress tensor**

$$\mathbf{T}_{\mathrm{r}} := \mathbf{R}^{T}\,\mathbf{T}\,\mathbf{R} \qquad \in \mathcal{Lin}$$

- the **material stress tensor**

$$\mathbf{S} := \mathbf{F}^{-1}\,\mathbf{T}\,\mathbf{F}^{-T} = J^{-1}\,\mathbf{T}^{2\mathrm{PK}} \qquad \in \mathcal{Lin}\,.$$

There are infinitely many other stress tensors. All of them have different properties, and it depends on the specific application which one is preferable. If we know the deformation gradient, then we can determine from one of them all the others uniquely.

CAUCHY´s 1st equation (3.20) can be brought into a material form

$$\int_{\mathcal{B}_t} div\,\mathbf{T}\,dv = \int_{\mathcal{B}_t} (\mathbf{a}-\mathbf{b})\,\rho\,dv = \int_{\mathcal{B}_0} (\mathbf{a}-\mathbf{b})\,\rho_0\,dv_0$$

$$= \int_{\partial\mathcal{B}_t} \mathbf{T}\,\mathbf{da} = \int_{\partial\mathcal{B}_0} \mathbf{T}\,\mathbf{E}^{\mathrm{F}}\,\mathbf{da}_0$$

$$= \int_{\partial\mathcal{B}_0} \mathbf{T}^{1\mathrm{PK}}\,\mathbf{da}_0 = \int_{\mathcal{B}_0} Div(\mathbf{T}^{1\mathrm{PK}})\,dv_0$$

and therefore locally

(3.25) $$Div\,\mathbf{T}^{1\mathrm{PK}} = Div(\mathbf{F}\,\mathbf{T}^{2\mathrm{PK}}) = \rho_0\,(\mathbf{a}-\mathbf{b})\,.$$

CAUCHY´s second equation (3.21) gives for non-polar media

$$\mathbf{T} = \mathbf{T}^{T} \qquad \mathbf{T}^{\mathrm{K}} = \mathbf{T}^{\mathrm{K}\,T}$$

$$\mathbf{T}^{1\mathrm{PK}}\,\mathbf{F}^{T} = \mathbf{F}\,\mathbf{T}^{1\mathrm{PK}\,T} \qquad \mathbf{T}^{2\mathrm{PK}} = \mathbf{T}^{2\mathrm{PK}\,T}$$

---

[64] Jean Mandel (1907-1982)

$$\mathbf{T}^{\mathrm{B}} = \mathbf{T}^{\mathrm{B}\,T} \qquad \mathbf{S} = \mathbf{S}^T$$

$$\mathbf{T}^{\mathrm{M}} = \mathbf{C}\,\mathbf{T}^{\mathrm{M}\,T}\,\mathbf{C}^{-1} \qquad \mathbf{T}_{\mathrm{r}} = \mathbf{T}_{\mathrm{r}}^{T}$$

$$\mathbf{T}_{\mathrm{k}} = \mathbf{T}_{\mathrm{k}}^{T}.$$

Hence, by (3.21) all of these different stress tensors are symmetric, with the exception of the 1st PIOLA-KIRCHHOFF tensor and MANDEL´s. However, these two exceptions also have for a given $\mathbf{F}$ only 6 independent components.

For the specific stress power[65] we obtain by (3.21)

$$l := \frac{1}{\rho}\,\mathbf{T}\cdot\mathbf{L} = \frac{1}{\rho}\,\mathbf{T}\cdot\mathbf{D} = \frac{1}{\rho_0}\,\mathbf{T}^{1\mathrm{PK}}\cdot\mathbf{F}^{\bullet}$$

$$= \frac{1}{\rho_0}\,\mathbf{T}^{2\mathrm{PK}}\cdot\mathbf{E}^{\mathrm{G}\bullet} = \frac{1}{2\rho_0}\,\mathbf{T}^{2\mathrm{PK}}\cdot\mathbf{C}^{\bullet}$$

$$(3.26) \qquad = \frac{1}{\rho_0}\,\mathbf{T}^{\mathrm{B}}\cdot\mathbf{U}^{\bullet} = \frac{1}{\rho}\,\mathbf{T}_{\mathrm{r}}\cdot\mathbf{U}^{\bullet}\,\mathbf{U}^{-1}$$

$$= \frac{1}{\rho}\,\mathbf{S}\cdot\mathbf{E}^{\mathrm{G}\bullet} = \frac{1}{2\rho}\,\mathbf{S}\cdot\mathbf{C}^{\bullet}.$$

A certain stress tensor $\mathbf{T}^x$ is called **work conjugate** to some strain tensor $\mathbf{E}^x$, or vice versa, if the stress power density in the reference placement is

$$\rho_0\,l = J\,\mathbf{T}\cdot\mathbf{D} = \mathbf{T}^x\cdot\mathbf{E}^{x\,\bullet}$$

for arbitrary $\mathbf{E}^{x\,\bullet}$ for all materials, and in particular for those where the stresses do not depend on the deformation rate. As the material time derivative makes only sense for material tensors, we restrict this definition to material stress tensors. According to this definition, $J\,\mathbf{S} = \mathbf{T}^{2\mathrm{PK}}$ is work conjugate to $\mathbf{E}^{\mathrm{G}}$. For CAUCHY´s stress tensor there is no work conjugate strain measure, since it is a spatial tensor[66].

The stress tensor $\mathbf{S}^{\mathrm{Gen}}$ that is work conjugate to a generalised strain tensor $\mathbf{E}^{\mathrm{Gen}}$, is called **generalised material stress tensor** [67]. For its determination, we bring all these tensors in spectral form by using the eigenbasis $\mathbf{P}_{ij} := \mathbf{u}_i \otimes \mathbf{u}_j$ of $\mathbf{U}$ and $\mathbf{E}^{\mathrm{G}}$

$$\mathbf{U} = \sum_{r=1}^{3} \mu_r\,\mathbf{P}_{rr}$$

$$\mathbf{E}^{\mathrm{G}} = \tfrac{1}{2}\sum_{r=1}^{3} (\mu_r^{\,2} - 1)\,\mathbf{P}_{rr} \qquad \mathbf{E}^{\mathrm{Gen}} = \sum_{r=1}^{3} f(\mu_r)\,\mathbf{P}_{rr}$$

[65] See MACVEAN (1968), KOROBEYNIKOV (2008).
[66] See HILL (1968), SANSOUR (2001).
[67] HILL (1968); see also HAUPT/ TSAKMAKIS (1989).

$$\mathbf{T}^{2PK} = \sum_{i,j=1}^{3} \tau^{ij}\, \mathbf{P}_{ij} \qquad\qquad \mathbf{S}^{Gen} = \sum_{i,j=1}^{3} \sigma^{ij}\, \mathbf{P}_{ij}\,.$$

The work conjugacy gives

$$\mathbf{T}^{2PK} \cdot \mathbf{E}^{G\bullet} = \tfrac{1}{2}\, \mathbf{T}^{2PK} \cdot \mathbf{C}^{\bullet}$$

$$= \mathbf{S}^{Gen} \cdot \mathbf{E}^{Gen\,\bullet} = \mathbf{S}^{Gen} \cdot \frac{d\mathbf{E}^{Gen}}{d\mathbf{C}}[\mathbf{C}^{\bullet}]\,.$$

Thus

$$\mathbf{S}^{Gen} = \tfrac{1}{2} \left(\frac{d\mathbf{E}^{Gen}}{d\mathbf{C}} \right)^{-T} [\mathbf{T}^{2PK}]\,.$$

With this we further obtain

$$\mathbf{S}^{Gen} \cdot \mathbf{E}^{Gen\,\bullet} = \sum_{i,j=1}^{3} \sigma^{ij}\, \mathbf{P}_{ij} \cdot \sum_{r=1}^{3} \{ f(\mu_r)'\, \mu_r^{\bullet}\, \mathbf{P}_{rr} + f(\mu_r)\, \mathbf{P}_{rr}^{\bullet} \}$$

$$= \mathbf{T}^{2PK} \cdot \mathbf{E}^{G\bullet} = \sum_{i,j=1}^{3} \tau^{ij}\, \mathbf{P}_{ij} \cdot \sum_{r=1}^{3} \{ \mu_r\, \mu_r^{\bullet}\, \mathbf{P}_{rr} + \tfrac{1}{2} (\mu_r^{2} - 1)\, \mathbf{P}_{rr}^{\bullet} \}\,.$$

By putting (see (1.43))

$$\mathbf{P}_{rr}^{\bullet} = \boldsymbol{\omega} \times \mathbf{P}_{rr} - \mathbf{P}_{rr} \times \boldsymbol{\omega} \qquad \text{with } \boldsymbol{\omega} = \omega^{i}\, \mathbf{u}_i$$

and comparing the coefficients in the independent virtual quantities $\mu_r^{\bullet}$, ω^1, ω^2, ω^3, we obtain the components of $\mathbf{S}^{Gen}$

$$\sigma^{ii} = \tau^{ii}\, \frac{\mu_i}{f(\mu_i)'}$$

and (no sum)

$$\sigma^{ij} = \frac{1}{2}\, \frac{\mu_i^{2} - \mu_j^{2}}{f(\mu_i) - f(\mu_j)}\, \tau^{ij} \qquad \text{for } i \neq j\,.$$

We get in particular for the SETH class with $f(\mu_r) = m^{-1}(\mu_r^{m} - 1)$ for $m \neq 0$

$$\sigma^{ii} = \tau^{ii}\, \mu_i^{2-m}$$

and (no sum)

$$\sigma^{ij} = \frac{m}{2}\, \frac{\mu_i^{2} - \mu_j^{2}}{\mu_i^{m} - \mu_j^{m}}\, \tau^{ij} \qquad \text{for } i \neq j\,.$$

By a limit process, we also obtain unique generalised stresses for multiple eigenvalues

$$\lim_{\mu_i \to \mu_j} \frac{m}{2} \frac{\mu_i^{\,2} - \mu_j^{\,2}}{\mu_i^{\,m} - \mu_j^{\,m}} = m/2 \; \mu_i^{\,2-m} .$$

For $m = 0$, we can also use this formula by applying the rule of L´HÔSPITAL

$$\sigma^{ii} = \tau^{ii} \mu_i^{\,2}$$

and (no sum)

$$\sigma^{ij} = \frac{1}{2} \frac{\mu_i^{\,2} - \mu_j^{\,2}}{\ln \mu_i - \ln \mu_j} \tau^{ij} \qquad \text{for } i \neq j .$$

This defines the stress tensor being work conjugate to the material HENCKY strain tensor.

If $\mathbf{U}$ and $\mathbf{T}^{2PK}$ are coaxial, which is the case in isotropic elasticity after an appropriate choice of the reference placement (see Chap. 6), then

$$\mathbf{S}^{Gen} = \mathbf{T}^{2PK} \, \mathbf{U}^{2-m} .$$

In the theory of small deformations, all these stress tensors coincide.

Stress Rates

For some purposes one needs **stress rates**[68], *i.e.*, incremental forms for the different stresses. For all material stress tensors like $\mathbf{S}$, $\mathbf{T}^{2PK}$, $\mathbf{S}^{Gen}$, etc., the derivative with respect to time is simply the material time derivative, such as $\mathbf{S}^{\bullet}$, $\mathbf{T}^{2PK\bullet}$, $\mathbf{S}^{Gen\,\bullet}$. All these material stress tensors are invariant under EUCLIDean transformations, and so are their rates (as we will see in Chap. 4.3). This transformation can either be interpreted as a change of observer, or as a superimposed rigid body motion.

If we express $\mathbf{S}$ by CAUCHY´s stress tensor, we obtain for its rate with the result of the example on page 63

$$\begin{aligned}
\mathbf{S}^{\bullet} &= (\mathbf{F}^{-1}\,\mathbf{T}\;\mathbf{F}^{-T})^{\bullet} \\
&= \mathbf{F}^{-1}\,\mathbf{T}^{\bullet}\mathbf{F}^{-T} + (\mathbf{F}^{-1})^{\bullet}\,\mathbf{T}\,\mathbf{F}^{-T} + \mathbf{F}^{-1}\,\mathbf{T}\,(\mathbf{F}^{-T})^{\bullet} \\
&= \mathbf{F}^{-1}\,\mathbf{T}^{\bullet}\mathbf{F}^{-T} - \mathbf{F}^{-1}\,\mathbf{F}^{\bullet}\,\mathbf{F}^{-1}\,\mathbf{T}\;\mathbf{F}^{-T} - \mathbf{F}^{-1}\,\mathbf{T}\,\mathbf{F}^{-T}\mathbf{F}^{T\bullet}\,\mathbf{F}^{-T} \\
&= \mathbf{F}^{-1}\,\mathbf{T}^{\bullet}\mathbf{F}^{-T} - \mathbf{F}^{-1}\,\mathbf{L}\,\mathbf{T}\;\mathbf{F}^{-T} - \mathbf{F}^{-1}\,\mathbf{T}\,\mathbf{L}^{T}\,\mathbf{F}^{-T} \\
&= \mathbf{F}^{-1}\,(\mathbf{T}^{\bullet} - \mathbf{L}\,\mathbf{T} - \mathbf{T}\,\mathbf{L}^{T})\,\mathbf{F}^{-T} .
\end{aligned}$$

The term in parenthesis

(3.27) $$\mathbf{T}^{\nabla} := \mathbf{T}^{\bullet} - \mathbf{L}\,\mathbf{T} - \mathbf{T}\,\mathbf{L}^{T}$$

---

[68] In WEGENER (1991) a rather detailed overview of different stress rates is given. See also KOROBEYNIKOV (2008).

is called the **OLDROYD rate** (1950) of CAUCHY's stress tensor. It transforms under a change of observer as an objective tensor and belongs to the *objective derivatives*, the same as the following ones. Mathematically, it is the **LIE**[69]-**derivative** of the stress tensor field along the velocity.

For the material time derivative of $\mathbf{S}/\rho$ we obtain analogously the **TRUESDELL**[70] **rate** (1955)

(3.28) $$\rho\,\mathbf{F}\,(\mathbf{S}/\rho)^\bullet\,\mathbf{F}^T = J^{-1}\,\mathbf{F}\,\mathbf{T}^{2PK\bullet}\,\mathbf{F}^T = \mathbf{T}^\bullet - \mathbf{L}\,\mathbf{T} - \mathbf{T}\,\mathbf{L}^T + \mathbf{T}\,div\,\mathbf{v}\,.$$

This corresponds to the OLDROYD rate of KIRCHHOFF's stress tensor

$$\mathbf{F}\,\mathbf{T}^{2PK\bullet}\,\mathbf{F}^T = J(\mathbf{T}^\bullet - \mathbf{L}\,\mathbf{T} - \mathbf{T}\,\mathbf{L}^T + \mathbf{T}\,div\,\mathbf{v}) = \mathbf{T}^{K\bullet} - \mathbf{L}\,\mathbf{T}^K - \mathbf{T}^K\,\mathbf{L}^T.$$

For the material time derivative of $\mathbf{C\,S\,C}$ one gets the **COTTER-RIVLIN rate** (1955)

(3.29) $$\mathbf{F}^{-T}\,(\mathbf{C\,S\,C})^\bullet\,\mathbf{F}^{-1} = \mathbf{T}^\bullet + \mathbf{L}^T\,\mathbf{T} + \mathbf{T}\,\mathbf{L}\,.$$

For the material time derivative of the non-symmetric material tensor $\mathbf{CS}$ one gets the **mixed OLDROYD rate**

(3.30) $$\mathbf{F}^{-T}\,(\mathbf{C\,S})^\bullet\,\mathbf{F}^T = \mathbf{T}^\bullet + \mathbf{L}^T\,\mathbf{T} - \mathbf{T}\,\mathbf{L}^T$$

which is neither symmetric. Analogously we obtain the other non-symmetric rate

$$\mathbf{F}\,(\mathbf{S\,C})^\bullet\,\mathbf{F}^{-1} = \mathbf{T}^\bullet + \mathbf{T}\,\mathbf{L} - \mathbf{L}\,\mathbf{T}.$$

The foregoing stress derivatives are called *convective*.

Quite often the **ZAREMBA-JAUMANN rate** (1903 and 1911, resp.)

(3.31) $$sym\{\mathbf{F}\,(\mathbf{S\,C})^\bullet\,\mathbf{F}^{-1}\} = \mathbf{T}^\bullet + \mathbf{T}\,\mathbf{W} - \mathbf{W}\,\mathbf{T}$$

or the rate after **GREEN/ NAGHDI**[71] (1965) - **GREEN/ McINNIS** (1967) - **DIENES** (1979), also called the *polar rate*,

(3.32) $$\mathbf{R}\,(\mathbf{U\,S\,U})^\bullet\,\mathbf{R}^T = \mathbf{T}^\bullet + \mathbf{T}\,\mathbf{R}^\bullet\,\mathbf{R}^T - \mathbf{R}^\bullet\,\mathbf{R}^T\,\mathbf{T}$$

are used.

In XIAO et al. (1996) the **logarithmic rate** of an objective tensor $\mathbf{T}$ is introduced as

(3.33) $$\mathbf{T}^\bullet + \mathbf{T}\,\mathbf{W}^{log} - \mathbf{W}^{log}\,\mathbf{T}$$

---

[69] James Gardner Oldroyd (1921-1982), Marius Sophus Lie (1842-1899)
[70] Clifford Ambrosius Truesdell (1919-2000)
[71] Paul Mansour Naghdi (1924-1994)

with the skew tensor

$$\mathbf{W}^{log} := \mathbf{W} + \sum_{r \neq s=1}^{K} \left(\frac{1+(\lambda_s / \lambda_r)}{1-(\lambda_s / \lambda_r)} + \frac{2}{ln(\lambda_s / \lambda_r)} \right) \mathbf{P}_{ss}\, \mathbf{D}\, \mathbf{P}_{rr}$$

(no sum) and with the eigenprojectors $\mathbf{P}_{ss}$ and the corresponding eigenvalues λ_s of $\mathbf{B}$, assumed to be distinct. For its properties see also XIAO et al. (1997, 2000) and FREED (2014).

These are *corrotational* objective rates because the skew terms stand for the rotational velocities.

All these stress rates in the brackets on the left-hand sides are invariant under EUCLIDean transformations, while those on the right-hand sides are objective (Chap. 4.3). Note that the time derivatives $\mathbf{T}^{\bullet}$, $\mathbf{T}^{K\bullet}$ of objective tensors like $\mathbf{T}$, $\mathbf{T}^{K}$ are in general neither objective nor invariant.[72]

3.5 Thermodynamical Balances

A Comment on the literature. Introductions to the thermodynamics of continuous media can be found in BERMUDEZ (2005), ERICKSEN (1991), GERMAIN/ NGUYEN/ SUQUET (1983), GREVE (2003), MAUGIN (1999), McLELLAN (1980), MÜLLER (1985), OTTOSEN/ RISTINMAA (2005), ŠILHAVÝ (1997), TRUESDELL (1969), WILMANSKI (1998), ZIEGLER (1983).

For the thermodynamic balances, we need some more variables. We introduce the following fields as primitive concepts.

- ε — the specific **internal energy** (a scalar field)
- r — the specific **heat source** (a scalar field)
- q — the **heat flux** per unit area and time (a scalar field on the surface).

The internal energy of the body is obtained by integration over the domain of the body

$$E = \int_{\mathscr{B}_t} \varepsilon \; dm \,.$$

One can supply the body with heat in two ways: by heat sources in its interior, or by heat fluxes through its surface. The global heat supply is thus assumed in the form

[72] see RUBIN (2021)

$$Q(\mathscr{B}, t) = \int_{\partial \mathscr{B}_t} q \; da + \int_{\mathscr{B}_t} r \; dm \,.$$

After the theorem of FOURIER[73], which is completely analogous to CAUCHY's stress principle, the heat flux is linear in the outer normal of the surface $\mathbf{n}$, and can therefore be represented by a scalar product with a vector, the **heat flux vector** $\mathbf{q}$, as $q = -\mathbf{q} \cdot \mathbf{n}$. We get by the GAUSS transformation

$$Q(\mathscr{B}, t) = \int_{\mathscr{B}_t} r \; dm - \int_{\partial \mathscr{B}_t} \mathbf{q} \cdot \mathbf{n} \; da = \int_{\mathscr{B}_t} (r \rho - div \, \mathbf{q}) \; dv \,.$$

Again, we can pull back the heat flux vector to the reference placement using NANSON's formula (2.18)

$$\int_{\mathscr{B}_t} div \, \mathbf{q} \; dv = \int_{\partial \mathscr{B}_t} \mathbf{q} \cdot \mathbf{n} \; da = \int_{\partial \mathscr{B}_t} \mathbf{q} \cdot \mathbf{da} = \int_{\partial \mathscr{B}_0} \mathbf{q} \cdot \mathbf{E}^{\mathrm{F}} \, \mathbf{da}_0$$

$$= \int_{\partial \mathscr{B}_0} \mathbf{q} \cdot (J \, \mathbf{F}^{-T} \mathbf{da}_0) = \int_{\partial \mathscr{B}_0} (J \, \mathbf{F}^{-1} \, \mathbf{q}) \cdot \mathbf{da}_0$$

$$= \int_{\partial \mathscr{B}_0} \mathbf{q}_0 \cdot \mathbf{da}_0 = \int_{\partial \mathscr{B}_0} \mathbf{q}_0 \cdot \mathbf{n}_0 \; da_0 = \int_{\mathscr{B}_0} Div \, \mathbf{q}_0 \; dv_0$$

with the **material heat flux vector**

$$\mathbf{q}_0 := det(\mathbf{F}) \; \mathbf{F}^{-1} \, \mathbf{q} \,.$$

The global **thermodynamic energy balance** is the

Assumption 3.9.
First Law of Thermodynamics (v. MAYER 1842, JOULE 1843, v. HELMHOLTZ 1847)[74]
The heat supply Q and the power of the forces L_e lead to a change of the kinetic energy K in an inertial frame and of the internal energy E of the body

(3.34) $$Q + L_e = K^\bullet + E^\bullet$$

The internal energy can be stored as strain energy (see Chap. 7) or as heat, but also in chemical connections, phase changes, changes of microstructure, etc. Together with the mechanical balance of power (3.24) this gives

$$Q + L = E^\bullet.$$

If we substitute the fields into (3.34), we obtain

---

[73] Jean Baptiste Joseph de Fourier (1768-1830)

[74] Julius Robert von Mayer (1814-1878), James Prescott Joule (1818-1889), Hermann Ludwig Ferdinand von Helmholtz (1821-1894)

$$\int_{\mathcal{B}_t} r\,dm - \int_{\partial\mathcal{B}_t} \mathbf{q}\cdot\mathbf{n}\,da + \int_{\partial\mathcal{B}_t} \mathbf{t}\cdot\mathbf{v}\,da + \int_{\mathcal{B}_t} \mathbf{b}\cdot\mathbf{v}\,dm$$

$$= \int_{\mathcal{B}_t} \mathbf{a}\cdot\mathbf{v}\,dm + \int_{\mathcal{B}_t} \varepsilon^\bullet\,dm$$

$$\Leftrightarrow \quad \int_{\mathcal{B}_t} \rho\,(\mathbf{a}\cdot\mathbf{v}+\varepsilon^\bullet)\,dv = \int_{\mathcal{B}_t} \rho\,(r+\mathbf{b}\cdot\mathbf{v})\,dv + \int_{\partial\mathcal{B}_t} (\mathbf{T}^T\,\mathbf{v}-\mathbf{q})\cdot\mathbf{n}\,da$$

$$= \int_{\mathcal{B}_t} \left\{\rho\,(r+\mathbf{b}\cdot\mathbf{v}) + div(\mathbf{T}^T\,\mathbf{v}-\mathbf{q})\right\}\,dv\,.$$

This holds for the body and all of its subbodies. With CAUCHY´s laws (3.20) and (3.21) we conclude the

Local version of the **energy balance** (first law of thermodynamics)

$$\rho\,\varepsilon^\bullet = \rho\,r - div\,\mathbf{q} + \mathbf{T}\cdot\mathbf{D}$$

The material version is then

(3.35) $$\rho_0\,\varepsilon^\bullet = \rho_0\,r - Div\,\mathbf{q}_0 + J\,\mathbf{S}\cdot\mathbf{E}^{G\,\bullet}\,.$$

For the *second law of thermodynamics* there are many suggestions in the literature. As this is not the place to discuss the foundations of thermodynamics, we will just briefly introduce the concepts of *Rational Thermodynamics* (TRUESDELL 1969). We need further fields.

θ	the absolute **temperature** (a scalar field)
$\mathbf{g} := grad\,\theta_E$	the spatial temperature gradient
$\mathbf{g}_0 := Grad\,\theta_L = \mathbf{F}^T\,\mathbf{g}$	the material temperature gradient
η	the specific **entropy** (a scalar field)

Also the introduction of

(3.36) $\psi := \varepsilon - \theta\,\eta$ — the specific v. HELMHOLTZ **free energy** (a scalar field)

turns out to be useful. The temperature and the entropy are introduced as primitive concepts, the others are deduced variables.

We now postulate as an axiom the

Assumption 3.10. Second Law of Thermodynamics
For all bodies at each time the following inequality holds

$$\int_{\mathcal{B}_t} \eta^\bullet\,dm \geq \int_{\mathcal{B}_t} \frac{r}{\theta}\,dm - \int_{\partial\mathcal{B}_t} \frac{\mathbf{q}\cdot\mathbf{n}}{\theta}\,da$$

The local form of the second law is obtained after applying the GAUSS transformation

$$\rho\,\eta^{\bullet} \geq \rho\,\frac{r}{\theta} - div(\frac{\mathbf{q}}{\theta}) = \rho\,\frac{r}{\theta} - \frac{1}{\theta}\,div\,\mathbf{q} + \frac{1}{\theta^2}\,\mathbf{q}\cdot\mathbf{g}$$

and by using (3.35) we finally get the

CLAUSIUS[75]-DUHEM[76] inequality in its spatial form

$$\frac{1}{\rho}\,\mathbf{T}\cdot\mathbf{D} - \frac{\mathbf{q}\cdot\mathbf{g}}{\rho\,\theta} \geq \psi^{\bullet} + \theta^{\bullet}\,\eta$$

or in its material form

(3.37)
$$\frac{1}{2\rho}\mathbf{S}\cdot\mathbf{C}^{\bullet} - \frac{\mathbf{q}_0\cdot\mathbf{g}_0}{\rho_0\theta} \geq \psi^{\bullet} + \theta^{\bullet}\,\eta$$

The specific **mechanical dissipation** is defined as

$$\delta := \theta\,\eta^{\bullet} - r + \frac{1}{\rho}\,div\,\mathbf{q} = \frac{1}{\rho}\mathbf{T}\cdot\mathbf{D} - \psi^{\bullet} - \theta^{\bullet}\,\eta\,.$$

This gives in equivalence to the CLAUSIUS-DUHEM inequality the **dissipation inequality**

$$\delta - \frac{\mathbf{q}\cdot\mathbf{g}}{\rho\theta} \geq 0\,.$$

The term $-\dfrac{\mathbf{q}\cdot\mathbf{g}}{\rho\theta}$ is the specific **thermal dissipation.**

For isothermal processes, the CLAUSIUS-DUHEM inequality reduces to the **CLAUSIUS-PLANCK[77] inequality**

(3.38)
$$\frac{1}{2\rho}\,\mathbf{S}\cdot\mathbf{C}^{\bullet} \geq \psi^{\bullet}$$

so that the rate of the free energy cannot be larger than the stress power. This is equivalent to the postulate, that the dissipation is non-negative $\delta \geq 0$.

These inequalities resulting from the second law must be fulfilled for all admissible processes the material can undergo. They are considered as restrictions on the material behaviour, which we will have to keep in mind when formulating material models.

[75] Rudolf Julius Emmanuel Clausius (1822-1888)
[76] Pierre Maurice Marie Duhem (1861-1916)
[77] Max Karl Ernst Ludwig Planck (1858-1947)

The internal energy and the entropy appear in the two laws only through their time rates. Therefore, they can principally be determined only up to additive constants, if at all. The free energy can only be determined up to the constant of the internal energy minus temperature times the constant of the entropy.

4 Principles of Material Theory

All the foregoing results are supposed to be valid for all material bodies and are thus labelled *universal*. In contrast to this, material theory aims to describe the individual behaviour of different materials. This is done by so-called *material equations, material laws* or *functionals, constitutive equations, material models* and whatever they are called. In going into material theory, we do not only intend to work out the differences in material behaviour, but primarily to find common features, so that we can make the form of a general material equation as precise as possible, of course, without restricting the range of validity or applicability of such theory. One way to do so is to assume certain *principles* of material theory, which are partly based on our experience, partly on plausibility. Further on we want to develop criteria to classify the universe of material behaviour, which shall help to get more concrete models for certain classes of materials. Finally, however, this will lead to a natural barrier, which we cannot trespass without experiments. But even for the design of appropriate experiments, the theory should give some general tools.

First of all we will introduce the principles of material theory, where we start with the pure mechanical variables.

4.1 Determinism

In mechanics one generally assumes that there is a deterministic relation between the stresses in a body and the motion of the body. In order to put this into a functional form, we have to decide which quantities to use as independent variables and which as dependent variables. There are also suggestions in the literature[78] to not use functions, but instead to use constitutive relations and, thus, circumvent this decision. This advantage, however, demands such a large conceptual effort that it did not gain much popularity. It has now become a more or less normal practise to consider the stresses as dependent variables, and motions as independent ones, although one might also consider the opposite.

Of course, we do not allow the future motion of a body to have any influence on the stresses in the presence. This is also expressed in the following

Assumption 4.1. Principle of determinism
The stresses in a material point at an instant of time are determined by the current and the past (but not the future) motion of the body.

[78] See PERZYNA/ KOSINSKI (1973), FRISCHMUTH/ KOSINSKI/ PERZYNA (1986), and GURTIN (1981).

A. Bertram, *Elasticity and Plasticity of Large Deformations*,
https://doi.org/10.1007/978-3-030-72328-6_4

The material theories in the literature principally differ about how to take the past motion into consideration. Some use the infinite history in *history functionals*, some introduce *internal variables* with evolution equations. We will instead use the past motion

$$\boldsymbol{\chi}(\mathbf{x}_0 \,,\, \tau)\Big|_{\tau=0}^{t}$$

in a finite time interval $[0\,,t]$.

4.2 Local Action

While the principle of determinism stands for a reduction of the temporal influence of deformations, the following will do this with the spatial aspect.

It is an empirical fact that the stresses at a material point do not depend on the motion of other points, provided those are sufficiently remote. Therefore, we assume in a first formulation of this principle that there is a (finite) neighbourhood of the point, which alone influences the stresses at that particular point, while the rest of the body has no direct effect on them.

Assumption 4.2. Principle of local action

The stresses at a material point depend on the motion of only a finite neighbourhood of that point.

According to this principle, it is still possible that the motion of points at a distance from the point under consideration have some influence. Theories with this property are called **non-local**, although *local action* in the above sense still holds.

A more restrictive formulation of this principle is obtained if we reduce the neighbourhood of influence to the smallest possible.

Assumption 4.3. Principle of local action for simple materials

The stresses at a material point depend on the motion of only its infinitesimal neighbourhood.

Infinitesimal means in the sense of differential geometry that the motion of the neighbourhood enters the material model only up to its differential. This means that only the deformation process of the tangent space at that point has influence, but not that of the entire body or of any subbody. As the differential is the first or the simple derivative of the motion, NOLL[79] called such materials **simple materials**. In contrast to this also theories exists which take the first n derivatives

---

[79] Walter Noll (1925-2017); see IGNATIEFF (1996) for the work and life of NOLL.

into account, what we would call **materials of grade n**[80]. Such material classes are described in the last chapter. However, in what follows here we will exclusively deal with simple materials. The ansatz for a **material functional** is

$$\mathbf{T}(\mathbf{x}_0 , t) = \mathcal{F}\{\boldsymbol{\chi}(\mathbf{x}_0 , \tau) , \mathbf{F}(\mathbf{x}_0 , \tau)\big|_{\tau=0}^{t}\} , \tag{4.1}$$

which assigns to each motion $\boldsymbol{\chi}$ and each **F**-process of that particular material point $\mathbf{x}_0$ in a time interval $[0 , t]$ the stress tensor $\mathbf{T}$ at its end.

However, even with these restrictions in the selection of the independent variables, we still have many choices within the variety of strain and stress measures which we introduced before. These choices will be limited by the following invariance requirements.

4.3 EUCLIDean Invariances

Our fundamental concepts like motions and stresses depend on observers. It is therefore necessary to work out the general rules for the observer-dependence of the material theory.

When measuring times (chronometry), the observer-dependence consists of the choice of the time origin and of the time unit. The latter can be overcome by an international unit (which is by no means an easy task). The first is not too important, as we mainly deal with time differences.

Much more relevant is the observer-dependence of kinematical and dynamical concepts. As we have already shown, the changes of observers form the EUCLIDean group, which acts on all kinematical quantities, some of which are classified as objective or as invariant. For the dynamical quantities we need further assumptions. And as a result, we obtain the observer-dependence of the material functional.

If one observer describes a simple material by the functional

$$\mathbf{T}(\mathbf{x}_0, t) = \mathcal{F}\{\boldsymbol{\chi}(\mathbf{x}_0 , \tau) , \mathbf{F}(\mathbf{x}_0, \tau)\big|_{\tau=0}^{t}\}$$

then with (3.15) some other observer possibly poses a different one like

$$\begin{aligned}\mathbf{T}^*(\mathbf{x}_0 , t) &= \mathcal{F}^*\{\boldsymbol{\chi}^*(\mathbf{x}_0, \tau) , \mathbf{F}^*(\mathbf{x}_0 , \tau)\big|_{\tau=0}^{t}\} \\ &= \mathcal{F}^*\{\mathbf{Q}(\tau)\, \boldsymbol{\chi}(\mathbf{x}_0 , \tau) + \mathbf{c}(\tau) , \mathbf{Q}(\tau)\, \mathbf{F}(\mathbf{x}_0 , \tau)\big|_{\tau=0}^{t}\} .\end{aligned}$$

[80] See BERTRAM/ FOREST (2007) and the references therein.

The transformation of the dynamical quantities is determined by the following assumption, which is also labelled *principle of EUCLIDean frame-indifference* in the literature.

Assumption 4.4. Principle of material objectivity *(PMO)*
The stress power is objective (and thus also invariant) under EUCLIDean transformations

$$l = l^*$$

This means with (3.26) and (3.18)

$$l = \frac{1}{\rho}\, \mathbf{T} \cdot \mathbf{L}$$

$$= l^* = \frac{1}{\rho^*}\, \mathbf{T}^* \cdot \mathbf{L}^* = \frac{1}{\rho}\, \mathbf{T}^* \cdot (\mathbf{Q}\,\mathbf{L}\,\mathbf{Q}^T + \mathbf{Q}^\bullet\mathbf{Q}^T) = \frac{1}{\rho}\, \mathbf{Q}^T\,\mathbf{T}^*\,\mathbf{Q} \cdot \mathbf{L}$$

as we have

$$\mathbf{T}^* \cdot \mathbf{Q}^\bullet\mathbf{Q}^T = 0$$

because of CAUCHY's second equation (3.21). The necessary and sufficient condition for this invariance to hold for all materials, is that CAUCHY's stress tensor is objective

(4.2)
$$\mathbf{T}^* = \mathfrak{F}^*\{\mathbf{Q}(\tau)\,\boldsymbol{\chi}(\mathbf{x}_0, \tau) + \mathbf{c}(\tau)\,,\ \mathbf{Q}(\tau)\,\mathbf{F}(\mathbf{x}_0, \tau)\big|_{\tau=0}^{t}\}$$

$$= \mathbf{Q}\,\mathbf{T}\,\mathbf{Q}^T = \mathbf{Q}(t)\,\mathfrak{F}\{\boldsymbol{\chi}(\mathbf{x}_0, \tau)\,,\ \mathbf{F}(\mathbf{x}_0, \tau)\big|_{\tau=0}^{t}\}\,\mathbf{Q}^T(t)\,.$$

This equation can be used to determine the material functional for any observer, once one observer has identified it.

By the Theorem 3.4. of CAUCHY (1823) and the objectivity of $\mathbf{n}$, the stress vector is also objective, and so is the contact force. If the body forces are also objective, then so are the resulting forces. This was already assumed in the preceding chapter.

So in the present context, this princple leads just to the objectivity of CAUCHY's stress tensor. One can, however, also start with this principle and use it to derive the balance laws from the assumption that the total power was objective. Such an approach is demonstrated in the last chapter of this book.

By the objectivity of the CAUCHY stresses and by the EUCLIDean transformation of the motion, we can determine the action of the EUCLIDean group on all other stress tensors

$\mathbf{S}^* = \mathbf{S}$	The material stress tensor is invariant.
$\mathbf{S}^{M}{}^* = \mathbf{S}^{M}$	MANDEL's stress tensor is invariant.
$\mathbf{S}^{Gen}{}^* = \mathbf{S}^{Gen}$	The generalised material stress tensor is invariant.

$\mathbf{T}^{\mathrm{K}}* = \mathbf{Q}\,\mathbf{T}^{\mathrm{K}}\,\mathbf{Q}^{T}$	KIRCHHOFF's stress tensor is objective.
$\mathbf{T}^{\mathrm{1PK}}* = \mathbf{Q}\,\mathbf{T}^{\mathrm{1PK}}$	The 1st PIOLA-KIRCHHOFF tensor is neither objective nor invariant.
$\mathbf{T}^{\mathrm{2PK}}* = \mathbf{T}^{\mathrm{2PK}}$	The 2nd PIOLA-KIRCHHOFF tensor is invariant.
$\mathbf{T}_{\mathrm{k}}* = \mathbf{T}_{\mathrm{k}}$	The convected stress tensor is invariant.
$\mathbf{T}_{\mathrm{r}}* = \mathbf{T}_{\mathrm{r}}$	The relative stress tensor is invariant.

The observer dependence of all kinematical and dynamical quantities are now completely determined.

Another important question, which has nothing to do with observer-dependence, is the behaviour of materials under superimposed rigid body motions. In the following principle we state that the CAUCHY stresses just rotate under such modifications.

Assumption 4.5.
Principle of invariance under superimposed rigid body motions (*PISM*)
If $\mathbf{T}(\mathbf{x}_0\,, t)$ *are the* CAUCHY *stresses after a motion*

$$\boldsymbol{\chi}(\mathbf{x}_0\,,\tau)\Big|_{\tau=0}^{t}$$

then

$$\mathbf{Q}(t)\,\mathbf{T}(\mathbf{x}_0\,,t)\,\mathbf{Q}(t)^T$$

are the stresses after superimposing a rigid body motion upon the original motion

$$\{\mathbf{Q}(\tau)\,\boldsymbol{\chi}(\mathbf{x}_0\,,\tau)+\mathbf{c}(\tau)\Big|_{\tau=0}^{t}\}$$

with arbitrary differentiable time functions $\mathbf{Q}(\tau)\in\mathcal{O}rth^{+}$ *and* $\mathbf{c}(\tau)\in\mathcal{V}$.

Note that here only proper orthogonal transformations are admissible, in contrast to the *PMO*, where also improper ones have been included.

This principle is a strong restriction on the material behaviour. By $\mathbf{Q}(\tau)$ and $\mathbf{c}(\tau)$ we can generate accelerations of arbitrary magnitude with respect to an inertial observer. However, after this assumption, inertia cannot directly affect the stresses, but only through the part $\rho(\mathbf{b}-\mathbf{a})$ in (3.20).

This principle is rather controversial (in contrast to the *PMO*, which seems to be universally accepted). There are counterexamples from mechanics, from thermodynamics, and from electrodynamics, which clearly show that this principle is not a general natural law[81]. However, for almost all materials under the conditions, under which they are technically used, deviations from this principle are

---

[81] see, *e.g.*, SPEZIALE (1998) giving counterexamples from statistical mechanics and turbulence where the micro-spin enters the constitutive equations.

negligible. On the other hand, the principle is a strong tool to make the general material functional more concrete, which is appreciated worldwide. We will next demonstrate this reduction for simple materials. For these it has the form

$$\mathbf{Q}(t)\ \mathfrak{F}\{\boldsymbol{\chi}(\mathbf{x}_0\,,\,\tau)\,,\,\mathbf{F}(\mathbf{x}_0\,,\,\tau)\Big|_{\tau=0}^{t}\}\ \mathbf{Q}(t)^T$$

$$= \mathfrak{F}\{\mathbf{Q}(\tau)\,\boldsymbol{\chi}(\mathbf{x}_0\,,\,\tau) + \mathbf{c}(\tau)\,,\,\mathbf{Q}(\tau)\,\mathbf{F}(\mathbf{x}_0\,,\,\tau)\Big|_{\tau=0}^{t}\}$$

for all time-dependent $\mathbf{Q}(\tau) \in \mathcal{O}rth^+$ and $\mathbf{c}(\tau) \in \mathcal{V}$. Because of the arbitrariness of the latter, the stresses cannot depend on the path $\boldsymbol{\chi}(\mathbf{x}_0\,,\,\tau)$ of the point under consideration, and we have only

$$\mathbf{Q}(t)\ \mathfrak{F}\{\mathbf{F}(\mathbf{x}_0\,,\,\tau)\Big|_{\tau=0}^{t}\}\ \mathbf{Q}(t)^T = \mathfrak{F}\{\mathbf{Q}(\tau)\,\mathbf{F}(\mathbf{x}_0\,,\,\tau)\Big|_{\tau=0}^{t}\} \quad \forall\ \mathbf{Q}(\tau) \in \mathcal{O}rth^+$$

which restricts the dependence on $\mathbf{F}$, what we will consider later. Abusing the notation, we have simply dropped $\boldsymbol{\chi}$ from the list of the arguments of $\mathfrak{F}$, without introducing a new symbol for this reduced functional.

In the literature, the *PMO* is sometimes labelled as the passive and the *PISM* as the active form of the principle. Also *EUCLIDean invariance* and *frame invariance* are other names for these principles. One should be always cautious if no difference is made between these two principles.

A comparison with this last equation and the *PMO* gives as a consequence the

Theorem 4.6. Principle of form invariance (*PFI*)[82]
The material functions are invariant under change of observer

$$\mathfrak{F}^*\{\cdot\} = \mathfrak{F}\{\cdot\}$$

We could have also used *PMO* and *PFI* to prove *PISM*. The three principles have the following logical implications:

$$PMO \wedge PISM \quad\Rightarrow\quad PFI$$

$$PMO \wedge PFI \quad\Rightarrow\quad PISM$$

$$PFI \wedge PISM \quad\Rightarrow\quad PMO$$

In other words, two of them imply the third. Or, if one of them holds, then the other two are equivalent.[83] In particular, if *PFI* holds, then *PMO* and *PISM* cannot be distinguished, although their physical meaning is quite different[84]. This

[82] See NOLL (2005).
[83] It shall be emphasized that this conclusion holds for simple materials, where the stresses and strains are described by second-order tensors. For gradient materials these conclusions do not hold, as we will see at the end of this book.
[84] See SVENDSEN/ BERTRAM (1999) and BERTRAM/ SVENDSEN (2001).

situation arises in TRUESDELL/ NOLL (1965, Sect. 19), where under the label of *material frame-indifference* firstly the *PFI* is assumed, and secondly the *PMO*. This has to be borne in mind, because in the literature the *PFI* is often used without further reasoning.[85]

Further on we will restrict our considerations to those materials, for which the three principles hold, and we will do this for simplicity under the label of *objectivity*. Accordingly, we will call a constitutive functional *objective* if it obeys *PMO* and *PISM*. The power of this kind of objectivity is demonstrated in the following example.

Example. *Viscous Fluids*

We consider a non-linear simple fluid. We assume that its stresses depend on the current mass density ρ, the velocity $\mathbf{v}$ and its gradient $\mathbf{L}$

$$\mathbf{T} = f(\rho, \mathbf{v}, \mathbf{L}) .$$

We can split the velocity gradient $\mathbf{L} = \mathbf{D} + \mathbf{W}$ into its symmetric part $\mathbf{D}$ and its skew one $\mathbf{W}$. If we apply the *PISM* to this, the following must hold

$$\begin{aligned} \mathbf{Q}\, f(\rho, \mathbf{v}, \mathbf{D} + \mathbf{W})\, \mathbf{Q}^T &= f(\rho^*, \mathbf{v}^*, \mathbf{D}^* + \mathbf{W}^*) \\ &= f(\rho, \mathbf{Q}\,\mathbf{v} + \mathbf{Q}^\bullet\,\mathbf{x} + \mathbf{c}^\bullet, \mathbf{Q}\,\mathbf{D}\,\mathbf{Q}^T + \mathbf{Q}\,\mathbf{W}\,\mathbf{Q}^T + \mathbf{Q}^\bullet\,\mathbf{Q}^T) \end{aligned}$$

for arbitrary time-dependent proper orthogonal tensors $\mathbf{Q}(t)$ and vectors $\mathbf{c}(t)$, where all values have to be taken at the present time. Because of the arbitrariness of $\mathbf{c}^\bullet$ we conclude that f cannot depend on $\mathbf{v}$, which we skip from the list of arguments (without introducing a new symbol for the reduced function). Similarly, because of the arbitrariness of the skew part $\mathbf{Q}^\bullet\,\mathbf{Q}^T$, f cannot depend on $\mathbf{W}$. Therefore, the last entry must be symmetric. From the *PISM* remains

$$\mathbf{Q}\, f(\rho, \mathbf{D})\, \mathbf{Q}^T = f(\rho, \mathbf{Q}\,\mathbf{D}\,\mathbf{Q}^T) .$$

Thus, f is a hemitropic tensor-function for which we have the general representation of Theorem 1.26

$$\mathbf{T} = \varphi_0\,\mathbf{I} + \varphi_1\,\mathbf{D} + \varphi_2\,\mathbf{D}^2 \tag{4.3}$$

with three scalar valued functions

$$\varphi_i(\rho, I_{\mathbf{D}}, II_{\mathbf{D}}, III_{\mathbf{D}}) \qquad i = 0, 1, 2$$

of the mass density and the three principal invariants of $\mathbf{D}$. This is the general form of an objective compressible non-linear **REINER**[86] **fluid** (1945). Within this class of materials, this representation is necessary and

[85] For an overview of the literature on this topic see FREWER (2009).
[86] Markus Reiner (1886-1976)

sufficient for the *PISM* to hold. By linearisation we obtain the linear compressible **NAVIER[87]-STOKES fluid**

(4.4) $$\mathbf{T} = k\,(1-\rho/\rho_0)\,\mathbf{I} + \mu_v\, I_{\mathbf{D}}\,\mathbf{I} + 2\mu\,\mathbf{D}'$$

with the material constants

k	compression modulus
μ_v	volumetric viscosity.
μ	deviatoric viscosity

This is the form of a compressible fluid, as it is often used in aerodynamics. If it is incompressible (see Chap. 5.1), it takes the form commonly used in hydrodynamics

$$\mathbf{T} = -p\,\mathbf{I} + 2\mu\,\mathbf{D}\,.$$

4.4 Extension of the Principles to Thermodynamics

If we want to include thermodynamics, then we have to extend the principles of material theory.

At first we have to decide again, which variables to choose as dependent ones and as independent ones. The following choice is quite common in thermodynamics, but by no means compulsory:

• **independent variables**: the motion $\boldsymbol{\chi}$ and the temperature field θ at the present and past times. We will call

$$\{\boldsymbol{\chi}(\mathbf{x}_0\,,\tau)\,,\ \theta(\mathbf{x}_0\,,\tau)\big|_{\tau=0}^{t}\}$$

a thermo-kinematical process.

• **dependent variables**: the stresses, the heat flux, the internal energy, and the entropy. We call

$$\{\mathbf{T}(\mathbf{x}_0\,,t)\,,\ \mathbf{q}(\mathbf{x}_0\,,t)\,,\ \varepsilon(\mathbf{x}_0\,,t)\,,\ \eta(\mathbf{x}_0\,,t)\}$$

a **caloro-dynamical state** at point $\mathbf{x}_0$ at time t. By the definition of the free energy (3.36), we can exchange the internal with the free energy

$$\{\mathbf{T}(\mathbf{x}_0\,,t)\,,\ \mathbf{q}(\mathbf{x}_0\,,t)\,,\ \psi(\mathbf{x}_0\,,t)\,,\ \eta(\mathbf{x}_0\,,t)\}\,.$$

We now transfer the principles to these variables.

Assumption 4.7. Principle of determinism for thermo-mechanical materials
The caloro-dynamical state of a material point at a certain time is determined by the present and past (but not future) thermo-kinematical process of the body.

[87] Claude Louis Marie Henri Navier (1785-1836)

Assumption 4.8. Principle of local action for thermo-mechanical materials
The caloro-dynamical state of a material point depends on the thermo-kinematical process of only a finite neighbourhood.

This principle can again be restricted to the minimal neighbourhood of influence.

Assumption 4.9.
Principle of local action for simple thermo-mechanical materials
The caloro-dynamical state of a material point depends on the thermo-kinematical process of only its infinitesimal neighbourhood.

By this latter principle, the independent variables can be reduced to the processes of the motion and temperature of that point

$$\boldsymbol{\chi}(\mathbf{x}_0\,,\,\tau)\Big|_{\tau=0}^{t}\,,\;\theta(\mathbf{x}_0\,,\,\tau)\Big|_{\tau=0}^{t}$$

and of their (material) gradients

$$\mathbf{F}(\mathbf{x}_0\,,\,\tau)\Big|_{\tau=0}^{t} := Grad\,\boldsymbol{\chi}(\mathbf{x}_0\,,\,\tau)\Big|_{\tau=0}^{t}$$

$$\mathbf{g}_0(\mathbf{x}_0\,,\,\tau)\Big|_{\tau=0}^{t} := Grad\,\theta_L(\mathbf{x}_0\,,\,\tau)\Big|_{\tau=0}^{t}\,.$$

It is commonly assumed that the temperature field and the spatial temperature gradient are objective fields under EUCLIDean transformations

$$\theta_E^* = \theta_E \qquad\qquad \mathbf{g}^* = \mathbf{Q}\,\mathbf{g}\,.$$

Consequently, the temperature in the LAGRANGEan description and the material temperature gradient can be considered as invariant fields

$$\theta_L^* = \theta_L \qquad\qquad \mathbf{g}_0^* = \mathbf{g}_0\,.$$

With these assumptions, we know how the independent variables transform under change of observer. The transformations of the dependent variables follow from the next principle.

Assumption 4.10.
Principle of material objectivity for thermo-mechanical materials
The total dissipation of all materials is objective (and thus also invariant) under EUCLIDean transformations

$$\delta - \frac{\mathbf{q}\cdot\mathbf{g}}{\rho\theta} = \delta^* - \frac{\mathbf{q}^*\cdot\mathbf{g}^*}{\rho^*\theta^*}$$

with the mechanical dissipation δ *after* (3.37).

Then, by an analogous argument which led to the objectivity of the CAUCHY stresses, we obtain in this context the objectivity of the entire caloro-dynamical state

$$(\mathbf{T}, \mathbf{q}, \psi, \eta)^* = (\mathbf{Q}\,\mathbf{T}\,\mathbf{Q}^T, \mathbf{Q}\,\mathbf{q}, \psi, \eta)$$

With this principle we assure that the first law (3.35) and the second law of thermodynamics hold for all inertial observers, if they hold for one.

In order to make this more precise, we specify the set of material functionals for simple thermo-mechanical materials in the form

$$\mathbf{T}(\mathbf{x}_0, t) = \mathfrak{F}_T\{\boldsymbol{\chi}(\mathbf{x}_0, \tau), \mathbf{F}(\mathbf{x}_0, \tau), \theta(\mathbf{x}_0, \tau), \mathbf{g}(\mathbf{x}_0, \tau)\Big|_{\tau=0}^{t}\}$$

$$\mathbf{q}(\mathbf{x}_0, t) = \mathfrak{F}_q\{\boldsymbol{\chi}(\mathbf{x}_0, \tau), \mathbf{F}(\mathbf{x}_0, \tau), \theta(\mathbf{x}_0, \tau), \mathbf{g}(\mathbf{x}_0, \tau)\Big|_{\tau=0}^{t}\}$$

$$\psi(\mathbf{x}_0, t) = \mathfrak{F}_\psi\{\boldsymbol{\chi}(\mathbf{x}_0, \tau), \mathbf{F}(\mathbf{x}_0, \tau), \theta(\mathbf{x}_0, \tau), \mathbf{g}(\mathbf{x}_0, \tau)\Big|_{\tau=0}^{t}\}$$

$$\eta(\mathbf{x}_0, t) = \mathfrak{F}_\eta\{\boldsymbol{\chi}(\mathbf{x}_0, \tau), \mathbf{F}(\mathbf{x}_0, \tau), \theta(\mathbf{x}_0, \tau), \mathbf{g}(\mathbf{x}_0, \tau)\Big|_{\tau=0}^{t}\}$$

with $0 \le \tau \le t$ as the time parameter. This ansatz, in which all material functionals are assumed to depend on the same set of independent variables, follows the *principle of equipresence.* Objectivity of these functions requires

$$\mathbf{Q}(t)\, \mathfrak{F}_T\{\boldsymbol{\chi}(\mathbf{x}_0, \tau), \mathbf{F}(\mathbf{x}_0, \tau), \theta(\mathbf{x}_0, \tau), \mathbf{g}(\mathbf{x}_0, \tau)\Big|_{\tau=0}^{t}\}\, \mathbf{Q}^T(t)$$

$$= \mathfrak{F}_T^*\{\mathbf{Q}(\tau)\,\boldsymbol{\chi}(\mathbf{x}_0, \tau) + \mathbf{c}(\tau), \mathbf{Q}(\tau)\,\mathbf{F}(\mathbf{x}_0, \tau), \theta(\mathbf{x}_0, \tau), \mathbf{Q}(\tau)\,\mathbf{g}(\mathbf{x}_0, \tau)\Big|_{\tau=0}^{t}\}$$

$$\mathbf{Q}(t)\, \mathfrak{F}_q\{\boldsymbol{\chi}(\mathbf{x}_0, \tau), \mathbf{F}(\mathbf{x}_0, \tau), \theta(\mathbf{x}_0, \tau), \mathbf{g}(\mathbf{x}_0, \tau)\Big|_{\tau=0}^{t}\}$$

$$= \mathfrak{F}_q^*\{\mathbf{Q}(\tau)\,\boldsymbol{\chi}(\mathbf{x}_0, \tau) + \mathbf{c}(\tau), \mathbf{Q}(\tau)\,\mathbf{F}(\mathbf{x}_0, \tau), \theta(\mathbf{x}_0, \tau), \mathbf{Q}(\tau)\,\mathbf{g}(\mathbf{x}_0, \tau)\Big|_{\tau=0}^{t}\}$$

$$\mathfrak{F}_\psi\{\boldsymbol{\chi}(\mathbf{x}_0, \tau), \mathbf{F}(\mathbf{x}_0, \tau), \theta(\mathbf{x}_0, \tau), \mathbf{g}(\mathbf{x}_0, \tau)\Big|_{\tau=0}^{t}\}$$

$$= \mathfrak{F}_\psi^*\{\mathbf{Q}(\tau)\,\boldsymbol{\chi}(\mathbf{x}_0, \tau) + \mathbf{c}(\tau), \mathbf{Q}(\tau)\,\mathbf{F}(\mathbf{x}_0, \tau), \theta(\mathbf{x}_0, \tau), \mathbf{Q}(\tau)\,\mathbf{g}(\mathbf{x}_0, \tau)\Big|_{\tau=0}^{t}\}$$

$$\mathfrak{F}_\eta\{\boldsymbol{\chi}(\mathbf{x}_0, \tau), \mathbf{F}(\mathbf{x}_0, \tau), \theta(\mathbf{x}_0, \tau), \mathbf{g}(\mathbf{x}_0, \tau)\Big|_{\tau=0}^{t}\}$$

$$= \mathfrak{F}_\eta^*\{\mathbf{Q}(\tau)\,\boldsymbol{\chi}(\mathbf{x}_0, \tau) + \mathbf{c}(\tau), \mathbf{Q}(\tau)\,\mathbf{F}(\mathbf{x}_0, \tau), \theta(\mathbf{x}_0, \tau), \mathbf{Q}(\tau)\,\mathbf{g}(\mathbf{x}_0, \tau)\Big|_{\tau=0}^{t}\}$$

for all thermo-kinematical processes. With this we know how to transform the material functionals from one observer to any other. Nothing else can be concluded from *PMO.* Much stronger are the restrictions of the following principle.

Assumption 4.11.

Principle of invariance under superimposed rigid body motions for thermo-mechanical simple materials

If

$$\{\mathbf{T}(\mathbf{x}_0, t), \mathbf{q}(\mathbf{x}_0, t), \psi(\mathbf{x}_0, t), \eta(\mathbf{x}_0, t)\}$$

is the caloro-dynamical state resulting from the thermo-kinematical process

$$\{\boldsymbol{\chi}(\mathbf{x}_0, \tau), \mathbf{F}(\mathbf{x}_0, \tau), \theta(\mathbf{x}_0, \tau), \mathbf{g}(\mathbf{x}_0, \tau)\Big|_{\tau=0}^{t}\},$$

then

$$\{\mathbf{Q}(t)\,\mathbf{T}(\mathbf{x}_0, t)\,\mathbf{Q}^T(t), \mathbf{Q}(t)\,\mathbf{q}(\mathbf{x}_0, t), \psi(\mathbf{x}_0, t), \eta(\mathbf{x}_0, t)\}$$

is the caloro-dynamical state resulting from the modified process

$$\{\mathbf{Q}(\tau)\,\boldsymbol{\chi}(\mathbf{x}_0, \tau) + \mathbf{c}(\tau), \mathbf{Q}(\tau)\,\mathbf{F}(\mathbf{x}_0, \tau), \theta(\mathbf{x}_0, \tau), \mathbf{Q}(\tau)\,\mathbf{g}(\mathbf{x}_0, \tau)\Big|_{\tau=0}^{t}\}$$

with arbitrary $\mathbf{Q}(\tau) \in \mathcal{O}rth^+$ *and* $\mathbf{c}(\tau) \in \mathcal{V}$, $\tau \in [0, t]$.

Because of the arbitrariness of $\mathbf{c}(\tau)$ we must conclude the independence of the caloro-dynamical state on the path $\boldsymbol{\chi}(\mathbf{x}_0, \tau)$, and the following invariance requirements remain

$$\mathbf{Q}(t)\,\mathfrak{F}_T\{\mathbf{F}(\mathbf{x}_0, \tau), \theta(\mathbf{x}_0, \tau), \mathbf{g}(\mathbf{x}_0, \tau)\Big|_{\tau=0}^{t}\}\,\mathbf{Q}(t)^T$$

$$= \mathfrak{F}_T\{\mathbf{Q}(\tau)\,\mathbf{F}(\mathbf{x}_0, \tau), \theta(\mathbf{x}_0, \tau), \mathbf{Q}(\tau)\,\mathbf{g}(\mathbf{x}_0, \tau)\Big|_{\tau=0}^{t}\}$$

$$\mathbf{Q}(t)\,\mathfrak{F}_q\{\mathbf{F}(\mathbf{x}_0, \tau), \theta(\mathbf{x}_0, \tau), \mathbf{g}(\mathbf{x}_0, \tau)\Big|_{\tau=0}^{t}\}$$

$$= \mathfrak{F}_q\{\mathbf{Q}(\tau)\,\mathbf{F}(\mathbf{x}_0, \tau), \theta(\mathbf{x}_0, \tau), \mathbf{Q}(\tau)\,\mathbf{g}(\mathbf{x}_0, \tau)\Big|_{\tau=0}^{t}\}$$

$$\mathfrak{F}_\psi\{\mathbf{F}(\mathbf{x}_0, \tau), \theta(\mathbf{x}_0, \tau), \mathbf{g}(\mathbf{x}_0, \tau)\Big|_{\tau=0}^{t}\}$$

$$= \mathfrak{F}_\psi\{\mathbf{Q}(\tau)\,\mathbf{F}(\mathbf{x}_0, \tau), \theta(\mathbf{x}_0, \tau), \mathbf{Q}(\tau)\,\mathbf{g}(\mathbf{x}_0, \tau)\Big|_{\tau=0}^{t}\}$$

$$\mathfrak{F}_\eta\{\mathbf{F}(\mathbf{x}_0, \tau), \theta(\mathbf{x}_0, \tau), \mathbf{g}(\mathbf{x}_0, \tau)\Big|_{\tau=0}^{t}\}$$

$$= \mathfrak{F}_\eta\{\mathbf{Q}(\tau)\,\mathbf{F}(\mathbf{x}_0, \tau), \theta(\mathbf{x}_0, \tau), \mathbf{Q}(\tau)\,\mathbf{g}(\mathbf{x}_0, \tau)\Big|_{\tau=0}^{t}\}$$

for all $\mathbf{Q}(\tau) \in \mathcal{O}rth^+$. We simple dropped $\boldsymbol{\chi}$ from the list of the arguments, without introducing new symbols for the functionals.

The *PMO* is again equivalent to the *principle of form invariance* in the forms

$$\mathfrak{F}_T^*\{\cdot\} = \mathfrak{F}_T\{\cdot\} \ ; \ \mathfrak{F}_q^*\{\cdot\} = \mathfrak{F}_q\{\cdot\} \ ; \ \mathfrak{F}_\psi^*\{\cdot\} = \mathfrak{F}_\psi\{\cdot\} \ ; \ \mathfrak{F}_\eta^*\{\cdot\} = \mathfrak{F}_\eta\{\cdot\}$$

and the logical structure between the three principles remains valid also in the thermo-mechanical context.

Reduced Forms for Simple Materials

We have already shown that for objective simple materials the following equality holds

$$\mathbf{Q}(t)\ \mathfrak{F}\{\mathbf{F}(\mathbf{x}_0\,,\,\tau)\Big|_{\tau=0}^{t}\}\ \mathbf{Q}(t)^T = \mathfrak{F}\{\mathbf{Q}(\tau)\ \mathbf{F}(\mathbf{x}_0\,,\,\tau)\Big|_{\tau=0}^{t}\}$$

for all time functions $\mathbf{Q}(\tau) \in \mathcal{O}rth^+$ for the material functional $\mathfrak{F}$ which gives the CAUCHY stresses. If we choose in particular $\mathbf{Q}(\tau) \equiv \mathbf{R}(\tau)^T \in \mathcal{O}rth^+$ from the polar decomposition $\mathbf{F}(\mathbf{x}_0\,,\,\tau) = \mathbf{R}(\mathbf{x}_0\,,\,\tau)\ \mathbf{U}(\mathbf{x}_0\,,\,\tau)$, we get

$$\mathbf{R}(\mathbf{x}_0\,,\,t)^T\ \mathfrak{F}\{\mathbf{F}(\mathbf{x}_0\,,\,\tau)\Big|_{\tau=0}^{t}\}\ \mathbf{R}(\mathbf{x}_0\,,\,t) = \mathfrak{F}\{\mathbf{U}(\mathbf{x}_0\,,\,\tau)\Big|_{\tau=0}^{t}\}$$

and thus

(4.5)
$$\mathbf{T}(\mathbf{x}_0\,,\,t) = \mathbf{R}(\mathbf{x}_0\,,\,t)\ \mathfrak{F}_1\{\mathbf{U}(\mathbf{x}_0\,,\,\tau)\Big|_{\tau=0}^{t}\}\ \mathbf{R}(\mathbf{x}_0\,,\,t)^T,$$

where $\mathfrak{F}_1$ is the restriction of $\mathfrak{F}$ on processes in symmetric variables. For the relative stress tensors we get

(4.6)
$$\mathbf{T}_\mathrm{r}(\mathbf{x}_0\,,\,t) = \mathfrak{F}_1\{\mathbf{U}(\mathbf{x}_0\,,\,\tau)\Big|_{\tau=0}^{t}\}.$$

The sufficiency of these forms for the *PISM* can easily be seen if we substitute $\mathbf{R}$ by $\mathbf{Q}\ \mathbf{R}$ in (4.5) or (4.6), while $\mathbf{U}$ is invariant under superimposed rigid body motions. Thus, a simple material fulfils *PISM*, if and only if it can be put into the **reduced form** (4.5) or (4.6), where $\mathfrak{F}_1$ is an arbitrary functional of the process of a symmetric tensor. Note that this reduction and the ones which follow in the sequel, are a direct consequence of *PISM* and have nothing to do with changes of observers or frames of reference.

These two, however, are not the only possible reduced forms which identically fulfil *PISM*. If we substitute the relative stress tensor by any other invariant stress tensor, and the right stretch tensor by the right CAUCHY-GREEN tensor $\mathbf{C}$ or GREEN's strain tensor $\mathbf{E}^G$, we obtain further reduced forms

(4.7)
$$\mathbf{T}_\mathrm{r} = \mathfrak{F}_2\{\mathbf{C}(\mathbf{x}_0\,,\,\tau)\Big|_{\tau=0}^{t}\}$$

(4.8)
$$\mathbf{T}_\mathrm{r} = \mathfrak{F}_3\{\mathbf{E}^G(\mathbf{x}_0\,,\,\tau)\Big|_{\tau=0}^{t}\}$$

(4.9) $$\mathbf{S} = \mathscr{K}_1\{\mathbf{U}(\mathbf{x}_0, \tau)\big|_{\tau=0}^{t}\}$$

(4.10) $$\mathbf{S} = \mathscr{K}\{\mathbf{C}(\mathbf{x}_0, \tau)\big|_{\tau=0}^{t}\}$$

(4.11) $$\mathbf{S} = \mathscr{K}_2\{\mathbf{E}^{\mathrm{G}}(\mathbf{x}_0, \tau)\big|_{\tau=0}^{t}\}$$

(4.12) $$\mathbf{T}^{2\mathrm{PK}} = \mathscr{H}_1\{\mathbf{U}(\mathbf{x}_0, \tau)\big|_{\tau=0}^{t}\}$$

(4.13) $$\mathbf{T}^{2\mathrm{PK}} = \mathscr{H}_2\{\mathbf{C}(\mathbf{x}_0, \tau)\big|_{\tau=0}^{t}\}$$

(4.14) $$\mathbf{T}^{2\mathrm{PK}} = \mathscr{H}_3\{\mathbf{E}^{\mathrm{G}}(\mathbf{x}_0, \tau)\big|_{\tau=0}^{t}\}$$

or generally

(4.15) $$\mathbf{S}^{\mathrm{Gen}} = \mathscr{G}\{\mathbf{E}^{\mathrm{Gen}}(\mathbf{x}_0, \tau)\big|_{\tau=0}^{t}\}$$

where these functionals are not further restricted by the above principles.

A reduced form of a thermo-mechanical simple material is, *e.g.*

(4.16)
$$\mathbf{T}(\mathbf{x}_0, t) = \mathbf{R}(\mathbf{x}_0, t)\ \mathscr{F}_T\{\mathbf{U}(\mathbf{x}_0, \tau), \theta(\mathbf{x}_0, \tau), \mathbf{g}_0(\mathbf{x}_0, \tau)\big|_{\tau=0}^{t}\}\ \mathbf{R}(\mathbf{x}_0, t)^T$$
$$\mathbf{q}(\mathbf{x}_0, t) = \mathbf{R}(\mathbf{x}_0, t)\ \mathscr{F}_q\{\mathbf{U}(\mathbf{x}_0, \tau), \theta(\mathbf{x}_0, \tau), \mathbf{g}_0(\mathbf{x}_0, \tau)\big|_{\tau=0}^{t}\}$$
$$\psi(\mathbf{x}_0, t) = \mathscr{F}_\psi\{\mathbf{U}(\mathbf{x}_0, \tau), \theta(\mathbf{x}_0, \tau), \mathbf{g}_0(\mathbf{x}_0, \tau)\big|_{\tau=0}^{t}\}$$
$$\eta(\mathbf{x}_0, t) = \mathscr{F}_\eta\{\mathbf{U}(\mathbf{x}_0, \tau), \theta(\mathbf{x}_0, \tau), \mathbf{g}_0(\mathbf{x}_0, \tau)\big|_{\tau=0}^{t}\}$$

or by use of material variables

(4.17)
$$\mathbf{S}(\mathbf{x}_0, t) = \mathscr{K}_S\{\mathbf{C}(\mathbf{x}_0, \tau), \theta(\mathbf{x}_0, \tau), \mathbf{g}_0(\mathbf{x}_0, \tau)\big|_{\tau=0}^{t}\}$$
$$\mathbf{q}_0(\mathbf{x}_0, t) = \mathscr{K}_q\{\mathbf{C}(\mathbf{x}_0, \tau), \theta(\mathbf{x}_0, \tau), \mathbf{g}_0(\mathbf{x}_0, \tau)\big|_{\tau=0}^{t}\}$$
$$\psi(\mathbf{x}_0, t) = \mathscr{K}_\psi\{\mathbf{C}(\mathbf{x}_0, \tau), \theta(\mathbf{x}_0, \tau), \mathbf{g}_0(\mathbf{x}_0, \tau)\big|_{\tau=0}^{t}\}$$
$$\eta(\mathbf{x}_0, t) = \mathscr{K}_\eta\{\mathbf{C}(\mathbf{x}_0, \tau), \theta(\mathbf{x}_0, \tau), \mathbf{g}_0(\mathbf{x}_0, \tau)\big|_{\tau=0}^{t}\}$$

with the material heat flux $\mathbf{q}_0 = det(\mathbf{F})\ \mathbf{F}^{-1}\,\mathbf{q}$.

A general theory of reduced forms is given by BERTRAM/ SVENDSEN (2001), which generalises the above procedure to arbitrary sets of variables.

It should be remarked that NOLL (1972)[88] suggested an approach which does not make use of reference placements. In it the body is described as a differentiable manifold, the properties of which are formulated by concepts from differential geometry. Within this *intrinsic theory*, NOLL introduces *intrinsic configurations* and *intrinsic stresses*, which automatically lead to reduced forms. Although this theory is much more comprehensive and elegant as the present one, only few authors adopted it[89]. One reason for this somehow regrettable fact is perhaps that engineers are normally no experts with geometrical concepts like manifolds.

---

[88] See also NOLL (2005).

[89] See, *e.g.*, DEL PIERO (1975), ŠILHAVÝ/ KRATOCHVÍL (1977), KRAWIETZ (1986), BERTRAM (1982, 1989).

5 Internal Constraints

5.1 Mechanical Internal Constraints

Until now we have assumed that the material functionals are defined for any **C**–process in $\mathcal{P}sym$. However, it is not very likely that there are materials which can be submitted to arbitrarily large deformations without destruction. In all practical cases the functionals can only be identified for a subset of $\mathcal{P}sym$. Without loss of generality they can be extrapolated to the whole space, so that our procedure does not lead to false conclusions.

Nevertheless, there are restrictions for certain materials on the accessible states of deformation, such that even small deformations in certain directions would lead to extreme stresses. If the stresses are not that large, then these deformations can be neglected. Such materials react on any attempt to impose such deformations by *reaction stresses*, and this makes it necessary to modify the principle of determinism. Such restrictions are called *internal constraints*, in contrast to *external constraints*, which can be understood as fixed displacement conditions on the boundary, and are not under the scope of this material theory.

Definition 5.1. An **internal constraint** is a material function

$$\gamma_{\mathbf{F}} : \mathcal{I}nv^+ \to \mathcal{R} \qquad | \qquad \mathbf{F} \mapsto \gamma_{\mathbf{F}}(\mathbf{F}),$$

such that for all admissible kinematical processes $\mathbf{F}(\tau)\Big|_{\tau=0}^{t}$

(5.1) $$\gamma_{\mathbf{F}}(\mathbf{F}(\tau)) = 0 \qquad \forall\, \tau \in [0\,, t]$$

holds.

The condition $\gamma_{\mathbf{F}}(\mathbf{F}) = 0$ defines a subset of $\mathcal{I}nv^+$, the *constraint manifold*, which one can interpret as an 8-dimensional hypersurface in the 9-dimensional space of all tensors[90].

As $\gamma_{\mathbf{F}}$ is a material function, it is subjected to the principles of material theory. In particular, by *PISM* it must obey

$$\gamma_{\mathbf{F}}(\mathbf{F}) = \gamma_{\mathbf{F}}(\mathbf{Q\,F}) \qquad \forall\, \mathbf{F} \in \mathcal{I}nv^+ , \mathbf{Q} \in \mathcal{O}rth^+.$$

If we again choose for $\mathbf{Q}^T$ the rotation tensor of the polar decomposition $\mathbf{R}$, we obtain the reduced forms

(5.2) $$\gamma_{\mathbf{F}}(\mathbf{F})\big|_{\mathcal{S}ym} =: \gamma_{\mathbf{U}}(\mathbf{U}) =: \gamma_{\mathbf{C}}(\mathbf{C}) \qquad \text{with } \mathbf{C} = \mathbf{U}^2.$$

---

[90] See VIANELLO (1990).

A. Bertram, *Elasticity and Plasticity of Large Deformations*,
https://doi.org/10.1007/978-3-030-72328-6_5

There are cases where not only one, but several, internal constraints hold simultaneously. As an extreme case we have the **rigid body**, which can be characterised by six independent constraints (see below). All following results can be easily applied to several internal constraints, but for simplicity, we will write them down for only one single constraint.

The compatibility of the deformation with the constraint is assured by the permanent validity of (5.2). If the constraint equation is differentiable, then we can equivalently formulate it in the rate form

(5.3) $$\frac{d\gamma_C(\mathbf{C})}{d\mathbf{C}} \cdot \mathbf{C}^\bullet = grad\, \gamma_C(\mathbf{C}) \cdot \mathbf{C}^\bullet = 0 \,.$$

By *grad* we mean the derivative of the function γ_C in the space of the symmetric tensors. The geometrical interpretation is as follows. In each moment only those deformation rates $\mathbf{C}^\bullet$ are admissible, which have no component in the direction of $grad\, \gamma_C(\mathbf{C})$. In the 6-dimensional space $\mathscr{Sym}$ the constraint $\gamma_C(\mathbf{C}) = 0$ defines a 5-dimensional manifold or hypersurface, the gradient $grad\, \gamma_C(\mathbf{C})$ of which is its normal. Thus, all admissible deformation rates $\mathbf{C}^\bullet$ must be tangential to the hypersurface. By defining the constraint equation, one must make sure, however, that $grad\, \gamma_C(\mathbf{C})$ lies in this space, *i.e.*, is symmetric. This will be further on assumed.

The reaction stresses, which are necessary to maintain the constraint, have the property of not working under all admissible deformations. As they cannot be determined by the deformations anymore, we have to modify the principle of determinism.

Assumption 5.2.
Principle of determinism for materials with internal constraints
The current stresses in a body are determined by the deformation process only up to an additive part that does no work in any admissible deformation process.

After this principle, we can decompose the CAUCHY stress tensor into

$$\mathbf{T} = \mathbf{T}_E + \mathbf{T}_R$$

or the material stress tensor into

$$\mathbf{S} = \mathbf{S}_E + \mathbf{S}_R$$

with the **reaction stresses** $\mathbf{T}_R$ and $\mathbf{S}_R$, which do no work

$$l = \frac{1}{\rho}\, \mathbf{T}_R \cdot \mathbf{D} = \frac{1}{2\rho}\, \mathbf{S}_R \cdot \mathbf{C}^\bullet = 0 \,,$$

and the **extra stress**, which can still be determined by a material functional such as

$$\mathbf{S}_E = \mathfrak{F}\{\mathbf{C}(\tau)\big|_{\tau=0}^{t}\}$$

or any other reduced form. By comparison with the rate form of the internal constraint, we conclude that the reaction stress must be directed in parallel to the normal to the constraint manifold, and thus is a scalar multiple of it

(5.4) $$\mathbf{S}_R = \alpha \, grad \, \gamma_{\mathbf{C}}(\mathbf{C})$$

or

$$\mathbf{T}_R = \alpha \, \mathbf{F} \, grad \, \gamma_{\mathbf{C}}(\mathbf{C}) \, \mathbf{F}^T \qquad \text{with } \alpha \in \mathscr{R}.$$

For the resulting stress to be symmetric after (3.21), both the reaction stresses $\mathbf{T}_R$ and $\mathbf{S}_R$ and the extra stresses $\mathbf{T}_E$ and $\mathbf{S}_E$ must be symmetric. In order to make the above decompositions unique, we can normalise the extra stresses by the conditions

(5.5) $$\mathbf{T}_R \cdot \mathbf{T}_E = 0 = \mathbf{S}_R \cdot \mathbf{S}_E \,.$$

The multiplier α, which forms a time-dependent scalar (real) field on the body, cannot be determined by constitutive functions, but only by the balance laws for a specific boundary-value problem, as we will see later in Chap. 8. For the different stress tensors, these factors are also different.

If more than one (K) internal constraints $\gamma_{\mathbf{C}\,i}$ hold, then $\mathbf{S}_R$ has equally many parts

(5.6) $$\mathbf{S}_R = \sum_{i=1}^{K} \alpha_i \, grad \, \gamma_{\mathbf{C}\,i}(\mathbf{C})$$

with K independent scalar fields α_i.

This shall be demonstrated next for different examples of internal constraints.

Incompressibility

Some materials resist to volumetric changes, such that the assumption of constant volume (or mass density) is justified

$$\rho(t) = \rho_0 \quad \Leftrightarrow \quad J = det(\mathbf{F}) = 1 \;\Leftrightarrow\; det(\mathbf{C}) = 1$$

for all admissible and otherwise arbitrary deformations $\mathbf{C}$. Therefore, we have the constraint

$$\gamma_{\mathbf{F}}(\mathbf{F}) \equiv det(\mathbf{F}) - 1 \quad \text{or} \quad \gamma_{\mathbf{C}}(\mathbf{C}) \equiv det(\mathbf{C}) - 1$$

and with (1.39)

$$grad \, \gamma_{\mathbf{C}}(\mathbf{C}) = det(\mathbf{C}) \, \mathbf{C}^{-1} = \mathbf{C}^{-1}.$$

The reaction stress is a hydrostatic pressure after (5.4)

$$\mathbf{S}_R = - \, p \, \mathbf{C}^{-1}$$

or

$$\mathbf{T}_R = - \, p \, \mathbf{I}$$

and the (normed) CAUCHY extra stress is traceless or deviatoric

$$tr(\mathbf{T}_E) = tr(\mathbf{T} - \mathbf{T}_R) = 0 .$$

The reaction stress in terms of the 1st PIOLA-KIRCHHOFF tensor lies in the direction of $\mathbf{F}^{-T}$

$$\mathbf{T}^{1PK}{}_R = -p\,\mathbf{F}^{-T}$$

and, thus, is not spherical.

If we again look at the **example** of a REINER fluid (4.3), we obtain its incompressible form as

$$\mathbf{T} = -p\,\mathbf{I} + \varphi_1\,\mathbf{D} + \varphi_2\,\mathbf{D}^2$$

with two real valued functions $\varphi_i\,(II_\mathbf{D}, III_\mathbf{D})$, $i = 1, 2$. The velocity field is in this case generally divergence free and $\mathbf{D}$ is deviatoric. This does not imply that $\mathbf{D}^2$ is also deviatoric, so that the above split into reaction pressure and extra stress is not yet normalised, and p is not the resulting pressure. The normalised version would contain the deviator $(\mathbf{D}^2)'$ instead.

This law contains the incompressible NAVIER-STOKES fluid

$$\mathbf{T} = -p\,\mathbf{I} + 2\,\mu\,\mathbf{D}$$

as a special case with the undetermined pressure p and only one material constant, the shear viscosity μ . In hydrostatics, the pressure can be determined by the equations of equilibrium

$$p = \rho_0\,g\,h + p_0 ,$$

with the water level h , the gravitational constant g , and the external pressure on the water surface p_0 , *i.e.*, as a result of the boundary-value problem.

Inextensibility in one Direction

In fibre-reinforced materials the situation may occur, that the fibre is almost inextensible compared to the stiffness of the matrix material. If $\mathbf{i} \in \mathcal{V}$ is tangent to the fibre at a material point $\mathbf{x}_0$ in the reference placement, then inextensibility of the fibre means locally that

$$(\mathbf{F\,i}) \cdot (\mathbf{F\,i}) = \mathbf{i} \cdot (\mathbf{C\,i}) = \mathbf{C} \cdot \mathbf{i} \otimes \mathbf{i}$$

is constant in all placements. If we norm this function to 1 , then the internal constraint is

$$\gamma_C(\mathbf{C}) \equiv \mathbf{C} \cdot \mathbf{i} \otimes \mathbf{i} - 1 .$$

The reaction stress is in direction of

$$grad\,\gamma_C(\mathbf{C}) = \mathbf{i} \otimes \mathbf{i} ,$$

i.e., a tensile stress in the direction of the fibre in the reference placement

$$\mathbf{S}_R = \alpha\, \mathbf{i} \otimes \mathbf{i}$$

or in the current placement

$$\mathbf{T}_R = \alpha\, (\mathbf{F}\, \mathbf{i}) \otimes (\mathbf{F}\, \mathbf{i}) = \alpha\, \mathbf{F} * (\mathbf{i} \otimes \mathbf{i}).$$

Shear Locking in one Plane

occurs, if the angle between two tangent vectors $\mathbf{i}_{1,\,2} \in \mathscr{V}$ in a certain material plane remains constant for all admissible placements. If we choose orthogonal ones in the reference placement, then they must remain so during the motion

$$(\mathbf{F}\, \mathbf{i}_1) \cdot (\mathbf{F}\, \mathbf{i}_2) = \mathbf{i}_1 \cdot (\mathbf{C}\, \mathbf{i}_2) = \mathbf{C} \cdot (\mathbf{i}_1 \otimes \mathbf{i}_2) = 0\,.$$

The constraint equation is in this case after symmetrisation

$$\gamma_C(\mathbf{C}) \equiv \mathbf{C} \cdot (\mathbf{i}_1 \otimes \mathbf{i}_2 + \mathbf{i}_2 \otimes \mathbf{i}_1) = 0$$

and the reaction stress

$$\mathbf{S}_R = \alpha\, (\mathbf{i}_1 \otimes \mathbf{i}_2 + \mathbf{i}_2 \otimes \mathbf{i}_1)$$

or

$$\mathbf{T}_R = \alpha\, (\mathbf{F}\, \mathbf{i}_1 \otimes \mathbf{F}\, \mathbf{i}_2 + \mathbf{F}\, \mathbf{i}_2 \otimes \mathbf{F}\, \mathbf{i}_1) = \alpha\, \mathbf{F} * (\mathbf{i}_1 \otimes \mathbf{i}_2 + \mathbf{i}_2 \otimes \mathbf{i}_1)\,,$$

which corresponds to the shear stresses in the plane.

Rigidity

occurs, if the material point can only exist in one single configuration $\mathbf{C}_0$, *i.e.*, if

$$\mathbf{C}_0^{-1}\, \mathbf{C}(t) = \mathbf{I}\,.$$

This corresponds to six internal constraints for the six components of $\mathbf{C}(t)$. In this case the entire stress tensor is indeterminate, and the extra stress can be normed to $\mathbf{0}$.

However, one must be carefull when applying this procedure since there cases where it leads to unrealistic results.

One example is the KRAWIETZ constraint

$$\gamma_C(\mathbf{C}) \equiv (\mathbf{C} - \mathbf{I}) \cdot (\mathbf{C} - \mathbf{I}) = 0$$

which obviously forces $\mathbf{C}$ to be $\mathbf{I}$, *i.e.*, rigidity. For it

$$grad\; \gamma_C(\mathbf{C} = \mathbf{I}) = \mathbf{O}$$

so that there are no reactive stresses at all.

Another example is the combination of incompressibility with BELL´s constraint (1973, 1985, 1996)[91]

$$\gamma_C(\mathbf{C}) \equiv tr\,\mathbf{C} - 3 .$$

In both cases the reaction stresses would be a hydrostatic pressure, while the combination of the two constraints also leads to rigidity, see BEATTY & HAYES (1992).

These two examples show that the theory of internal constraints needs some refinement. In particular, one should make sure that the constraint equation really defines a constraint *manifold* (which is not the case for the KRAIWETZ constraint), and in the case of multiple constraints that they are really independent.

[91] BELL uses the square-root of **C**, which however does not make much difference here.

5.2 Thermo-Mechanical Internal Constraints

Not only in mechanics, but also more generally in thermo-mechanics, the introduction of internal constraints can be reasonable, if not all thermo-kinematical processes are admissible. In the literature, several suggestions have been made to generalise the mechanical concepts[92], which we will here adopt in a modified way.

Definition 5.3. A **thermo-mechanical internal constraint** consists of material functions

$$\mathbf{J} : \mathscr{Psym} \times \mathscr{R}^+ \to \mathscr{Sym} \quad | \quad (\mathbf{C}, \theta) \mapsto \mathbf{J}(\mathbf{C}, \theta)$$

$$\mathbf{j} : \mathscr{Psym} \times \mathscr{R}^+ \to \mathscr{V} \quad | \quad (\mathbf{C}, \theta) \mapsto \mathbf{j}(\mathbf{C}, \theta)$$

$$j : \mathscr{Psym} \times \mathscr{R}^+ \to \mathscr{R} \quad | \quad (\mathbf{C}, \theta) \mapsto j(\mathbf{C}, \theta)$$

such that for all admissible thermo-kinematical processes the constraint equation

(5.7) $$\mathbf{J}(\mathbf{C}, \theta) \cdot \mathbf{C}^{\bullet} + \mathbf{j}(\mathbf{C}, \theta) \cdot \mathbf{g}_0 + j(\mathbf{C}, \theta)\, \theta^{\bullet} = 0$$

holds at each instant.

Note that the first term, to which the equation is reduced in the isothermal case, corresponds to the mechanical constraint in its rate form (5.3). Since we made use of material variables, this constraint is already in a reduced form, *i.e.* objective.

Once again, we have to modify the principle of determinism.

Assumption 5.4. Principle of determinism for materials with thermo-mechanical internal constraints
The current values of stress, heat flux, internal energy, and entropy are determined by the thermo-kinematical process only up to additive parts that are not dissipative during all admissible processes.

Thus, we have the decompositions of the

stress $\quad \mathbf{S} = \mathbf{S}_E + \mathbf{S}_R$

heat flux $\quad \mathbf{q}_0 = \mathbf{q}_{0E} + \mathbf{q}_{0R}$

internal energy $\quad \varepsilon = \varepsilon_E + \varepsilon_R$

entropy $\quad \eta = \eta_E + \eta_R$

and, consequently, also for the

free energy $\quad \psi = \varepsilon_E + \varepsilon_R - \theta\, \eta_E - \theta\, \eta_R =: \psi_E + \psi_R$

---

92 GREEN/ NAGHDI/ TRAPP (1970), GURTIN/ PODIO GUIDUGLI (1973), BERTRAM/ HAUPT (1974)

where only the extra-terms depend on the thermo-kinematical process. The reaction parts do not dissipate in the sense of the CLAUSIUS-DUHEM inequality (3.37)

$$\frac{1}{2\rho}\mathbf{S}_R \cdot \mathbf{C}^\bullet - \frac{1}{\theta\,\rho_0}\mathbf{q}_{0R} \cdot \mathbf{g}_0 - \psi_R^\bullet - \eta_R\,\theta^\bullet = 0$$

for all admissible thermo-kinematical processes. If we subtract from this equation an α-fold of the constraint equation (5.7), we get

$$(\frac{1}{2\rho}\mathbf{S}_R - \alpha\,\mathbf{J}) \cdot \mathbf{C}^\bullet - (\frac{1}{\theta\,\rho_0}\mathbf{q}_{0R} + \alpha\,\mathbf{j}) \cdot \mathbf{g}_0$$

$$- \psi_R^\bullet - (\eta_R + \alpha j)\,\theta^\bullet = 0$$

for any real α. Because of the arbitrariness of $\mathbf{C}^\bullet$, $\mathbf{g}_0$, and $\theta^\bullet$, this is solved for all constrained materials only by

$$\mathbf{S}_R = 2\,\alpha\,\rho\,\mathbf{J}(\mathbf{C}, \theta)$$

$$\mathbf{q}_{0R} = -\,\alpha\,\rho_0\,\theta\,\mathbf{j}(\mathbf{C}, \theta)$$

$$\psi_R^\bullet = 0$$

$$\eta_R = -\,\alpha\,j(\mathbf{C}, \theta)$$

or spatially

$$\mathbf{T}_R = 2\,\alpha\,\rho\,\mathbf{F}\,\mathbf{J}(\mathbf{C}, \theta)\,\mathbf{F}^T \qquad \text{with } \mathbf{S}_R = \mathbf{F}^{-1}\,\mathbf{T}_R\,\mathbf{F}^{-T}$$

$$\mathbf{q}_R = -\,\alpha\,\rho\,\theta\,\mathbf{F}\,\mathbf{j}(\mathbf{C}, \theta) \qquad \text{with } \mathbf{q}_{0R} = J\,\mathbf{F}^{-1}\,\mathbf{q}_R\,.$$

With this form, for no $\alpha \in \mathscr{R}$ can a contradiction to the CLAUSIUS-DUHEM inequality occur, if the extra terms already fulfil it alone.

As the free energy is only determined up to a constant, we can principally assume

$$\psi_R = 0\,.$$

If more than one constraint is active, then the reactive parts are simply additive superpositions of those resulting from each constraint alone.

This shall be demonstrated by some examples.

Perfect Heat Conduction in one Direction

means that in this direction, indicated by a tangent vector $\mathbf{j} \in \mathscr{V}$ in the reference placement, no temperature gradient can be achieved

$$\mathbf{g}_0 \cdot \mathbf{j} = 0\,.$$

In this case, $\mathbf{J} \equiv \mathbf{0}$ and $j \equiv 0$, and thus

$$\mathbf{S}_R = \mathbf{0}$$

$$\mathbf{q}_R = \alpha\,\rho\,\theta\,\mathbf{F}\,\mathbf{j}$$

$$\eta_R = 0\,.$$

The heat flux in the direction of $\mathbf{F\,j}$ is a reactive part, which is determined by the heat supply on the surface and/or heat sources in the interior of the body through the first law of thermodynamics.

Temperature-Dependent Incompressibility

Some materials cannot be compressed by pressure, but they can still expand their volume under changes of temperature. For them, the change of mass density is a function of temperature alone

$$det(\mathbf{C}) = f(\theta) \quad \text{with } f : \mathscr{R}^+ \to \mathscr{R}^+.$$

Its time derivative is by the chain rule

$$det(\mathbf{C})\, \mathbf{C}^{-1} \cdot \mathbf{C}^\bullet = f(\theta)'\, \theta^\bullet.$$

This forms the constraint equation with the identifications

$$\mathbf{J} \equiv \mathbf{C}^{-1}$$

$$\mathbf{j} \equiv \mathbf{o}$$

$$j(\theta) \equiv -\frac{f(\theta)'}{f(\theta)}\,.$$

The reaction stress is again a pressure

$$\mathbf{S}_R = 2\,\alpha\,\rho\,\mathbf{C}^{-1}$$

and

$$\mathbf{q}_R = \mathbf{o}$$

$$\eta_R = -\,\alpha\, j(\theta)\,.$$

Temperature-Dependent Inextensibility in one Direction

being indicated by the tangent vector $\mathbf{i} \in \mathscr{V}$ in the reference placement, is established by the constraint

$$(\mathbf{F\,i}) \cdot (\mathbf{F\,i}) = \mathbf{i} \cdot (\mathbf{C\,i}) = \mathbf{C} \cdot (\mathbf{i} \otimes \mathbf{i}) = f(\theta) \quad \text{with } f : \mathscr{R}^+ \to \mathscr{R}^+$$

or, equivalently, by the rate form

$$(\mathbf{i} \otimes \mathbf{i}) \cdot \mathbf{C}^\bullet = f(\theta)'\, \theta^\bullet.$$

By the identifications

$$\mathbf{J} \equiv \mathbf{i} \otimes \mathbf{i}$$

$$\mathbf{j} \equiv \mathbf{o}$$

$$j(\theta) \equiv -f(\theta)'$$

this again leads to the constraint equation. The reaction stress is a tensile stress in the direction of $\mathbf{i}$

$$\mathbf{S}_R = 2\,\rho\,\alpha\,\mathbf{i} \otimes \mathbf{i}$$

and

$$\mathbf{q}_R = \mathbf{o}$$

$$\eta_R = -\alpha\, j(\theta)\,.$$

In Chap. 8.2 an exact solution for a material with two thermo-mechanical constraints is given.

6 Elasticity

A Comment on the Literature. The following books and articles can be considered as introductions to elasticity: ATKIN/ FOX (1980), CARLSON/ SHIELD (1982), CIARLET (1983 and 1988)), DOYLE/ ERICKSEN (1956), FRAEIJS DE VEUBEKE (1979), GREEN/ ZERNA (1954), GREEN/ ADKINS (1960, 1970), OGDEN (1984a), PODIO-GUIDUGLI (2000), TABER (2004).

If one prefers a more geometrical introduction, we suggest MARSDEN/ HUGHES (1983) and HANYGA (1985).

The following books also contain a detailed part on elasticity in a more general context of material theory or continuum mechanics: BAŞAR/ WEICHERT (2000), BETTEN (1993, 2001), DOGHRI (2000), DROZDOV (1996), HAUPT (2000), HOLZAPFEL (2000), KRAWIETZ (1986), SALENÇON (2001), ŠILHAVÝ (1997).

The following books contain much more than a mere introduction and can serve as handbooks: ANTMAN (1995), LUBARDA (2002), LURIE (1990), WANG/ TRUESDELL (1973), TRUESDELL/ NOLL (1965). In particular, this last book is a very rich source in the field and has been extensively used for the present text.

The following works focus on more specific aspects of finite elasticity: BALL (1977), BEATTY (1987), BONET/ WOOD (1997), HAYES/ SACCOMANDI (2001), LEMAITRE (2001), MÜLLER/ STREHLOW (2004), OGDEN (2003), TRELOAR (1975), VALENT (1988).

The following works by pioneers in finite elasticity are recommended for reading, not only for historical reasons: NEUMANN (1860), FINGER (1894), KAPPUS (1939), RICHTER (1948).

Elasticity is the most simple of all material theories, and, at the same time, the most frequently applied one. HOOKE´s[93] famous statement "*ut tensio sic vis*" (1675) hidden in an anagram, was the starting point for a tremendous development soon passing from success to success.

In the classical (geometrically and physically linear) elasticity theory one assumes a linear relation (HOOKE´s law) between the (linear) deformation tensor **E** and CAUCHY´s stress tensor **T**

(6.1) $$\mathbf{T} = \mathbf{C}[\mathbf{E}] ,$$

[93] Robert Hooke (1635-1703)

A. Bertram, *Elasticity and Plasticity of Large Deformations*,
https://doi.org/10.1007/978-3-030-72328-6_6

and material theory is only concerned by the properties of the *4*th-order elasticity tensors $\mathbf{C}$, as this has been done by CAUCHY, NAVIER, POISSON, and others at the beginning of the 19th century.

The extension of elasticity theory to large deformations, the *finite elasticity theory*, was almost simultaneously developed by CAUCHY, GREEN, KIRCHHOFF, KELVIN, FINGER, and others, but did not begin to blossom before the middle of the 20th century, when TRELOAR, RIVLIN, NOLL and others developed a theory applicable to rubber polymers. The foundations of finite elasticity have already been worked out half a century ago, so the main task that remains is to invent specific material models within this framework, and to identify the material parameters by appropriate experiments and/or by micro-physical considerations.

6.1 Reduced Elastic Forms

Elasticity is based on the strongest possible restriction of the influence of the past deformations on the present stresses.

Definition 6.1. A simple material is called **elastic** if the current stress is determined by the current placement

$$\mathbf{T}(\mathbf{x}_0, t) = f(\boldsymbol{\chi}(\mathbf{x}_0, t), \mathbf{F}(\mathbf{x}_0, t))$$

by a function

$$f : \mathscr{V} \times \mathscr{Inv}^+ \rightarrow \mathscr{Sym}$$

In the sequel, we consider all variables at a certain instant t and at a certain material point $\mathbf{x}_0$. For brevity, we suppress the arguments $(\mathbf{x}_0, t)$ of all field variables.

By applying the same objectivity arguments as in Chap. 4, we immediately see that the stresses are independent of $\boldsymbol{\chi}(\mathbf{x}_0, t)$, so that only the second argument remains

(6.2) $$\mathbf{T} = f(\mathbf{F}) \qquad \text{with} \qquad f : \mathscr{Inv}^+ \rightarrow \mathscr{Sym},$$

where we incorrectly used the same notation for the reduced function. A reduced form is

(6.3) $$\mathbf{T} = \mathbf{R}\, f_1(\mathbf{U})\, \mathbf{R}^T$$

with f_1 being the restriction of f to $\mathscr{Psym}$. Other reduced forms are

(6.4) $$\mathbf{T}_r = f_1(\mathbf{U})$$

(6.5) $$\mathbf{T}_r = f_2(\mathbf{C})$$

(6.6) $$\mathbf{T}_r = f_3(\mathbf{E}^G)$$

(6.7) $$\mathbf{S} = k_1(\mathbf{U})$$

(6.8) $\mathbf{S} = k(\mathbf{C})$

(6.9) $\mathbf{S} = k_2(\mathbf{E}^{\mathrm{G}})$

(6.10) $\mathbf{T}^{2\mathrm{PK}} = h_1(\mathbf{U})$

(6.11) $\mathbf{T}^{2\mathrm{PK}} = h(\mathbf{C})$

(6.12) $\mathbf{T}^{2\mathrm{PK}} = h_2(\mathbf{E}^{\mathrm{G}})$

(6.13) $\mathbf{S}^{\mathrm{Gen}} = g(\mathbf{E}^{\mathrm{Gen}})$

etc. with

$$f_1, f_2, k_1, k, h_1, h : \mathcal{Psym} \rightarrow \mathcal{Sym}$$

and

$$f_3, k_2, h_2, g : \mathcal{Sym} \rightarrow \mathcal{Sym},$$

as special versions of (4.5) to (4.17), which identically fulfil the *PISM*. Because of the rate-independence of elasticity (see Chap. 9), it is sufficient to require the invariance under GALILEIan transformations instead of all EUCLIDean ones.

In contrast to these reduced forms, the geometrically linear HOOKE´s law (6.1) violates *PISM*, and, thus, is not valid for large deformations and rotations.

6.2 Thermo-Elasticity

Analogously we extend the definition to thermo-elasticity.

Definition 6.2. A simple material is called **thermo-elastic** if the current caloro-dynamical state is determined by its current thermo-kinematical state, *i.e.*, if there exist functions

$$f_T : \mathcal{V} \times \mathcal{Inv}^+ \times \mathcal{R}^+ \times \mathcal{V} \rightarrow \mathcal{Sym} \mid (\boldsymbol{\chi}, \mathbf{F}, \theta, \mathbf{g}) \mapsto \mathbf{T}$$

$$f_q : \mathcal{V} \times \mathcal{Inv}^+ \times \mathcal{R}^+ \times \mathcal{V} \rightarrow \mathcal{V} \mid (\boldsymbol{\chi}, \mathbf{F}, \theta, \mathbf{g}) \mapsto \mathbf{q}$$

$$f_\psi : \mathcal{V} \times \mathcal{Inv}^+ \times \mathcal{R}^+ \times \mathcal{V} \rightarrow \mathcal{R} \mid (\boldsymbol{\chi}, \mathbf{F}, \theta, \mathbf{g}) \mapsto \psi$$

$$f_\eta : \mathcal{V} \times \mathcal{Inv}^+ \times \mathcal{R}^+ \times \mathcal{V} \rightarrow \mathcal{R} \mid (\boldsymbol{\chi}, \mathbf{F}, \theta, \mathbf{g}) \mapsto \eta$$

where all field-variables are taken at the same $(\mathbf{x}_0, t)$.

Possible reduced forms are then

$$\mathbf{T} = \mathbf{R}\, f_T(\mathbf{U}, \theta, \mathbf{g}_0)\, \mathbf{R}^T \quad \mid \quad f_T : \mathcal{Psym} \times \mathcal{R}^+ \times \mathcal{V} \rightarrow \mathcal{Sym}$$

$$\mathbf{q} = \mathbf{R}\, f_q(\mathbf{U}, \theta, \mathbf{g}_0) \quad \mid \quad f_q : \mathcal{Psym} \times \mathcal{R}^+ \times \mathcal{V} \rightarrow \mathcal{V}$$

(6.14) $$\psi = f_\psi(\mathbf{U}, \theta, \mathbf{g}_0) \quad \mid \quad f_\psi : \mathcal{Psym} \times \mathcal{R}^+ \times \mathcal{V} \rightarrow \mathcal{R}$$

$$\eta = f_\eta(\mathbf{U}, \theta, \mathbf{g}_0) \qquad | \quad f_\eta: \mathscr{Psym} \times \mathscr{R}^+ \times \mathscr{V} \to \mathscr{R}$$

or in material representation

$$\mathbf{S} = k_S(\mathbf{C}, \theta, \mathbf{g}_0) \qquad | \quad k_T: \mathscr{Psym} \times \mathscr{R}^+ \times \mathscr{V} \to \mathscr{Sym}$$

$$\mathbf{q}_0 = k_q(\mathbf{C}, \theta, \mathbf{g}_0) \qquad | \quad k_q: \mathscr{Psym} \times \mathscr{R}^+ \times \mathscr{V} \to \mathscr{V}$$

(6.15)
$$\psi = k_\psi(\mathbf{C}, \theta, \mathbf{g}_0) \qquad | \quad k_\psi: \mathscr{Psym} \times \mathscr{R}^+ \times \mathscr{V} \to \mathscr{R}$$

$$\eta = k_\eta(\mathbf{C}, \theta, \mathbf{g}_0) \qquad | \quad k_\eta: \mathscr{Psym} \times \mathscr{R}^+ \times \mathscr{V} \to \mathscr{R}.$$

6.3 Change of the Reference Placement

If we use an elastic law (6.2)

$$\mathbf{T} = f(\mathbf{F}) \qquad \text{with } \mathbf{F} = Grad\, \boldsymbol{\chi} = Grad(\kappa\, \kappa_0^{-1}) \in \mathscr{Inv},$$

then this function f evidently depends on the used reference placement κ_0. One should therefore keep κ_0 in the list of arguments

$$\mathbf{T} = f(\kappa_0, \mathbf{F})$$

in this section.

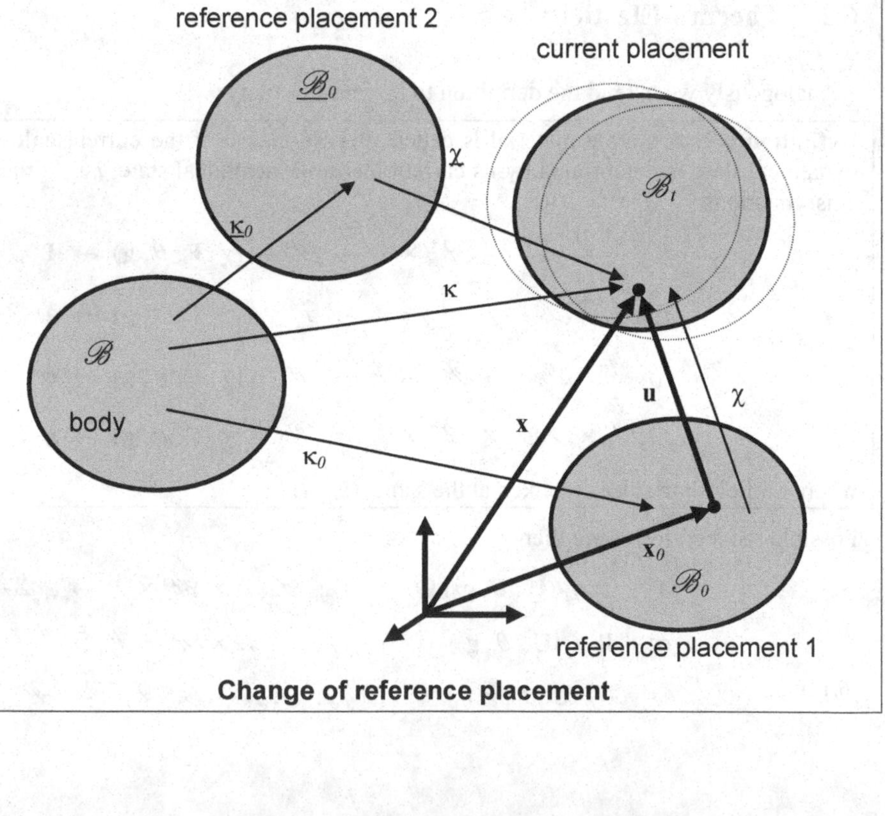

Change of reference placement

A change of the reference placement from κ_0 to $\underline{\kappa}_0$ also changes the deformation gradient because of the chain rule

$$\underline{\mathbf{F}} = Grad(\kappa\, \underline{\kappa}_0^{-1})$$
$$= Grad(\kappa\, \kappa_0^{-1}\, \kappa_0\, \underline{\kappa}_0^{-1})$$
$$= Grad(\kappa\, \kappa_0^{-1})\, Grad(\kappa_0\, \underline{\kappa}_0^{-1}) =$$

(6.16) $$\underline{\mathbf{F}} = \mathbf{F}\,\mathbf{K} \qquad \in \mathcal{I}nv$$

with

$$\mathbf{K} := Grad(\kappa_0\, \underline{\kappa}_0^{-1}) \qquad \in \mathcal{I}nv$$

and

$$\mathbf{F} := Grad(\kappa\, \kappa_0^{-1}) \qquad \in \mathcal{I}nv\,.$$

If the two involved reference placements are admissible placements, then $\kappa_0\, \underline{\kappa}_0^{-1}$ must be orientation preserving, and, thus, $\mathbf{F}$, $\underline{\mathbf{F}}$, $\mathbf{K} \in \mathcal{I}nv^+$. This shall be assumed further on. If we can obtain one of the reference placements by shifting and rotating the other (superimposed rigid body motion), then $\mathbf{K}$ is a versor and $\mathbf{F}$ and $\underline{\mathbf{F}}$ differ only by their rotation tensors. This, however, shall not be assumed in general.

For the relation of the mass densities in the reference placements we obtain the relation

$$\underline{\rho}_0 / \rho_0 = det(\mathbf{K}) > 0\,.$$

With the transformation of $\mathbf{F}$ under a change of the reference placement, those of all other deformation tensors are determined, like the right CAUCHY-GREEN tensor

(6.17) $$\underline{\mathbf{C}} = \underline{\mathbf{F}}^T\, \underline{\mathbf{F}} = \mathbf{K}^T\, \mathbf{C}\, \mathbf{K} \qquad \in \mathcal{P}sym$$

or GREEN's strain tensor

$$\underline{\mathbf{E}}^G = \tfrac{1}{2}\,(\underline{\mathbf{C}} - \mathbf{I}) = \tfrac{1}{2}\,(\mathbf{K}^T\, \mathbf{C}\, \mathbf{K} - \mathbf{I})$$

(6.18) $$= \mathbf{K}^T\, \mathbf{E}^G\, \mathbf{K} + \tfrac{1}{2}\,(\mathbf{K}^T\, \mathbf{K} - \mathbf{I}) \qquad \in \mathcal{S}ym\,.$$

The last term vanishes for the special case of an orthogonal $\mathbf{K}$. Even more complicated transformations result for the generalised material strain tensors for non-orthogonal $\mathbf{K}$, as we have to solve the eigenvalue problem for both reference placements.

As CAUCHY's stresses do not depend on the reference placement, we must conclude that

$$\mathbf{T} = f(\kappa_0, \mathbf{F}) = f(\underline{\kappa}_0, \underline{\mathbf{F}}) = f(\underline{\kappa}_0, \mathbf{F}\,\mathbf{K}) \qquad \forall\, \mathbf{F} \in \mathcal{L}in\,.$$

For the material stress tensor in two reference placements we obtain

$$\underline{\mathbf{S}} = k(\underline{\kappa}_0, \underline{\mathbf{C}}) \\ = \underline{\mathbf{F}}^{-1} f(\underline{\kappa}_0, \underline{\mathbf{F}})\, \underline{\mathbf{F}}^{-T}$$

and

$$\begin{aligned} \mathbf{S} &= k(\kappa_0, \mathbf{C}) \\ &= \mathbf{F}^{-1} f(\kappa_0, \mathbf{F})\, \mathbf{F}^{-T} \\ &= \mathbf{K}\, \underline{\mathbf{F}}^{-1} f(\underline{\kappa}_0, \underline{\mathbf{F}})\, \underline{\mathbf{F}}^{-T}\, \mathbf{K}^T \\ &= \mathbf{K}\, k(\underline{\kappa}_0, \underline{\mathbf{C}})\, \mathbf{K}^T \\ &= \mathbf{K}\, \underline{\mathbf{S}}\, \mathbf{K}^T. \end{aligned}$$

With this we know how the reduced form k behaves under a change of the reference placement

(6.19) $$k(\kappa_0, \mathbf{C}) = \mathbf{K}\, k(\underline{\kappa}_0, \mathbf{K}^T \mathbf{C}\, \mathbf{K})\, \mathbf{K}^T.$$

Also the 2nd PIOLA-KIRCHHOFF tensor depends on the reference placement

$$\underline{\mathbf{T}}^{2PK} = h(\underline{\kappa}_0, \underline{\mathbf{C}}) = h(\underline{\kappa}_0, \mathbf{K}^T \mathbf{C}\, \mathbf{K})$$

and

$$\begin{aligned} \mathbf{T}^{2PK} &= h(\kappa_0, \mathbf{C}) \\ &= det(\mathbf{F})\, \mathbf{F}^{-1} f(\kappa_0, \mathbf{F})\, \mathbf{F}^{-T} \\ &= det(\mathbf{K}^{-1})\, det(\mathbf{F}\, \mathbf{K})\, \mathbf{K}\, \mathbf{K}^{-1}\, \mathbf{F}^{-1} f(\underline{\kappa}_0, \mathbf{F}\, \mathbf{K})\, \mathbf{F}^{-T} \mathbf{K}^{-T}\, \mathbf{K}^T \\ &= det(\mathbf{K}^{-1})\, \mathbf{K}\, h(\underline{\kappa}_0, \underline{\mathbf{C}})\, \mathbf{K}^T \\ &= det^{-1}(\mathbf{K})\, \mathbf{K}\, \underline{\mathbf{T}}^{2PK}\, \mathbf{K}^T. \end{aligned}$$

This determines the transformation of the elastic law h

$$h(\kappa_0, \mathbf{C}) = det^{-1}(\mathbf{K})\, \mathbf{K}\, h(\underline{\kappa}_0, \mathbf{K}^T \mathbf{C}\, \mathbf{K})\, \mathbf{K}^T.$$

If the material is submitted to an internal constraint $\gamma_{\mathbf{F}}(\mathbf{F}) = 0$ or in a reduced form (5.2) as $\gamma_{\mathbf{C}}(\mathbf{C}) = 0$, then this form also depends on the reference placement. Under a change we get

$$\gamma_{\mathbf{F}}(\kappa_0, \mathbf{F}) = \gamma_{\mathbf{F}}(\underline{\kappa}_0, \underline{\mathbf{F}}) = \gamma_{\mathbf{F}}(\underline{\kappa}_0, \mathbf{F}\, \mathbf{K})$$

or

$$\gamma_{\mathbf{C}}(\kappa_0, \mathbf{C}) = \gamma_{\mathbf{C}}(\underline{\kappa}_0, \underline{\mathbf{C}}) = \gamma_{\mathbf{C}}(\underline{\kappa}_0, \mathbf{K}^T \mathbf{C}\, \mathbf{K}).$$

It seems to be natural to choose the reference placements such that they are compatible with the internal constraint

$$\gamma_{\mathbf{F}}(\kappa_0, \mathbf{I}) = 0 = \gamma_{\mathbf{C}}(\kappa_0, \mathbf{I}).$$

6.4 Elastic Isomorphy

Changes of the reference placement become important when we want to compare the elastic laws of two different material points. Let

$$f_X(\underline{\kappa}_X, \underline{\mathbf{F}}_X) \qquad \text{and} \qquad f_Y(\underline{\kappa}_Y, \underline{\mathbf{F}}_Y)$$

be two elastic laws for the material points X , $Y \in \mathscr{B}$. The important question of whether the two laws describe *identical material behaviour* leads to the following criterion.

Definition 6.3. Two elastic material points X and Y are called **elastically isomorphic**, if we can find (local) reference placements κ_X for X and κ_Y for Y such that the following two conditions hold.

- In κ_X and κ_Y the mass densities are equal

$$\rho_{0X} = \rho_{0Y}$$

- With respect to κ_X and κ_Y the elastic laws are identical

$$f_X(\kappa_X, \cdot) = f_Y(\kappa_Y, \cdot)$$

If two elastic laws are given with respect to arbitrary reference placements $\underline{\kappa}_X$ and $\underline{\kappa}_Y$, then we must probably first transform them to κ_X and κ_Y

$$f_X(\kappa_X, \mathbf{F}) = f_X(\underline{\kappa}_X, \mathbf{F}\,\mathbf{K}_X) \quad \text{with } \mathbf{K}_X := Grad(\kappa_X\, \underline{\kappa}_X^{-1}) \in \mathscr{Inv}^+$$

$$f_Y(\kappa_Y, \mathbf{F}) = f_Y(\underline{\kappa}_Y, \mathbf{F}\,\mathbf{K}_Y) \quad \text{with } \mathbf{K}_Y := Grad(\kappa_Y\, \underline{\kappa}_Y^{-1}) \in \mathscr{Inv}^+$$

as well as the mass densities in them

$$\underline{\rho}_{0X} / \rho_{0X} = det(\mathbf{K}_X) \qquad \text{and} \qquad \underline{\rho}_{0Y} / \rho_{0Y} = det(\mathbf{K}_Y) ,$$

so that the above isomorphy conditions hold if

$$\underline{\rho}_{0X}\, det(\mathbf{K}_X)^{-1} = \underline{\rho}_{0Y}\, det(\mathbf{K}_Y)^{-1}$$

$$f_X(\underline{\kappa}_X, \mathbf{F}\,\mathbf{K}_X) = f_Y(\underline{\kappa}_Y, \mathbf{F}\,\mathbf{K}_Y) .$$

However, it is generally sufficient to only change the reference placement for just one of the two points, and to keep the other, like, *e.g.*, $\kappa_Y \equiv \underline{\kappa}_Y$. We obtain with the identifications $\mathbf{K}_Y \equiv \mathbf{I}$ and an appropriate $\mathbf{K}_X =: \mathbf{K}$ the following equivalent, but simpler isomorphy conditions.

Theorem 6.4. *Two elastic material points X and Y with elastic laws f_X and f_Y with respect to arbitrary reference placements are elastically isomorphic, if and only if there is an* **elastic isomorphism** $\mathbf{K} \in \mathscr{Inv}^+$ *such that*

- $\rho_{0X} = \rho_{0Y}\, det(\mathbf{K})$
- $f_X(\mathbf{F}\,\mathbf{K}) = f_Y(\mathbf{F}) \qquad \forall\, \mathbf{F} \in \mathscr{Inv}^+$

hold with ρ_{0X} and ρ_{0Y} being the mass densities in the reference placements of X and Y, respectively.

The practical advantage of these conditions is that we can use the elastic laws with respect to arbitrary reference placements. Only the tensor $\mathbf{K}$ is needed. Because of its invertibility, it can be interpreted as an identification of the line elements $\mathbf{dx}_0$ in X with $\mathbf{dy}_0$ in Y by

$$\mathbf{dy}_0 \equiv \mathbf{K}\,\mathbf{dx}_0$$

which play the same role in the elastic behaviour. Its natural representation is similar to that of the deformation gradient (2.5). After choosing a material COOS $\{\Psi^i\}$ in one reference placement $\underline{\kappa}_X$ with natural bases $\{\mathbf{r}_{\Psi i}\}$ and $\{\mathbf{r}_\Psi^i\}$, and another COOS $\{\Phi^i\}$ in the other $\underline{\kappa}_Y$ with $\{\mathbf{r}_{\Phi i}\}$ and $\{\mathbf{r}_\Phi^i\}$, we obtain

$$\mathbf{K} = \frac{\partial \Phi^k}{\partial \Psi^i}\, \mathbf{r}_{\Phi k} \otimes \mathbf{r}_\Psi^i \,.$$

If we transform the two elastic laws into a reduced form such as (6.8)

$$\mathbf{S}_Y = k_Y(\mathbf{C}_Y) = \mathbf{F}_Y^{-1}\, f_Y(\mathbf{F}_Y)\, \mathbf{F}_Y^{-T}$$
$$\mathbf{S}_X = k_X(\mathbf{C}_X) = \mathbf{F}_X^{-1}\, f_X(\mathbf{F}_X)\, \mathbf{F}_X^{-T},$$

then we obtain with the identifications $\mathbf{F}_Y \equiv \mathbf{F}$, $\mathbf{C} \equiv \mathbf{F}^T\,\mathbf{F}$, and $\mathbf{F}_X \equiv \mathbf{F}\,\mathbf{K}$ for

$$\mathbf{C}_Y = \mathbf{F}_Y^{\,T}\,\mathbf{F}_Y = \mathbf{F}^T\,\mathbf{F} = \mathbf{C}$$
$$\mathbf{C}_X = \mathbf{F}_X^{\,T}\,\mathbf{F}_X = \mathbf{K}^T\,\mathbf{C}\,\mathbf{K}$$

and for the second isomorphy condition

$$\begin{aligned} k_Y(\mathbf{C}) &= \mathbf{F}^{-1}\, f_Y(\mathbf{F})\, \mathbf{F}^{-T} \\ &= \mathbf{F}^{-1}\, f_X(\mathbf{F}\,\mathbf{K})\, \mathbf{F}^{-T} \\ &= \mathbf{F}^{-1}\, (\mathbf{F}\,\mathbf{K})\, k_X(\mathbf{C}_X)\, (\mathbf{F}\,\mathbf{K})^T\, \mathbf{F}^{-T} \\ &= \mathbf{K}\, k_X(\mathbf{K}^T\,\mathbf{C}\,\mathbf{K})\, \mathbf{K}^T \qquad \forall\, \mathbf{C} \in \mathcal{P}sym\,, \end{aligned} \tag{6.20}$$

where we can again interpret the tensor $\mathbf{K}$ as a suitable change of the reference placement for k_X.

If we choose another reduced form, like, *e.g.*, (6.12), then the second isomorphy condition is with (6.18)

$$h_{2Y}(\mathbf{E}^G) = det^{-1}(\mathbf{K})\, \mathbf{K}\, h_{2X}(\mathbf{K}^T\,\mathbf{E}^G\,\mathbf{K} + \tfrac{1}{2}\,(\mathbf{K}^T\,\mathbf{K} - \mathbf{I}))\, \mathbf{K}^T \qquad \forall\, \mathbf{E}^G \in \mathcal{S}ym\,.$$

Only in the special case of orthogonal isomorphisms, this reduces to the simple form

$$h_{2Y}(\mathbf{E}^G) = \mathbf{K}\, h_{2X}(\mathbf{K}^T\,\mathbf{E}^G\,\mathbf{K})\, \mathbf{K}^T \qquad \forall\, \mathbf{E}^G \in \mathcal{S}ym\,.$$

If two elastic points are elastically isomorphic, then they have the same elastic properties. If, for example, $\mathbf{C}_{Yu}$ is a stress-free configuration for k_Y, then

$$\mathbf{C}_{Xu} := \mathbf{K}^T\,\mathbf{C}_{Yu}\,\mathbf{K}$$

is stress-free for k_X, and vice versa.

The elastic isomorphy defines an equivalence relation between elastic laws, the equivalence classes of which are the **elastic materials**. If the whole body consists of one material, then all its points are mutually elastically isomorphic, and we call it (elastically) **uniform**. If there is, moreover, a (global) reference placement such that the isomorphisms between all points are simply $\mathbf{K} \equiv \mathbf{I}$, then we call it **homogeneous**. For two of its points $X, Y \in \mathscr{B}$ we have in this case

$$f_X(\cdot) = f_Y(\cdot) \quad \text{and} \quad k_X(\cdot) = k_Y(\cdot)$$

with respect to this particular reference placement. And for spatially constant (homogeneous) states of deformation, the stresses are also homogeneous.

If both material points are submitted to internal constraints $\gamma_{\mathbf{F}X}(\mathbf{F}) = 0 = \gamma_{\mathbf{F}Y}(\mathbf{F})$ or in the reduced form $\gamma_{\mathbf{C}X}(\mathbf{C}) = 0 = \gamma_{\mathbf{C}Y}(\mathbf{C})$, an isomorphism $\mathbf{K}$ between them, if existent, would give

$$\gamma_{\mathbf{F}X}(\mathbf{F}) = \gamma_{\mathbf{F}Y}(\mathbf{F\,K}) \qquad \forall\, \mathbf{F} \in \mathscr{I}nv^{+}$$

or

$$\gamma_{\mathbf{C}X}(\mathbf{C}) = \gamma_{\mathbf{C}Y}(\mathbf{K}^T\,\mathbf{C}\,\mathbf{K}) \qquad \forall\, \mathbf{C} \in \mathscr{P}sym\,,$$

respectively.

6.5 Elastic Symmetry

Quite often we know that the properties of a material do or do not depend on certain directions. Such information can be used to specify the elastic law. This shall be made more precise in the sequel.

We consider an arbitrary, but fixed elastic point $X \in \mathscr{B}$ with elastic law (6.2) $\mathbf{T} = f_X(\mathbf{F})$. For brevity, we will further on suppress the suffix X.

Clearly, each point is isomorphic to itself (automorphic). Trivially the identity $\mathbf{I} \in \mathscr{I}nv$ serves here as an automorphism. For most materials, there are others, non-trivial automorphisms. Such an automorphy is called *symmetry*.

Definition 6.5. Let $\mathbf{T} = f(\mathbf{F})$ be an elastic law. A tensor $\mathbf{A} \in \mathscr{I}nv$ is called (elastic) **symmetry-transformation** of f, if

(6.21)
$$f(\mathbf{F}) = f(\mathbf{FA}) \qquad \forall\, \mathbf{F} \in \mathscr{I}nv.$$

According to this definition, we can interpret a symmetry-transformation as a change of the reference placement, which does not alter the material law.[94]

In the first isomorphy condition we assumed equal mass density in the two reference placements, the change of which we locally describe by $\mathbf{A} \equiv \mathbf{K}$. With the identification $X \equiv Y$ and, consequently, $\underline{\kappa}_X \equiv \underline{\kappa}_Y$ and $\rho_{0X} \equiv \rho_{0Y}$, we conclude that symmetry transformations are generally unimodular

$$det\, \mathbf{A} = 1 .$$

Theorem and Definition 6.6.
The set $\mathcal{G}$ *of all symmetry transformations of an elastic law* $\mathbf{T} = f(\mathbf{F})$ *forms a group under composition, the* **symmetry group** *of* f.

It is easy to prove that $\mathcal{G}$ fulfils the axioms of a group (see Definition 1.7.). The symmetry group is a subgroup of the special unimodular group

$$\mathcal{G} \subseteq \mathcal{U}nim .$$

For the reduced form k the symmetry transformation is

$$(6.22) \qquad \boxed{k(\mathbf{C}) = \mathbf{A}\, k(\mathbf{A}^T \mathbf{C}\, \mathbf{A})\, \mathbf{A}^T} \qquad \forall\ \mathbf{C} \in \mathcal{P}sym .$$

after (6.20). The symmetry transformation can also be notated by the RAYLEIGH product as

$$k(\mathbf{C}) = \mathbf{A} * k(\mathbf{A}^T * \mathbf{C}) .$$

In the way the symmetry transformation $\mathbf{A}$ appears in (6.22), it would always hold for $-\mathbf{A}$ too. However, symmetry transformations with negative determinant do not contain additional information in the mechanical theory, and will therefore not be considered here.

If we choose another reduced form, such as (6.12), the symmetry condition is

$$h_2(\mathbf{E}^G) = \mathbf{A}\, h_2\big(\mathbf{A}^T \mathbf{E}^G \mathbf{A} + \tfrac{1}{2}\,(\mathbf{A}^T \mathbf{A} - \mathbf{I})\big)\, \mathbf{A}^T \qquad \forall\ \mathbf{E}^G \in \mathcal{S}ym .$$

Only in the special case of orthogonal symmetry transformations do we obtain the simple form

$$h_2(\mathbf{E}^G) = \mathbf{A}\, h_2(\mathbf{A}^T \mathbf{E}^G \mathbf{A})\, \mathbf{A}^T \qquad \forall\ \mathbf{E}^G \in \mathcal{S}ym .$$

For all these forms the symmetry group $\mathcal{G}$ is identical.

If the material is submitted to an internal constraint $\gamma_{\mathbf{F}}(\mathbf{F}) = 0$ or in the reduced form (5.2) $\gamma_{\mathbf{C}}(\mathbf{C}) = 0$, then a symmetry transformation $\mathbf{A}$ of the internal constraint requires

---

[94] However, one should not be too dogmatic here, as one can introduce symmetry transformations also in theories, which do not make use of any reference placement at all, by means of a so-called intrinsic description (see NOLL 1972, 2006).

$$\gamma_F(\mathbf{F}) = \gamma_F(\mathbf{F\,A}) \qquad \forall\, \mathbf{F} \in \mathscr{Inv}^+$$

or

$$\gamma_C(\mathbf{C}) = \gamma_C(\mathbf{A}^T\mathbf{C\,A}) \qquad \forall\, \mathbf{C} \in \mathscr{Psym}\,,$$

respectively. For the incompressibility, the symmetry group of the internal constraint is the special unimodular group.

The symmetry group of the internal constraint is conceptually independent of that of the elastic law of the extra stresses. The *symmetry group of the material* is then the section of the two groups.

For simplicity, however, we will now consider only simple elastic materials without internal constraints. Let $\mathscr{G}$ be the symmetry group of the elastic law. As the elastic law depends on the reference placement, then so does $\mathscr{G}$. The dependence is clarified by NOLL's rule.

Theorem 6.7. (NOLL 1958)
Let $k(\kappa_0, \mathbf{C})$ *be the elastic law with respect to a reference placement* κ_0 *with symmetry group* $\mathscr{G}$ *, and* $k(\underline{\kappa}_0, \underline{\mathbf{C}})$ *the elastic law of the same material with respect to* $\underline{\kappa}_0$ *with symmetry group* $\underline{\mathscr{G}}$. *Then*

$$\mathscr{G} = \mathbf{K}\,\underline{\mathscr{G}}\,\mathbf{K}^{-1}$$

i.e.,
$$\mathbf{A} \in \underline{\mathscr{G}} \;\Leftrightarrow\; \mathbf{K\,A\,K}^{-1} \in \mathscr{G} \qquad \textit{with } \mathbf{K} := \mathit{Grad}(\kappa_0\,\underline{\kappa}_0^{\;-1})\,.$$

Proof. Let $\mathbf{A} \in \underline{\mathscr{G}}$. Then

$$k(\underline{\kappa}_0, \underline{\mathbf{C}}) = \mathbf{A}\,k(\underline{\kappa}_0, \mathbf{A}^T\underline{\mathbf{C}}\,\mathbf{A})\,\mathbf{A}^T \qquad \forall\, \underline{\mathbf{C}} \in \mathscr{Psym}$$

holds. Under a change of the reference placement with (6.19) we obtain

$$\begin{aligned}
k(\kappa_0, \mathbf{C}) &= \mathbf{K}\,k(\underline{\kappa}_0, \mathbf{K}^T\mathbf{C\,K})\,\mathbf{K}^T \\
&= \mathbf{K\,A}\,k(\underline{\kappa}_0, \mathbf{A}^T\mathbf{K}^T\mathbf{C\,K\,A})\,\mathbf{A}^T\mathbf{K}^T \\
&= \mathbf{K\,A\,K}^{-1}\,\mathbf{K}\,k(\underline{\kappa}_0, \mathbf{K}^T\mathbf{K}^{-T}\mathbf{A}^T\mathbf{K}^T\mathbf{C\,K\,A\,K}^{-1}\mathbf{K})\,\mathbf{K}^T\mathbf{K}^{-T}\mathbf{A}^T\mathbf{K}^T \\
&= (\mathbf{K\,A\,K}^{-1})\,k\big(\kappa_0, (\mathbf{K\,A\,K}^{-1})^T\mathbf{C}\,(\mathbf{K\,A\,K}^{-1})\big)\,(\mathbf{K\,A\,K}^{-1})^T.
\end{aligned}$$

Therefore, $\mathbf{K\,A\,K}^{-1} \in \mathscr{G}$; *q.e.d.*

As changes of the reference placement do not need to be unimodular, we can also consider dilatations $\mathbf{K} = \alpha\,\mathbf{I}$, $\alpha \neq 0$. By the foregoing theorem we conclude, that dilatations do not alter the symmetry group.

The symmetry group characterises the material. The next theorem explains exhaustively the relation between symmetry transformations and elastic isomorphisms.

Theorem 6.8. *Let k_X and k_Y be two elastic laws with symmetry groups $\mathcal{G}_X$ and $\mathcal{G}_Y$, respectively.*

1) If $\mathbf{K}$ *is an elastic isomorphism between* k_X *and* k_Y*, then*

$$\mathcal{G}_Y = \mathbf{K}\,\mathcal{G}_X\,\mathbf{K}^{-1},$$

i.e., $$\mathbf{A}_X \in \mathcal{G}_X \quad \Leftrightarrow \quad \mathbf{K}\,\mathbf{A}_X\,\mathbf{K}^{-1} \in \mathcal{G}_Y .$$

2) If $\mathbf{K}$ *is an elastic isomorphism between* k_X *and* k_Y*, then so is* $\mathbf{A}_Y\,\mathbf{K}\,\mathbf{A}_X$ *for all* $\mathbf{A}_X \in \mathcal{G}_X$ *and all* $\mathbf{A}_Y \in \mathcal{G}_Y$.

3) If $\mathbf{K}_1$ *and* $\mathbf{K}_2$ *are elastic isomorphisms between* k_X *and* k_Y*, then*

$$\mathbf{K}_1\,\mathbf{K}_2^{-1} \in \mathcal{G}_Y \quad \text{and} \quad \mathbf{K}_1^{-1}\,\mathbf{K}_2 \in \mathcal{G}_X .$$

Proof of *1)*. If $\mathbf{K}$ is an isomorphism between k_X and k_Y, then with (6.20)

$$k_Y(\mathbf{C}) = \mathbf{K}\,k_X(\mathbf{K}^T\,\mathbf{C}\,\mathbf{K})\,\mathbf{K}^T \qquad \forall\, \mathbf{C} \in \mathcal{P}sym$$

or vice versa

$$k_X(\mathbf{C}) = \mathbf{K}^{-1}\,k_Y(\mathbf{K}^{-T}\,\mathbf{C}\,\mathbf{K}^{-1})\,\mathbf{K}^{-T} \qquad \forall\, \mathbf{C} \in \mathcal{P}sym$$

and with $\mathbf{A}_X \in \mathcal{G}_X$ further

$$\begin{aligned} k_Y(\mathbf{C}) &= \mathbf{K}\,\mathbf{A}_X\,k_X(\mathbf{A}_X^T\,\mathbf{K}^T\,\mathbf{C}\,\mathbf{K}\,\mathbf{A}_X)\,\mathbf{A}_X^T\,\mathbf{K}^T \\ &= \mathbf{K}\,\mathbf{A}_X\mathbf{K}^{-1}\,k_Y(\mathbf{K}^{-T}\mathbf{A}_X^T\,\mathbf{K}^T\,\mathbf{C}\,\mathbf{K}\,\mathbf{A}_X\mathbf{K}^{-1})\,\mathbf{K}^{-T}\mathbf{A}_X^T\,\mathbf{K}^T \end{aligned}$$

and therefore $\mathbf{K}\,\mathbf{A}_X\,\mathbf{K}^{-1} \in \mathcal{G}_Y$; *q. e. d.*

The proofs of the other parts are quite similar and thus suppressed here for brevity.

We are now able to classify elastic materials according to their symmetry groups. We already saw that the group $\{\mathbf{I}\}$ is contained in all symmetry groups. Therefore, the set $\{\mathbf{I}\}$ is the *minimal symmetry group*. A material with this minimal symmetry group is called **triclinic**.

The *maximal symmetry group* is the special unimodular group $\mathcal{U}nim^+$, which defines (elastic) **fluids**. So all symmetry groups lie in between these extremes: $\{\mathbf{I}\} \subseteq \mathcal{G} \subseteq \mathcal{U}nim^+$.

For elastic fluids one shows the following representation theorem.

Theorem 6.9. *An elastic material is a fluid, if and only if the following equivalent representations hold*

$$\mathbf{T} = -\,p(\rho)\,\mathbf{I}$$

$$\mathbf{S} = -\,p(\rho)\,\mathbf{C}^{-1}$$

where the pressure p *is a scalar function of the current mass density* ρ.

Proof. For elastic fluids we have by definition

$$\mathbf{S} = k(\mathbf{C}) = \mathbf{A}\, k(\mathbf{A}^T \mathbf{C}\, \mathbf{A})\, \mathbf{A}^T \quad \forall\, \mathbf{C} \in \mathscr{Psym}\,, \forall\, \mathbf{A} \in \mathscr{Unim}^+.$$

If we decompose an arbitrary $\mathbf{C} \in \mathscr{Psym}$ into

$$\mathbf{C} = \mathbf{C}^\circ \underline{\mathbf{C}}$$

after Theorem 1.10. into its unimodular part $\underline{\mathbf{C}} := J^{-2/3}\, \mathbf{C}$ and its spherical part $\mathbf{C}^\circ := J^{2/3}\, \mathbf{I}$, choose an arbitrary orthogonal tensor $\mathbf{Q}$, and identify $\mathbf{A} \equiv \underline{\mathbf{C}}^{-1/2}\, \mathbf{Q} \in \mathscr{Unim}^+$, then the symmetry conditions gives

$$\begin{aligned} k(\mathbf{C}) &= \underline{\mathbf{C}}^{-1/2}\, \mathbf{Q}\, k(\mathbf{Q}^T \underline{\mathbf{C}}^{-1/2}\, \mathbf{C}\, \underline{\mathbf{C}}^{-1/2}\, \mathbf{Q})\, \mathbf{Q}^T\, \underline{\mathbf{C}}^{-1/2} \\ &= J^{2/3}\, \mathbf{C}^{-1/2}\, \mathbf{Q}\, k(J^{2/3}\, \mathbf{I})\, \mathbf{Q}^T\, \mathbf{C}^{-1/2} \\ &= \mathbf{U}^{-1}\, \mathbf{Q}\, \underline{k}(J)\, \mathbf{Q}^T\, \mathbf{U}^{-1} \qquad \text{with } \underline{k}(J) := J^{2/3}\, k(J^{2/3}\, \mathbf{I})\,. \end{aligned}$$

This must hold for arbitrary rotations $\mathbf{Q}$, which leads to the condition

$$\underline{k}(J) = \mathbf{Q}\, \underline{k}(J)\, \mathbf{Q}^T.$$

The only function with this property has spherical values like $\underline{k}(J) = -\,p(\rho)\, \mathbf{I}$ with an arbitrary scalar-valued function $p(\rho)$ because of (3.1) $J = \rho_0/\rho$. Thus

$$k(\mathbf{C}) = \mathbf{U}^{-1}\, \underline{k}(J)\, \mathbf{U}^{-1} = -p(\rho)\, \mathbf{C}^{-1}.$$

The second representation is obtained by substituting $\mathbf{T} = \mathbf{F}\, \mathbf{S}\, \mathbf{F}^T$.

If, vice versa, one of the two representations holds, then it is easy to see that all unimodular tensors are symmetry transformations, and, thus, the material is an (elastic) fluid; *q. e. d.*

Such a stress state is called *hydrostatic stress* or *pressure*. Friction and viscosity are not possible in such fluids, as they are elastic. One can show, however, that for all fluids, even the viscous ones, the above laws hold, if they are at rest. *Incompressible* elastic fluids have the representations

$$\begin{aligned} \mathbf{T} &= -p\, \mathbf{I} \\ \mathbf{S} &= -p\, \mathbf{C}^{-1} \end{aligned}$$

with an undetermined hydrostatic pressure p .

After this short excursion into fluidity, we come back to the general case. A reference placement is called **undistorted**, if the symmetry group of the material with respect to it is contained in the special orthogonal group

$$\mathscr{G} \subseteq \mathscr{Orth}^+.$$

In it the symmetry transformations can be interpreted as (metric conserving) rotations of the tangent space, which do not alter the material behaviour. Note that this property can be lost for another reference placement after NOLL´s rule Theorem 6.7. , *i.e.*, $\mathbf{K}\, \mathscr{G}\, \mathbf{K}^{-1}$ may also contain non-orthogonal elements.

If a material possesses an undistorted reference placement, then it is a **solid**. In extreme cases, all rotations are symmetry transformations.

If

$$\mathscr{G} \supseteq \mathscr{O}rth^{+}$$

for a reference placement, then this is called **isotropic**. A material, which possesses an isotropic reference placement, is called **isotropic material**, otherwise it is **anisotropic** (or *aelotropic*). Accordingly, an isotropic solid has a reference placement with symmetry group $\mathscr{G} \equiv \mathscr{O}rth^{+}$.

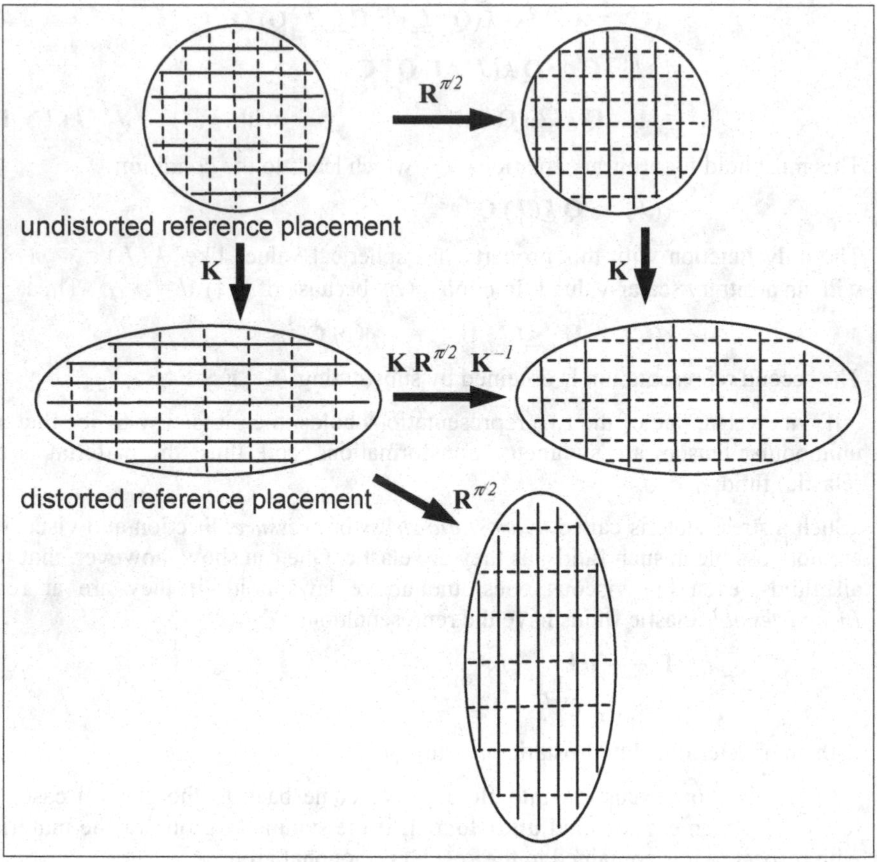

Example. We consider a material, which is characterised in its undistorted reference placement by two orthogonal layers of some substructure. If these layers are identical with respect to their elastic properties, then a rotation of 90° (denoted by a versor $\mathbf{R}^{\pi/2}$) constitutes a symmetry transformation.

If we now change the reference placement by straining the material as sketched in the figure (denoted by $\mathbf{K} \in \mathcal{Inv}^+$), then $\mathbf{R}^{\pi/2}$ is no longer a symmetry transformation with respect to this distorted reference placement. But, instead, the non-orthogonal $\mathbf{K}\,\mathbf{R}^{\pi/2}\,\mathbf{K}^{-1}$ is a symmetry transformation.

As the orthogonal group is contained in the unimodular group, *all fluids are isotropic.*

The question arises of whether the orthogonal group can be contained in the symmetry group as a proper subgroup. For fluids this is principally the case, as their symmetry group is maximal. The question after other examples is negatively answered by the following theorem from group theory.

Theorem 6.10. of BRAUER and NOLL (1965)
The special orthogonal group is maximal in the special unimodular group, i.e. if

$$\mathcal{Orth}^+ \subseteq \mathcal{G} \subseteq \mathcal{Unim}^+$$

holds, then either $\mathcal{G} \equiv \mathcal{Orth}^+$ *or* $\mathcal{G} \equiv \mathcal{Unim}^+$.

Proof (after NOLL 1965).
1) *Sketch of the proof.* If $\mathbf{A}$ is a proper-unimodular, non-orthogonal element in $\mathcal{G}$, then each $\mathbf{M}$ in $\mathcal{Unim}^+$ can be generated by products of powers of $\mathbf{A}$ and orthogonal tensors.

2) Let $\mathbf{M} \in \mathcal{Unim}^+$. By the polar decomposition we get

$$\mathbf{M} = \mathbf{R}\,\mathbf{H} \qquad \text{with } \mathbf{R} \in \mathcal{Orth}^+ ,\ \mathbf{H} \in \mathcal{Psym} \cap \mathcal{Unim}^+.$$

Thus we can immediately restrict our consideration to a symmetric, positive-definite and unimodular transformation $\mathbf{H}$. Similarly, we can also start with a symmetric, positive-definite $\mathbf{A}$.

3) Let $\mathbf{H}$ have the positive eigenvalues h, k, $\frac{1}{hk}$, and $\mathbf{A}$ the positive eigenvalues t, s, $\frac{1}{ts}$, two of which are at least different, *e.g.*, $t \neq s$. If all eigenvalues were equal, then it could only be the unit tensor.

4) An appropriate versor $\mathbf{Q} \in \mathcal{Orth}^+$ rotates the eigenbasis of $\mathbf{H}$ into that of $\mathbf{A}$. Then $\mathbf{H}' := \mathbf{Q}\,\mathbf{H}\,\mathbf{Q}^T$ has the same eigenvalues as $\mathbf{H}$ and the same eigendirections as $\mathbf{A}$.

5) We factorise $\mathbf{H}' = \mathbf{H}_1\,\mathbf{H}_2$ with

$$\mathbf{H}_1 \cong \begin{bmatrix} h & 0 & 0 \\ 0 & \frac{1}{h} & 0 \\ 0 & 0 & 1 \end{bmatrix} \qquad \text{and} \qquad \mathbf{H}_2 \cong \begin{bmatrix} 1 & 0 & 0 \\ 0 & hk & 0 \\ 0 & 0 & \frac{1}{hk} \end{bmatrix}$$

with respect to the eigenbasis of $\mathbf{A}$, which are also contained in $\mathcal{Psym} \cap \mathcal{Unim}^+$.

6) With another appropriate versor $\mathbf{Q}_2 \in \mathcal{Orth}^+$ we rotate $\mathbf{H}_2$ into

$$\mathbf{H}_2' := \mathbf{Q}_2\, \mathbf{H}_2\, \mathbf{Q}_2^T \cong \begin{bmatrix} hk & 0 & 0 \\ 0 & \frac{1}{hk} & 0 \\ 0 & 0 & 1 \end{bmatrix}$$

with respect to the eigenbasis of $\mathbf{A}$, and set $\mathbf{H}_1' \equiv \mathbf{H}_1$. Therefore, it is sufficient to generate the tensor

$$\mathbf{H}_i' \cong \begin{bmatrix} x_i & 0 & 0 \\ 0 & \frac{1}{x_i} & 0 \\ 0 & 0 & 1 \end{bmatrix}$$

with respect to the eigenbasis of $\mathbf{A}$ for $x_i > 1$ by products of powers of $\mathbf{A}$ and elements of $\mathcal{Orth}^+$.

7) For an arbitrary angle φ, we define an orthogonal tensor by

$$\mathbf{R}(\varphi) \cong \begin{bmatrix} \cos\varphi & -\sin\varphi & 0 \\ \sin\varphi & \cos\varphi & 0 \\ 0 & 0 & 1 \end{bmatrix}$$

with respect to the eigenbasis of $\mathbf{A}$. Then

$$\mathbf{T}(\varphi) := \mathbf{A}\, \mathbf{R}(\varphi)\, \mathbf{A}^{-2}\, \mathbf{R}(\varphi)^T\, \mathbf{A}$$

$$\cong \begin{bmatrix} \cos^2\varphi + \left(\frac{s}{t}\right)^2 \sin^2\varphi & \left(\frac{t}{s} - \frac{s}{t}\right)\sin\varphi\cos\varphi & 0 \\ \left(\frac{t}{s} - \frac{s}{t}\right)\sin\varphi\cos\varphi & \cos^2\varphi + \left(\frac{t}{s}\right)^2 \sin^2\varphi & 0 \\ 0 & 0 & 1 \end{bmatrix}$$

is positive-definite and symmetric, and in particular, $\mathbf{T}(\varphi \equiv 0) = \mathbf{I}$ and

$$\mathbf{T}(\varphi \equiv \pi/2) \cong \begin{bmatrix} \left(\frac{s}{t}\right)^2 & 0 & 0 \\ 0 & \left(\frac{t}{s}\right)^2 & 0 \\ 0 & 0 & 1 \end{bmatrix}.$$

8) Without loss of generality, we assume $s > t$ and $x_i > 1$. The first eigenvalue of $\mathbf{T}(\varphi)$ grows with $0 \leq \varphi \leq \pi/2$ from 1 to $(s/t)^2$. The second is the inverse of the first and shrinks from 1 to $(t/s)^2$. The third is always 1. We next choose a natural number m_i such that $x_i \leq (s/t)^{2m_i}$. Because of continuity, there exists a φ_i (see 9), such that $\mathbf{T}(\varphi_i)^{m_i}$ has the eigenvalues x_i, $1/x_i$, 1. With an appropriate $\mathbf{O}_i \in \mathcal{O}rth^+$ we get

$$\mathbf{H}_i{}' = \mathbf{O}_i\, \mathbf{T}(\varphi_i)^{m_i}\, \mathbf{O}_i{}^T.$$

9) Determination of φ_i. Let us assume that already $x_i \leq (s/t)^2$ holds. We want to obtain

$$\begin{aligned} tr(\mathbf{T}(\varphi_i)) &= tr(\mathbf{O}_i\, \mathbf{T}(\varphi_i)\, \mathbf{O}_i{}^T) = x_i + 1/x_i + 1 \\ &= 2\cos^2\varphi_i + [(s/t)^2 + (t/s)^2]\sin^2\varphi_i + 1 \\ &= 2\cos^2\varphi_i + [(s/t) - (t/s)]^2 \sin^2\varphi_i + 2\sin^2\varphi_i + 1 \\ &= [(s/t) - (t/s)]^2 \sin^2\varphi_i + 3\,. \end{aligned}$$

This equation is solved by

$$\varphi_i = arc\,sin \frac{\sqrt{x_i} - \sqrt{x_i^{-1}}}{\frac{s}{t} - \frac{t}{s}}.$$

10) Summary. We obtained for an arbitrary $\mathbf{M}$ in $\mathcal{U}nim^+$ the representation

$$\begin{aligned} \mathbf{M} &= \mathbf{R}\,\mathbf{H} \\ &= \mathbf{R}\,\mathbf{Q}^T\,\mathbf{H}'\,\mathbf{Q} \\ &= \mathbf{R}\,\mathbf{Q}^T\,\mathbf{H}_1\,\mathbf{H}_2\,\mathbf{Q} \\ &= \mathbf{R}\,\mathbf{Q}^T\,\mathbf{H}_1{}'\,\mathbf{Q}_2{}^T\,\mathbf{H}_2{}'\,\mathbf{Q}_2\,\mathbf{Q} \\ &= \mathbf{R}\,\mathbf{Q}^T\,\mathbf{O}_1\,\mathbf{T}(\varphi_1)^{m_1}\,\mathbf{O}_1{}^T\,\mathbf{Q}_2{}^T\,\mathbf{O}_2\,\mathbf{T}(\varphi_2)^{m_2}\,\mathbf{O}_2{}^T\,\mathbf{Q}_2\,\mathbf{Q} \\ &= \mathbf{R}\,\mathbf{Q}^T\,\mathbf{O}_1\,(\mathbf{A}\,\mathbf{R}(\varphi_1)\,\mathbf{A}^{-2}\,\mathbf{R}(\varphi_1)^T\,\mathbf{A})^{m_1}\,\mathbf{O}_1{}^T \\ &\qquad \mathbf{Q}_2{}^T\,\mathbf{O}_2\,(\mathbf{A}\,\mathbf{R}(\varphi_2)\,\mathbf{A}^{-2}\,\mathbf{R}(\varphi_2)^T\,\mathbf{A})^{m_2}\,\mathbf{O}_2{}^T\,\mathbf{Q}_2\,\mathbf{Q} \quad ; q.e.d. \end{aligned}$$

After this theorem, *isotropic materials are either solids or fluids*.

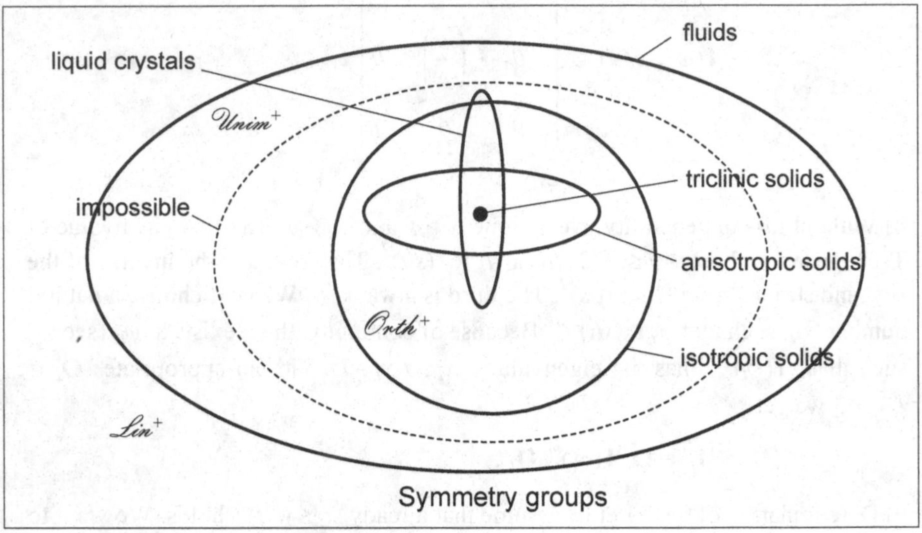

On the other hand, there are (anisotropic) materials, which are neither fluids nor solids. Such materials are called **liquid crystals** (also *anisotropic fluids* or *semi-liquids*) and were investigated by ERICKSEN (since 1960), WANG (1965), MUSCHIK (since 1990) and others.

For elasticity theory, mainly the isotropic and anisotropic solids are of interest. For them, the symmetry group with respect to an undistorted reference configuration contains just rotations (proper orthogonal transformations). Theoretically, there are infinitely many such subgroups of the orthogonal group. In crystallography, only 32 crystal classes are used. In the mechanical context, many of them turn out to be equivalent, so that only 11 classes have to be distinguished, and in the linear elastic theory only 7.

In the Table these 11 groups are characterised by their generators. The corresponding group is generated by these generators under the group operations. There, $\{\mathbf{i}, \mathbf{j}, \mathbf{k}\}$ is an ONB and $\mathbf{d} := 1/\sqrt{3}\,(\mathbf{i} + \mathbf{j} + \mathbf{k})$ the unit diagonal vector. $\mathbf{R}_{\mathbf{n}}^{\varphi}$ is that particular versor which describes a right-handed rotation through an angle $\varphi \in [0, 2\pi)$ about an axis with direction of $\mathbf{n}$.

For all of these groups, only rotations through discrete angels are permitted. This is why they are called *discrete groups*. Only the isotropic group and the following transversely isotropic group are *continuous*. The **transversely isotropic** group is generated by rotations $\mathbf{R}_{\mathbf{k}}^{\varphi}$ about a fixed axis $\mathbf{k}$ through arbitrary angles φ.

Table. The symmetry groups for the crystal classes [95]

No.	crystal classes	generators in $\mathcal{O}rth^+$	number of elements in $\mathcal{O}rth^+$
1.	**triclinic systems**	$\mathbf{I}$	1
2.	**monoclinic systems**	$\mathbf{R}_\mathbf{k}^{\pi}$	2
3.	**rhombic systems**	$\mathbf{R}_\mathbf{i}^{\pi}$, $\mathbf{R}_\mathbf{j}^{\pi}$	4
4.	**tetragonal systems** tetragonal-disphenoidal tetragonal-pyramidal tetragonal-dipyramidal	$\mathbf{R}_\mathbf{k}^{\pi/2}$	4
5.	tetragonal-skalenohedral ditetragonal-pyramidal tetragonal-trapezohedral ditetragonal-dipyramidal	$\mathbf{R}_\mathbf{k}^{\pi/2}$, $\mathbf{R}_\mathbf{i}^{\pi}$	8
6.	**cubic systems** tetratoidal diploidal	$\mathbf{R}_\mathbf{i}^{\pi}$, $\mathbf{R}_\mathbf{j}^{\pi}$, $\mathbf{R}_\mathbf{d}^{2\pi/3}$	12
7.	hextetrahedral gyroidal hexoktahedral	$\mathbf{R}_\mathbf{i}^{\pi/2}$, $\mathbf{R}_\mathbf{j}^{\pi/2}$	24
8.	**hexagonal systems** trigonal-pyramidal rhombo-hedral	$\mathbf{R}_\mathbf{k}^{2\pi/3}$	3
9.	ditrigonal-pyramidal trigonal-trapezohedral hexagonal-skalenohedral	$\mathbf{R}_\mathbf{k}^{2\pi/3}$, $\mathbf{R}_\mathbf{i}^{\pi}$	6
10.	trigonal-dipyramidal hexagonal-pyramidal hexagonal-dipyramidal	$\mathbf{R}_\mathbf{k}^{\pi/3}$	6
11.	ditrigonal-dipyramidal dihexagonal-pyramidal hexagonal-trapezohedral dihexagonal-dipyramidal	$\mathbf{R}_\mathbf{k}^{\pi/3}$, $\mathbf{R}_\mathbf{i}^{\pi}$	12

[95] after COLEMAN/ NOLL (1964)

6.6 Isotropic Elasticity

is characterised by the existence of an isotropic reference placement, with respect to which the reduced form of the elastic law (6.8) is an isotropic tensor-function, *i.e.*,

$$\mathbf{S} = k(\mathbf{C}) = \mathbf{Q}\, k(\mathbf{Q}^T \mathbf{C}\, \mathbf{Q})\, \mathbf{Q}^T \qquad \forall\, \mathbf{C} \in \mathscr{P}sym$$

or equivalently denoted by the RAYLEIGH product

$$\mathbf{Q} * k(\mathbf{C}) = k(\mathbf{Q} * \mathbf{C})$$

holds for all proper-orthogonal transformations $\mathbf{Q} \in \mathscr{O}rth^+$. Therefore, we can apply the representation of Theorem 1.26 to obtain

- the RICHTER representation

(6.23) $$k(\mathbf{C}) = \alpha_0\, \mathbf{I} + \alpha_1\, \mathbf{C} + \alpha_2\, \mathbf{C}^2$$

with three real functions $\alpha_i\,(I_\mathbf{C}, II_\mathbf{C}, III_\mathbf{C})$ of the principal invariants of $\mathbf{C}$

or, as $\mathbf{C}$ is invertible,

- the alternative RICHTER representation

(6.24) $$k(\mathbf{C}) = \underline{\alpha}_0\, \mathbf{I} + \underline{\alpha}_1\, \mathbf{C} + \underline{\alpha}_{-1}\, \mathbf{C}^{-1}$$

with three real functions $\underline{\alpha}_i\,(I_\mathbf{C}, II_\mathbf{C}, III_\mathbf{C})$ of the principal invariants of $\mathbf{C}$

- or the spectral representation

(6.25) $$\begin{aligned} k(\mathbf{C}) = {} & \sigma(\lambda_1, \lambda_2, \lambda_3)\, \mathbf{u}_1 \otimes \mathbf{u}_1 \\ & + \sigma(\lambda_2, \lambda_3, \lambda_1)\, \mathbf{u}_2 \otimes \mathbf{u}_2 \\ & + \sigma(\lambda_3, \lambda_1, \lambda_2)\, \mathbf{u}_3 \otimes \mathbf{u}_3 \end{aligned}$$

with one real function σ of three scalar arguments, being symmetric in the latter two, if $\mathbf{C}$ has the eigenvalues λ_i and the normed eigenvectors $\mathbf{u}_i$.

This symmetry can be identically fulfilled by introducing intermediate symmetrising variables like

(6.26) $$x_i := \lambda_i \qquad y_i := \lambda_{i+1} + \lambda_{i+2} \qquad z_i := |\,\lambda_{i+1} - \lambda_{i+2}\,|$$

$$i = 1, 2, 3 \text{ modulo } 3$$

for this real function

$$\sigma(\lambda_i, \lambda_{i+1}, \lambda_{i+2}) = \underline{\sigma}(\lambda_i, \lambda_{i+1} + \lambda_{i+2}, |\,\lambda_{i+1} - \lambda_{i+2}\,|)\,,$$

as one can solve the system of equations (6.26) for given x_i, y_i, and z_i uniquely up to changes of λ_{i+1} and λ_{i+2} by

$$\lambda_i = x_i$$

$$\lambda_{i+1} = y_i / 2 \pm z_i / 2$$

$$\lambda_{i+2} = y_i / 2 \mp z_i / 2 .$$

With this we get for the material stresses

$$
\begin{aligned}
\mathbf{S} &= \underline{\sigma}(\lambda_1, \lambda_2 + \lambda_3, |\lambda_2 - \lambda_3|)\, \mathbf{u}_1 \otimes \mathbf{u}_1 \\
&+ \underline{\sigma}(\lambda_2, \lambda_1 + \lambda_3, |\lambda_3 - \lambda_1|)\, \mathbf{u}_2 \otimes \mathbf{u}_2 \qquad (6.27)\\
&+ \underline{\sigma}(\lambda_3, \lambda_1 + \lambda_2, |\lambda_1 - \lambda_2|)\, \mathbf{u}_3 \otimes \mathbf{u}_3 .
\end{aligned}
$$

Of course, this is by no means the only possible choice of such intermediate variables. It can be made in different ways under practical aspects. Disadvantage of this above choice is that the third variable is not differentiable in the eigenvalues (because of the norm). One can also use any other intermediate variable there which is symmetric in λ_{i+1} and λ_{i+2} and independent of the other two intermediate variables. *E.g.*, the determinant $\lambda_1 \lambda_2 \lambda_3$ is such an alternative choice, being differentiable and having a clear physical interpretation as the change of the mass density after (2.17) and (3.1).

In order to show the connection between the two representations, we will first calculate the different sets of variables. For that purpose, we number the eigenvalues such that $\lambda_{i+1} \geq \lambda_i$ holds. With the above intermediate variables we obtain

$$
\begin{aligned}
I_{\mathbf{C}} &= I_{\mathbf{B}} = x_i + y_i = \lambda_i + (\lambda_{i+1} + \lambda_{i+2}) \\
II_{\mathbf{C}} &= II_{\mathbf{B}} = x_i\, y_i + \tfrac{1}{4}\,(y_i^2 - z_i^2) \\
&= \lambda_i(\lambda_{i+1} + \lambda_{i+2}) + \tfrac{1}{4}\,(\lambda_{i+1} + \lambda_{i+2})^2 - \tfrac{1}{4}\,|\lambda_{i+1} - \lambda_{i+2}|^2 \qquad (6.28)\\
III_{\mathbf{C}} &= III_{\mathbf{B}} = \tfrac{1}{4}\, x_i\,(y_i^2 - z_i^2) \\
&= \tfrac{1}{4}\,\lambda_i(\lambda_{i+1} + \lambda_{i+2})^2 - \tfrac{1}{4}\,\lambda_i\,|\lambda_{i+1} - \lambda_{i+2}|^2 .
\end{aligned}
$$

By comparison with (6.27) we obtain for the first component

$$
\begin{aligned}
\sigma_{11} &= \underline{\sigma}(x_1, y_1, z_1) \\
&= \alpha_0 + \alpha_1\, x_1 + \alpha_2\, x_1^2
\end{aligned}
$$

with

$$\alpha_i(x_1 + y_1,\, x_1\, y_1 + \tfrac{1}{4}\,(y_1^2 - z_1^2),\, \tfrac{1}{4}\, x_1\,(y_1^2 - z_1^2))$$

and for the others similar forms.

Analogous representations can be used for all material stress tensors as (isotropic) functions of material strain tensors, like

$$\mathbf{S} = \gamma_0\, \mathbf{I} + \gamma_1\, \mathbf{E}^{\mathrm{G}} + \gamma_2\, \mathbf{E}^{\mathrm{G}\,2} \qquad \text{with } \gamma_i(I_{\mathbf{E}^{\mathrm{G}}}, II_{\mathbf{E}^{\mathrm{G}}}, III_{\mathbf{E}^{\mathrm{G}}})$$

$$\mathbf{S} = \delta_0\, \mathbf{I} + \delta_1\, \mathbf{U} + \delta_2\, \mathbf{U}^2 \qquad \text{with } \delta_i(I_{\mathbf{U}}, II_{\mathbf{U}}, III_{\mathbf{U}})$$

$$\mathbf{S} = \underline{\delta}_0\, \mathbf{I} + \underline{\delta}_1\, \mathbf{U} + \underline{\delta}_{-1}\, \mathbf{U}^{-1} \qquad \text{with } \underline{\delta}_i(I_{\mathbf{U}}, II_{\mathbf{U}}, III_{\mathbf{U}})$$

$$\mathbf{T}^{2PK} = \varepsilon_0\,\mathbf{I} + \varepsilon_1\,\mathbf{C} + \varepsilon_2\,\mathbf{C}^2 \qquad \text{with } \varepsilon_i(I_\mathbf{C}, II_\mathbf{C}, III_\mathbf{C})$$

$$\mathbf{T}^{2PK} = \underline{\varepsilon}_0\,\mathbf{I} + \underline{\varepsilon}_1\,\mathbf{C} + \underline{\varepsilon}_{-1}\,\mathbf{C}^{-1} \qquad \text{with } \underline{\varepsilon}_i(I_\mathbf{C}, II_\mathbf{C}, III_\mathbf{C})$$

$$\mathbf{T}^{2PK} = \gamma_0\,\mathbf{I} + \gamma_1\,\mathbf{E}^G + \gamma_2\,\mathbf{E}^{G\,2} \qquad \text{with } \gamma_i(I_{\mathbf{E}^G}, II_{\mathbf{E}^G}, III_{\mathbf{E}^G})$$

$$\mathbf{T}^{2PK} = \mu_0\,\mathbf{I} + \mu_1\,\mathbf{U} + \mu_2\,\mathbf{U}^2 \qquad \text{with } \mu_i(I_\mathbf{U}, II_\mathbf{U}, III_\mathbf{U})$$

$$\mathbf{T}^{2PK} = \underline{\mu}_0\,\mathbf{I} + \underline{\mu}_1\,\mathbf{U} + \underline{\mu}_{-1}\,\mathbf{U}^{-1} \qquad \text{with } \underline{\mu}_i(I_\mathbf{U}, II_\mathbf{U}, III_\mathbf{U})$$

etc. Some of these kinematical tensors are not invertible. Therefore, they do not allow for the alternative RICHTER representation.

For the generalised material strain and stress tensors we obtain the isotropic RICHTER representation

(6.29) $$\mathbf{S}^{Gen} = \nu_0\,\mathbf{I} + \nu_1\,\mathbf{E}^{Gen} + \nu_2\,\mathbf{E}^{Gen\,2} \qquad \text{with } \nu_i(I_{\mathbf{E}^{Gen}}, II_{\mathbf{E}^{Gen}}, III_{\mathbf{E}^{Gen}})$$

and the spectral representation

(6.30) $$\mathbf{S}^{Gen} = \sum_{i=1}^{3} \underline{\sigma}_{Gen}(x_i, y_i, z_i)\,\mathbf{u}_i \otimes \mathbf{u}_i$$

with the symmetrising intermediate variables

$$x_i := f(\lambda_i) \qquad y_i := f(\lambda_{i+1}) + f(\lambda_{i+2}) \qquad z_i := |f(\lambda_{i+1}) - f(\lambda_{i+2})|$$

and f after (2.14). As f is invertible, we can bring this into the following form

$$\mathbf{S}^{Gen} = \sum_{i=1}^{3} \sigma_{iso}(\lambda_i, \lambda_{i+1} + \lambda_{i+2}, |\lambda_{i+1} - \lambda_{i+2}|)\,\mathbf{u}_i \otimes \mathbf{u}_i$$

with

$$\sigma_{iso}(\lambda_i, \lambda_{i+1} + \lambda_{i+2}, |\lambda_{i+1} - \lambda_{i+2}|)$$
$$:= \underline{\sigma}_{Gen}(f(\lambda_i), f(\lambda_{i+1}) + f(\lambda_{i+2}), |f(\lambda_{i+1}) - f(\lambda_{i+2})|)\,.$$

If we use spatial variables instead of material ones, then this would also lead to isotropic tensor-functions, not because of material isotropy, but because of objectivity. In the following table we list the different strain and stress tensors and their transformations under EUCLIDean transformations and under orthogonal symmetry transformations.

	EUCLIDean transformations	orthogonal symmetry transf.
deformation gradient	$\mathbf{Q\,F}$	$\mathbf{F\,Q}^T$
right CAUCHY-GREEN tensor	$\mathbf{C}$	$\mathbf{Q\,C\,Q}^T$
GREEN´s strain tensor	$\mathbf{E}^G$	$\mathbf{Q\,E}^G\,\mathbf{Q}^T$
right stretch tensor	$\mathbf{U}$	$\mathbf{Q\,U\,Q}^T$
generalised material strain tensor	$\mathbf{E}^{Gen}$	$\mathbf{Q\,E}^{Gen}\,\mathbf{Q}^T$
left CAUCHY-GREEN tensor	$\mathbf{Q\,B\,Q}^T$	$\mathbf{B}$
ALMANSI´s strain tensor	$\mathbf{Q\,E}^a\,\mathbf{Q}^T$	$\mathbf{E}^a$
generalised spatial strain tensor	$\mathbf{Q\,E}^{gen}\,\mathbf{Q}^T$	$\mathbf{E}^{gen}$
left stretch tensor	$\mathbf{Q\,V\,Q}^T$	$\mathbf{V}$
CAUCHY´s stress tensor	$\mathbf{Q\,T\,Q}^T$	$\mathbf{T}$
generalised material stress tensor	$\mathbf{S}^{Gen}$	$\mathbf{Q\,S}^{Gen}\,\mathbf{Q}^T$
material stress tensor	$\mathbf{S}$	$\mathbf{Q\,S\,Q}^T$
1st PIOLA-KIRCHHOFF tensor	$\mathbf{Q\,T}^{1PK}$	$\mathbf{T}^{1PK}\,\mathbf{Q}^T$
2nd PIOLA-KIRCHHOFF tensor	$\mathbf{T}^{2PK}$	$\mathbf{Q\,T}^{2PK}\,\mathbf{Q}^T$

We have to principally differentiate between spatial tensors, which are objective, and material ones, which are invariant. Only the deformation gradient and the 1st PIOLA-KIRCHHOFF tensor are as two-point tensors neither invariant nor objective.

If both objectivity and isotropy are valid, then all laws with spatial variables must be isotropic tensor-functions, as well as all laws including material variables. As an example, we consider the CAUCHY stress as a function of the left CAUCHY-GREEN tensor

$$\mathbf{T} = l(\mathbf{B})\,.$$

This is not a reduced form. Such a law can only describe isotropic behaviour. And it is objective, if and only if l is an isotropic tensor-function. Thus, a RICHTER representation can be used

(6.31) $$\mathbf{T} = \beta_0\,\mathbf{I} + \beta_1\,\mathbf{B} + \beta_2\,\mathbf{B}^2 \qquad \text{with } \beta_i\,(I_\mathbf{B}\,,\,II_\mathbf{B}\,,\,III_\mathbf{B})$$

(6.32) $$\mathbf{T} = \underline{\beta}_0\,\mathbf{I} + \underline{\beta}_1\,\mathbf{B} + \underline{\beta}_{-1}\,\mathbf{B}^{-1} \qquad \text{with } \underline{\beta}_i\,(I_\mathbf{B}\,,\,II_\mathbf{B}\,,\,III_\mathbf{B})$$

or the spectral representation

(6.33) $$\mathbf{T} = \sigma(\lambda_1\,,\,\lambda_2\,,\,\lambda_3)\,\mathbf{v}_1\otimes\mathbf{v}_1 + \sigma(\lambda_2\,,\,\lambda_3\,,\,\lambda_1)\,\mathbf{v}_2\otimes\mathbf{v}_2$$

$$+ \ \sigma(\lambda_3, \lambda_1, \lambda_2)\, \mathbf{v}_3 \otimes \mathbf{v}_3$$

$$= \ \underline{\sigma}(\lambda_1, \lambda_2 + \lambda_3, |\lambda_2 - \lambda_3|)\, \mathbf{v}_1 \otimes \mathbf{v}_1$$

$$+ \ \underline{\sigma}(\lambda_2, \lambda_1 + \lambda_3, |\lambda_3 - \lambda_1|)\, \mathbf{v}_2 \otimes \mathbf{v}_2$$

$$+ \ \underline{\sigma}(\lambda_3, \lambda_1 + \lambda_2, |\lambda_1 - \lambda_2|)\, \mathbf{v}_3 \otimes \mathbf{v}_3$$

if $\mathbf{B}$ has the eigenvalues λ_i and the eigenvectors $\mathbf{v}_i$.

Analogous representations also hold for the other spatial stress and strain tensors, like

$$\mathbf{T} = \theta_0\, \mathbf{I} + \theta_1\, \mathbf{E}^a + \theta_2\, \mathbf{E}^{a\,2} \qquad \text{with } \theta_i(I_{\mathbf{E}^a}, II_{\mathbf{E}^a}, III_{\mathbf{E}^a})$$

$$\mathbf{T} = \kappa_0\, \mathbf{I} + \kappa_1\, \mathbf{V} + \kappa_2\, \mathbf{V}^2 \qquad \text{with } \kappa_i(I_{\mathbf{V}}, II_{\mathbf{V}}, III_{\mathbf{V}})$$

$$\mathbf{T} = \underline{\kappa}_0\, \mathbf{I} + \underline{\kappa}_1\, \mathbf{V} + \underline{\kappa}_{-1}\, \mathbf{V}^{-1} \qquad \text{with } \underline{\kappa}_i(I_{\mathbf{V}}, II_{\mathbf{V}}, III_{\mathbf{V}})\,.$$

XIAO et al. (2002) considered problems of the RICHTER representation with regularity and suggested alternative forms.

For **incompressible** isotropic elastic materials $\underline{\alpha}_{-1}$, β_0, $\underline{\beta}_0$, $\underline{\varepsilon}_{-1}$, θ_0, κ_0, $\underline{\kappa}_0$ are reactions and not given by material functions. And the dependence of the determinants vanishes with

$$III_{\mathbf{C}} = III_{\mathbf{U}} = III_{\mathbf{B}} = III_{\mathbf{V}} \equiv 1\,.$$

Two examples are

$$\mathbf{S} = \underline{\alpha}_0\, \mathbf{I} + \underline{\alpha}_1\, \mathbf{C} + \underline{\alpha}_{-1}\, \mathbf{C}^{-1} \qquad \text{with } \underline{\alpha}_i(I_{\mathbf{C}}, II_{\mathbf{C}}),\ i = 0, 1$$

$$\mathbf{T} = \underline{\beta}_0\, \mathbf{I} + \underline{\beta}_1\, \mathbf{B} + \underline{\beta}_{-1}\, \mathbf{B}^{-1} \qquad \text{with } \underline{\beta}_i(I_{\mathbf{B}}, II_{\mathbf{B}}),\ i = 1, -1$$

none of which is yet normalised according to (5.5).

6.7 Incremental Elastic Laws

For certain purposes, one uses incremental or rate forms of the elastic law. Rate forms are obtained of the finite law by calculating the differential. Of course, this is only feasible for differentiable material functions. By using the material reduced forms such as (6.11) or (6.12), we obtain

$$\mathbf{dT}^{2PK} = dh(\mathbf{C}, \mathbf{dC}) = \frac{dh(\mathbf{C})}{d\mathbf{C}}[\mathbf{dC}] = 2\,\frac{dh(\mathbf{C})}{d\mathbf{C}}[\mathbf{dE}^G]$$

(6.34)

$$= \frac{dh_2(\mathbf{E}^G)}{d\mathbf{E}^G}[\mathbf{dE}^G] =: \boldsymbol{H}[\mathbf{dE}^G] = \tfrac{1}{2}\,\boldsymbol{H}[\mathbf{dC}]$$

or for (6.8) or (6.9)

$$\mathbf{dS} = dk(\mathbf{C}, \mathbf{dC}) = \frac{dk(\mathbf{C})}{d\mathbf{C}}[\mathbf{dC}] = 2\,\frac{dk(\mathbf{C})}{d\mathbf{C}}[\mathbf{dE}^{\mathrm{G}}]$$

(6.35)
$$= \frac{dk_2(\mathbf{E}^{\mathrm{G}})}{d\mathbf{E}^{\mathrm{G}}}[\mathbf{dE}^{\mathrm{G}}] =: \boldsymbol{K}[\mathbf{dE}^{\mathrm{G}}] = \tfrac{1}{2}\,\boldsymbol{K}[\mathbf{dC}]$$

or for (6.13)

(6.36)
$$\mathbf{dS}^{\mathrm{Gen}} = \frac{dg(\mathbf{E}^{\mathrm{Gen}})}{d\mathbf{E}^{\mathrm{Gen}}}[\mathbf{dE}^{\mathrm{Gen}}] =: \boldsymbol{H}^{\mathrm{Gen}}[\mathbf{dE}^{\mathrm{Gen}}].$$

By interpreting the differentials in a temporal sense, we can likewise substitute them by time differentials such as

$$\mathbf{T}^{2\mathrm{PK}\bullet} = \boldsymbol{H}[\mathbf{E}^{\mathrm{G}\bullet}] = \tfrac{1}{2}\,\boldsymbol{H}[\mathbf{C}^{\bullet}]$$
$$= (J\,\mathbf{S})^{\bullet} = J\,\mathbf{S}^{\bullet} + J^{\bullet}\,\mathbf{S} = J\,\boldsymbol{K}[\mathbf{E}^{\mathrm{G}\bullet}] + J\,\mathbf{S} \otimes \mathbf{C}^{-1}[\mathbf{E}^{\mathrm{G}\bullet}]$$
$$= J\,\boldsymbol{K}[\mathbf{E}^{\mathrm{G}\bullet}] + J\,div(\mathbf{v})\,\mathbf{S}$$

with (2.20). Thus

$$\boldsymbol{H} = J\,\boldsymbol{K} + J\,\mathbf{S} \otimes \mathbf{C}^{-1} = J\,\boldsymbol{K} + \mathbf{T}^{2\mathrm{PK}} \otimes \mathbf{C}^{-1}.$$

These *4*th-order tensors or tetrads are called tangential **stiffness tetrads**, which are for non-linear elastic laws again functions of $\mathbf{C}$ or $\mathbf{E}^{\mathrm{G}}$, respectively. For symmetric independent and dependent variables we can generally assume for them the left and right subsymmetry and use a VOIGT representation. Their (major) symmetry, however, depends on the material. We will later show (Theorem 7.5.), that $\boldsymbol{H}$ is symmetric for hyperelastic materials only.

These rate forms would be incomplete without initial values for the integration with respect to time. These can be given by the information that a certain configuration $\mathbf{C}_0$ or $\mathbf{E}^{\mathrm{G}}{}_0$ is stress free, like, *e.g.*, the reference placement. Together with such information, the rate forms are then equivalent to the finite elastic laws.

If one, however, does it vice versa and starts with a rate form by defining a configuration-dependent tetrad, then the integrability to a finite elastic law is not generally assured. One can formulate integrability conditions for such cases. We will not do this here, as it is always preferable and easier to start with a finite law and to derive the rate forms, if needed.

The advantage of using material variables when generating rate forms is, that all quantities, including the stiffness tetrads, are invariant under EUCLIDean transformations. This is not valid, if we use spatial or mixed variables.

For the rate of the CAUCHY stress tensor we obtain with (6.3)

$$\mathbf{T}^{\bullet} = \mathbf{R}\,f_1(\mathbf{U})^{\bullet}\,\mathbf{R}^T + \mathbf{R}^{\bullet}\,f_1(\mathbf{U})\,\mathbf{R}^T + \mathbf{R}\,f_1(\mathbf{U})\,\mathbf{R}^{T\bullet}$$
$$= \mathbf{R}\,(\frac{df_1(\mathbf{U})}{d\mathbf{U}}[\mathbf{U}^{\bullet}])\,\mathbf{R}^T + \mathbf{R}^{\bullet}\,\mathbf{R}^T\,\mathbf{T} - \mathbf{T}\,\mathbf{R}^{\bullet}\,\mathbf{R}^T$$

with $\mathbf{R}^{\bullet}\,\mathbf{R}^{T}$ being skew. This stress rate is neither objective nor invariant. It contains the GREEN-NAGHDI rate, see (3.32).

If one prefers spatial representations, then one can push all material rate forms forward into space by the deformation gradient. If we do so with (6.34), we obtain the TRUESDELL rate (3.28) of CAUCHY´s stress tensor

$$\begin{aligned} \mathbf{F}\,\mathbf{T}^{2PK\bullet}\,\mathbf{F}^{T} &= J(\mathbf{T}^{\bullet} - \mathbf{L}\,\mathbf{T} - \mathbf{T}\,\mathbf{L}^{T} + \mathbf{T}\,div\,\mathbf{v}) \\ &= \mathbf{F}\,\boldsymbol{H}[\mathbf{E}^{G\bullet}]\,\mathbf{F}^{T} \\ &= \mathbf{F}\,\boldsymbol{H}[\mathbf{F}^{T}\,\mathbf{D}\,\mathbf{F}]\,\mathbf{F}^{T} \qquad (6.37) \\ &= \boldsymbol{H}_{E}[\mathbf{D}] \qquad \text{with } \boldsymbol{H}_{E} := \mathbf{F} * \boldsymbol{H}. \end{aligned}$$

Thus, the push forward of the stiffness tetrad is the RAYLEIGH product of the material one.

If we push (6.35) forward

$$\begin{aligned} \mathbf{F}\,\mathbf{S}^{\bullet}\,\mathbf{F}^{T} &= \mathbf{T}^{\bullet} - \mathbf{L}\,\mathbf{T} - \mathbf{T}\,\mathbf{L}^{T} = \mathbf{T}^{\nabla} \\ &= \mathbf{F}\,\boldsymbol{K}[\mathbf{E}^{G\bullet}]\,\mathbf{F}^{T} \\ &= \mathbf{F}\,\boldsymbol{K}[\mathbf{F}^{T}\,\mathbf{D}\,\mathbf{F}]\,\mathbf{F}^{T} = \\ \mathbf{T}^{\nabla} &= \boldsymbol{K}_{E}[\mathbf{D}] \qquad \text{with } \boldsymbol{K}_{E} := \mathbf{F} * \boldsymbol{K} \end{aligned}$$

then this gives the OLDROYD rate (3.27) of the CAUCHY stress. Thus, the stiffness tetrads $\boldsymbol{H}_{E}$ and $\boldsymbol{K}_{E}$ are objective and develop even under pure rotations of the body. They can no longer be interpreted as differentials of finite elastic laws.

If, for such a rate form of the elastic law, the stiffness tetrad does not depend on the configuration, but is instead constant, then we have a **physically linear** elastic law. When using this notion, we must keep in mind that this linearity does depend on the particular choice of the stress and strain measures. If we express the same linear law in other tensors, then it generally becomes non-linear. Thus, there is not a single linear law, but infinitely many. All of them still obtain the geometrical non-linearity, and therefore generally lead to a non-linear field problem. Nevertheless, linear elastic laws in material variables are quite popular, as they are reduced forms, and thus applicable for small strains, but large rotations. Their isotropic or anisotropic forms are well known and contain only few material constants. If the strains become large, however, one must be extremely careful when applying linear laws, as they all have regions of instability in their domains[96].

An **example** for such a physically linear law is the **ST. VENANT[97]-KIRCHHOFF law**, which is given with respect to a stress-free reference placement by

[96] See BERTRAM/ BÖHLKE/ ŠILHAVY (2007).
[97] Adhémar Jean Claude Barré de Saint-Venant (1797-1886)

(6.38) $$\mathbf{T}^{2PK} = \boldsymbol{H}[\mathbf{E}^G]$$

with a constant stiffness tetrad $\boldsymbol{H}$.

Another **example** of a linear law is given by

(6.39) $$\mathbf{S} = k(\mathbf{C}) = \tfrac{1}{2}\boldsymbol{K}[\mathbf{C} - \mathbf{C}_u]$$

where $\boldsymbol{K}$ is a positive-definite tetrad with both subsymmetries, and $\mathbf{C}_u \in \mathcal{P}sym$ is the unloaded or stress-free configuration, which can be considered as a material constant. Because of objectivity, we postulate the invariance of $\boldsymbol{K}$ and $\mathbf{C}_u$ under EUCLIDean transformations.

If two of such laws (with indices *0* and *1*) are isomorphic, then by the isomorphy condition (6.20)

$$\boldsymbol{K}_1[\mathbf{C} - \mathbf{C}_{u1}] = \mathbf{K}\,\boldsymbol{K}_0[\mathbf{K}^T\,\mathbf{C}\,\mathbf{K} - \mathbf{C}_{u0}]\,\mathbf{K}^T.$$

with the stress-free configurations $\mathbf{C}_{u1}$ and $\mathbf{C}_{u0}$, respectively. The stiffness tetrads transform as

$$\boldsymbol{K}_1 = \mathbf{K} * \boldsymbol{K}_0$$

and the stress-free configurations as

$$\mathbf{C}_{u0} = \mathbf{K}^T * \mathbf{C}_{u1} = \mathbf{K}^T\,\mathbf{C}_{u1}\,\mathbf{K}.$$

$\mathbf{K}$ transforms the characteristic directions (anisotropy directions such as crystal lattices) of the elastic behaviour of the two laws into another.

We will next focus our considerations on *isotropic* rate forms of the elastic laws. The time-rate of, *e.g.*, (6.29) is then

(6.40) $$\mathbf{S}^{Gen\,\bullet} = \nu_0^{\bullet}\,\mathbf{I} + \nu_1^{\bullet}\,\mathbf{E}^{Gen} + \nu_1\,\mathbf{E}^{Gen\,\bullet} + \nu_2^{\bullet}\,\mathbf{E}^{Gen\,2} + \nu_2\,(\mathbf{E}^{Gen\,2})^{\bullet}$$

with

$$\nu_i(I_{\mathbf{E}^G}, II_{\mathbf{E}^G}, III_{\mathbf{E}^G})^{\bullet} = (\alpha_{i0}\,\mathbf{I} + \alpha_{i1}\,\mathbf{E}^{Gen} + \alpha_{i2}\,\mathbf{E}^{Gen\,2}) \cdot \mathbf{E}^{Gen\,\bullet}$$

for $i = 0, 1, 2$ with the three scalar functions

$$\alpha_{i0}(I_{\mathbf{T}}, II_{\mathbf{T}}, III_{\mathbf{T}}) := \frac{\partial \nu_i}{\partial I_{\mathbf{E}^{Gen}}} + \frac{\partial \nu_i}{\partial II_{\mathbf{E}^{Gen}}}\, I_{\mathbf{E}^{Gen}} + \frac{\partial \nu_i}{\partial III_{\mathbf{E}^{Gen}}}\, II_{\mathbf{E}^{Gen}}$$

$$\alpha_{i1}(I_{\mathbf{T}}, II_{\mathbf{T}}, III_{\mathbf{T}}) := -\frac{\partial \nu_i}{\partial II_{\mathbf{E}^{Gen}}} - \frac{\partial \nu_i}{\partial III_{\mathbf{E}^{Gen}}}\, I_{\mathbf{E}^G}$$

$$\alpha_{i2}(I_{\mathbf{T}}, II_{\mathbf{T}}, III_{\mathbf{T}}) := \frac{\partial \nu_i}{\partial III_{\mathbf{E}^{Gen}}}$$

after (1.40). We substitute this into (6.40) and obtain

$$\begin{aligned} \mathbf{S}^{\mathrm{Gen}\,\bullet} &= \{(\alpha_{00}\,\mathbf{I} + \alpha_{01}\,\mathbf{E}^{\mathrm{Gen}} + \alpha_{02}\,\mathbf{E}^{\mathrm{Gen}\,2}) \cdot \mathbf{E}^{\mathrm{Gen}\,\bullet}\}\,\mathbf{I} \\ &+ \{(\alpha_{10}\,\mathbf{I} + \alpha_{11}\,\mathbf{E}^{\mathrm{Gen}} + \alpha_{12}\,\mathbf{E}^{\mathrm{Gen}\,2}) \cdot \mathbf{E}^{\mathrm{Gen}\,\bullet}\}\,\mathbf{E}^{\mathrm{Gen}} + \nu_1\,\mathbf{E}^{\mathrm{Gen}\,\bullet} \\ &+ \{(\alpha_{20}\,\mathbf{I} + \alpha_{21}\,\mathbf{E}^{\mathrm{Gen}} + \alpha_{22}\,\mathbf{E}^{\mathrm{Gen}\,2}) \cdot \mathbf{E}^{\mathrm{Gen}\,\bullet}\}\,\mathbf{E}^{\mathrm{Gen}\,2} \\ &+ \nu_2\,(\mathbf{E}^{\mathrm{Gen}\,\bullet}\,\mathbf{E}^{\mathrm{Gen}} + \mathbf{E}^{\mathrm{Gen}}\,\mathbf{E}^{\mathrm{Gen}\,\bullet}) \\ &= \boldsymbol{H}^{\mathrm{Gen}}\,[\mathbf{E}^{\mathrm{Gen}\,\bullet}] \end{aligned}$$

with the stiffness tetrad according to (1.42)

$$\begin{aligned} \boldsymbol{H}^{\mathrm{Gen}} &= \mathbf{I} \otimes (\alpha_{00}\,\mathbf{I} + \alpha_{01}\,\mathbf{E}^{\mathrm{Gen}} + \alpha_{02}\,\mathbf{E}^{\mathrm{Gen}\,2}) \\ &+ \mathbf{E}^{\mathrm{Gen}} \otimes (\alpha_{10}\,\mathbf{I} + \alpha_{11}\,\mathbf{E}^{\mathrm{Gen}} + \alpha_{12}\,\mathbf{E}^{\mathrm{Gen}\,2}) + \nu_1\,\boldsymbol{I} \\ &+ \mathbf{E}^{\mathrm{Gen}\,2} \otimes (\alpha_{20}\,\mathbf{I} + \alpha_{21}\,\mathbf{E}^{\mathrm{Gen}} + \alpha_{22}\,\mathbf{E}^{\mathrm{Gen}\,2}) \\ &+ \nu_2\,(G_{ki}\,E^{\mathrm{Gen}}{}_{lj} + G_{lj}\,E^{\mathrm{Gen}}{}_{ik})\,\mathbf{g}^i \otimes \mathbf{g}^j \otimes \mathbf{g}^k \otimes \mathbf{g}^l \end{aligned}$$

with respect to an arbitrary basis $\{\mathbf{g}^i\}$. The initial stiffness tetrad can be determined by linearising in the reference placement, *i.e.*, by putting $\mathbf{E}^{\mathrm{Gen}} \equiv \mathbf{0}$. Here we have

$$\mathbf{S}^{\mathrm{Gen}\,\bullet} = (\lambda\,\mathbf{I} \otimes \mathbf{I} + 2\mu\,\boldsymbol{I})[\mathbf{E}^{\mathrm{Gen}\,\bullet}]\,,$$

with two constants

$$\lambda := \frac{\partial \nu_0}{\partial I_{\mathbf{E}^{\mathrm{Gen}}}}\Big|_{(0,\,0,\,0)} \qquad \text{and} \qquad \mu := \tfrac{1}{2}\,\nu_1(0\,,0\,,0)\,.$$

If we, moreover, linearise geometrically, then we can substitute the stress increment by the CAUCHY stress tensor $\mathbf{T}$ and the deformation rate by the linearised GREEN strain tensor $\mathbf{E}$ obtaining the isotropic **HOOKE´s law** (CAUCHY 1823)

(6.41) $$\mathbf{T} = \lambda\,tr(\mathbf{E})\,\mathbf{I} + 2\mu\,\mathbf{E}$$

with the LAMÉ[98] constants λ and μ. This law is no longer objective. And there is no way to bring this law into a reduced form[99]. Objective *and* geometrically linear elastic laws do not exist.

The stiffness tetrad of HOOKE´s law is

$$\lambda\,\mathbf{I} \otimes \mathbf{I} + 2\mu\,\boldsymbol{I}$$

wherein one can also use the *4*th-order symmetriser $\boldsymbol{I}^S$ instead of the identity $\boldsymbol{I}$.

A comparison of (6.30) with the isotropic HOOKE´s law (6.41) gives under the assumption of small deformations with (2.25) for the linear isotropic law

$$\sigma_{iso}(\lambda_i\,,\,\lambda_{i+1} + \lambda_{i+2}\,,\,|\,\lambda_{i+1} - \lambda_{i+2}\,|)$$

---

[98] Gabriel Lamé (1795-1870)
[99] See FOSDICK/ SERRIN (1979).

$$\equiv (\lambda + 2\,\mu)\,½\,(\lambda_i - 1) + ½\,\lambda\,(\lambda_{i+1} + \lambda_{i+2} - 2)$$

with LAMÉ's constants λ and μ, *i.e.*, a linear dependence on the first two arguments and no dependence on the third. These are the values of the principal CAUCHY stresses.

We can also start with the isotropic spectral representation of (6.30) with the eigenprojectors $\mathbf{P}_{ij} := \mathbf{u}_i \otimes \mathbf{u}_j$ of $\mathbf{C}$ and $\mathbf{S}^{Gen}$, *i.e.*,

$$\mathbf{C} = \sum_{i=1}^{3} \lambda_i\, \mathbf{P}_{ii} \quad \text{and} \quad \mathbf{S}^{Gen} = \sum_{i=1}^{3} \sigma_i\, \mathbf{P}_{ii} = \sum_{i=1}^{3} \underline{\sigma}_{Gen}(x_i\,, y_i\,, z_i)\, \mathbf{P}_{ii}$$

with (6.26) and f after (2.14), where we use the eigenvalues λ_i of $\mathbf{C}$ instead of $\mathbf{U}$ as variables. The rate form of this representation is by use of the product and the chain rule

$$\begin{aligned}
\mathbf{S}^{Gen\,\bullet} &= ½\, \underline{\boldsymbol{H}}^{Gen}[\mathbf{C}^\bullet] \\
&= (\sigma_i^\bullet\, \mathbf{P}_{ii} + \sigma_i\, \mathbf{P}_{ii}^\bullet) \\
&= \sum_{i=1}^{3} \Big(\sum_{j=1}^{3} \sigma_{i\,,\,j}\, f(\lambda_j)'\, \lambda_j^\bullet\, \mathbf{P}_{ii} + \sigma_i\, \frac{d\mathbf{P}_{ii}}{d\mathbf{C}}[\mathbf{C}^\bullet]\Big) \\
&= \sum_{i=1}^{3} \Big(\sum_{j=1}^{3} \sigma_{i\,,\,j}\, f(\lambda_i)'\, (\mathbf{P}_{jj} \cdot \mathbf{C}^\bullet)\, \mathbf{P}_{ii} + \sigma_i\, \frac{d\mathbf{P}_{ii}}{d\mathbf{C}}[\mathbf{C}^\bullet]\Big)
\end{aligned}$$

with the partial derivatives

$$\sigma_{i\,,\,i} := \frac{\partial \underline{\sigma}_{Gen}(x_i, y_i, z_i)}{\partial x_i}$$

$$\sigma_{i\,,\,i+1} := \frac{\partial \underline{\sigma}_{Gen}(x_i, y_i, z_i)}{\partial y_i} \pm \frac{\partial \underline{\sigma}_{Gen}(x_i, y_i, z_i)}{\partial z_i}$$

$$\sigma_{i\,,\,i+2} := \frac{\partial \underline{\sigma}_{Gen}(x_i, y_i, z_i)}{\partial y_i} \mp \frac{\partial \underline{\sigma}_{Gen}(x_i, y_i, z_i)}{\partial z_i}\,,$$

where the upper signs hold for $\lambda_{i+1} > \lambda_{i+2}$ and the lower for $\lambda_{i+2} > \lambda_{i+1}$, and the tetrads after (1.44). The stiffness tetrad is then

$$\begin{aligned}
\underline{\boldsymbol{H}}^{Gen}(\mathbf{C}) = 2 \sum_{i=1}^{3} \Big\{ &\sum_{j=1}^{3} \sigma_{i\,,\,j}\, f(\lambda_j)'\, \mathbf{P}_{ii} \otimes \mathbf{P}_{jj} \\
&+ \sigma_i\, (\lambda_i - \lambda_{i+1})^{-1}\, (\mathbf{P}_{i\,i+1} \otimes \mathbf{P}_{i\,i+1} + \mathbf{P}_{i+1\,i} \otimes \mathbf{P}_{i+1\,i}) \\
&+ \sigma_i\, (\lambda_i - \lambda_{i+2})^{-1}\, (\mathbf{P}_{i\,i+2} \otimes \mathbf{P}_{i\,i+2} + \mathbf{P}_{i+2\,i} \otimes \mathbf{P}_{i+2\,i})\Big\}\,.
\end{aligned}$$

The symmetry of this tensor depends on the condition

$$\sigma_{i\,,\,j}\, f(\lambda_j)' \;=\; \sigma_{j\,,\,i}\, f(\lambda_i)'$$

for all indices and all arguments, or whether

$$\sigma_{i\,,\,2}\,(\lambda_j\,,\, \lambda_{j+1} + \lambda_{j+2}\,, \lvert\, \lambda_{j+1} - \lambda_{j+2} \rvert) \pm \sigma_{i\,,\,3}\,(\lambda_j\,,\, \lambda_{j+1} + \lambda_{j+2}\,, \lvert\, \lambda_{j+1} - \lambda_{j+2} \rvert)$$

$$= \sigma_{j\,,\,2}\,(\lambda_i\,,\, \lambda_{i+1} + \lambda_{i+2}\,, \lvert\, \lambda_{i+1} - \lambda_{i+2} \rvert) \pm \sigma_{j\,,\,3}\,(\lambda_i\,,\, \lambda_{i+1} + \lambda_{i+2}\,, \lvert\, \lambda_{i+1} - \lambda_{i+2} \rvert)$$

holds for all i and j with the same rule for the signs as before.

In a stress-free reference placement all $\lambda_i \equiv 1$ and all $f(\lambda_j)' \equiv \frac{1}{2}$ and all $\sigma_i \equiv 0$, and therefore

$$\underline{\boldsymbol{H}}^{\text{Gen}} = \sum_{i=1}^{3} \sum_{j=1}^{3} \sigma_{i\,,\,j}\big|_{(1,2,0)}\, \mathbf{P}_{ii} \otimes \mathbf{P}_{jj}\,.$$

6.8 Symmetries in Thermo-Elasticity

As the independent and dependent variables in thermo-elasticity, as we chose them, depend on the reference placement, we have to determine their transformation behaviour.

We again consider a change of the (local) reference placement

$$\mathbf{K} := \mathit{Grad}(\kappa_0\, \underline{\kappa}_0^{-1}) \qquad \in \mathscr{Inv}^{+}.$$

Then we obtain the following transformations of the variables:

$\underline{\mathbf{F}} = \mathbf{F}\,\mathbf{K}$	deformation gradient
$\underline{\mathbf{C}} = \mathbf{K}^T * \mathbf{C} = \mathbf{K}^T\,\mathbf{C}\,\mathbf{K}$	right CAUCHY-GREEN tensor
$\underline{\theta} = \mathbf{K}^T * \theta = \theta$	temperature
$\underline{\mathbf{g}}_0 = \mathbf{K}^T * \mathbf{g}_0 = \mathbf{K}^T\,\mathbf{g}_0$	material temperature gradient
$\mathbf{T} = \underline{\mathbf{T}}$	CAUCHY´s stress tensor
$\mathbf{S} = \mathbf{K} * \underline{\mathbf{S}} = \mathbf{K}\,\underline{\mathbf{S}}\,\mathbf{K}^T$	material stress tensor
$\mathbf{T}^{2\text{PK}} = \mathit{det}(\mathbf{K}^{-1})\,\mathbf{K}\,\underline{\mathbf{S}}\,\mathbf{K}^T$	2nd PIOLA-KIRCHHOFF tensor
$\mathbf{q}_0 = \mathit{det}(\mathbf{K}^{-1})\,\mathbf{K}\,\underline{\mathbf{q}}_0$	material heat flux vector
$\varepsilon = \mathbf{K} * \underline{\varepsilon} = \underline{\varepsilon}$	specific internal energy
$\psi = \mathbf{K} * \underline{\psi} = \underline{\psi}$	specific free energy
$\eta = \mathbf{K} * \underline{\eta} = \underline{\eta}$	specific entropy.

By analogy to the mechanical case, we define the isomorphy for two thermo-elastic material points.

Definition 6.11. Two thermo-elastic points X and Y with material laws $k_{SX,Y}$, $k_{qX,Y}$, $k_{\psi X,Y}$ and $k_{\eta X,Y}$ are called **thermo-elastically isomorphic**, if a $\mathbf{K} \in \mathcal{Inv}^+$ and two real constants ψ_c and η_c exist such that

$$\rho_{0X} = \rho_{0Y}\ det(\mathbf{K})$$

$$k_{SY}(\mathbf{C}, \theta, \mathbf{g}_0) = \mathbf{K}\ k_{SX}(\mathbf{K}^T \mathbf{C} \mathbf{K}, \theta, \mathbf{K}^T \mathbf{g}_0)\ \mathbf{K}^T$$

$$k_{qY}(\mathbf{C}, \theta, \mathbf{g}_0) = det(\mathbf{K}^{-1})\ \mathbf{K}\ k_{qX}(\mathbf{K}^T \mathbf{C} \mathbf{K}, \theta, \mathbf{K}^T \mathbf{g}_0)$$

$$k_{\psi Y}(\mathbf{C}, \theta, \mathbf{g}_0) = k_{\psi X}(\mathbf{K}^T \mathbf{C} \mathbf{K}, \theta, \mathbf{K}^T \mathbf{g}_0) + \psi_c - \theta\, \eta_c$$

$$k_{\eta Y}(\mathbf{C}, \theta, \mathbf{g}_0) = k_{\eta X}(\mathbf{K}^T \mathbf{C} \mathbf{K}, \theta, \mathbf{K}^T \mathbf{g}_0) + \eta_c$$

hold for all $\mathbf{C} \in \mathcal{Psym}$, $\theta \in \mathcal{R}^+$, $\mathbf{g}_0 \in \mathcal{V}$.

The two constants are due to the fact, that additive constants in the internal energy and the entropy cannot be determined in principle. These constants can be chosen differently for the two points under consideration,

If we consider automorphisms ($X \equiv Y$), we obtain the following definition of symmetry.

Definition 6.12. A mapping $\mathbf{A} \in \mathcal{Unim}$ is called **symmetry transformation** of a thermo-elastic point with material laws k_S, k_q, k_ψ and k_η, if

$$k_S(\mathbf{C}, \theta, \mathbf{g}_0) = \mathbf{A}\ k_S(\mathbf{A}^T \mathbf{C} \mathbf{A}, \theta, \mathbf{A}^T \mathbf{g}_0)\ \mathbf{A}^T$$

$$k_q(\mathbf{C}, \theta, \mathbf{g}_0) = \mathbf{A}\ k_q(\mathbf{A}^T \mathbf{C} \mathbf{A}, \theta, \mathbf{A}^T \mathbf{g}_0)$$

$$k_\psi(\mathbf{C}, \theta, \mathbf{g}_0) = k_\psi(\mathbf{A}^T \mathbf{C} \mathbf{A}, \theta, \mathbf{A}^T \mathbf{g}_0)$$

$$k_\eta(\mathbf{C}, \theta, \mathbf{g}_0) = k_\eta(\mathbf{A}^T \mathbf{C} \mathbf{A}, \theta, \mathbf{A}^T \mathbf{g}_0)$$

holds for all $\mathbf{C} \in \mathcal{Psym}$, $\theta \in \mathcal{R}^+$, $\mathbf{g}_0 \in \mathcal{V}$.

The set of all symmetry transformations forms the **symmetry group** of the thermo-elastic point. This serves for characterising solids and fluids, and isotropic and anisotropic materials, as we did in the mechanical context.

Note that in this context with some $\mathbf{A} \in \mathcal{Unim}$ in the symmetry group, $-\mathbf{A} \in \mathcal{Unim}$ is *not* automatically in it, as is the case in the mechanical theory.

7 Hyperelasticity

In this chapter, we will embed elastic theory in a thermo-mechanical setting and draw conclusions from the thermodynamical restrictions on the mechanical theory of elasticity. The starting point is thermo-elasticity.

7.1 Thermodynamical Restrictions

We will next consider the restrictions imposed by the second law of thermodynamics on the reduced forms of thermo-elasticity (without internal constraints). As a possible set of reduced thermo-elastic laws we use (6.15)

$$\mathbf{S} = k_S(\mathbf{C}, \theta, \mathbf{g}_0) \qquad \text{material stress tensor}$$

$$\mathbf{q}_0 = k_q(\mathbf{C}, \theta, \mathbf{g}_0) \qquad \text{material heat flux vector}$$

$$\psi = k_\psi(\mathbf{C}, \theta, \mathbf{g}_0) \qquad \text{specific free energy}$$

$$\eta = k_\eta(\mathbf{C}, \theta, \mathbf{g}_0) \qquad \text{specific entropy.}$$

Further on, it is assumed that all these functions are differentiable with respect to all arguments.

For exploiting the CLAUSIUS-DUHEM inequality (3.37), the rate of the free energy is needed. By the chain rule we obtain

$$\psi^\bullet = k_\psi(\mathbf{C}, \theta, \mathbf{g}_0)^\bullet = \frac{\partial k_\psi}{\partial \mathbf{C}} \cdot \mathbf{C}^\bullet + \frac{\partial k_\psi}{\partial \theta}\, \theta^\bullet + \frac{\partial k_\psi}{\partial \mathbf{g}_0} \cdot \mathbf{g}_0^\bullet$$

wherein $\dfrac{\partial k_\psi}{\partial \mathbf{C}}$ is assumed symmetric. Substituting this into (3.37), we obtain

$$0 \le \frac{1}{2\rho}\mathbf{S} \cdot \mathbf{C}^\bullet - \frac{1}{\rho_0\,\theta}\,\mathbf{q}_0 \cdot \mathbf{g}_0 - \eta\,\theta^\bullet - \frac{\partial k_\psi}{\partial \mathbf{C}} \cdot \mathbf{C}^\bullet - \frac{\partial k_\psi}{\partial \theta}\,\theta^\bullet - \frac{\partial k_\psi}{\partial \mathbf{g}_0} \cdot \mathbf{g}_0^\bullet$$

$$= \Big[\frac{1}{2\rho} k_S(\mathbf{C}, \theta, \mathbf{g}_0) - \frac{\partial k_\psi}{\partial \mathbf{C}}\Big] \cdot \mathbf{C}^\bullet - \frac{1}{\rho_0\,\theta}\, k_q(\mathbf{C}, \theta, \mathbf{g}_0) \cdot \mathbf{g}_0$$

$$+ \Big[- k_\eta(\mathbf{C}, \theta, \mathbf{g}_0) - \frac{\partial k_\psi}{\partial \theta}\Big]\, \theta^\bullet - \frac{\partial k_\psi}{\partial \mathbf{g}_0} \cdot \mathbf{g}_0^\bullet .$$

This must be fulfilled for all admissible thermo-kinematical processes, *i.e.*, for all those which satisfy the balances of mass, linear momentum (3.20), moment of momentum (3.21), and energy (3.35). If one is able to control the supply terms of body force $\mathbf{b}$ and heat r arbitrarily, at least in principle, then (3.20) and (3.35) can be satisfied for all thermo-kinematical processes. In particular, at each instant

A. Bertram, *Elasticity and Plasticity of Large Deformations*,
https://doi.org/10.1007/978-3-030-72328-6_7

the following forward rates can be arbitrarily prescribed: $\mathbf{C}^\bullet \in \mathscr{Sym}$, $\mathbf{g}_0{}^\bullet \in \mathscr{V}$, and $\theta^\bullet \in \mathscr{R}$. As $\mathbf{S}$, $\mathbf{q}_0$, and η do not depend on $\mathbf{C}^\bullet$, $\mathbf{g}_0{}^\bullet$, and $\theta^\bullet$, we conclude the

Theorem 7.1. of COLEMAN and NOLL (1963)

The CLAUSIUS-DUHEM inequality is fulfilled for all admissible thermo-kinematical processes, if and only if the following conditions hold for the thermo-elastic laws.

1) The free energy (and consequently also the entropy and the stresses) do not depend on the temperature gradient

$$\psi = k_\psi(\mathbf{C}, \theta)$$

2) The free energy is a potential for the stresses

$$k_S(\mathbf{C}, \theta) = 2\rho \frac{\partial k_\psi(\mathbf{C}, \theta)}{\partial \mathbf{C}}$$

3) and for the entropy

$$k_\eta(\mathbf{C}, \theta) = -\frac{\partial k_\psi(\mathbf{C}, \theta)}{\partial \theta}$$

4) The heat fluxes obey the ***heat conduction inequality***

$$\mathbf{q}_0 \cdot \mathbf{g}_0 \leq 0$$

If we bring this latter inequality into the EULERean representation, then we obtain equivalently

$$\mathbf{q} \cdot \mathbf{g} \leq 0 .$$

The potential forms of *2)* and *3)* are known as *GIBBS*[100] *relations.*

If we had interchanged the roles of entropy and temperature as dependent and independent variables in the principle of determinism, then we could have concluded similarly that the internal energy is a potential for the stresses and the temperature

$$\mathbf{S} = 2\rho \frac{\partial \varepsilon(\mathbf{C}, \eta)}{\partial \mathbf{C}}$$

$$\theta = \frac{\partial \varepsilon(\mathbf{C}, \eta)}{\partial \eta} .$$

The following special cases lead to mechanical elasticity.

---

[100] Josiah Willard Gibbs (1839-1903)

1) Isothermal case

If the temperature is constant in time $\theta(t) \equiv \theta_0$ in a thermo-elastic point, then it is only a parameter in the thermo-elastic laws

$$\mathbf{S} = 2\rho \frac{\partial k_\psi(\mathbf{C}, \theta_0)}{\partial \mathbf{C}} \qquad \psi = k_\psi(\mathbf{C}, \theta_0).$$

2) Isentropic case

If the entropy in a thermo-elastic point is constant in time $\eta(t) \equiv \eta_0$, then we obtain for the stresses the potential

$$\mathbf{S} = 2\rho \frac{\partial \varepsilon(\mathbf{C}, \eta_0)}{\partial \mathbf{C}}.$$

3) The free energy does not depend on the temperature

$$\mathbf{S} = 2\rho \frac{dk_\psi(\mathbf{C})}{d\mathbf{C}}.$$

4) The internal energy does not depend on the entropy

$$\mathbf{S} = 2\rho \frac{d\varepsilon(\mathbf{C})}{d\mathbf{C}}.$$

However, all of these cases are rather unrealistic, as they prohibit all thermo-mechanical couplings. We see that the transition from the thermo-elastic theory to the mechanical elasticity is neither trivial nor unique. In all of these cases, we obtain potentials for the stresses (hyperelasticity), but four different ones.

7.2 Hyperelastic Materials

We have just seen that thermo-elasticity leads under special conditions (isothermy, isentropy) to mechanical elasticity, where the thermodynamic energy serves as a potential for the stresses. This assumption leads in pure mechanics to the notion of *hyperelasticity*, also called *perfect elasticity* or *GREEN elasticity*.

Definition 7.2. A material is called **hyperelastic,** if a specific **strain energy** $w_\mathbf{F}$ exists as a differentiable function of the deformation gradient $\mathbf{F}$

$$w_\mathbf{F} : \mathcal{I}nv^+ \rightarrow \mathcal{R}$$

the rate of which equals the specific stress power

$$w_\mathbf{F}(\mathbf{F})^\bullet = l$$

If the stress power is a complete differential, then the work along a deformation process does not depend on the path (in the strain space), but only on the initial and final value of it. The stress power is generally

$$l = \rho^{-1}\,\mathbf{T}\cdot\mathbf{L} = \rho^{-1}\,\mathbf{T}\cdot\mathbf{F}^{\bullet}\,\mathbf{F}^{-1} = \rho^{-1}\,\mathbf{T}\,\mathbf{F}^{-T}\cdot\mathbf{F}^{\bullet}$$

and for hyperelastic materials by the chain rule

$$l = w_{\mathbf{F}}(\mathbf{F})^{\bullet} = \frac{dw_{\mathbf{F}}}{d\mathbf{F}}\cdot\mathbf{F}^{\bullet}\,.$$

For $\mathbf{F}^{\bullet}$ being arbitrary, we conclude the potential relation for the stresses (NEUMANN[101] 1860)

(7.1)
$$\mathbf{T} = \rho\,\frac{dw_{\mathbf{F}}}{d\mathbf{F}}\,\mathbf{F}^{T}.$$

Obviously, the reverse is also true: If a potential for the stresses of this form exists, then the material is hyperelastic.

As in this expression, the stress tensor is a function of only the current deformation, we have the following implication.

Theorem 7.3. *Every hyperelastic material is elastic.*

The inverse, however, does not hold. The hyperelastic materials form a proper subset of the elastic ones.

If we transform the potential relation into other stress tensors, we obtain

(7.2)
$$\mathbf{T}^{1PK} = \rho_0\,\frac{dw_{\mathbf{F}}}{d\mathbf{F}} \qquad \mathbf{S} = \rho\,\mathbf{F}^{-1}\,\frac{dw_{\mathbf{F}}}{d\mathbf{F}} \qquad \mathbf{T}^{2PK} = \rho_0\,\mathbf{F}^{-1}\,\frac{dw_{\mathbf{F}}}{d\mathbf{F}}\,.$$

Note that neither in the power nor in the stress functions does the strain energy appear, but only its derivative. If we are able to measure only kinematical quantities and stresses, then the strain energy remains determined only up to an additive constant. If the material possesses a natural configuration, as we would expect in the case of elasticity through an unloading process, then we could use it for normalising the energy to zero.

By (3.21) we conclude that these potentials must give symmetric tensors for $\mathbf{T}$ and for $\mathbf{S}$, etc. under all deformations. Furthermore, we require objectivity for the CAUCHY stresses. The following theorem shows that these postulates are not only equivalent, but also include the objectivity of the strain energy function.

Theorem 7.4. of NOLL (1955)
For hyperelastic materials, for which the strain energy remains bounded under arbitrary rotations, the following conditions are equivalent.

1) objectivity of the strain energy function

$$w_{\mathbf{F}}(\mathbf{F}) = w_{\mathbf{F}}(\mathbf{Q}\,\mathbf{F}) \qquad \forall\,\mathbf{F}\in\mathcal{Inv}^{+},\ \forall\,\mathbf{Q}\in\mathcal{Orth}^{+}$$

2) objectivity of the stress function (PISM)

[101] Franz Ernst Neumann (1798-1895)

$$\mathbf{Q}\, f(\mathbf{F})\, \mathbf{Q}^T = f(\mathbf{Q\, F}) \qquad \forall\, \mathbf{F} \in \mathcal{Inv}^+ ,\ \forall\, \mathbf{Q} \in \mathcal{Orth}^+$$

3) *local balance of moment of momentum* (3.21)

$$\mathbf{T} = \mathbf{T}^T \quad \text{with } \mathbf{T} = \rho\, \frac{dw_\mathbf{F}}{d\mathbf{F}}\, \mathbf{F}^T \qquad \forall\, \mathbf{F} \in \mathcal{Inv}^+$$

Proof.

2) ⇒ *1)* We set $\mathbf{F}^* := \mathbf{Q\, F}$. Then $det(\mathbf{F}^*) = det(\mathbf{F}) \Leftrightarrow \rho^* = \rho$, and with *2)*

$$\mathbf{Q}\, f(\mathbf{F})\, \mathbf{Q}^T = \rho\, \mathbf{Q}\, \frac{dw_\mathbf{F}}{d\mathbf{F}}\, \mathbf{F}^T \mathbf{Q}^T = f(\mathbf{F}^*) = \rho\, \frac{dw_\mathbf{F}}{d\mathbf{F}^*}\, \mathbf{F}^{*T} = \rho\, \frac{dw_\mathbf{F}}{d\mathbf{F}^*}\, \mathbf{F}^T \mathbf{Q}^T$$

$$\Leftrightarrow \qquad \mathbf{Q}\, \frac{dw_\mathbf{F}}{d\mathbf{F}} = \frac{dw_\mathbf{F}}{d\mathbf{F}^*} \qquad \Leftrightarrow \qquad \frac{dw_\mathbf{F}}{d\mathbf{F}} = \mathbf{Q}^T\, \frac{dw_\mathbf{F}}{d\mathbf{F}^*}.$$

By integrating over $\mathbf{dF}$ und using $\mathbf{dF}^* = \mathbf{Q\, dF}$

$$\int \frac{dw_\mathbf{F}}{d\mathbf{F}} \cdot \mathbf{dF} = \int \mathbf{Q}^T\, \frac{dw_\mathbf{F}}{d\mathbf{F}^*} \cdot \mathbf{dF} = \int \frac{dw_\mathbf{F}}{d\mathbf{F}^*} \cdot \mathbf{Q\, dF}$$

$$= \int \frac{dw_\mathbf{F}}{d\mathbf{F}^*} \cdot \mathbf{dF}^*,$$

we conclude

$$w_\mathbf{F}(\mathbf{F}) = w_\mathbf{F}(\mathbf{F}^*) + w_\mathbf{Q}(\mathbf{Q}) = w_\mathbf{F}(\mathbf{Q\, F}) + w_\mathbf{Q}(\mathbf{Q})$$

with an unknown function $w_\mathbf{Q}$ of $\mathbf{Q}$. By putting $\mathbf{F} \equiv \mathbf{I}$, we get

$$w_\mathbf{F}(\mathbf{I}) = w_\mathbf{F}(\mathbf{Q}) + w_\mathbf{Q}(\mathbf{Q})$$

$$\Rightarrow \qquad w_\mathbf{F}(\mathbf{F}) = w_\mathbf{F}(\mathbf{Q\, F}) - w_\mathbf{F}(\mathbf{Q}) + w_\mathbf{F}(\mathbf{I}).$$

Further, by $\mathbf{F} \equiv \mathbf{Q}$ we obtain

$$w_\mathbf{F}(\mathbf{Q}) = w_\mathbf{F}(\mathbf{Q}^2) - w_\mathbf{F}(\mathbf{Q}) + w_\mathbf{F}(\mathbf{I})$$

$$\Rightarrow \qquad 2\, w_\mathbf{F}(\mathbf{Q}) = w_\mathbf{F}(\mathbf{Q}^2) + w_\mathbf{F}(\mathbf{I}).$$

For $\mathbf{F} \equiv \mathbf{Q}^2$

$$w_\mathbf{F}(\mathbf{Q}^2) = w_\mathbf{F}(\mathbf{Q}^3) - w_\mathbf{F}(\mathbf{Q}) + w_\mathbf{F}(\mathbf{I}) = 2\, w_\mathbf{F}(\mathbf{Q}) - w_\mathbf{F}(\mathbf{I})$$

$$\Rightarrow \qquad w_\mathbf{F}(\mathbf{Q}^3) = 3\, w_\mathbf{F}(\mathbf{Q}) - 2\, w_\mathbf{F}(\mathbf{I})$$

and in general for $\mathbf{F} \equiv \mathbf{Q}^n$, $n \geq 1$

$$w_\mathbf{F}(\mathbf{Q}^n) = n\, \{w_\mathbf{F}(\mathbf{Q}) - w_\mathbf{F}(\mathbf{I})\} + w_\mathbf{F}(\mathbf{I}).$$

So only by rotations the strain energy would grow unboundedly, which would contradict our assumption, unless

$$w_\mathbf{F}(\mathbf{Q}) = w_\mathbf{F}(\mathbf{I}).$$

Thus,

$$w_{\mathbf{F}}(\mathbf{F}) \;=\; w_{\mathbf{F}}(\mathbf{Q\,F})$$

thus, *1)* is proven.

If we invert the first part of the proof, we see the inclusion *1)* $\Rightarrow$ *2)*.

For showing *1)* $\Rightarrow$ *3)*, we use *1)* with $\mathbf{Q} \equiv \mathbf{R}^T$ and obtain the reduced form

$$w_{\mathbf{F}}(\mathbf{F}) \;=\; w_{\mathbf{F}}(\mathbf{U}) \;=:\; w_{\mathbf{C}}(\mathbf{C})\,.$$

The differential is

$$dw_{\mathbf{C}}(\mathbf{C}, \mathbf{dC}) \;=\; \frac{dw_{\mathbf{C}}}{d\mathbf{C}} \cdot \mathbf{dC} \;=\; \frac{dw_{\mathbf{C}}}{d\mathbf{C}} \cdot (\mathbf{dF}^T\,\mathbf{F} + \mathbf{F}^T\,\mathbf{dF})\,.$$

Because of the symmetry of the right factor, only the symmetric part of the derivative enters this expression

$$dw_{\mathbf{C}}(\mathbf{C}, \mathbf{dC}) \;=\; \{\frac{dw_{\mathbf{C}}}{d\mathbf{C}} + (\frac{dw_{\mathbf{C}}}{d\mathbf{C}})^T\} \cdot \mathbf{F}^T\,\mathbf{dF}$$

$$=\; \mathbf{F}\,\{\frac{dw_{\mathbf{C}}}{d\mathbf{C}} + (\frac{dw_{\mathbf{C}}}{d\mathbf{C}})^T\} \cdot \mathbf{dF}$$

$$=\; dw_{\mathbf{F}}(\mathbf{F}, \mathbf{dF}) \;=\; \frac{dw_{\mathbf{F}}}{d\mathbf{F}} \cdot \mathbf{dF}$$

$$\Rightarrow \qquad \frac{dw_{\mathbf{F}}}{d\mathbf{F}} \;=\; \mathbf{F}\,\{\frac{dw_{\mathbf{C}}}{d\mathbf{C}} + (\frac{dw_{\mathbf{C}}}{d\mathbf{C}})^T\}$$

$$\Rightarrow \text{with (7.1)} \qquad \mathbf{T} \;=\; \rho\,\frac{dw_{\mathbf{F}}}{d\mathbf{F}}\,\mathbf{F}^T \;=\; \rho\,\mathbf{F}\,\{\frac{dw_{\mathbf{C}}}{d\mathbf{C}} + (\frac{dw_{\mathbf{C}}}{d\mathbf{C}})^T\}\,\mathbf{F}^T$$

being therefore symmetric, *i.e.*, *3)*.

3) $\Rightarrow$ *1)*. We set $\mathbf{F}(t) \equiv \mathbf{Q}(t)\,\mathbf{F}_0$ with $\mathbf{F}_0 \in \mathcal{Inv}^+$ arbitrary, but constant. Then the stress power is

$$w_{\mathbf{F}}(\mathbf{F})^{\bullet} \;=\; w_{\mathbf{F}}(\mathbf{Q\,F}_0)^{\bullet}$$

$$=\; \rho^{-1}\,\mathbf{T}\cdot\mathbf{L} \;=\; \rho^{-1}\,\mathbf{T}\cdot\mathbf{F}^{\bullet}\,\mathbf{F}^{-1} \;=\; \rho^{-1}\,\mathbf{T}\cdot\mathbf{Q}^{\bullet}\,\mathbf{F}_0\,\mathbf{F}_0^{-1}\,\mathbf{Q}^{-1}$$

$$=\; \rho^{-1}\,\mathbf{T}\cdot\mathbf{Q}^{\bullet}\,\mathbf{Q}^T \;=\; 0$$

for $\mathbf{T} \in \mathcal{Sym}$ because of CAUCHY's second law (3.21), as $\mathbf{Q}^{\bullet}\,\mathbf{Q}^T \in \mathcal{Skw}$. Consequently, arbitrary rotations do not contribute to the stress power

$$w_{\mathbf{F}}(\mathbf{Q\,F}) \;=\; w_{\mathbf{F}}(\mathbf{F})\,.$$

Thus *1)*; *q. e. d.*

According to this theorem, we postulate the objectivity of $w_{\mathbf{F}}(\mathbf{F})$. By again setting $\mathbf{Q} \equiv \mathbf{R}^T$ (from the polar decomposition), we obtain the *reduced forms for the strain energy function*

$$w_{\mathbf{U}}(\mathbf{U}) \;=\; w_{\mathbf{C}}(\mathbf{C}) \;=\; w_{\mathbf{E}^{\mathrm{Gen}}}(\mathbf{E}^{\mathrm{Gen}})$$

with $w_{\mathbf{U}}(\cdot)$ being the restriction of $w_{\mathbf{F}}(\cdot)$ to $\mathscr{Psym}$ and

$$w_{\mathbf{U}}(\cdot) \;=\; w_{\mathbf{C}}(\cdot\,^2) \;=\; w_{\mathbf{E}G}(\tfrac{1}{2}(\cdot\,^2 - \mathbf{I}))\,.$$

The power is then

$$l \;=\; \frac{1}{2\rho}\,\mathbf{S}\cdot\mathbf{C}^{\bullet} \;=\; w_{\mathbf{C}}{}^{\bullet} \;=\; \frac{dw_{\mathbf{C}}}{d\mathbf{C}}\cdot\mathbf{C}^{\bullet}$$

and we conclude the *reduced forms of the hyperelastic material laws*

$$\boxed{\begin{array}{lll} \mathbf{S} \;=\; 2\rho\,\dfrac{dw_{\mathbf{C}}}{d\mathbf{C}} & \mathbf{T}^{\mathrm{B}} \;=\; \rho_0\,\dfrac{dw_{\mathbf{U}}}{d\mathbf{U}} & \mathbf{T}^{2\mathrm{PK}} \;=\; 2\rho_0\,\dfrac{dw_{\mathbf{C}}}{d\mathbf{C}} \\[2ex] \mathbf{T}^{1\mathrm{PK}} \;=\; 2\rho_0\,\mathbf{F}\,\dfrac{dw_{\mathbf{C}}}{d\mathbf{C}} & \mathbf{S}^{\mathrm{Gen}} \;=\; \rho_0\,\dfrac{dw_{\mathbf{E}^{\mathrm{Gen}}}}{d\mathbf{E}^{\mathrm{Gen}}} & \mathbf{T} \;=\; 2\rho\,\mathbf{F}\,\dfrac{dw_{\mathbf{C}}}{d\mathbf{C}}\,\mathbf{F}^{T} \end{array}} \tag{7.3}$$

These are partly the formulae of DOYLE/ ERICKSEN (1956). We again take the derivative of a scalar function such as $w_{\mathbf{C}}$ with respect to a symmetric tensor $\mathbf{C}$ as symmetric (to assure (3.21)).

For the same purposes as in Chap. 6.7, one needs the rate forms of the hyperelastic laws, such as (6.34)

$$d\mathbf{T}^{2\mathrm{PK}} \;=\; \frac{dh(\mathbf{C})}{d\mathbf{C}}\,[d\mathbf{C}] \;=\; 2\rho_0\,\frac{d^2w_{\mathbf{C}}}{d\mathbf{C}^2}\,[d\mathbf{C}] \;=\; \tfrac{1}{2}\,\boldsymbol{H}[d\mathbf{C}]$$

or generally (6.36)

$$d\mathbf{S}^{\mathrm{Gen}} = \frac{dg\left(\mathbf{E}^{\mathrm{Gen}}\right)}{d\mathbf{E}^{\mathrm{Gen}}}\,[d\mathbf{E}^{\mathrm{Gen}}] = \rho_0\,\frac{d^2w_{\mathbf{E}^{\mathrm{Gen}}}}{d\mathbf{E}^{\mathrm{Gen}\,2}}\,[d\mathbf{E}^{\mathrm{Gen}}] =: \boldsymbol{H}^{\mathrm{Gen}}[d\mathbf{E}^{\mathrm{Gen}}]\,.$$

The tangential stiffness tetrads $\boldsymbol{H}$ and $\boldsymbol{H}^{\mathrm{Gen}}$ are in the case of hyperelasticity double derivatives and therefore symmetric. This is true for all rate forms with conjugate stress and strain variables. By transforming this form into the EULERean or spatial version, we obtain in the TRUESDELL rate (3.28) as in (6.37) the (time-dependent) objective elasticity tetrad

$$\boldsymbol{H}_{\mathrm{E}} := 4\,\rho_0\,\mathbf{F} * \frac{d^2w_{\mathbf{C}}}{d\mathbf{C}^2}\,,$$

the symmetry of which is again given, if and only if the material is hyperelastic. For the rate form (6.35), the elasticity tensor $\boldsymbol{K}$ is non-symmetric even for hyperelastic materials, but then restricted by other integrability conditions.

From potential theory, we know that fields, for which the curl vanishes, have a potential, and vice versa. In addition, the curl is zero, if and only if the gradient of the field is symmetric everywhere.

Theorem 7.5. *The following conditions are equivalent for all elastic materials.*

1) The material is hyperelastic.

2) The stiffness tetrad $\boldsymbol{H}^{\text{Gen}}$ *(or* $\boldsymbol{H}_{\text{E}}$ *or* $\boldsymbol{H}$*) is symmetric for all* $\mathbf{F} \in \mathscr{I}nv^{+}$ *.*

3) The stress work is path-independent.

4) The stress work is zero for cyclic processes.

The last two statements are referred to paths in the strain space.

After the 4th part of the above theorem, one can find cyclic processes for elastic materials that are not hyperelastic, which add energy to the body, or, if processed the other way round, set energy free. By repeatedly performing this cycle, one can extract any amount of energy. Such a behaviour can be considered as non-physical. This argument is a strong objection against any theory of elasticity that does not automatically include hyperelasticity.

On the other hand, elasticity (the same as hyperelasticity) is based on a clear physical definition, which already serves as a foundation of a well-developed mathematical theory.

It should be noted here that also in linear elasticity using HOOKE´s law, hyper-elasticity coincides with the symmetry of the stiffness tetrad, which is, however, always assured if the material is isotropic.

If also the external loads (forces) on the body allow for a potential U

$$L_e = -U^{\bullet}$$

then the balance of power (3.24) of a hyperelastic body turns into the **conservation law of the total energy**

$$(K + W + U)^{\bullet} = 0$$

with the global strain energy of the body

$$W := \int_{\mathscr{B}} w \, dm \, .$$

This law should be distinguished from the first law of thermodynamics, which holds independently for all materials.

7.3 Hyperelastic Isomorphy and Symmetry

If we define the strain energy function on the deformations, then it automatically depends on the chosen reference configuration. Its value, however, should be independent of such choice. We therefore have to consider invariance under change of reference placements. By (6.16) a change of the reference placement gives

$$w_{\mathbf{F}}(\kappa_0, \mathbf{F}) = w_{\mathbf{F}}(\underline{\kappa}_0, \mathbf{F}\,\mathbf{K})$$
$$= w_{\mathbf{C}}(\kappa_0, \mathbf{C}) = w_{\mathbf{C}}(\underline{\kappa}_0, \mathbf{K}^T\,\mathbf{C}\,\mathbf{K})\,.$$

With this, we can define the isomorphy for hyperelastic materials.

Theorem 7.6. *Two hyperelastic material points* X *and* Y *with strain energy functions* $w_{\mathbf{F}X}$ *and* $w_{\mathbf{F}Y}$ *with respect to arbitrary reference placements* κ_X *and* κ_Y *, respectively, are isomorphic, if and only if there exists an* **isomorphism** $\mathbf{K} \in \mathcal{Inv}^+$ *and a constant* $w_c \in \mathcal{R}$ *such that*

(I1) $\qquad \rho_{0X} = \rho_{0Y}\, det(\mathbf{K})$

and

(I2) $\qquad w_{\mathbf{F}X}(\underline{\kappa}_X, \mathbf{F}\,\mathbf{K}) = w_{\mathbf{F}Y}(\underline{\kappa}_Y, \mathbf{F}) + w_c \qquad \forall\, \mathbf{F} \in \mathcal{Inv}^+$

hold. Here, ρ_{0X} *and* ρ_{0Y} *are the mass densities in the reference placements* κ_X *and* κ_Y*, respectively.*

For the reduced form $w_{\mathbf{C}}$ the second isomorphy condition becomes

(7.4) $$w_{\mathbf{C}X}(\kappa_X, \mathbf{K}^T\,\mathbf{C}\,\mathbf{K}) = w_{\mathbf{C}Y}(\kappa_Y, \mathbf{C}) + w_c \qquad \forall\, \mathbf{C} \in \mathcal{Psym}^+.$$

For the following considerations of symmetries, we immediately work with the reduced form $w_{\mathbf{C}}(\mathbf{C})$, although one can obtain similar results also for $w_{\mathbf{F}}(\mathbf{F}) = w_{\mathbf{U}}(\mathbf{U})$.

A **symmetry transformation** of the strain energy function is a tensor $\mathbf{A} \in \mathcal{Unim}^+$, for which

$$\boxed{w_{\mathbf{C}}(\mathbf{C}) = w_{\mathbf{C}}(\mathbf{A}^T\,\mathbf{C}\,\mathbf{A})} \qquad \forall\, \mathbf{C} \in \mathcal{Psym}$$

holds. The symmetry transformations form a group under composition, the **symmetry group of the strain energy** $\mathcal{G}_w$. This should be kept distinct from the symmetry group $\mathcal{G}_\sigma$ of the elastic law k . For investigating the relation between these two groups, we start from the above symmetry condition of the strain energy function. If two functions of $\mathbf{C}$ are equal, then this surely also holds for their differentials

$$dw_{\mathbf{C}}(\mathbf{C}, d\mathbf{C}) = \frac{dw_{\mathbf{C}}}{d\mathbf{C}} \cdot d\mathbf{C}$$

$$= dw_{\mathbf{C}}(\mathbf{A}^T\,\mathbf{C}\,\mathbf{A}) = \frac{dw_{\mathbf{C}}\left(\mathbf{A}^T\mathbf{C}\,\mathbf{A}\right)}{d\mathbf{C}} \cdot d(\mathbf{A}^T\,\mathbf{C}\,\mathbf{A}) \qquad \forall\, \mathbf{A} \in \mathcal{G}_w$$

$$= \frac{dw_{\mathbf{C}}\left(\mathbf{A}^T\mathbf{C}\,\mathbf{A}\right)}{d\mathbf{C}} \cdot (\mathbf{A}^T\,d\mathbf{C}\,\mathbf{A}) = (\mathbf{A}\,\frac{dw_{\mathbf{C}}\left(\mathbf{A}^T\mathbf{C}\,\mathbf{A}\right)}{d\mathbf{C}}\,\mathbf{A}^T) \cdot d\mathbf{C}$$

with the notation

$$\frac{dw_{\mathbf{C}}\left(\mathbf{A}^T\mathbf{C}\,\mathbf{A}\right)}{d\mathbf{C}} := \frac{dw_{\mathbf{C}}(\underline{\mathbf{C}})}{d\underline{\mathbf{C}}} \quad \text{at} \quad \underline{\mathbf{C}} \equiv \mathbf{A}^T\,\mathbf{C}\,\mathbf{A}\,.$$

By comparison with (7.3,1) we obtain

$$\frac{k(\mathbf{C})}{2\rho} = \frac{dw_{\mathbf{C}}}{d\mathbf{C}} = \mathbf{A}\,\frac{dw_{\mathbf{C}}\left(\mathbf{A}^T\mathbf{C}\,\mathbf{A}\right)}{d\mathbf{C}}\,\mathbf{A}^T = \mathbf{A}\,\frac{k\left(\mathbf{A}^T\mathbf{C}\mathbf{A}\right)}{2\rho}\,\mathbf{A}^T.$$

For $\mathbf{A}$ being unimodular, we get

$$det(\mathbf{C}) = det(\mathbf{A}^T\,\mathbf{C}\,\mathbf{A})$$

and, hence, the mass densities in the configurations $\mathbf{C}$ and $\mathbf{A}^T\,\mathbf{C}\,\mathbf{A}$ are equal. Therefore,

$$k(\mathbf{C}) = \mathbf{A}\,k(\mathbf{A}^T\,\mathbf{C}\,\mathbf{A})\,\mathbf{A}^T$$

and $\mathbf{A} \in \mathscr{G}_\sigma$.

The reverse of this part of the proof gives for all $\mathbf{A} \in \mathscr{G}_\sigma$

$$dw_{\mathbf{C}}(\mathbf{C}) = dw_{\mathbf{C}}(\mathbf{A}^T\,\mathbf{C}\,\mathbf{A})\,,$$

which means that the integrals are only equal up to a function $w_{\mathbf{A}}(\mathbf{A})$

$$w_{\mathbf{C}}(\mathbf{C}) = w_{\mathbf{C}}(\mathbf{A}^T\,\mathbf{C}\,\mathbf{A}) + w_{\mathbf{A}}(\mathbf{A})\,.$$

Only for hyperelastic *solids*, which possess an undistorted reference placement, for which

$$\mathbf{A} \in \mathscr{G}_\sigma \subset \mathscr{O}\!rth^+,$$

can we conclude that

$$w_{\mathbf{C}}(\mathbf{I}) = w_{\mathbf{C}}(\mathbf{I}) + w_{\mathbf{A}}(\mathbf{A}) \quad \Rightarrow \quad w_{\mathbf{A}}(\mathbf{A}) = 0$$

and, hence, also $\mathbf{A} \in \mathscr{G}_w$.

Theorem 7.7.[102] *For hyperelastic materials the symmetry group of the strain energy function* $w_{\mathbf{C}}$ *is contained in the symmetry group for the elastic law* k

$$\mathscr{G}_w \subset \mathscr{G}_\sigma\,.$$

For hyperelastic solids, both groups coincide

$$\mathscr{G}_w \equiv \mathscr{G}_\sigma\,.$$

---

[102] See HUILGOL (1971) and WANG (1966).

7.4 Isotropic Hyperelasticity

In the sequel we will consider isotropic hyperelastic solids and compare them with isotropic elastic ones. Further on, we will use an isotropic undistorted reference placement and use isotropic tensor-functions.

Theorem 7.8. *An objective hyperelastic solid is isotropic, if and only if the strain energy function* $w_\mathbf{F}$ *is an isotropic function of* $\mathbf{U}$, $\mathbf{C}$, $\mathbf{V}$, $\mathbf{B}$ *or* $\mathbf{E}^G$ *with respect to an isotropic reference placement*

(*1*) $$w_\mathbf{F}(\mathbf{F}) = w_\mathbf{U}(\mathbf{U}) = w_\mathbf{U}(\mathbf{Q\,U\,Q}^T) \qquad \forall\, \mathbf{U} \in \mathscr{Psym}$$

(*2*) $$w_\mathbf{F}(\mathbf{F}) = w_\mathbf{C}(\mathbf{C}) = w_\mathbf{C}(\mathbf{Q\,C\,Q}^T) \qquad \forall\, \mathbf{C} \in \mathscr{Psym}$$

(*3*) $$w_\mathbf{F}(\mathbf{F}) = w_\mathbf{V}(\mathbf{V}) = w_\mathbf{V}(\mathbf{Q\,V\,Q}^T) \qquad \forall\, \mathbf{V} \in \mathscr{Psym}$$

(*4*) $$w_\mathbf{F}(\mathbf{F}) = w_\mathbf{B}(\mathbf{B}) = w_\mathbf{B}(\mathbf{Q\,B\,Q}^T) \qquad \forall\, \mathbf{B} \in \mathscr{Psym}$$

(*5*) $$w_\mathbf{F}(\mathbf{F}) = w_{\mathbf{E}^G}(\mathbf{E}^G) = w_{\mathbf{E}^G}(\mathbf{Q\,E}^G\,\mathbf{Q}^T) \qquad \forall\, \mathbf{E}^G \in \mathscr{Sym}$$

for all $\mathbf{Q} \in \mathscr{Orth}^+$ *with*

$$w_\mathbf{U}(\mathbf{U}) = w_\mathbf{C}(\mathbf{U}^2) = w_\mathbf{V}(\mathbf{U}) = w_\mathbf{B}(\mathbf{U}^2) = w_{\mathbf{E}^G}(\tfrac{1}{2}(\mathbf{U}^2 - \mathbf{I})) := w_\mathbf{F}(\mathbf{U})\,|_{\mathscr{Psym}} .$$

Proof. Let $w_\mathbf{F}(\mathbf{F})$ be the strain energy of a hyperelastic material with respect to an isotropic reference placement. By objectivity (as already shown) we obtain the reduced form

$$w_\mathbf{U}(\mathbf{U}) := w_\mathbf{F}(\mathbf{U})\,|_{\mathscr{Psym}}$$

and for an (orthogonal) symmetry transformation

$$w_\mathbf{U}(\mathbf{U}) = w_\mathbf{U}(\mathbf{Q\,U\,Q}^T) \qquad \forall\, \mathbf{Q} \in \mathscr{Orth}^+.$$

Hence, $w_\mathbf{U}$ is a real isotropic tensor-function of $\mathbf{U}$ (*1*). In what follows, we will always use the function $w_\mathbf{F}$ as restricted to $\mathscr{Psym}$. We get

$$w_\mathbf{C}(\mathbf{C}) = w_\mathbf{C}(\mathbf{U}^2) := w_\mathbf{U}(\mathbf{U}) = w_\mathbf{U}(\mathbf{Q\,U\,Q}^T)$$

$$= w_\mathbf{C}(\mathbf{Q\,U\,Q}^T\,\mathbf{Q\,U\,Q}^T) = w_\mathbf{C}(\mathbf{Q\,C\,Q}^T)$$

(*2*) $$= w_{\mathbf{E}^G}(\tfrac{1}{2}(\mathbf{C} - \mathbf{I})) = w_{\mathbf{E}^G}\{\tfrac{1}{2}(\mathbf{Q\,C\,Q}^T - \mathbf{Q\,Q}^T)\}$$

(*5*) $$= w_{\mathbf{E}^G}(\mathbf{Q\,E}^G\,\mathbf{Q}^T)$$

$$= w_\mathbf{C}(\mathbf{Q\,F}^T\,\mathbf{F\,Q}^T) = w_\mathbf{C}(\mathbf{Q\,R}^T\,\mathbf{V\,V\,R\,Q}^T)$$

and with $\mathbf{R} \equiv \mathbf{Q}$ $$= w_\mathbf{C}(\mathbf{V}^2) = w_\mathbf{C}(\mathbf{B}) =: w_\mathbf{B}(\mathbf{B})$$

$$= w_\mathbf{U}(\mathbf{R\,U\,R}^T) = w_\mathbf{U}(\mathbf{V}) =: w_\mathbf{V}(\mathbf{V})$$

with (*1*) we get (*3*) $\qquad = w_{\mathbf{V}}(\mathbf{Q\,V\,Q}^T)$

and (*4*) $\qquad = w_{\mathbf{B}}(\mathbf{V}^2) = w_{\mathbf{B}}(\mathbf{Q\,B\,Q}^T)\,;$ *q. e. d.*

With

$$\mathbf{B}^\bullet = \mathbf{B}^{\bullet T} = (\mathbf{F\,F}^T)^\bullet = \mathbf{F}^\bullet\,\mathbf{F}^T + \mathbf{F\,F}^{T\bullet} \qquad \in \mathscr{Sym}$$

and using the symmetry of the CAUCHY stress tensor (3.21), we obtain

$$l = w_{\mathbf{B}}(\mathbf{B})^\bullet = \frac{dw_{\mathbf{B}}}{d\mathbf{B}} \cdot \mathbf{B}^\bullet = \frac{dw_{\mathbf{B}}}{d\mathbf{B}} \cdot (\mathbf{F}^\bullet\,\mathbf{F}^T + \mathbf{F\,F}^{T\bullet})\,.$$

Only the symmetric part of $\dfrac{dw_{\mathbf{B}}}{d\mathbf{B}}$ affects this expression, which is coaxial to $\mathbf{B}$ in the isotropic case. We assume its symmetry and continue

$$l = \frac{dw_{\mathbf{B}}}{d\mathbf{B}} 2 \cdot \mathbf{F\,F}^{T\bullet} = 2\,\mathbf{F}^T \frac{dw_{\mathbf{B}}}{d\mathbf{B}} \cdot \mathbf{F}^{T\bullet}$$

$$= \rho^{-1}\,\mathbf{T}\cdot\mathbf{L} = \rho^{-1}\,\mathbf{T}\cdot\mathbf{F}^\bullet\,\mathbf{F}^{-1} = \rho^{-1}\,\mathbf{T}\cdot\mathbf{F}^{-T}\,\mathbf{F}^{T\bullet}$$

$$= \rho^{-1}\,\mathbf{F}^{-1}\,\mathbf{T}\cdot\mathbf{F}^{T\bullet}\,.$$

By comparison, we conclude with Theorem 1.16.

(7.5)
$$\boxed{\mathbf{T} = 2\,\rho\,\frac{dw_{\mathbf{B}}}{d\mathbf{B}}\,\mathbf{B} = 2\,\rho\,\mathbf{B}\,\frac{dw_{\mathbf{B}}}{d\mathbf{B}}}$$

By Theorem 1.24, any real isotropic tensor-function of a symmetric tensor can be represented as a function of the three principal invariants.

Theorem 7.9. *A hyperelastic solid is isotropic, if and only if the strain energy can be represented as a function of the three principal invariants of their symmetric arguments*

$$w_{\mathbf{U}}(\mathbf{U}) = w_{\mathbf{U}iso}(I_{\mathbf{U}}\,, II_{\mathbf{U}}\,, III_{\mathbf{U}}) \qquad \forall\ \mathbf{U} \in \mathscr{Psym}$$

$$w_{\mathbf{C}}(\mathbf{C}) = w_{\mathbf{C}iso}(I_{\mathbf{C}}\,, II_{\mathbf{C}}\,, III_{\mathbf{C}}) \qquad \forall\ \mathbf{C} \in \mathscr{Psym}$$

$$w_{\mathbf{V}}(\mathbf{V}) = w_{\mathbf{V}iso}(I_{\mathbf{V}}\,, II_{\mathbf{V}}\,, III_{\mathbf{V}}) \qquad \forall\ \mathbf{V} \in \mathscr{Psym}$$

$$w_{\mathbf{B}}(\mathbf{B}) = w_{\mathbf{B}iso}(I_{\mathbf{B}}\,, II_{\mathbf{B}}\,, III_{\mathbf{B}}) \qquad \forall\ \mathbf{B} \in \mathscr{Psym}$$

Note that the principal invariants of $\mathbf{U}$ and $\mathbf{V}$ are identical, the same as those of $\mathbf{B}$ and $\mathbf{C}$. Thus

$$w_{\mathbf{U}iso} = w_{\mathbf{V}iso} \quad \text{and} \quad w_{\mathbf{C}iso} = w_{\mathbf{B}iso} =: w_{iso}\,.$$

The link between these strain energy functions and RICHTER´s representation of isotropic elasticity is established, *e.g.*, by

Theorem 7.10. of FINGER (1894)[103]
An isotropic elastic solid with the representation

$$\mathbf{T} = \beta_0\,\mathbf{I} + \beta_1\,\mathbf{B} + \beta_2\,\mathbf{B}^2$$

is hyperelastic, if and only if the following relations hold

$$\beta_0(I_\mathbf{B}, II_\mathbf{B}, III_\mathbf{B}) = 2\rho\, III_\mathbf{B}\,\frac{\partial w_{iso}}{\partial III_\mathbf{B}}$$

$$\beta_1(I_\mathbf{B}, II_\mathbf{B}, III_\mathbf{B}) = 2\rho\Big(\frac{\partial w_{iso}}{\partial I_\mathbf{B}} + I_\mathbf{B}\,\frac{\partial w_{iso}}{\partial II_\mathbf{B}}\Big)$$

$$\beta_2(I_\mathbf{B}, II_\mathbf{B}, III_\mathbf{B}) = -2\rho\,\frac{\partial w_{iso}}{\partial II_\mathbf{B}}$$

or equivalently for

$$\mathbf{T} = \underline{\beta}_0\,\mathbf{I} + \underline{\beta}_1\,\mathbf{B} + \underline{\beta}_{-1}\,\mathbf{B}^{-1}$$

the relations

$$\underline{\beta}_0(I_\mathbf{B}, II_\mathbf{B}, III_\mathbf{B}) = 2\rho\Big(II_\mathbf{B}\,\frac{\partial w_{iso}}{\partial II_\mathbf{B}} + III_\mathbf{B}\,\frac{\partial w_{iso}}{\partial III_\mathbf{B}}\Big)$$

$$\underline{\beta}_1(I_\mathbf{B}, II_\mathbf{B}, III_\mathbf{B}) = 2\rho\,\frac{\partial w_{iso}}{\partial I_\mathbf{B}}$$

$$\underline{\beta}_{-1}(I_\mathbf{B}, II_\mathbf{B}, III_\mathbf{B}) = -2\rho\, III_\mathbf{B}\,\frac{\partial w_{iso}}{\partial II_\mathbf{B}}$$

Proof. If we determine the differential of $w_{iso}(I_\mathbf{B}, II_\mathbf{B}, III_\mathbf{B})$ by use of (1.40) and the chain rule, and compare it with the representations of isotropic tensor-functions (6.31), then

$$\mathbf{T} = \beta_0\,\mathbf{I} + \beta_1\,\mathbf{B} + \beta_2\,\mathbf{B}^2 = 2\rho\,\mathbf{B}\,\frac{dw_\mathbf{B}}{d\mathbf{B}}$$

with

$$\mathbf{B}\,\frac{dw_\mathbf{B}}{d\mathbf{B}} = \mathbf{B}\Big(\frac{\partial w_{iso}}{\partial I_\mathbf{B}}\,\frac{dI_\mathbf{B}}{d\mathbf{B}} + \frac{\partial w_{iso}}{\partial II_\mathbf{B}}\,\frac{dII_\mathbf{B}}{d\mathbf{B}} + \frac{\partial w_{iso}}{\partial III_\mathbf{B}}\,\frac{dIII_\mathbf{B}}{d\mathbf{B}}\Big)$$

$$= \mathbf{B}\Big(\frac{\partial w_{iso}}{\partial I_\mathbf{B}}\,\mathbf{I} + \frac{\partial w_{iso}}{\partial II_\mathbf{B}}\,(I_\mathbf{B}\,\mathbf{I} - \mathbf{B}) + \frac{\partial w_{iso}}{\partial III_\mathbf{B}}\,III_\mathbf{B}\,\mathbf{B}^{-1}\Big)$$

[103] JOSEPH FINGER (1894) derived these conditions from NEUMANN´s potentials solely based on geometrical arguments, still unaware of RICHTER´s representation.

$$= \frac{\partial w_{iso}}{\partial I_\mathbf{B}} \mathbf{B} + \frac{\partial w_{iso}}{\partial II_\mathbf{B}} (I_\mathbf{B}\, \mathbf{B} - \mathbf{B}^2) + \frac{\partial w_{iso}}{\partial III_\mathbf{B}} III_\mathbf{B}\, \mathbf{I}\,.$$

By comparing the powers of $\mathbf{B}$, we obtain the relations of the first part of the theorem. The second part can be proven analogously by using the CAYLEY-HAMILTON equation of Theorem 1.13.; *q. e. d.*

If the reference placement is stress-free, then for $\mathbf{B} \equiv \mathbf{I}$, $I_\mathbf{B} \equiv 3$, $II_\mathbf{B} \equiv 3$, $III_\mathbf{B} \equiv 1$, $\rho \equiv \rho_0$ we have

$$\mathbf{T} = \mathbf{0} = 2\rho_0 \Big(\frac{\partial w_{iso}}{\partial I_\mathbf{B}} + 2\, \frac{\partial w_{iso}}{\partial II_\mathbf{B}} + \frac{\partial w_{iso}}{\partial III_\mathbf{B}} \Big)\, \mathbf{I}$$

taking all derivatives at $\{3, 3, 1\}$. This leads to the conditions on $w_{iso}(I_\mathbf{B}, II_\mathbf{B}, III_\mathbf{B})$ at $\{3, 3, 1\}$

$$\frac{\partial w_{iso}}{\partial I_\mathbf{B}} + 2\, \frac{\partial w_{iso}}{\partial II_\mathbf{B}} + \frac{\partial w_{iso}}{\partial III_\mathbf{B}} = 0\,.$$

If, inversely, the RICHTER representation of isotropic elasticity is given, then the following theorem gives necessary and sufficient conditions for hyperelasticity.

Theorem 7.11. of GOLDENBLAT (1950)
An isotropic elastic solid with the representation

$$\mathbf{T} = \beta_0\, \mathbf{I} + \beta_1\, \mathbf{B} + \beta_2\, \mathbf{B}^2 \qquad \textit{with } \beta_i(I_\mathbf{B}, II_\mathbf{B}, III_\mathbf{B})$$

is hyperelastic, if and only if the following conditions of integrability hold

$$\frac{\partial \beta_0}{\partial II_\mathbf{B}} + III_\mathbf{B} \frac{\partial \beta_2}{\partial III_\mathbf{B}} + \frac{\beta_2}{2} = 0$$

$$III_\mathbf{B} \frac{\partial \beta_1}{\partial III_\mathbf{B}} + \frac{\beta_1}{2} - I_\mathbf{B} \frac{\partial \beta_0}{\partial II_\mathbf{B}} - \frac{\partial \beta_0}{\partial I_\mathbf{B}} = 0$$

$$\frac{\partial \beta_1}{\partial II_\mathbf{B}} + I_\mathbf{B} \frac{\partial \beta_2}{\partial II_\mathbf{B}} + \frac{\partial \beta_2}{\partial I_\mathbf{B}} = 0$$

Proof. By taking the derivatives of the relations of FINGER's Theorem 7.10., we obtain the expressions

(1) $$\frac{\partial(\beta_0 / 2\rho)}{\partial I_\mathbf{B}} = III_\mathbf{B} \frac{\partial^2 w_{iso}}{\partial I_\mathbf{B}\, \partial III_\mathbf{B}}$$

(2) $$\frac{\partial(\beta_0/2\rho)}{\partial II_{\mathbf{B}}} = III_{\mathbf{B}} \frac{\partial^2 w_{iso}}{\partial II_{\mathbf{B}}\, \partial III_{\mathbf{B}}}$$

(3) $$\frac{\partial(\beta_1/2\rho)}{\partial II_{\mathbf{B}}} = \frac{\partial^2 w_{iso}}{\partial I_{\mathbf{B}}\, \partial II_{\mathbf{B}}} + I_{\mathbf{B}} \frac{\partial^2 w_{iso}}{\partial II_{\mathbf{B}}{}^2}$$

(4) $$\frac{\partial(\beta_1/2\rho)}{\partial III_{\mathbf{B}}} = \frac{\partial^2 w_{iso}}{\partial I_{\mathbf{B}}\, \partial III_{\mathbf{B}}} + I_{\mathbf{B}} \frac{\partial^2 w_{iso}}{\partial II_{\mathbf{B}}\, \partial III_{\mathbf{B}}}$$

(5) $$\frac{\partial(\beta_2/2\rho)}{\partial I_{\mathbf{B}}} = -\frac{\partial^2 w_{iso}}{\partial I_{\mathbf{B}}\, \partial II_{\mathbf{B}}}$$

(6) $$\frac{\partial(\beta_2/2\rho)}{\partial II_{\mathbf{B}}} = -\frac{\partial^2 w_{iso}}{\partial II_{\mathbf{B}}{}^2}$$

(7) $$\frac{\partial(\beta_2/2\rho)}{\partial III_{\mathbf{B}}} = -\frac{\partial^2 w_{iso}}{\partial II_{\mathbf{B}}\, \partial III_{\mathbf{B}}} .$$

By (2) and (7) we conclude

$$\frac{\partial(\beta_0/2\rho)}{\partial II_{\mathbf{B}}} = -III_{\mathbf{B}} \frac{\partial(\beta_2/2\rho)}{\partial III_{\mathbf{B}}} .$$

By expressing the mass density by the third principal invariant

$$\rho = \rho_0\, III_{\mathbf{B}}{}^{-1/2}$$

after (2.17), we obtain the first condition of integrability. By (1), (2), and (4) we analogously conclude the second condition, and by (3), (5), and (6) the third; *q.e.d.*

Similar conditions can be derived for the other scalar functions of the invariants[104].

If we particularise these results to *incompressible* solids with respect to an isotropic configuration being physically admissible, then

$$det\, \mathbf{B} = III_{\mathbf{B}} = 1 .$$

Consequently, w_{iso} cannot depend on the third argument $III_{\mathbf{B}}$

$$w_{iso}(I_{\mathbf{B}}, II_{\mathbf{B}}, 1) .$$

[104] See TRUESDELL/NOLL 1965 p. 318 f.

Theorem 7.12. (ARIANO 1930, RIVLIN 1948)
An incompressible isotropic hyperelastic solid has the representation

$$\mathbf{T} = -p_1\,\mathbf{I} + 2\rho\Big(\frac{\partial w_{iso}(I_\mathbf{B}, II_\mathbf{B}, 1)}{\partial I_\mathbf{B}} + I_\mathbf{B}\,\frac{\partial w_{iso}(I_\mathbf{B}, II_\mathbf{B}, 1)}{\partial II_\mathbf{B}}\Big)\,\mathbf{B} - 2\rho\,\frac{\partial w_{iso}(I_\mathbf{B}, II_\mathbf{B}, 1)}{\partial II_\mathbf{B}}\,\mathbf{B}^2$$

or

$$\mathbf{T} = -p_2\,\mathbf{I} + 2\rho\,\frac{\partial w_{iso}(I_\mathbf{B}, II_\mathbf{B}, 1)}{\partial I_\mathbf{B}}\,\mathbf{B} - 2\rho\,\frac{\partial w_{iso}(I_\mathbf{B}, II_\mathbf{B}, 1)}{\partial II_\mathbf{B}}\,\mathbf{B}^{-1}$$

with the pressures p_1 *and* p_2 *and* **B** *unimodular and positive-definite and symmetric.*

Herein, we did not yet normalise the extra stresses in the sense of (5.5), which still contain spherical components, and p_1 and p_2 are not the complete pressures.

For the other reduced forms, we find corresponding representations for isotropic incompressible hyperelastic materials.

A different representation for isotropic hyperelasticity is obtained by the following invariants of $\mathbf{B} \in \mathcal{Lin}$

$$\underline{I}_i := tr(\mathbf{B}^i) \qquad i = 1, 2, 3.$$

Their derivatives are simply

$$\frac{\partial \underline{I}_i}{\partial \mathbf{B}} = i\,\mathbf{B}^{i-1} \qquad i = 1, 2, 3$$

and their traces

$$tr\left(\frac{\partial \underline{I}_i}{\partial \mathbf{B}}\right) = i\,\underline{I}_{i-1} \qquad i = 1, 2, 3.$$

If we now use the ansatz $\underline{w}_{iso}(\underline{I}_1, \underline{I}_2, \underline{I}_3)$, we obtain the representation

$$\mathbf{T} = 2\rho\Big(\frac{\partial \underline{w}_{iso}}{\partial \underline{I}_1}\,\mathbf{B} + 2\,\frac{\partial \underline{w}_{iso}}{\partial \underline{I}_2}\,\mathbf{B}^2 + 3\,\frac{\partial \underline{w}_{iso}}{\partial \underline{I}_3}\,\mathbf{B}^3\Big)$$

by analogy to FINGER´s Theorem 7.10.

PENN (1970), LURIE (1990), LAINE et al. (1999), and CRISCIONE *et al.* (2000) suggested other, more physics-based systems of invariants.

Alternatively to the use of invariants, we can again make use of the eigenvalues λ_i of **B** and **C**.

Theorem 7.13. *A hyperelastic solid is isotropic, if and only if the strain energy can be expressed as a symmetric function* w_λ *of the three eigenvalues of its argument*

$$w_{\mathbf{C}}(\mathbf{C}) = w_\lambda(\lambda_1, \lambda_2, \lambda_3) = w_\lambda(\lambda_3, \lambda_2, \lambda_1) = w_\lambda(\lambda_2, \lambda_1, \lambda_3)$$

The CAUCHY stresses are then

$$\mathbf{T} = 2\rho \frac{dw_{\mathbf{B}}}{d\mathbf{B}} \mathbf{B}$$

$$= 2\rho \left(\frac{\partial w_\lambda}{\partial \lambda_1} \frac{d\lambda_1}{d\mathbf{B}} + \frac{\partial w_\lambda}{\partial \lambda_2} \frac{d\lambda_2}{d\mathbf{B}} + \frac{\partial w_\lambda}{\partial \lambda_3} \frac{d\lambda_3}{d\mathbf{B}} \right) \mathbf{B}$$

$$= 2\rho \sum_{i=1}^{3} \frac{\partial w_\lambda}{\partial \lambda_i} \mathbf{v}_i \otimes \mathbf{v}_i \mathbf{B} = 2\rho \sum_{i=1}^{3} \frac{\partial w_\lambda}{\partial \lambda_i} \lambda_i \mathbf{v}_i \otimes \mathbf{v}_i .$$

This can be put into the form

$$\mathbf{T} = \sum_{i=1}^{3} \sigma(\lambda_i, \lambda_{i+1} + \lambda_{i+2}, |\lambda_{i+1} - \lambda_{i+2}|) \mathbf{v}_i \otimes \mathbf{v}_i$$

with

$$\sigma(\lambda_i, \lambda_{i+1} + \lambda_{i+2}, |\lambda_{i+1} - \lambda_{i+2}|) := \frac{\partial w_\lambda}{\partial \lambda_i} \lambda_i .$$

The relation between the two representations is

$$w_\lambda(\lambda_1, \lambda_2, \lambda_3) = w_{iso}(I_{\mathbf{B}}, II_{\mathbf{B}}, III_{\mathbf{B}})$$

$$= w_{iso}(\lambda_1 + \lambda_2 + \lambda_3, \lambda_1 \lambda_2 + \lambda_2 \lambda_3 + \lambda_3 \lambda_1, \lambda_1 \lambda_2 \lambda_3)$$

with (1.20).

If one prefers the material representation, we obtain with (6.26) and (7.3)

$$\mathbf{S} = \sum_{i=1}^{3} \underline{\sigma}(\lambda_i, \lambda_{i+1} + \lambda_{i+2}, |\lambda_{i+1} - \lambda_{i+2}|) \mathbf{u}_i \otimes \mathbf{u}_i$$

$$= 2\rho \frac{dw_{\mathbf{C}}}{d\mathbf{C}} = 2\rho \left(\sum_{i=1}^{3} \frac{\partial w_\lambda}{\partial \lambda_i} \mathbf{u}_i \otimes \mathbf{u}_i \right) .$$

A comparison of the coefficients gives

$$2\rho \frac{\partial w_\lambda}{\partial \lambda_i} = \underline{\sigma}(\lambda_i, \lambda_{i+1} + \lambda_{i+2}, |\lambda_{i+1} - \lambda_{i+2}|) .$$

This leads to the integrability conditions

$$2\rho \frac{\partial^2 w_\lambda}{\partial \lambda_i \partial \lambda_k} = \frac{\partial}{\partial \lambda_i} \underline{\sigma}(\lambda_k, \lambda_{k+1} + \lambda_{k+2}, |\lambda_{k+1} - \lambda_{k+2}|)$$

$$= \frac{\partial}{\partial \lambda_k} \underline{\sigma}(\lambda_i, \lambda_{i+1} + \lambda_{i+2}, |\lambda_{i+1} - \lambda_{i+2}|) .$$

Theorem 7.14. *An isotropic elastic solid with the representation*

$$\mathbf{S} = \sum_{i=1}^{3} \underline{\sigma}(\lambda_i, \lambda_{i+1} + \lambda_{i+2}, |\lambda_{i+1} - \lambda_{i+2}|)\, \mathbf{u}_i \otimes \mathbf{u}_i$$

with one scalar function $\underline{\sigma}$ *of the three eigenvalues* λ_i *of* $\mathbf{C}$ *is hyperelastic, if and only if the following conditions of integrability hold for all positive* $\lambda_1 \neq \lambda_2 \neq \lambda_3$*:*

$$\underline{\sigma}_{,2}(\lambda_1, \lambda_2 + \lambda_3, |\lambda_2 - \lambda_3|) + sgn(\lambda_2 - \lambda_3)\, \underline{\sigma}_{,3}(\lambda_1, \lambda_2 + \lambda_3, |\lambda_2 - \lambda_3|)$$
$$= \underline{\sigma}_{,2}(\lambda_2, \lambda_3 + \lambda_1, |\lambda_3 - \lambda_1|) + sgn(\lambda_1 - \lambda_3)\, \underline{\sigma}_{,3}(\lambda_2, \lambda_3 + \lambda_1, |\lambda_3 - \lambda_1|)$$

$$\underline{\sigma}_{,2}(\lambda_1, \lambda_2 + \lambda_3, |\lambda_2 - \lambda_3|) + sgn(\lambda_3 - \lambda_2)\, \underline{\sigma}_{,3}(\lambda_1, \lambda_2 + \lambda_3, |\lambda_2 - \lambda_3|)$$
$$= \underline{\sigma}_{,2}(\lambda_3, \lambda_1 + \lambda_2, |\lambda_1 - \lambda_2|) + sgn(\lambda_1 - \lambda_2)\, \underline{\sigma}_{,3}(\lambda_3, \lambda_1 + \lambda_2, |\lambda_1 - \lambda_2|)$$

$$\underline{\sigma}_{,2}(\lambda_3, \lambda_1 + \lambda_2, |\lambda_1 - \lambda_2|) + sgn(\lambda_2 - \lambda_1)\, \underline{\sigma}_{,3}(\lambda_3, \lambda_1 + \lambda_2, |\lambda_1 - \lambda_2|)$$
$$= \underline{\sigma}_{,2}(\lambda_2, \lambda_3 + \lambda_1, |\lambda_3 - \lambda_1|) + sgn(\lambda_3 - \lambda_1)\, \underline{\sigma}_{,3}(\lambda_2, \lambda_3 + \lambda_1, |\lambda_3 - \lambda_1|)$$

Herein, $\underline{\sigma}_{,i}$ *stands for the partial derivative with respect to the i-th argument.*

Note that the dependence of the stresses on the first argument is not restricted.

The following **examples** for hyperelastic models can be found in the literature[105]:

Compressible Models

- BLATZ-KO model (1962)

$$w_{iso} = \alpha \{\gamma [(I_{\mathbf{B}} - 3) - \beta^{-1}(III_{\mathbf{B}}^{\beta} - 1)]$$
$$+ (1-\gamma)[II_{\mathbf{B}} / III_{\mathbf{B}} - 3 - \beta^{-1}(III_{\mathbf{B}}^{-\beta} - 1)]\}$$

[105] For hyperelastic models see also LURIE (1990), EHLERS/ EIPPER (1992), ŠILHAVÝ (1997), HOLZAPFEL (2000), CRISCIONE (2004), TRELOAR (2005).

with real constants γ with $0 \leq \gamma \leq 1$, α and β. This energy gives the following stresses

$$\mathbf{T} = 2\rho\,\alpha\,\{[-\gamma\, III_\mathbf{B}{}^{\beta} - (1-\gamma)\, II_\mathbf{B} / III_\mathbf{B} + (1-\gamma)\, III_\mathbf{B}{}^{-\beta}]\,\mathbf{I} + [\gamma + (1-\gamma)\, I_\mathbf{B} / III_\mathbf{B}]\,\mathbf{B} - (1-\gamma) / III_\mathbf{B}\,\mathbf{B}^2\}.$$

- CIARLET model (1988)

$$w_{iso} = \lambda/4\,(III_\mathbf{B} - 1) - (\lambda/2 + \mu)/2\,\ln(III_\mathbf{B}) + \mu/2\,(I_\mathbf{B} - 3)$$

with the positive LAMÉ constants λ and μ. This energy gives the following stresses

$$\mathbf{T} = \rho\,[\mu\,(\mathbf{B} - \mathbf{I}) + \lambda/2\,(III_\mathbf{B} - 1)\,\mathbf{I}]\,.$$

- HILL model (1978)

$$w_{iso} = \sum_{i=1}^{N} \frac{\beta_i}{\alpha_i}\,\{\lambda_1{}^{\alpha_i} + \lambda_2{}^{\alpha_i} + \lambda_3{}^{\alpha_i} - 3 + n^{-1}\,(III_\mathbf{B}{}^{-\alpha_i{}^n} - 1)\}$$

with real constants α_i and β_i with $\alpha_i\,\beta_i > 0$ and natural constants n and N. By use of (1.44) we obtain for the stresses

$$\mathbf{T} = 2\rho \sum_{i=1}^{N} \beta_i\,(\mathbf{B}^{\alpha_i} - \alpha_i^{n-1}/n\; III_\mathbf{B}{}^{-\alpha_i{}^n}\,\mathbf{I})\,.$$

- ATTARD model (2003)

$$w_{iso} = \sum_{i=1}^{N} \{\frac{\alpha_i}{2i}\,tr(\mathbf{B}^{i} - \mathbf{I}) + \frac{\beta_i}{2i}\,tr(\mathbf{B}^{-i} - \mathbf{I})\} + \sum_{i=1}^{M} \frac{\gamma_i}{2i}\,(\ln J)^{2i} - \{\sum_{i=1}^{N} (\alpha_i - \beta_i)\}\,\ln J$$

with real constants α_i, β_i, and γ_i and natural N and M. With (1.38) we obtain

$$\mathbf{T} = \rho \sum_{i=1}^{N} (\alpha_i\,\mathbf{B}^{i} - \beta_i\,\mathbf{B}^{-i}) + \{\sum_{i=1}^{M} \gamma_i\,(\ln J)^{2i-1} - \sum_{i=1}^{N} (\alpha_i - \beta_i)\}\,\mathbf{I}\,.$$

- MURNAGHAN model (1937)

$$w_{iso} = \gamma\, I_{\mathbf{E}^G} + (2\lambda + \eta)\, I_{\mathbf{E}^G}{}^2 + \varepsilon\, I_{\mathbf{E}^G}{}^3 - 2\eta\, II_{\mathbf{E}^G} + \xi\, I_{\mathbf{E}^G}\, II_{\mathbf{E}^G} + \zeta\, III_{\mathbf{E}^G}$$

with real constants γ, λ, η, ε, ξ and ζ. This is a rather versatile but also complicated model. The corresponding law for the stresses can be found in the original work.

Incompressible Models

- MOONEY-RIVLIN model (1940)

$$w_{iso} = ½\,\alpha\,[(½ + \beta)\,(I_\mathbf{B} - 3) + (½ - \beta)\,(II_\mathbf{B} - 3)]$$

$$= \tfrac{1}{2}\,\alpha\,[(\tfrac{1}{2}+\beta)\,(\lambda_1 + \lambda_2 + \lambda_3 - 3) + (\tfrac{1}{2} - \beta)\,(\lambda_1^{-1} + \lambda_2^{-1} + \lambda_3^{-1} - 3)]$$

with two real constants α and β. Note that for incompressible materials we obtain the relation

$$\lambda_1\,\lambda_2 + \lambda_2\,\lambda_3 + \lambda_3\,\lambda_1 = \lambda_1^{-1} + \lambda_2^{-1} + \lambda_3^{-1}$$

by use of the incompressibility condition $\lambda_1\,\lambda_2\,\lambda_3 = 1$.

The corresponding stresses are

$$\mathbf{T} = -p\,\mathbf{I} + \rho_0\,\alpha\,[(\tfrac{1}{2} + \beta)\,\mathbf{B} + (\tfrac{1}{2} - \beta)\,\mathbf{B}^{-1}]$$

As a special case if this we obtain for $\beta \equiv \tfrac{1}{2}$ the

- neo-HOOKE model (RIVLIN 1948)

$$w_{iso} = \tfrac{1}{2}\,\alpha\,(I_{\mathbf{B}} - 3) = \tfrac{1}{2}\,\alpha\,(\lambda_1 + \lambda_2 + \lambda_3 - 3)$$

with only one real constant α which leads to the stresses

$$\mathbf{T} = -p\,\mathbf{I} + \rho_0\,\alpha\,\mathbf{B}\,.$$

- VALANIS-LANDEL model (1967)

$$w_{iso} = g(\lambda_1) + g(\lambda_2) + g(\lambda_3)$$

with a real-valued scalar-function g. The corresponding stresses are

$$\mathbf{T} = -p\,\mathbf{I} + 2\rho_0 \sum_{i=1}^{3} g'(\lambda_i)\,\lambda_i\,\mathbf{v}_i \otimes \mathbf{v}_i\,.$$

- RIVLIN-SAUNDERS model (1951)

$$w_{iso} = \alpha/2\,(I_{\mathbf{B}} - 3) + g(II_{\mathbf{B}} - 3)$$

with the real constant α and a function g, for which $g(0) = 0$ holds. The according stresses are

$$\mathbf{T} = -p\,\mathbf{I} + \rho_0\,[\alpha\,\mathbf{B} - 2\,g'(II_{\mathbf{B}} - 3)\,\mathbf{B}^{-1}]\,.$$

- OGDEN model (1972)

$$w_{iso} = \sum_{i=1}^{N} \frac{\beta_i}{\alpha_i}\,(\lambda_1^{\alpha_i} + \lambda_2^{\alpha_i} + \lambda_3^{\alpha_i} - 3) = \sum_{i=1}^{N} \frac{\beta_i}{\alpha_i}\,tr(\mathbf{B}^{\alpha_i} - \mathbf{I})$$

with real constants α_i and β_i with $\alpha_i\,\beta_i > 0$ and a natural N. The stresses are then

$$\mathbf{T} = -p\,\mathbf{I} + 2\rho_0 \sum_{i=1}^{N} \beta_i\,\mathbf{B}^{\alpha_i}.$$

- YEOH model (1990,1993)

$$w_{iso} = \sum_{i=1}^{N} \alpha_i\,(I_{\mathbf{B}} - 3)^i$$

with real constants α_i and a natural N. The stresses are then

$$\mathbf{T} = -p\,\mathbf{I} + 2\rho_0\,[\sum_{i=1}^{N} i\,\alpha_i\,(I_\mathbf{B} - 3)^{i-1}]\,\mathbf{B}\,.$$

A special case of this is the following

- RAGHAVAN-VORP model (2000)

$$w_{iso} = \alpha\,(I_\mathbf{B} - 3) + \beta\,(I_\mathbf{B} - 3)^2$$

with real constants α and C. This gives the stresses

$$\mathbf{T} = -p\,\mathbf{I} + 2\rho_0\,[\alpha + 2\beta\,(I_\mathbf{B} - 3)]\,\mathbf{B}\,.$$

This model has been applied to biological tissues.

- KNOWLES model (1977)

$$w_{iso} = \frac{\mu}{2b}\{[1 + \frac{b}{n}\;(I_\mathbf{B} - 3)]^n - 1\}$$

with three positive constants b, μ, and n. The stresses are

$$\mathbf{T} = -p\,\mathbf{I} + \rho_0\,\mu\,[1 + b/n\;(I_\mathbf{B} - 3)]^{n-1}\;\mathbf{B}\,.$$

With growing n this converges to the

- FUNG (1967) model[106]

$$w_{iso} = \frac{\mu}{2b}\{exp[b\;(I_\mathbf{B} - 3)] - 1\}$$

with real constants μ and b. This gives the stresses

$$\mathbf{T} = -p\,\mathbf{I} + \rho_0\,\mu\,exp[b\,(I_\mathbf{B} - 3)]\,\mathbf{B}\,.$$

This model has also been applied to biological tissues.

- GENT (1996) model

$$w_{iso} = -\frac{E}{6}\;b\;ln[1 - (I_\mathbf{B} - 3)\,/\,b]$$

with real constants E, μ and b. This gives the stresses

$$\mathbf{T} = -p\,\mathbf{I} + \rho_0\,\frac{E}{3}\;[1 - (I_\mathbf{B} - 3)\,/\,b]^{-1}\,\mathbf{B}\,.$$

This model was originally suggested for rubber, but has also been applied to biological tissues.

- ARRUDA-BOYCE model (1993)

$$w_{iso} = \alpha\,[\frac{1}{2}(I_\mathbf{B} - 3) + \frac{1}{20N}(I_\mathbf{B}^{\,2} - 9) + \frac{11}{1050\,N^2}(I_\mathbf{B}^{\,3} - 27)$$

---

[106] See also TABER (2004).

$$+ \frac{19}{7000\,N^3}\,(I_{\mathbf{B}}^{\,4} - 81) + \frac{519}{673750\,N^4}\,(I_{\mathbf{B}}^{\,5} - 243)]$$

with real constants α and N, which possess a microphysical interpretation[107]. Note that in the last models only the trace of $\mathbf{B}$ enters, but not the other invariants.[108] The corresponding stresses are

$$\mathbf{T} = -p\,\mathbf{I} + \rho_0\,\alpha\,[1 + \frac{1}{5N}\,I_{\mathbf{B}} + \frac{11}{175N^2}\,I_{\mathbf{B}}^{\,2} + \frac{19}{875N^3}\,I_{\mathbf{B}}^{\,3} + \frac{519}{67375N^4}\,I_{\mathbf{B}}^{\,4}]\,\mathbf{B}\,.$$

In HOSS/ MARCZAK (2010) and MARCKMANN/ VERRON (2006) one finds an overviews of hyperelastic models applied to rubber-like materials. The invertibility of isotropic hyperelastic laws is investigated in BLUME (1992).

---

[107] See HORGAN/ SACCOMANDI (2002).

[108] See also WU/ v. d. GIESSEN (1993) and BEATTY (2004).

8 Solutions

8.1 Elastic Boundary Value Problems

After considering in the previous chapters almost exclusively the material theory of elasticity, we will now focus on the boundary-value problem. We assume an elastic law and establish the system of field-equations which define a boundary and initial value problem. Mathematically, this leads to a non-linear coupled system of partial differential equations with spatial and temporal boundary conditions. As we do not intend to further restrict the class of materials under consideration, we will neither be able to further specify the manifold of solutions. Despite this generality, we will still be able to give some non-trivial solutions, which enable us to study the material behaviour.

We will primarily try to find analytical solutions. However, these are rather limited, and it will be unavoidable to also consider approximate solutions.

The **initial and boundary value problem of elasticity** for the case without internal constraints is posed in the following way.

We consider an elastic body $\mathscr{B}$ consisting of a defined or identified elastic material, *i.e.*, for each of its points $X \in \mathscr{B}$ an *elastic law* is given as a reduced form like (6.8)

$$\mathbf{S} = k_X(\mathbf{C}) \, .$$

The body can be uniform or non-uniform. In the first case, we have elastic isomorphisms between all body points, which transform their elastic laws according to (6.20). In the latter case, the material laws must be given for all points.

All of the other equations in this section are universal, *i.e.*, they hold for all materials.

The *motion of the body* in the EUCLIDean space in a time interval $[t_a, t_e]$ is a time-dependent sequence of placements

$$\kappa : \mathscr{B} \times [t_a, t_e] \rightarrow \mathscr{E} .$$

After choosing a reference placement $\kappa_0 : \mathscr{B} \rightarrow \mathscr{E}$, we obtain the description of the motion in the EUCLIDean space as

$$\boldsymbol{\chi}(\mathbf{x}_0 , t) = \kappa(\kappa_0^{-1}(\mathbf{x}_0) , t) \, .$$

After choosing a material COOS $\{\Psi^k\}$ and a spatial one $\{\varphi^i\}$, we can represent the motion of the body by means of a time-dependent coordinate transformation

$$\varphi^i = \chi^i(\Psi^1, \Psi^2, \Psi^3, t) \, .$$

A. Bertram, *Elasticity and Plasticity of Large Deformations*,
https://doi.org/10.1007/978-3-030-72328-6_8

The deformation gradient $\mathbf{F} = Grad\,\boldsymbol{\chi}$ can then be determined for each point and each instant by (2.5)

$$\mathbf{F} = \frac{\partial \varphi^k}{\partial \Psi^i}\,\mathbf{r}_{\varphi k} \otimes \mathbf{r}_{\Psi}{}^i \qquad \in \mathscr{Inv}^+$$

and with it the field of the right CAUCHY-GREEN tensor

$$\mathbf{C} = \mathbf{F}^T\,\mathbf{F} \qquad \in \mathscr{Psym}^+.$$

We can then use the elastic law to determine the material stress tensor $\mathbf{S}$ at each point. The transition to the CAUCHY stress tensor is

$$\mathbf{T} = \mathbf{F}\,\mathbf{S}\,\mathbf{F}^T.$$

If the material law k_X generally gives a symmetric stress tensor - which shall be assumed -, then also $\mathbf{T}$ is symmetric and CAUCHY´s second equation (3.21) is identically fulfilled. CAUCHY´s first equation (3.20) still remains to be satisfied at each point and all instants

$$div\,\mathbf{T} + \rho\,(\mathbf{b} - \mathbf{a}) = \mathbf{o}\,,$$

where the field of the specific body force $\mathbf{b} \in \mathscr{V}$ should be known (*e.g.*, as the gravitational field), and the field of the acceleration is related to the motion by (2.1)

$$\mathbf{a}(\mathbf{x}_0\,,t) = \frac{\partial^2 \boldsymbol{\chi}(\mathbf{x}_0,t)}{\partial t^2} \qquad \in \mathscr{V}.$$

This completes the set of field equations. We can now substitute the stresses $\mathbf{T}$ by the elastic law in (3.20), as well as $\mathbf{a}$ by (2.1), so that finally the only unknown is the motion $\boldsymbol{\chi}(\mathbf{x}_0\,,t)$ (*elimination of the stresses*) as a solution of the vectorial differential equation

$$div\{(Grad\,\boldsymbol{\chi})\,k_X((Grad^T\boldsymbol{\chi})\,(Grad\,\boldsymbol{\chi}))\,(Grad^T\boldsymbol{\chi})\} + \rho\,\mathbf{b} = \rho\,\frac{\partial^2 \boldsymbol{\chi}(\mathbf{x}_0,t)}{\partial t^2}.$$

This is all valid when no internal constraints are active. If this is the case, the system changes principally. If we consider the case of $K \leq 6$ internal constraints in reduced form (5.2)

$$\gamma_{Ci}(\mathbf{C}) = 0\,, \quad i = 1, \dots, K > 0\,,$$

then the stresses in each point are decomposed

$$\mathbf{S} = \mathbf{S}_R + \mathbf{S}_E$$

into reaction stresses

$$\mathbf{S}_R = \sum_{i=1}^{K} \alpha_i \frac{d\gamma_{Ci}}{d\mathbf{C}}$$

and extra stresses being given by the elastic law

$$\mathbf{S}_E = k_X(\mathbf{C}) .$$

Thus, there are $K \leq 6$ scalar fields α_i and the same number of equations, namely the internal constraints.

Nevertheless, the problem, with or without internal constraints, is not yet well posed, as *boundary values* and *initial values* are needed. There are many possibilities of posing such conditions. We will here only consider the most simple kinds of such conditions, which, however, already open a wide variety of applications.

We distribute the boundary $\partial\mathscr{B}_0$ of the body into two disjoint surfaces $\partial\mathscr{B}_g$ and $\partial\mathscr{B}_d$, on which the following shall be prescribed:

- on $\partial\mathscr{B}_g$ the geometrical or DIRICHLET or **displacement boundary conditions** for

$$\boldsymbol{\chi}(\mathbf{x}_0 , t) \qquad \forall\, \mathbf{x}_0 \in \partial\mathscr{B}_g ,\ \forall\, t \in [t_a , t_e]$$

- on $\partial\mathscr{B}_d$ the dynamical or NEUMANN[109] or **traction boundary conditions** for

$$\mathbf{t}(\mathbf{x}_0 , t) = \mathbf{T}(\mathbf{x}_0 , t)\, \mathbf{n}(\mathbf{x}_0 , t) \qquad \forall\, \mathbf{x}_0 \in \partial\mathscr{B}_d ,\ \forall\, t \in [t_a , t_e]$$

  with the outer normal $\mathbf{n}$.

For some applications, one may prefer an EULERean condition instead of this LAGRANGEan form.

With respect to the initial conditions, we also have many choices. For example, the initial placement of the body

$$\boldsymbol{\chi}(\mathbf{x}_0 , t_a) \qquad \forall\, \mathbf{x}_0 \in \mathscr{B}_0$$

and the initial velocities

$$\boldsymbol{\chi}(\mathbf{x}_0 , t_a)^{\bullet} \qquad \forall\, \mathbf{x}_0 \in \mathscr{B}_0$$

can be prescribed.

With such boundary conditions, we have defined a complete initial and boundary problem of finite elasticity.

If the problem is time-dependent, then it belongs to *elastodynamics*, if not, it belongs to *elastostatics*. If the inertia part $\rho\, \mathbf{a}$ in (3.20) is negligible, then the problem is called *quasi-static*.

A motion $\boldsymbol{\chi}$, which satisfies all the above field equations and conditions, is called an *exact* or *strong solution*. The question of existence and uniqueness of strong solutions is a complicated matter within such a non-linear theory, and is subject of many investigations (see, *e.g.*, BALL 1977, CIARLET 1983). Beyond the existence of solutions, however, stability of solutions is of great concern,

[109] Carl Gottfried Neumann (1832-1925)

because such a solution may have no physical importance at all, as it turns out to be unstable. Stability of solutions in finite elasticity is also an important field of research (KNOPS/ WILKES 1973, HANYGA 1985, DE BORST/ v. d. GIESSEN 1998, REESE 2000, NGUYEN 2000). Only for classical, *i.e.*, linear elasticity, can questions of existence, uniqueness and stability be answered for a broad class of problems (KNOPS/ PAYNE 1973).

In most cases we will not be able to find exact solutions for a given boundary-value problem. In such cases one tries to find at least approximate solutions. These do not exactly satisfy all field equations at each point, but only in a mean sense. For that purpose, one uses variational principles instead of differential equations, the (weak) solutions of which are determined in finite-dimensional subspaces of the functional space of displacements. The most important method to construct approximate solutions is the Finite Element Method (FEM).

Example: Homogenisation by the RVE-Technique

Many materials possess a microstructure, which strongly influences the mechanical behaviour. As an example, most metallic materials consist of grains, in which the material is in crystalline form. These grains are so small compared with construction parts that a simulation, which takes all these grains into account, would lead to prohibitive computational costs. In order to still investigate the interaction between the microstructure and the macro-behaviour, homogenisation methods have been developed. The most versatile of these methods is perhaps the **representative volume element** (RVE), which was already suggested by BISHOP/ HILL (1951) under the label *unit cube*. For this, one cuts a small cubic element out of the body, which has to be large enough to reflect all the essential parts of the microstructure so as to be representative, while also being small enough for computation of its boundary value problem to still be feasible. This RVE is then submitted to the relevant load modes, and the generally inhomogeneous stress and deformation fields are determined. These fields are then homogenised by appropriate averages and interpreted as macro-variables. This procedure shall be made more precise in the sequel.

We consider an RVE, which is cubic in the reference placement. We submit it to a motion in the form

(8.1) $$\boldsymbol{\chi}(\mathbf{x}_0\,, t) \;=\; \underline{\mathbf{F}}(t)\,\mathbf{x}_0 \;+\; \mathbf{w}(\mathbf{x}_0\,, t)$$

with

- the *mean deformation* $\underline{\mathbf{F}}(t)$ depending only on time, but not on the point,
- the *fluctuation of the motion* $\mathbf{w}(\mathbf{x}_0\,, t)$ depending on time and point.

This motion is induced by prescribed displacements on the boundary. For this we have different choices:

- *homogeneous boundary-conditions*

We prescribe on all boundaries spatially constant or linear fields of displacements. For a homogeneous material (which is not interesting in this context) the resulting deformation of the RVE would be constant in space or homogeneous $\mathbf{w}(\mathbf{x}_0, t) \equiv \mathbf{o}$.

- *periodic boundary conditions*

We prescribe in the above decomposition of the motion only the part $\underline{\mathbf{F}}(t)\ \mathbf{x}_0$ which describes the mean deformation. The fluctuation is permitted on the boundary and set equal on opposite boundaries like $\mathscr{A}_0^-$ and $\mathscr{A}_0^+$

$$\mathbf{w}(\mathbf{x}_0, t)\ |_{\mathscr{A}_0^+} \equiv \mathbf{w}(\mathbf{x}_0, t)\ |_{\mathscr{A}_0^-} \qquad \forall\ t \in [t_a, t_e]\,.$$

In addition, the traction vector field on opposite boundaries is assumed to be equilibrated

$$\mathbf{t}(\mathbf{x}_0, t)\ |_{\mathscr{A}_0^+} \equiv -\,\mathbf{t}(\mathbf{x}_0, t)\ |_{\mathscr{A}_0^-} \qquad \forall\ t \in [t_a, t_e]\,.$$

The fluctuation is then determined by the solution of the entire boundary value problem for the RVE.

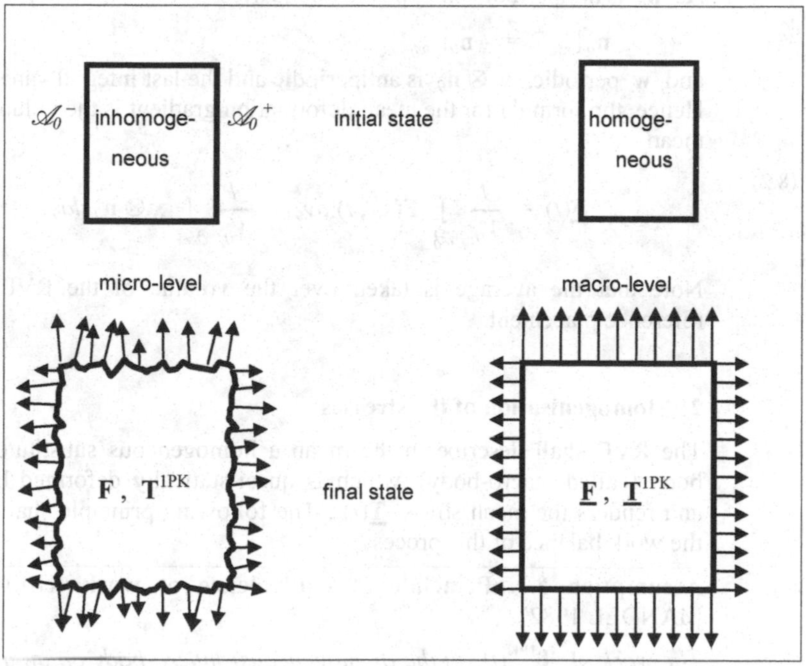

As the homogeneous boundary conditions lead to systematic errors in the boundary layer of the RVE, one prefers the more realistic periodic boundary conditions, although they are more complicated than the homogeneous ones.

We now come to the homogenisation of the micro-fields.

1) Homogenisation of the deformations

The (inhomogeneous) deformation gradient on the micro-(RVE)-level is with (8.1)

$$\mathbf{F}(\mathbf{x}_0\,,t) \;=\; Grad\,\boldsymbol{\chi}(\mathbf{x}_0\,,t) \;=\; \underline{\mathbf{F}}(t) + Grad\,\mathbf{w}(\mathbf{x}_0\,,t)\,.$$

By the GAUSS transformation we get

$$\int_{\mathcal{B}_0} \mathbf{F}\,dv_0 = \int_{\mathcal{B}_0} \underline{\mathbf{F}}\,dv_0 + \int_{\mathcal{B}_0} Grad\,\mathbf{w}\,dv_0$$

$$= \underline{\mathbf{F}}\,V_0 + \int_{\partial\mathcal{B}_0} \mathbf{w}\otimes\mathbf{n}_0\;da_0\,.$$

As $\mathbf{n}_0$ is antiperiodic on opposite boundaries

$$\mathbf{n}_0\big|_{\mathcal{A}_0^+} = -\,\mathbf{n}_0\big|_{\mathcal{A}_0^-}$$

and $\mathbf{w}$ periodic, $\mathbf{w}\otimes\mathbf{n}_0$ is antiperiodic and the last integral vanishes. Hence, the formula for the mean deformation gradient is the volumetric mean

(8.2)
$$\underline{\mathbf{F}}(t) \;=\; \frac{1}{V_0}\int_{\mathcal{B}_0} \mathbf{F}(\mathbf{x}_0\,,t)\;dv_0 \;=\; \frac{1}{V_0}\int_{\partial\mathcal{B}_0} \mathbf{x}\otimes\mathbf{n}_0\;da_0\,.$$

Note that the average is taken over the volume of the RVE in the reference placement.

2) Homogenisation of the stresses

The RVE shall describe in the mean a homogeneous substitute of the body (called macro-body), which is quasi-statically deformed by $\underline{\mathbf{F}}(t)$ and renders the mean stress $\underline{\mathbf{T}}(t)$. The following principle shall govern the work balance of this process.

Assumption 8.1. Principle of equivalence of work (HILL 1963, MANDEL 1982)

The work of $\underline{\mathbf{T}}^{1PK}(t)$ *in the (homogeneous) macro-body on an arbitrary deformation process* $\underline{\mathbf{F}}(t)$ *is equal to that of the (inhomogeneous) RVE*

$$\int_0^T \int_{\mathcal{B}_0} \underline{\mathbf{T}}^{1PK}\cdot\underline{\mathbf{F}}^{\bullet}\;dv_0\;dt \;=\; \int_0^T \int_{\mathcal{B}_0} \mathbf{T}^{1PK}\cdot\mathbf{F}^{\bullet}\;dv_0\;dt$$

As the work equivalence must hold for all processes, it must also hold for the power at each instant

$$\int_{\mathscr{B}_0} \underline{\mathbf{T}}^{1PK} \cdot \underline{\mathbf{F}}^{\bullet} \, dv_0 = \int_{\mathscr{B}_0} \mathbf{T}^{1PK} \cdot \mathbf{F}^{\bullet} \, dv_0$$

with

$\underline{\mathbf{F}}^{\bullet}$ the time-derivative of $\underline{\mathbf{F}}(t)$ at the end of the (arbitrary) process

$\mathbf{F}^{\bullet}$ the resulting field of the deformation rates on the micro level (as a solution of the boundary-value problem for the RVE).

We restrict our considerations to the quasi-static case (neglecting inertia) in the absence of body forces. The stress power equals

$$\int_{\mathscr{B}_0} \underline{\mathbf{T}}^{1PK} \cdot \underline{\mathbf{F}}^{\bullet} \, dv_0 = V_0 \, \underline{\mathbf{T}}^{1PK} \cdot \underline{\mathbf{F}}^{\bullet}$$

$$= \int_{\mathscr{B}_0} \mathbf{T}^{1PK} \cdot \mathbf{F}^{\bullet} \, dv_0 = \int_{\mathscr{B}_0} \mathbf{T}^{1PK} \cdot \{\underline{\mathbf{F}}^{\bullet} + (Grad\, \mathbf{w})^{\bullet}\} \, dv_0$$

$$= \Big(\int_{\mathscr{B}_0} \mathbf{T}^{1PK} \, dv_0\Big) \cdot \underline{\mathbf{F}}^{\bullet} + \int_{\mathscr{B}_0} \mathbf{T}^{1PK} \cdot Grad\, \mathbf{w}^{\bullet} \, dv_0 .$$

By the product rule we have

$$\mathbf{T}^{1PK} \cdot Grad\, \mathbf{w}^{\bullet} = Div(\mathbf{T}^{1PK\,T} \mathbf{w}^{\bullet}) - (Div\, \mathbf{T}^{1PK}) \cdot \mathbf{w}^{\bullet}.$$

The last term vanishes because of equilibrium after (3.25). By the GAUSS transformation we get

$$\int_{\mathscr{B}_0} Div\, (\mathbf{T}^{1PK\,T} \mathbf{w}^{\bullet}) \, dv_0 = \int_{\partial\mathscr{B}_0} (\mathbf{T}^{1PK\,T} \mathbf{w}^{\bullet}) \cdot \mathbf{n}_0 \, da_0$$

$$= \int_{\partial\mathscr{B}_0} \mathbf{w}^{\bullet} \cdot (\mathbf{T}^{1PK} \mathbf{n}_0) \, da_0 = \int_{\partial\mathscr{B}_t} \mathbf{w}^{\bullet} \cdot \mathbf{t} \, da$$

with the traction vector

$$\mathbf{t} := \mathbf{T}^{1PK} \mathbf{n}_0$$

on the boundary of the deformed RVE. $\mathbf{w}^{\bullet}$ is periodic and $\mathbf{t}$ is anti-periodic, so that the entire integral vanishes. Thus, we obtain the average of the stresses

$$\boxed{\underline{\mathbf{T}}^{1PK} = \frac{1}{V_0} \int_{\mathscr{B}_0} \mathbf{T}^{1PK} \, dv_0}$$

as a necessary and sufficient condition for the work equivalence. By $\underline{\mathbf{F}}$ we can transform the mean stress into any other stress tensor without violating the work equivalence.

If the RVE is submitted to homogeneous boundary conditions, then the same homogenisation formula can be used with $\mathbf{w}(\mathbf{x}_0, t) \equiv \mathbf{o}$ on the boundary of the body.

If we set the fluctuation $\mathbf{w}(\mathbf{x}_0, t) \equiv \mathbf{o}$ on the whole body, then the deformations become constant in space, which results in the VOIGT (1910) or TAYLOR (1938) model. Because of the heterogeneity of the material, the stress field will not be constant in space, which leads to a violation of the equilibrium condition.

If the RVE contains voids, then these can also be taken into account. One thinks of these voids as filled with a material having no or negligible stiffness, like air or gas. For the averaging of the deformation by the volume integral in (8.2), one must also integrate over these voids. Therefore, the surface averaging is preferable, as the voids enter it only through V_0. In a deformation-controlled process, $\underline{\mathbf{F}}$ is prescribed.

For the determination of the mean stress $\underline{\mathbf{T}}^{1PK}$ the voids do not directly contribute, as the stresses are (almost) zero in them. However, for the volume V_0 of $\mathscr{B}_0$ the voids have to be taken into account anyway.

Further information on homogenisation can be found in HILL (1963, 1965, 1972, 1984), HAVNER (1971), RICE (1971), KRAWIETZ (1986), NEMAT-NASSER/ HORI (1993), PETRYK (1999), MIEHE/ SCHRÖDER/ SCHOTTE (1999), BESSON *et al.* (2001, 2010), LUBARDA (2002), MIEHE (2003), RAABE *et al.* (2004), BERTRAM/ TOMAS (2008).

8.2 Universal Solutions

In some cases the boundary-value problem is posed differently, namely if we do not investigate a specific material, but a class of materials like, *e.g.*, all homogeneous elastic materials. In this case, we are looking for motions or displacements of the body, which satisfy the field equations for all materials of this class by posing appropriate loads on the boundary. Such solutions do not only give results that hold for a whole bunch of materials, but also ones that open up the possibility, at least in principle, to use them as test deformations, to measure the necessary loads, and to identify the material functions and constants.

A motion, which satisfies the field equations for all materials of a certain material class, is called a **universal solution** of this class. Often one looks only for static solutions for homogeneous bodies. And as one has little possibilities to manipulate the body force field, we restrict our considerations to the case of negligible body forces.

This means precisely, to find a placement of the body which leads to a field **C** of local configurations such that the equilibrium equation after (3.25)

$$Div(J\,\mathbf{F}\,\mathbf{S}) \;=\; Div(J\,\mathbf{F}\,\mathfrak{k}(\mathbf{C})) \;=\; \mathbf{o}$$

is satisfied for all material laws $\mathfrak{k}$ of this class.

Homogeneous Deformations

Let κ_0 and κ be two placements. κ is called *homogeneously deformed relative to* κ_0 if for all body points $X, Y \in \mathscr{B}$

$$\mathbf{F}_X \;=\; Grad\,(\kappa\,\kappa_0^{-1}(X)) \;\equiv\; \mathbf{F}_Y \;=\; Grad\,(\kappa\,\kappa_0^{-1}(Y))$$

holds. If the body consists of a homogeneous material, then also the CAUCHY stresses are constant fields and thus the equilibrium equation $div\,\mathbf{T} = \mathbf{o}$ is fulfilled everywhere.

Theorem 8.2. *All homogeneous deformations are universal solutions for all bodies of homogeneous material.*

ERICKSEN (1955) has shown that beyond the homogeneous deformations (**Family 0**) there are no more universal solutions for simple materials without internal constraints, even if we restrict the class of materials to

- elastic
- hyperelastic
- isotropic

ones. Only for materials with internal constraints are there universal solutions other than the homogeneous ones.

However, one can entirely identify the behaviour of a simple material by studying its behaviour under all homogeneous deformations alone.

Universal Solutions for Incompressible Elastic Materials

We will now restrict our considerations to *incompressible elastic* matcrials. By substituting the decomposition of CAUCHY's stress tensor into a pressure part and a deviatoric extra stress

$$\mathbf{T} \;=\; \mathbf{T}_R + \mathbf{T}_E \;=\; -p\,\mathbf{I} + \mathbf{F}\,k(\mathbf{C})\,\mathbf{F}^T$$

into (3.20), we obtain

$$div(\mathbf{T}_R + \mathbf{T}_E) \;=\; -\,grad(p) + div(\mathbf{T}_E) \;=\; \rho\,(\mathbf{a} - \mathbf{b})$$

with a constant mass density ρ. For a conservative body force $\mathbf{b} = -\,grad\ U$ with potential U we further obtain

$$div\ \mathbf{T}_E \;=\; \rho\ grad(p + U) + \rho\,\mathbf{a}\,.$$

If both the deformation and the material are homogeneous, then the left-hand side of the equation is again zero, and so is the right-hand side for all acceleration fields of the form

$$\mathbf{a} \;=\; -\,grad(U + p)\,,$$

i.e., for all motions with a conservative acceleration field, which are called *circulation-preserving*. As p is arbitrary, also the conservative body force can be given arbitrarily, *e.g.*, as $\mathbf{b} \equiv \mathbf{o}$. Consequently, these motions can be maintained by surface forces alone. This reasoning does not only hold for elastic materials, but for all simple materials.

Theorem 8.3. *For incompressible homogeneous bodies all circulation-preserving homogeneous motions are universal solutions.*

For static universal solutions, the above equation can be solved for conservative fields $div\ \mathbf{T}_E$, *i.e.*, by the condition

$$rot\ div\ \mathbf{T}_E \;=\; \mathbf{o}\,.$$

This fact can be used for the construction of inhomogeneous universal solutions for incompressible *isotropic* elastic materials. In MÜLLER (1970), an interesting method for this task was developed, which uses the fact that the invariance of $\mathbf{C}$ under certain rotations $\mathbf{Q}_i \in \mathcal{O}rth$

$$\mathbf{Q}_i * \mathbf{C} \;=\; \mathbf{C}$$

implies the same invariance for the stresses

$$\mathbf{Q}_i * \mathbf{S} \;=\; \mathbf{S}\,.$$

By cleverly using this property of isotropic materials, MÜLLER could find forms of the stress field, the divergence of which is conservative. However, this method only gives sufficient conditions, but not necessary ones for a solution to be universal.

Until today, five families of such universal solutions are known, which we briefly list in the sequel.

Family 1. *Bending, stretching, and shearing of a rectangular block*

We use a homogeneous reference placement with a Cartesian COOS $\{X, Y, Z\}$ and for the deformed placement cylindrical COOS $\{r, \theta, z\}$. The deformation is given by

$$r = \sqrt{(2AX)} \qquad \theta = BY \qquad z = Z/(AB) - BC\,Y$$

with three real constants A, B, C with $AB \neq 0$.

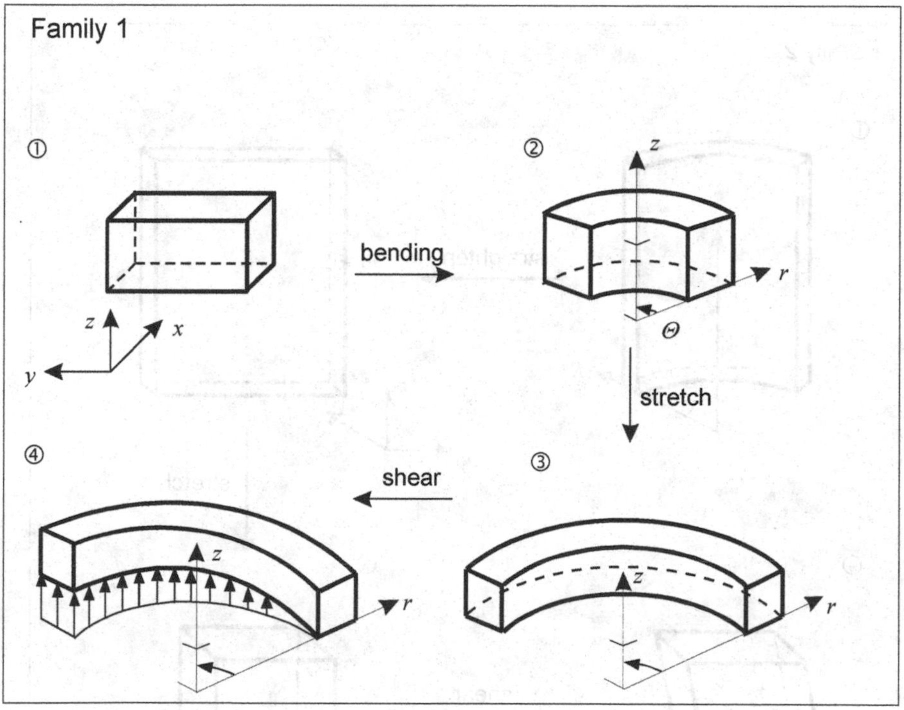

Family 2. *Straightening, stretching, and shearing a sector of a hollow cylinder*

We use a cylindrical COOS $\{R\,,\Theta\,,Z\}$ for the reference placement and a Cartesian COOS $\{x\,,y\,,z\}$ for the deformed placement. The deformation is given by

$$x = \tfrac{1}{2}\,A\,B^2\,R^2 \quad y = \Theta(AB)^{-1} \qquad z = Z/B + C\,\Theta/(AB)$$

with three real constants $A\,,B\,,C$ with $AB \neq 0$.

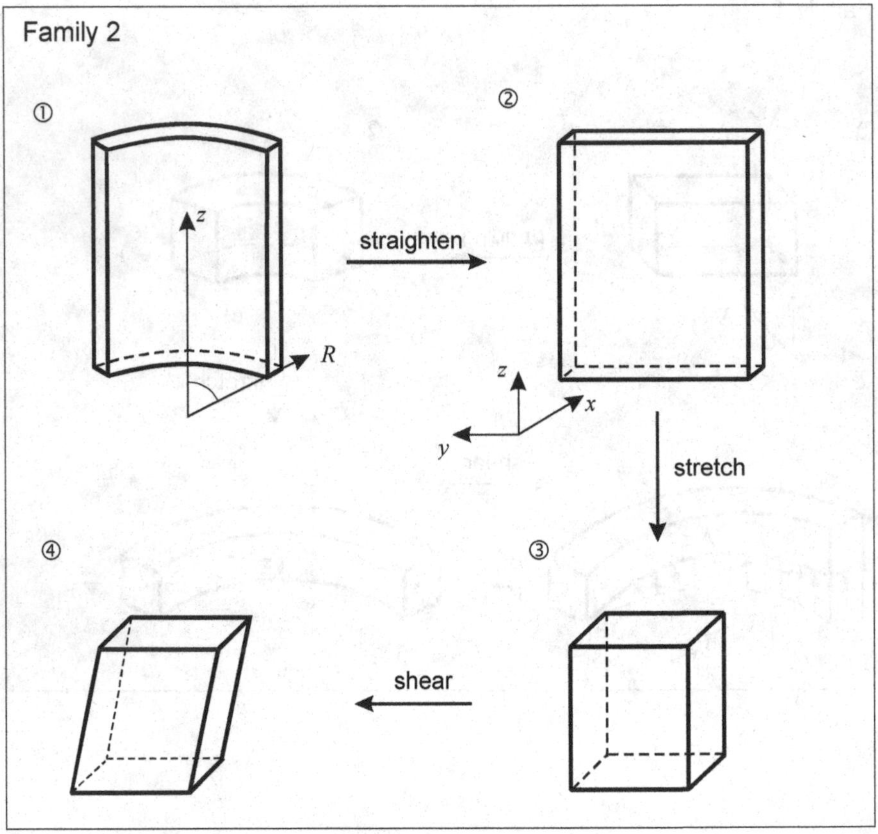

Family 3. *Inflation, bending, torsion, extension, and shearing of an annular wedge*

We use a cylindrical COOS $\{R, \Theta, Z\}$ for the reference placement and another cylindrical COOS $\{r, \theta, z\}$ for the deformed placement. The deformation is given by

$$r = \sqrt{(A R^2 + B)} \qquad \theta = C\Theta + DZ \qquad z = E\Theta + FZ$$

with the real constants A, B, C, D, E, F obeying

$$A(CF - DE) = 1 \qquad A R^2 + B > 0 \qquad \text{with } R_i < R < R_a .$$

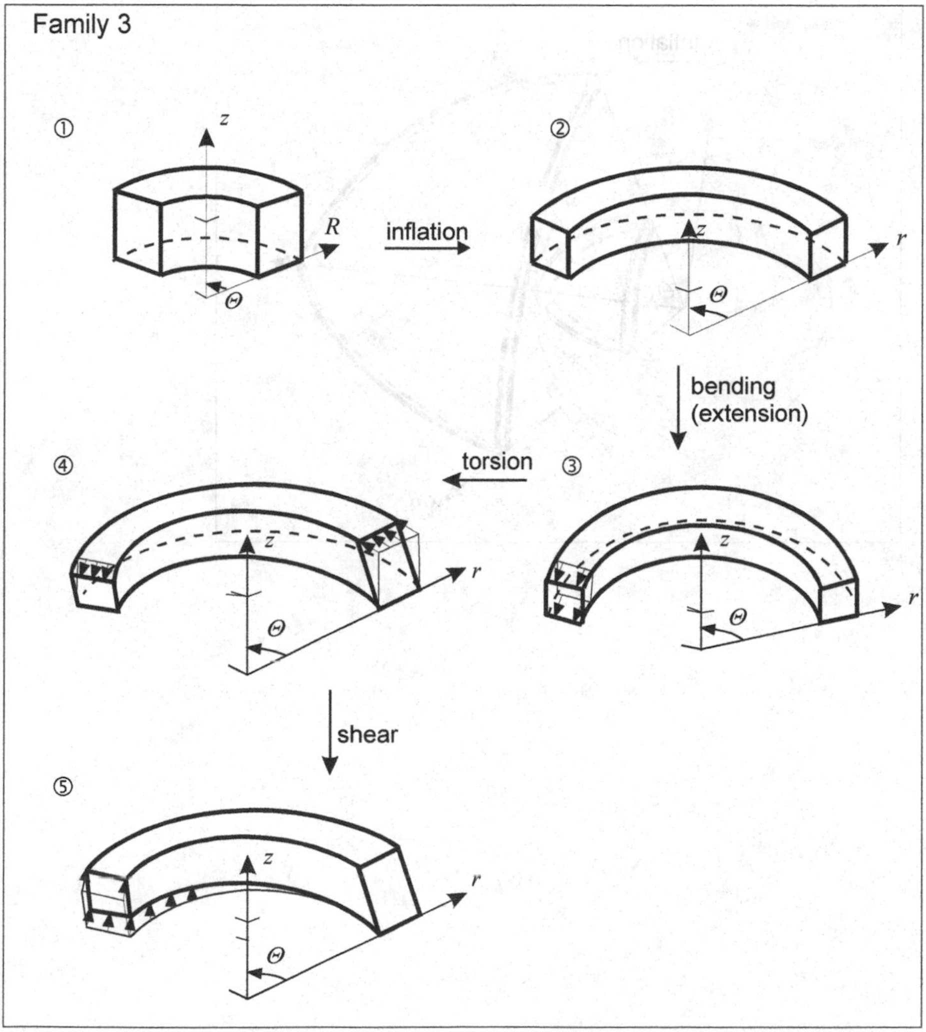

Family 4. *Inflation and/or inversion of a sector of a spherical shell*

We use a spherical COOS $\{R\,,\,\Theta\,,\,\Phi\}$ for the reference placement and another spherical COOS $\{r\,,\,\theta\,,\,\varphi\}$ for the deformed placement. The deformation is given by

$$r = (\pm R^3 + A)^{1/3} \qquad \theta = \pm\,\Theta \qquad \varphi = \Phi$$

with a real constant A.

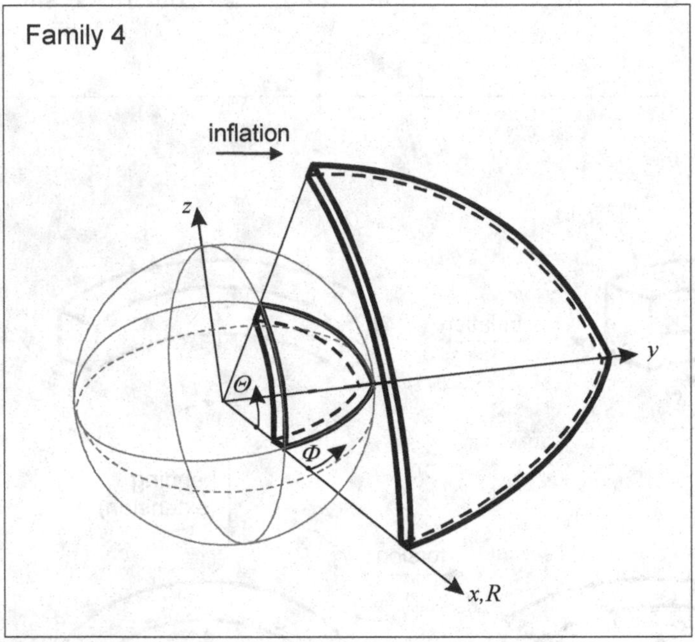

Family 5. *Inflation, bending, and azimuthal shearing of an annular wedge*

We use a cylindrical COOS $\{R, \Theta, Z\}$ for the reference placement and another cylindrical COOS $\{r, \theta, z\}$ for the deformed placement. The deformation is given by

$$r = A\,R \qquad \theta = B\,ln(R) + C\,\Theta \qquad z = Z/(A^2\,B)$$

with three real constants $A \neq 0, B, C$.

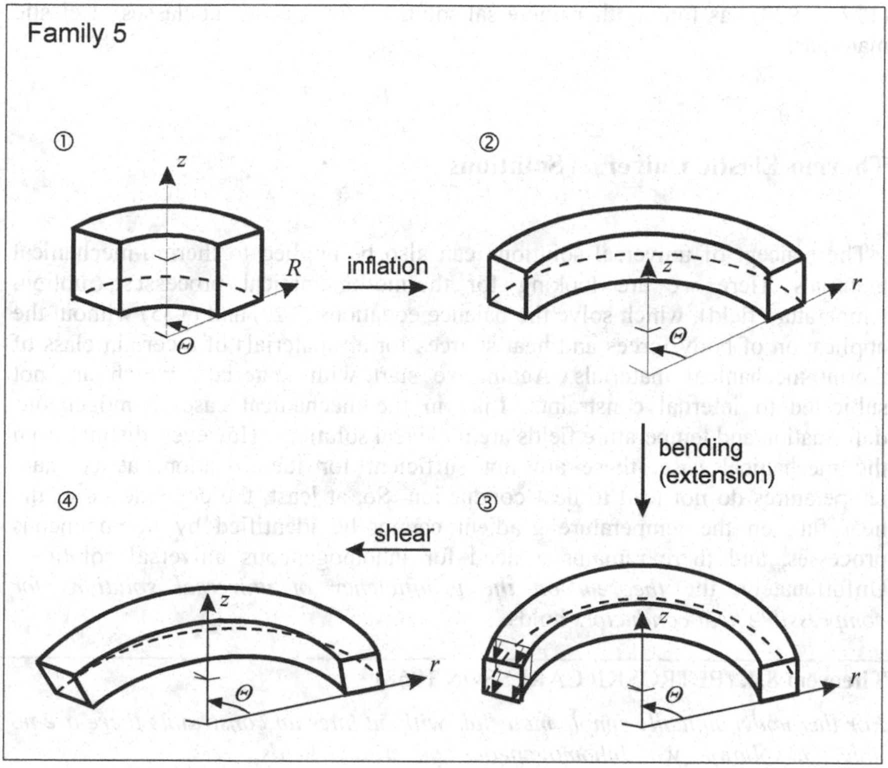

A detailed description of the first four families, which are mainly due to ERICKSEN and RIVLIN, can be found in TRUESDELL/NOLL (1965), p. 186 ff. The fifth family was not discovered until 1966 by KLINGBEIL and SYNGH (see TRUESDELL 1968, BEATTY 1987).

The question of whether there are universal solutions for incompressible isotropic elastic materials other than these five families, is the **ERICKSEN**[110] **problem** (1954), which has only been solved for special cases[111]. CARROLL (1971, 1988) has found other universal solutions for special subclasses of elastic materials.

Thermo-Elastic Universal Solutions

The concept of universal solutions can also be applied to thermo-mechanical materials. Here we are looking for thermo-kinematical processes (motion, temperature field), which solve the balance equations (3.20) and (3.35) without the application of body forces and heat sources for all materials of a certain class of thermo-mechanical materials. Again, we start with materials, which are not subjected to internal constraints. Like in the mechanical case, homogeneous deformation and temperature fields are universal solutions. However, distinct from the mechanical case, these are not sufficient for identification, as constant temperatures do not lead to heat conduction. So, at least, the dependence of the heat flux on the temperature gradient cannot be identified by homogeneous processes, and there remains a need for inhomogeneous universal solutions. Unfortunately, the *theorem on the insufficiency of universal solutions for compressible heat conductors* holds.

Theorem 8.4. (PETROSKI/ CARLSSON 1968)

For thermodynamically simple materials without internal constraints there are no universal solutions with inhomogeneous temperature fields.

This theorem still remains valid, if we restrict our consideration to thermo-elastic isotropic materials without internal constraints. In SACCOMANDI (1999) universal solutions with inhomogeneous temperature fields are presented for a special class of heat conductors (called "restricted Fourier law materials")[112].

Again, by introducing internal thermo-mechanical constraints we open the door for further universal solutions. This is one important reason to consider internal constraints.

---

110 Jerald Laverne Ericksen (1924-)

111 See MARRIS/ SHIAU (1970), MARRIS (1982), HILL/ ARRIGO (1999), BEATTY (2001) and SACCOMANDI (2001).

112 See also DUNWOODY (2003).

As an **example**, we consider a thermo-elastic material with two internal constraints, namely

1) temperature-dependent inextensibility in one direction indicated by $\mathbf{i} \in \mathscr{V}$ in the reference placement

$$\mathbf{i} \otimes \mathbf{i} \cdot \mathbf{C}^{\bullet} - f(\theta)' \, \theta^{\bullet} = 0$$

2) perfect heat conduction in the same direction

$$\mathbf{i} \cdot \mathbf{g}_0 = 0 \, .$$

These constraints maybe caused by an axial reinforcement of an isotropic compressible matrix material like a rubber foam.

We obtain the reaction parts of the dependent variables as

reaction stresses	$\mathbf{S}_R = \alpha_1 \, 2\rho \, \mathbf{i} \otimes \mathbf{i}$
reaction heat flux	$\mathbf{q}_R = \alpha_2 \, \mathbf{F} \, \mathbf{i}$
reaction entropy	$\eta_R = \alpha_1 f'(\theta)$

(see Chap. 5.2) with two scalar fields α_1 and α_2. For this material the following thermo-kinematical process is an

Exact thermo-elastic solution (BERTRAM/ HAUPT 1974)
Static torsion and radial deformation of a circular cylinder followed by a homogeneous axial stretching and a radial heat flux

We use the same cylindrical COOS $\{R \, , \Theta \, , Z\}$ for the homogeneous reference placement and another cylindrical COOS $\{r \, , \vartheta \, , z\}$ for the deformed one, as we did in the COUETTE-flow example in Chap. 2.3 and the same natural bases as we did already there. The motion is given by

$$r = r(R \, , t) \qquad \text{invertible in } R$$

$$\vartheta = \Theta + B \, Z$$

$$z = D(t) \, Z$$

with the function $D(t) := A + B \, t$ and two real constants $A \neq 0$ and B. The temperature field is of a radial form

$$\theta = \theta(R \, , t)$$

and leads to radial heat fluxes. The material direction $\mathbf{i}$ coincides with the Z-axis in the reference placement.

The deformation gradient has the components

$$\begin{bmatrix} r' & 0 & 0 \\ 0 & 1 & B \\ 0 & 0 & D \end{bmatrix}$$

with respect to the natural bases $\{\mathbf{r}_{\varphi k} \otimes \mathbf{r}_\Psi{}^i\}$ and

$$\begin{bmatrix} r' & 0 & 0 \\ 0 & r/R & rB \\ 0 & 0 & D \end{bmatrix}$$

with respect to the normed bases $\{\mathbf{e}_{\varphi k} \otimes \mathbf{e}_{\Psi i}\}$ with

$$\mathbf{e}_{\varphi k} := \mathbf{r}_{\varphi k} / |\mathbf{r}_{\varphi k}| \qquad k = 1, 2, 3$$

and

$$\mathbf{e}_{\Psi i} := \mathbf{r}_{\Psi i} / |\mathbf{r}_{\Psi i}| \qquad i = 1, 2, 3.$$

The prime $'$ at a function denotes the derivative with respect to R. Note that these components are *not* physical components. The JACOBIan of this deformation is

$$J = det\,\mathbf{F} = D(t)\, r'\, r/R\,.$$

The acceleration is radial $\mathbf{a} = r^{\bullet\bullet}\, \mathbf{e}_{\varphi\, 1}$. With respect to the basis $\{\mathbf{e}_{\Psi k} \otimes \mathbf{e}_{\Psi i}\}$, the right CAUCHY-GREEN tensor $\mathbf{C}$ has the components

$$\begin{bmatrix} r'^2 & 0 & 0 \\ 0 & r^2/R^2 & Br^2/R \\ 0 & Br^2/R & D^2 + r^2 B^2 \end{bmatrix}.$$

The first constraint can be alternatively written as

$$\mathbf{i} \otimes \mathbf{i} \cdot \mathbf{C} = D^2 + r^2\, B^2 = f(\theta)$$

which is satisfied for

$$r(R\,, t) = B^{-1} \sqrt{\{f(\theta(R\,, t)) - D^2\}}\,.$$

Thus, the motion depends on the material function f. Therefore, this solution is not a universal solution in a strict sense. However, it fulfils the balance laws for all materials within this class, as we will see below.

The second constraint is already fulfilled by the choice of the temperature field since its material gradient

$$\mathbf{g}_0 = Grad\, \theta = \theta'\, \mathbf{e}_{\Psi\, 1}$$

is radial and, thus, perpendicular to $\mathbf{i}$. If the material laws for the extra variables are isotropic, then the orthogonal tensor $\mathbf{A}$ with components

$$\begin{bmatrix} 1 & 0 & 0 \\ 0 & -1 & 0 \\ 0 & 0 & -1 \end{bmatrix}$$

with respect to $\{\mathbf{e}_{\Psi k} \otimes \mathbf{e}_{\Psi i}\}$ is a symmetry transformation. It gives

$$\mathbf{A}^T \mathbf{C}\, \mathbf{A} = \mathbf{C}$$

so that the extra stresses must also obey

$$\mathbf{A}^T \mathbf{S}_E\, \mathbf{A} = \mathbf{S}_E$$

following (6.22). This is only possible, if the stresses σ_{12} and σ_{13} vanish. Accordingly, the extra-stresses have the component matrix

$$\begin{bmatrix} \sigma_{E11} & 0 & 0 \\ 0 & \sigma_{E22} & \sigma_{E23} \\ 0 & \sigma_{E23} & \sigma_{E33} \end{bmatrix}$$

with respect to $\{\mathbf{e}_{\Psi k} \otimes \mathbf{e}_{\Psi i}\}$, and the extra CAUCHY stresses $\mathbf{T}_E = \mathbf{F}\, \mathbf{S}_E\, \mathbf{F}^T$ have the *same* components with respect to $\{\mathbf{e}_{\varphi k} \otimes \mathbf{e}_{\varphi i}\}$. The extra stresses can only depend on the radius and on the time, but not on the other coordinates. The reaction CAUCHY stresses have the form

$$\begin{aligned} \mathbf{T}_R &= \alpha_1\, 2\rho\, (\mathbf{F}\, \mathbf{i}) \otimes (\mathbf{F}\, \mathbf{i}) \\ &= \alpha_1\, 2\rho\, (r\, B\, \mathbf{e}_{\varphi 2} + D\, \mathbf{e}_{\varphi 3}) \otimes (r\, B\, \mathbf{e}_{\varphi 2} + D\, \mathbf{e}_{\varphi 3}) \\ &= \alpha_1\, 2\rho\, \{r^2 B^2\, \mathbf{e}_{\varphi 2} \otimes \mathbf{e}_{\varphi 2} + D^2\, \mathbf{e}_{\varphi 3} \otimes \mathbf{e}_{\varphi 3} \\ &\qquad + r\, B\, D\, (\mathbf{e}_{\varphi 3} \otimes \mathbf{e}_{\varphi 2} + \mathbf{e}_{\varphi 2} \otimes \mathbf{e}_{\varphi 3})\}\,. \end{aligned}$$

With the divergence of a tensor field after (1.48) we obtain

$$div\, \mathbf{T}_E = (\sigma_{E\,11\,,\,1} + \sigma_{E\,11}/r - \sigma_{E\,22}/r)\, \mathbf{e}_{\varphi 1}$$

$$\begin{aligned} div\, \mathbf{T}_R &= -\,\sigma_{R\,22}/r\ \mathbf{e}_{\varphi 1} + (\sigma_{R\,22\,,\,2}/r + \sigma_{R\,23\,,\,3})\, \mathbf{e}_{\varphi 2} \\ &\quad + (\sigma_{R\,23\,,\,2}/r + \sigma_{R\,33\,,\,3})\, \mathbf{e}_{\varphi 3} \\ &= 2\rho\, \{-\,\alpha_1\, r\, B^2\, \mathbf{e}_{\varphi 1} + (\alpha_{1\,,\,2}\, r\, B^2 + \alpha_{1\,,\,3}\, r\, B\, D)\, \mathbf{e}_{\varphi 2} \\ &\quad + (\alpha_{1,\,2}\, B\, D + \alpha_{1\,,\,3}\, D^2)\, \mathbf{e}_{\varphi 3}\}. \end{aligned}$$

This gives in CAUCHY's first equation (3.20)

$$\begin{aligned} & div\, \mathbf{T}_E + div\, \mathbf{T}_R - \rho\, \mathbf{a} \\ &= (\sigma_{E\,11\,,\,1} + \sigma_{E\,11}/r - \sigma_{E\,22}/r - 2\rho\, \alpha_1\, r\, B^2 - \rho\, r^{\bullet\bullet})\, \mathbf{e}_{\varphi 1} \\ &\quad + 2\rho\, (\alpha_{1\,,\,2}\, r\, B^2 + \alpha_{1\,,\,3}\, r\, B\, D)\, \mathbf{e}_{\varphi 2} \\ &\quad + 2\rho\, (\alpha_{1\,,\,2}\, B\, D + \alpha_{1\,,\,3}\, D^2)\, \mathbf{e}_{\varphi 3} \end{aligned}$$

which can always be fulfilled if α_1 depends appropriately on r and does not depend at all on ϑ and or z.

Also the first law of thermodynamics (3.35) can always be fulfilled by an appropriate choice of α_2, which enters (only) through

$$Div\, \mathbf{q}_{0R} = J\, \alpha_{2\,,\,3}\,.$$

8.3 Trajectory Method

In the case of *isotropic* elasticity, an interesting geometrical interpretation can be given which is based on the concept of stress and strain trajectories.

In what follows we consider a body consisting of an isotropic elastic material in a particular (static) current placement κ and an unloaded stress-free reference placement κ_0. Then we can determine the deformation gradient $\mathbf{F}$ as a second-order tensor field in all points of κ and calculate the field of the left CAUCHY-GREEN tensor

$$\mathbf{B} := \mathbf{F}\,\mathbf{F}^T = \sum_{i=1}^{3} \lambda_i\, \mathbf{v}_i \otimes \mathbf{v}_i$$

after (2.11) in its spectral form with eigenvalues λ_i and with respect to a normed eigenbasis $\{\mathbf{v}_i\}$. We will call a curve defined on κ a **strain trajectory** if it is at all points tangent to $\mathbf{v}_i$ for $i = 1, 2, 3$. These trajectories will be unique in body regions where the three eigenvalues λ_i are different. If two or even three of them are equal then there is some arbitrariness in the determination of the strain trajectories. In each case, however, we can assume that such trajectories exist as a grid of three sets of smooth curves that are mutually orthogonal at each point. They can therefore be interpreted as an orthogonal coordinate system on κ which we denote by $\{\varphi^i\}$.

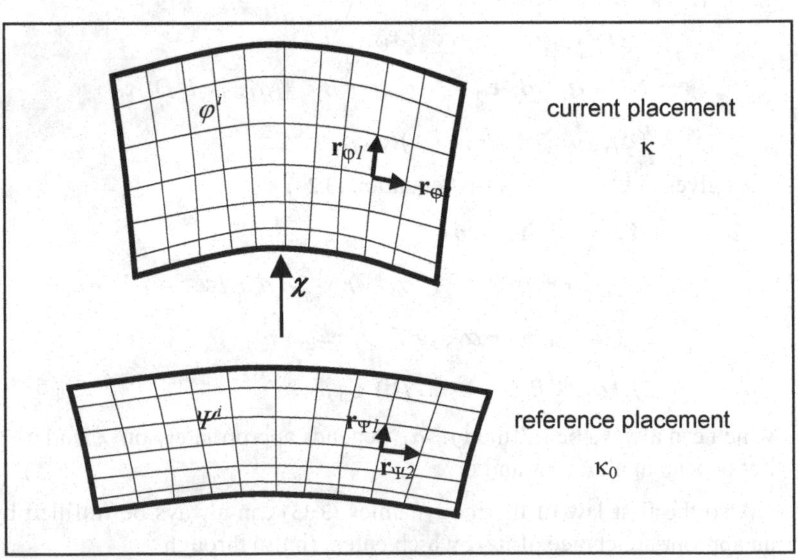

Next we consider the stress field in the current placement. The CAUCHY stresses at each point can be also represented in their spectral form

$$\mathbf{T} = \sum_{i=1}^{3} \sigma_i \, \mathbf{v}_i \otimes \mathbf{v}_i$$

with three principal stresses σ_i and normed eigenvectors $\mathbf{v}_i$. Here we already used the fact that for an isotropic elastic material the eigenvectors of $\mathbf{B}$ and $\mathbf{T}$ coincide.

We next introduce the **stress trajectories** in the current placement as curves being tangent to the eigendirections of $\mathbf{T}$. We can immediately conclude that they coincide with the strain trajectories in the isotropic case. This is an important property of isotropic bodies for what follows.

In the reference placement all stresses and all strains are assumed to be zero. So here the principal stresses and strains are all equal to zero, and every orthogonal coordinate system can be considered as stress or strain trajectories.

If we use these (spatial) coordinates $\{\varphi^i\}$ in the current placement as convected coordinates (see Sect. 2.3), then their pull-back into the reference placement by the motion χ produces a second (material) coordinate system $\{\Psi^i\}$. This is also orthogonal, since the deformations of material line elements aligned to the principal strain directions remain orthogonal in isotropic elasticity. They are just rotated by the orthogonal part $\mathbf{R}$ of the polar decomposition (2.6).

Next, we introduce at some material point the natural bases to these two coordinate systems

- the tangent bases $\{\mathbf{r}_{\varphi i}\}$ and $\{\mathbf{r}_{\Psi i}\}$
- the gradient bases $\{\mathbf{r}_\varphi{}^i\}$ and $\{\mathbf{r}_\Psi{}^i\}$
- the metric coefficients (no sum over i)

$$G_{\varphi ii} := \mathbf{r}_{\varphi i} \cdot \mathbf{r}_{\varphi i} \quad \text{and} \quad G_{\Psi ii} := \mathbf{r}_{\Psi i} \cdot \mathbf{r}_{\Psi i}$$

$$G_\varphi{}^{ii} := \mathbf{r}_\varphi{}^i \cdot \mathbf{r}_\varphi{}^i \quad \text{and} \quad G_\Psi{}^{ii} := \mathbf{r}_\Psi{}^i \cdot \mathbf{r}_\Psi{}^i$$

so that $$G_{\varphi ii} = (G_\varphi{}^{ii})^{-1} \quad \text{and} \quad G_{\Psi ii} = (G_\Psi{}^{ii})^{-1}$$

while all $G_{\varphi ij}$, $G_\varphi{}^{ij}$, $G_{\Psi ij}$, $G_\Psi{}^{ij}$ are zero for $i \neq j$ because of orthogonality.

We know already that the components of $\mathbf{F}$ and $\mathbf{F}^{-1}$ become trivial for convected coordinates with respect to these natural bases

$$\frac{\partial \varphi_t{}^k}{\partial \Psi^i} = \delta^k{}_i$$

so that we have the representaton of the deformation gradient as

$$\mathbf{F} = \mathbf{r}_{\varphi i} \otimes \mathbf{r}_\Psi{}^i \quad \Leftrightarrow \quad \mathbf{F}^{-1} = \mathbf{r}_{\Psi i} \otimes \mathbf{r}_\varphi{}^i$$

and for the left CAUCHY-GREEN tensor

$$\mathbf{B} = (\mathbf{r}_\Psi{}^i \cdot \mathbf{r}_\Psi{}^j)\, \mathbf{r}_{\varphi i} \otimes \mathbf{r}_{\varphi j} = \sum_{i=1}^{3} G_\Psi{}^{ii}\, \mathbf{r}_{\varphi i} \otimes \mathbf{r}_{\varphi i} .$$

The normed eigenvectors are

$$\mathbf{v}_i = \mathbf{r}_{\varphi i} / \sqrt{G_{\varphi ii}} = \sqrt{G_\varphi{}^{ii}}\, \mathbf{r}_{\varphi i} \qquad i = 1, 2, 3.$$

Comparison with the presentation above gives

$$\lambda_i\, \mathbf{v}_i \otimes \mathbf{v}_i = G_\Psi{}^{ii}\, \mathbf{r}_{\varphi i} \otimes \mathbf{r}_{\varphi i} \qquad i = 1, 2, 3$$

so that the principal strains are

$$\lambda_i = G_\Psi{}^{ii}\, G_{\varphi\, ii} = G_\Psi{}^{ii} / G_\varphi{}^{ii} \qquad i = 1, 2, 3$$

being completely determined by the two coordinate systems.

We can now apply the representation of an isotropic elastic law in analogy to (6.27)

$$\sigma_i = \sigma(\lambda_i\,, \lambda_{i+1} + \lambda_{i+2}\,, |\, \lambda_{i+1} - \lambda_{i+2}\,|)$$

with only one scalar function σ of three scalar variables, which can be easily evaluated once the two coordinate systems are known.

The appealing property of this approach is that the two coordinate systems do not only determine the displacements but also the three-dimensonal stress field in the body. At each point, they give the direction and value of the extremal normal stresses, of the shear stresses, etc. So they offer an illustrative picture of how the stresses are distributed in the body.

For the local balance law (3.20) we will need the divergence of the CAUCHY stresses $\mathbf{T}$. By use of the chain rule this gives an expression where the CHRISTOFFEL symbols of the convected coordinate systems from Chapt. 1.4.1 appear.

Let us next consider boundary conditions. Displacement boundary conditions can be directly fulfilled by employing the representation for convected coordinates

$$\varphi^i = \chi^i(\Psi^1, \Psi^2, \Psi^3) .$$

But also NEUMANN conditions can be easily evaluated, since the tensions on a surface point in the current placement with outer normal $\mathbf{n}$ are

$$\mathbf{t} = \mathbf{T}\,\mathbf{n} = \sum_{i=1}^{3} (\mathbf{v}_i \cdot \mathbf{n})\, \sigma_i\, \mathbf{v}_i\,,$$

i.e,. a linear combination of the normed eigenvectors $\mathbf{v}_i$.

We can use these findings to construct an approximation method to solve boundary value problems in isotropic elasticity. This would surely be a challenging object of further investigations. For this aim, one needs a representation for orthogononal grids. Conformal mappings of Cartesian coordinates are an example for such a representation, although rather particular.

By the way, this approach shares some common properties with the slip line approach to plasticity. In contrast to the stress trajectories, however, the slip lines occur under an angle of 45° to the trajectories.

9 Inelasticity

A Comment on the Literature. A more detailed description of a framework for inelasticity can be found in BERTRAM (1982, 1989), or KRAWIETZ (1986). A comparison of different frameworks is given in BERTRAM (1993, 1999, 2014).

Elastic material behaviour, as studied it in Chap. 6, is characterised by the assumption that the stresses depend exclusively on the current state of deformation with respect to a specific reference placement. In contrast to this, inelastic materials have a memory of the way this current state was reached. In such a context, it is no longer sufficient to just consider specific deformations (as points in the strain space), but instead we have to consider sequences of deformations, *i.e.*, **deformation processes**. By this we understand a continuous time function

$$\mathbf{F}(\tau)\Big|_{\tau=0}^{t} : [0\,,t] \to \mathscr{Inv}^{+} \quad \Big| \quad \tau \mapsto \mathbf{F}(\tau)\,,$$

which assigns to each instant τ in a certain closed time-interval between 0 and $t \geq 0$ a deformation tensor $\mathbf{F}(\tau)$. Without restriction of generality, we set the initial time to 0. Besides continuity, we will also assume that the time-derivative $\mathbf{F}^{\bullet}$ exists at each instant and is piecewise continuous. At each instant in $[0\,,t]$, $\mathbf{F}$ is the gradient of a time-dependent sequence of placements and, thus, depends on point and time. In material theory, we consider local quantities. Therefore, the dependence of $\mathbf{F}$ on the point is not important here.

If we truncate a process to a shorter interval, then we obtain a **subprocess**. In contrast, if we extend a process into the future by a **continuation process**, then the resulting process consists of the two subprocesses, namely the original one and its continuation. Naturally, because of continuity, not all processes can be combined, as they must be compatible at the connection instant.

As we are not able to know the history of a certain material within an indeterminately long past, we assume, that at a freely chosen initial instant $t_0 \equiv 0$, the material was in a certain **initial state**. Starting from this, we can submit the material to deformation processes that have to meet the following initial conditions

- they all begin at a time $t_0 \equiv 0$;
- they all start at a certain initial deformation $\mathbf{F}_0 = \mathbf{F}(0)$.

Often this initial deformation is selected as a reference placement, so that $\mathbf{F}_0 \equiv \mathbf{I}$ holds. This choice, however, is not compulsory and will not be assumed, but instead we leave the reference placement arbitrary.

A. Bertram, *Elasticity and Plasticity of Large Deformations*,
https://doi.org/10.1007/978-3-030-72328-6_9

After the *principle of determinism*, the stresses $\mathbf{T}(t)$ at the end of such a process are determined by this process. Hence, there exists a **material functional** $\mathfrak{F}$, which assigns to each **F**-process the stresses $\mathbf{T}(t)$ at its end

$$\mathfrak{F}:\ \mathbf{F}(\tau)\Big|_{\tau=0}^{t} \mapsto \mathbf{T}(t)\,,$$

where the deformation process lasts from $t_0 \equiv 0$ until t and starts with $\mathbf{F}_0$.

Note that we only consider processes of finite duration, and not half-infinite *deformation histories*, as was preferred by TRUESDELL/ NOLL (1965), WANG (1967), MÜLLER (1985), OWEN (1970), HAUPT (1977, 2000), GREVE (2003), and many others.

As we only take the local deformation process into account, the *principle of local action* is already fulfilled by this ansatz for a material functional.

The remaining principles *PMO* and *PISM* are identically fulfilled by reduced forms of the functional like

$$\mathfrak{K}:\ \mathbf{C}(\tau)\Big|_{\tau=0}^{t} \mapsto \mathbf{S}(t)\,, \tag{9.1}$$

which assigns to each process of the right CAUCHY-GREEN tensor the material stress tensor. The proof of this statement is entirely analogous to that which led to the elastic reduced form (4.10). The following transformation between these two functional holds

$$\begin{aligned}\mathbf{S}(t) &= \mathfrak{K}\{\mathbf{C}(\tau)\Big|_{\tau=0}^{t}\}\\ &= \mathbf{F}(t)^{-1}\,\mathbf{T}(t)\,\mathbf{F}(t)^{-T}\\ &= \mathbf{F}(t)^{-1}\,\mathfrak{F}\{\mathbf{F}(\tau)\Big|_{\tau=0}^{t}\}\,\mathbf{F}(t)^{-T}.\end{aligned}$$

$\mathbf{C}$, $\mathbf{S}$, and the material functional $\mathfrak{K}$ are principally invariant under change of observer and under superimposed rigid body motions (EUCLIDean transformations). In this sense, we chose a *material description.*

However, all these quantities may depend on the choice of the reference placement. Under a change of the reference placement from κ_0 to $\underline{\kappa}_0$, CAUCHY's stress tensor must remain unaltered

$$\mathfrak{F}\{\kappa_0,\, \mathbf{F}(\tau)\Big|_{\tau=0}^{t}\} = \mathfrak{F}\{\underline{\kappa}_0,\, \mathbf{F}(\tau)\Big|_{\tau=0}^{t}\,\mathbf{K}\}\,,$$

while the material one changes according to

$$\begin{aligned}\mathbf{S}_{\kappa_0}(t) &= \mathfrak{K}\{\kappa_0,\, \mathbf{C}(\tau)\Big|_{\tau=0}^{t}\}\\ &= \mathbf{F}(t)^{-1}\,\mathfrak{F}\{\kappa_0,\, \mathbf{F}(\tau)\Big|_{\tau=0}^{t}\}\,\mathbf{F}(t)^{-T}\end{aligned}$$

$$= \mathbf{K}\,\mathbf{K}^{-1}\,\mathbf{F}(t)^{-1}\,\mathfrak{F}\{\underline{\kappa}_0\,,\mathbf{F}(\tau)\big|_{\tau=0}^{t}\,\mathbf{K}\}\;\mathbf{F}(t)^{-T}\,\mathbf{K}^{-T}\,\mathbf{K}^{T}$$

$$= \mathbf{K}\,\underline{\mathbf{F}}(t)^{-1}\,\mathfrak{F}\{\underline{\kappa}_0\,,\underline{\mathbf{F}}(\tau)\big|_{\tau=0}^{t}\,\}\;\underline{\mathbf{F}}^{-T}\cdot\mathbf{K}^{T}$$

$$= \mathbf{K}\,\mathfrak{K}\{\underline{\kappa}_0\,,\underline{\mathbf{C}}(\tau)\big|_{\tau=0}^{t}\,\}\;\mathbf{K}^{T} = \mathbf{K}\,\underline{\mathbf{S}}_{\underline{\kappa}_0}\,(t)\,\mathbf{K}^{T}$$

with

$$\mathbf{K} := Grad(\kappa_0\,\underline{\kappa}_0^{-1})$$

(time-independent), the $\underline{\mathbf{F}}$-process

$$\underline{\mathbf{F}}(\tau) := \mathbf{F}(\tau)\,\mathbf{K}$$

and the $\underline{\mathbf{C}}$-process

$$\underline{\mathbf{C}}(\tau) := \mathbf{K}^{T}\,\mathbf{C}(\tau)\,\mathbf{K}$$

and the material stress tensor in $\underline{\kappa}_0$

$$\underline{\mathbf{S}}_{\underline{\kappa}_0} = \mathbf{K}^{-1}\,\mathbf{S}_{\kappa_0}\,(t)\,\mathbf{K}^{-T}.$$

By analogy with the definition in Chap. 6.5 we define a **symmetry transformation** $\mathbf{A} \in \mathcal{U}nim^{+}$ of the functional $\mathfrak{K}$ by the condition

$$\boxed{\mathfrak{K}\{\mathbf{C}(\tau)\big|_{\tau=0}^{t}\,\} = \mathbf{A}\,\mathfrak{K}\{\mathbf{A}^{T}\,\mathbf{C}(\tau)\big|_{\tau=0}^{t}\,\mathbf{A}\}\,\mathbf{A}^{T}}$$

for all admissible $\mathbf{C}$–processes.

The set of all symmetry transformations is the **initial symmetry group** $\mathcal{G}$ of the material, *i.e.*, the one of the initial state of the material.

With respect to the dependence of the symmetry group upon the reference placement, NOLL's rule from Theorem 6.7. holds analogously.

We call an inelastic material (initially) **isotropic**, if there exists a reference placement such that the inclusion $\mathcal{G} \supseteq \mathcal{O}rth^{+}$ holds. If, however $\mathcal{G} \subset \mathcal{O}rth^{+}$ (excluding equality), we have an **anisotropic** inelastic solid.

State Space Formulation

In certain cases, but by no means in all cases, it is possible to describe the influence of the deformation process on the present response of the material by a finite set of variables, being tensors of arbitrary order. We will denote these by $\mathbf{Z} \in \mathcal{L}in$ without further specifying this space, as it depends on the individual material model. These variables are called **state-variables, internal variables**, or **hidden variables**, as they cannot be directly determined by measuring the stresses and strains, but only indirectly by comparing the stress-response after different deformation processes. Moreover, they are usually not uniquely determined, but there are alternative choices. In the state space format, there exists a function

$$f : \mathcal{I}nv^+ \times \underline{\mathcal{L}in} \to \mathcal{S}ym \mid \{\mathbf{F}, \mathbf{Z}\} \mapsto \mathbf{T}$$

such that the functional can be substituted by this function of present values only

$$\mathfrak{F}\{\mathbf{F}(\tau)\big|_{\tau=0}^{t}\} = f(\mathbf{F}(t), \mathbf{Z}(t)) .$$

Of course, we need a functional to up-date the value of the state variables like

$$\mathbf{Z}(t) = z\{\mathbf{F}(\tau)\big|_{\tau=0}^{t}\}$$

which is called an **evolution function**. In many cases this is assumed to be given by an ordinary differential equation in time

$$\mathbf{Z}^{\bullet} = z(\mathbf{F}, \mathbf{Z}, \mathbf{F}^{\bullet})$$

which shall be uniquely integrable along a deformation process with appropriate initial values. The advantage of such a procedure is that the general material functional already obtains a certain concrete form, although it is still valid for a large class of materials. In practical cases one will of course use appropriate reduced forms for the variables appearing in these equations. For an axiomatic approach to state space concepts see NOLL (1972), BERTRAM (1982, 1989).

Rate-Independence

An important classification of material models is their *rate-(in)dependence.* For that purpose, we consider retarded and accelerated processes. This can be achieved by a **time-transformation**

$$\alpha : \mathcal{R} \to \mathcal{R} \mid \tau \mapsto \alpha(\tau)$$

which shall be monotonously increasing. Then rate-independence characterises that material property, that the response (in terms of stresses) to a deformation-process is independent of the rate of this process.

Definition 9.1. A material functional $\mathfrak{F}$ is called **rate-independent**, if

$$\mathfrak{F}\{\mathbf{F}(\tau)\big|_{\tau=0}^{t}\} = \mathfrak{F}\{\mathbf{F}(\alpha(\tau))\big|_{\alpha(0)}^{\alpha(t)}\}$$

holds for all time-transformations α.

In particular, for such materials constant deformation-processes (freezes) lead to constant stress-processes (no stress relaxation).

In the literature, this property is sometimes called *time-independence*. This is, however, misleading, as the time explicitly appears in the functionals only for aging materials, which are only a subtype of rate-dependent materials.

Elastic materials are rate-independent. For an inelastic and rate-independent material with memory, we will consider an important example in the next chapter.

10 Plasticity

A Comment on the Literature. The following papers will often be cited in this chapter, as they are pioneering original works: GREEN/ NAGHDI (1965, 1971), LEE/ LIU (1967), LEE (1969), MANDEL (1971). As *finite plasticity* is not yet a mature science, only few text-books on the subject are available, like POZDEYEV/ TRUSOV/ NYASHIN (1986), KHAN/ HUANG (1995), LUBARDA (2002), NEMAT-NASSER (2004), WU (2005). In some other books on *non-linear continuum mechanics* one also finds chapters on *finite plasticity*, like in BESSELING/ DIMITRIENKO (2011), v. d. GIESSEN (1994), LEMAITRE (2001), KRAUSZ/ KRAUSZ (1996), KRAWIETZ (1986), HASHIGUCHI (2014), HAUPT (2000, 2002), NEGAHBAN (2012), or specific aspects of it, as in TEODOSIU/ RAPHANEL/ SIDOROFF (1993), YANG (1991), HUTTER/ BAASER (2003), GURTIN/ FRIED/ ANAND (2013). Some books on *classical continuum mechanics* contain chapters on *finite plasticity*, like BESSON/ CAILLETAUD/ CHABOCHE/ FOREST (2001, 2010), DOGHRI (2000), LUBLINER (1990), OTTOSEN/ RISTINMAA (2005). There are many textbooks describing the material mechanisms of plasticity, like FRANCOIS/ PINEAU/ ZAOUI (1998), KETTUNEN/ KUOKKALA (2002), RÖSLER/ HARDERS/ BÄKER (2003), TOLEDANO (2012), YANG/ LEE (1993). Aspects of applications of plasticity in metal forming are given in BANABIC/ BUNGE/ PÖHLANDT/ TEKKAYA (2000), HOSFORD/ CADDELL (1983, 1993), HONEYCOMBE (1968, 1984), PAGLIETTI (2007), RAABE *et al.* (2004), SLUZALEC (2004), SZCZEPIŃSKI (1979), TALBERT/ AVITZUR (1996). Numerical aspects of *finite plasticity* are considered in ANANDARAJAH (2010), BORJA (2013), DUNNE/ PETRINIC (2005), SIMO/ HUGHES (1998), SIMO (1998), HAN/ REDDY (1999), NEMAT-NASSER (2004). The history of finite plasticity is briefly reported in NAGHDI (1960) and BRUHNS (2020). The theory of *finite plasticity*, which will be presented in the sequel, was originally published in BERTRAM (1992, 1998, 2003), see also ŠILHAVÝ/ KRATOCHVÍL (1977) and DEL PIERO (1975).

Historical Notes

Under small deformations, most metallic and many non-metallic solid materials can be described as elastic. Only after reaching a certain threshold, can permanent and, thus, inelastic deformations occur. For steel this threshold strain lies typically in a range around or even below one percent of strain. Beyond this *yield limit* the material undergoes plastic deformations or yields. These plastic deformations can become rather large, typically up to 4 or 5 or even more orders of magnitude larger than the threshold strain. Finally, the sample fails because of instabilities like shear -bands, necking, rupture or fracture.

As the material is still far from failure when reaching the yield limit, it is interesting to also use this large range of plastic behaviour for engineering

A. Bertram, *Elasticity and Plasticity of Large Deformations*,
https://doi.org/10.1007/978-3-030-72328-6_10

purposes. For this reason the classical plasticity theory was developed in the first half of the 20th century, and is associated with names like LEVY, PRANDTL, REUSS, v. MISES, HENCKY, PRAGER, ZIEGLER and others.

Classical plasticity theory, however, is geometrically linear, and, thus, limited to small deformations, such as strains below *10%* and rotations below *10 °*, or even less. For many technological applications the deformations and rotations are much larger. In metal forming processes such as rolling, extrusion moulding, deep drawing, etc. the deformations may reach some thousands of percent. In this range, *classical plasticity theory* would produce absurd results. One is therefore rather interested in generalising the *classical* (*i.e.*, geometrically linear) *theory of plasticity* to a *finite*, *i.e.* geometrically non-linear theory.

The first theories of *finite plasticity* were suggested in the 1960s or even later (GREEN/ NAGHDI 1965, LEE 1969, MANDEL 1971), when *finite elasticity* had already achieved a certain level of maturity. Such theories were based on suggestions of KRÖNER, BILBY, KONDO, ECKART, BESSELING, FOX, amongst others. These early theories were conceptually quite different, causing an everlasting controversy among the different schools[113]. Four or five decades later, still none of these theories is universally accepted and considered as established textbook knowledge.

One of the fundamental problems in *finite plasticity theory* is the definition of the internal variables and, in particular, that of plastic or inelastic strain. Although the term *plastic deformation* seems to be rather customary in engineering, it turns out in the context of large deformations that it is extremely difficult to define it precisely. "There is no general agreement on how it is to be identified, either conceptually or experimentally, in the presence of finite deformations." (CASEY/ NAGHDI 1992). And, if this is not accurately done, it leaves space for controversial interpretations.

If we try to develop a general theory of *finite plasticity* in the remainder of this book, then this can only be a compromise between generality on the one hand and necessary concreteness and applicability on the other hand.

For this purpose, we will first of all work out the main features of plastic behaviour and condense these notions into a mathematically precise foundation of *elasto-plasticity*. Since we use different concepts as many other authors, we have to relate the current format to others from literature. Finally, we will exemplify this framework for an important class of plastic materials, namely for *crystals*.

[113] See NAGHDI (1990).

10.1 Elastic Ranges

Plasticity is characterised by the following features:

- *elastic ranges* in the stress or strain space, which are bounded by a *yield limit.* Within these ranges, the behaviour is assumed to be elastic,
- *yielding* or *plastic flow*, which can occur only at the yield limit and causes a change of the current elastic range,
- an alteration of the yield limit (only) during yielding, which leads to *hardening* or *softening*

This shall now be made more precise. We firstly introduce the concept of an *elastic range* in the configuration space.

Definition 10.1. An **elastic range** is a pair $\{\mathscr{Ela}_p, k_p\}$ consisting of

1) a path-connected submanifold with boundary $\mathscr{Ela}_p \subset \mathscr{Psym}$ in the space of the right CAUCHY-GREEN tensors

2) an elastic law (taken as a reduced form (6.8))

$$k_p : \mathscr{Ela}_p \to \mathscr{Sym} \quad \Big| \quad \mathbf{C} \mapsto \mathbf{S} \,,$$

such that the stresses after any continuation process $\mathbf{C}(\tau)\big|_{t_o}^{t}$, which remains entirely in $\mathscr{Ela}_p$

$$\mathbf{C}(\tau) \in \mathscr{Ela}_p \qquad \forall\, \tau \in [t_o, t]$$

are determined by its ultimate configuration

(10.1)
$$\mathbf{S}(t) = k_p(\mathbf{C}(t)) \qquad \forall\; \mathbf{C}(t) \in \mathscr{Ela}_p \,.$$

The choice of these special stress and configuration tensors is immaterial, *i.e.*, any other choice would lead to the same material class. Because of objectivity, however, it is recommended to use only material variables and invariant laws.

One could also assume more strongly the existence of *hyperelastic ranges*. Then there would exist an elastic energy $w_p(\mathbf{C})$ as a reduced form such that (10.1) becomes more specifically

$$k_p(\mathbf{C}(t)) = 2\rho\, \frac{dw_p(\mathbf{C}(t))}{d\mathbf{C}} \,.$$

If we write briefly about a (hyper-) elastic range $\mathscr{Ela}_p$, then we tacitly assume the existence of an associated k_p (or w_p). We further assume that both functions are continuously differentiable and as such extendible on the whole space $\mathscr{Psym}$.

Yield Criteria

We decompose $\mathscr{E}la_p$ topologically into its interior $\mathscr{E}la_p{}^o$ and its boundary $\partial\mathscr{E}la_p$. The latter is called **yield surface** or **yield limit** or **yield locus** (in the strain space). In order to describe it more easily, we introduce a real-valued tensor-function in the strain space

$$\Phi_p : \mathscr{P}sym \to \mathscr{R} \quad | \quad \mathbf{C} \mapsto \Phi_p(\mathbf{C})$$

the kernel of which coincides with the yield limit

$$\Phi_p(\mathbf{C}) = 0 \qquad \Leftrightarrow \qquad \mathbf{C} \in \partial\mathscr{E}la_p .$$

For distinguishing points in the interior and in the exterior of the elastic ranges, we postulate

$$\Phi_p(\mathbf{C}) < 0 \qquad \Leftrightarrow \qquad \mathbf{C} \in \mathscr{E}la_p{}^o$$

and, consequently,

$$\Phi_p(\mathbf{C}) > 0 \qquad \Leftrightarrow \qquad \mathbf{C} \in \mathscr{P}sym \setminus \mathscr{E}la_p .$$

We call such an indicator function (or level set function) a **yield criterion,** and assume further on that Φ_p is at least piecewise differentiable.

Clearly, for each yield limit there exist arbitrarily many yield criteria. A scalar multiple of a yield criterion by a positive real is again a yield criterion, as the kernel remains unaltered by such multiplication. In this sense all yield criteria of an elastic range $\mathscr{E}la_p$ are equivalent and we will not distinguish between them.

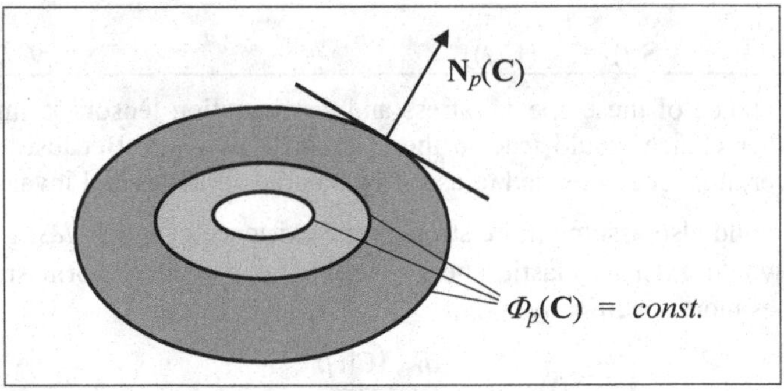

The gradient or the derivative of Φ_p (where existent) normed by its value

$$\mathbf{N}_p(\mathbf{C}) := \frac{grad\, \Phi_p(\mathbf{C})}{|grad\, \Phi_p(\mathbf{C})|} \in \mathcal{Sym}$$

at the yield limit $\mathbf{C} \in \partial\mathcal{Ela}_p$ can be interpreted geometrically as the outer normal to the yield surface in $\mathcal{P}sym$.

Often it is postulated that the elastic ranges have to be convex. A subset of a linear space is called *convex* if the linear connection of two points of this subset also belongs to it

$$\nu \mathbf{C}_1 + (1-\nu)\, \mathbf{C}_2 \in \mathcal{Ela}_p \qquad \forall\, \nu \in [0\,,1]\,,\ \forall\, \mathbf{C}_1, \mathbf{C}_2 \in \mathcal{Ela}_p\,.$$

This convexity-condition can be equivalently expressed by the yield criterion as

$$\Phi_p(\nu \mathbf{C}_1 + (1-\nu)\, \mathbf{C}_2) \leq 0 \qquad \forall\, \nu \in [0\,,1]\,,\ \forall\, \mathbf{C}_1, \mathbf{C}_2 \in \mathcal{Ela}_p$$

or, for a differentiable Φ_p, as

$$grad\, \Phi_p(\mathbf{C}_1) \cdot (\mathbf{C}_2 - \mathbf{C}_1) \leq 0 \qquad \forall\, \mathbf{C}_1, \mathbf{C}_2 \in \partial\mathcal{Ela}_p\,.$$

Although we will not postulate this convexity, for most of our examples it will be fulfilled.

Sometimes it is more practical or convenient to formulate the yield criterion in the stress space, such as a function $\Phi_{\sigma p}(\mathbf{CS})$. By the elastic law we can easily transform it again into the strain space

$$\Phi_{\sigma p}(\mathbf{C}\, k_p(\mathbf{C})) = \Phi_p(\mathbf{C}) \qquad \forall\, \mathbf{C} \in \mathcal{P}sym\,.$$

Yield criteria are often generated by defining an **equivalent stress,** *i.e.*, a function

$$\sigma_{\mathbf{T}}: \mathcal{Sym} \to \mathcal{R} \quad | \quad \mathbf{T} \mapsto \sigma_{\mathbf{T}}(\mathbf{T})$$

in the stress space, which assigns to each CAUCHY stress tensor a non-negative real value, which stands for a stress intensity of the stress state. The yield criterion is then the difference of this equivalent stress and a critical equivalent stress called **yield stress** σ_Y

(10.2) $$\Phi_{\mathbf{T}p}(\mathbf{T}) := \sigma_{\mathbf{T}}(\mathbf{T}) - \sigma_Y$$

which is identified as the yield limit in a tensile test in a certain material direction. The yield criterion as well as $\sigma_{\mathbf{T}}$ are material functions and thus submitted to the general principles of material theory. Because of objectivity, $\sigma_{\mathbf{T}}$ must be an isotropic tensor-function and, therefore, presentable as a function of the principal invariants of $\mathbf{T}$

$$\sigma_{\mathbf{T}}(\mathbf{T}) = \sigma_{iso}(I_{\mathbf{T}}\,, II_{\mathbf{T}}\,, III_{\mathbf{T}})$$

or as a symmetric function of the principal stresses

$$\sigma_{\mathbf{T}}(\mathbf{T}) = \underline{\sigma}_{iso}(\sigma_1\,, \sigma_2\,, \sigma_3) = \underline{\sigma}_{iso}(\sigma_2\,, \sigma_1\,, \sigma_3) = \underline{\sigma}_{iso}(\sigma_3\,, \sigma_2\,, \sigma_1)$$

after Theorem 1.24. If one wants to introduce anisotropic equivalent stresses for the CAUCHY stresses, then this can be done by (EULERean) structure tensors. Here again a material description is preferable. For this purpose, we can use the fact that **T** and **CS** (the same as **SC**) have the same eigenvalues and principal invariants. The tensor **CS** (the same as **SC**) is non-symmetric with respect to the EUCLIDean metric, but diagonalisable. With (3.22) it allows for a spectral representation

$$\mathbf{CS} = (\mathbf{SC})^T = \sum_{i=1}^{3} \sigma_i \, \underline{\mathbf{s}}^i \otimes \underline{\mathbf{s}}_i$$

with respect to the basis of the right eigenvectors $\{\underline{\mathbf{s}}^j := \mathbf{F}^T \mathbf{s}_j\}$ and its dual $\{\underline{\mathbf{s}}_k := \mathbf{F}^{-1} \mathbf{s}_k\}$, the left eigenbasis, with $\mathbf{s}_j$ being the eigenvectors of the CAUCHY stresses. This stress tensor is proportional to MANDEL´s stress tensor. If the elastic law is isotropic, then **C** and **S** are coaxial, and $\mathbf{CS} \equiv \mathbf{SC}$ is symmetric after Theorem 1.16.

The material version of the equivalent stress is

$$\sigma_{\mathbf{CS}}(\mathbf{CS}) = \sigma_{iso}(I_{\mathbf{CS}}, II_{\mathbf{CS}}, III_{\mathbf{CS}}) = \sigma_{iso}(I_{\mathbf{T}}, II_{\mathbf{T}}, III_{\mathbf{T}}) = \sigma_{\mathbf{T}}(\mathbf{T})$$
$$= \underline{\sigma}_{iso}(\sigma_1, \sigma_2, \sigma_3) \, .$$

A symmetry transformation **A** of this equivalent stress is given by the symmetry condition

$$\sigma_{\mathbf{CS}}(\mathbf{CS}) = \sigma_{\mathbf{CS}}(\mathbf{A}^T \mathbf{C}\, \mathbf{A}\, \mathbf{A}^{-1} \mathbf{S}\, \mathbf{A}^{-T}) = \sigma_{\mathbf{CS}}(\mathbf{A}^T \mathbf{CS}\, \mathbf{A}^{-T}) \, .$$

One should keep in mind that the general (isotropic or anisotropic) form of a material equivalent stress $\sigma_{\mathbf{CS}}(\mathbf{CS})$ is automatically objective (because of its invariance under EUCLIDean transformations). Its isotropy as a tensor-function would correspond to the material isotropy. A material form of the yield criteria analogous to the above EULERean one (10.2) is then

$$\Phi_{\mathbf{CS}p}(\mathbf{CS}) := \sigma_{\mathbf{CS}}(\mathbf{CS}) - \sigma_Y \, .$$

One of the first anisotropic yield criteria was suggested by v. MISES[114] (1928) where the equivalent stress

$$\sigma_{\mathbf{T}}(\mathbf{T}) = \mathbf{T}' \cdot \boldsymbol{K}[\mathbf{T}'] = \boldsymbol{K} \cdot (\mathbf{T}' \otimes \mathbf{T}')$$

is a quadratic form with a tetrad $\boldsymbol{K}$, which meets the symmetry of the material. Such tetrads are well known for all symmetry-classes from *linear elasticity*. This form of an equivalent stress, although seldom cited, can be considered as the root of a large quantity of yield criteria. If the material is orthotropic, then this criterion reduces to HILL´s (1948) yield criterion[115]. If the material is isotropic, then $\boldsymbol{K}$ coincides with the forth-order identity and

[114] Richard von Mises (1883-1953)
[115] See also HSU (1966).

$$\sigma_T(\mathbf{T}) = \mathbf{T}' \cdot \mathbf{T}'$$

which gives rise to the isotropic HUBER-v. MISES equivalent stress (10.4) to be considered later.

VOYIADJIS/ THIAGARAJAN/ PETRAKIS (1995) use two of such quadratic forms for modelling a piecewise yield criterion to describe distortional hardening[116].

Another group of anisotropic yield criteria is based on the KELVIN modes from *linear elasticity*. These are the (symmetric) projections of the strains into the eigenspaces of HOOKE's elasticity tetrad of (6.1). If

$$\boldsymbol{C} = \sum_{r=1}^{K} \Lambda_r \, \boldsymbol{P}_r$$

is the spectral form of HOOKE's tetradin analogy to (1.23) with $K \leq 6$ being the number of different eigenvalues Λ_r (principal stiffnesses) of $\boldsymbol{C}$, and $\boldsymbol{P}_r$ its eigenprojectors, then the elastic energy can be additively decomposed into the partial energies

$$e_r := \tfrac{1}{2} \Lambda_r^{-1} \, \mathbf{T}_r \cdot \mathbf{T}_r \qquad \text{with the partial stresses } \mathbf{T}_r := \boldsymbol{P}_r[\mathbf{T}] \, .$$

The yield criterion can now depend on a linear combination of such energies

$$\Phi_{Tp}(\mathbf{T}) := \Big(\sum_{r=1}^{K} c_r \, e_r \Big) - \sigma_Y$$

with different weight factors c_r, or on a nonlinear combination like

$$\Phi_{Tp}(\mathbf{T}) := \Big(\sum_{r=1}^{K} c_r \, e_r^{\,n_r} \Big) - \sigma_Y$$

(DESMORAT/ MARULL 2011) with exponents n_r, or as a multi-limit criterion on each of them separately

$$\Phi_{Tp}(\mathbf{T}) := \max_r \, \{ e_r - \sigma_{Y\,r} \} \, .$$

BIEGLER/ MEHRABADI (1995) used this as a damage criterion, SCHREYER/ ZUO (1995) as a yield criterion[117]. All these criteria are anisotropic according to the symmetry group of the elasticity tensor from *linear elasticity*.

---

[116] See also VOYIADJIS/ FOROOZESH (1990).
[117] See ARRAMON *et al.* (2000).

Another interesting extension of the quadratic yield criterion is due to FEIGENBAUM/ DAFALIAS (2007)[118]. They include distortional hardening by composing the tetrad of two parts

$$\boldsymbol{K} = \boldsymbol{K_0} + \left(\frac{\mathbf{T}' - \mathbf{A}}{|\mathbf{T}' - \mathbf{A}|} \cdot \mathbf{H}\right) \boldsymbol{A}$$

with two hardening variables $\mathbf{H}$ (second order tensor) and $\boldsymbol{A}$ (fourth order tensor), for which evolution equations are suggested in a thermodynamically consistent way.

A more general form of an anisotropic yield criterion comes from BETTEN (1979) and can be considered as a generalisation of GOLDENBLAT/ KOPNOV (1965) and TSAI/ WU (1971) who introduced it as a strength criterion. BETTEN introduced three scalar parameters defined by

$$\alpha_1(\mathbf{T}') = \overset{\langle 2 \rangle}{\mathbf{K}} \cdot \mathbf{T}'$$
$$\alpha_2(\mathbf{T}') = \overset{\langle 4 \rangle}{\mathbf{K}} \cdot (\mathbf{T}' \otimes \mathbf{T}')$$
$$\alpha_3(\mathbf{T}') = \overset{\langle 6 \rangle}{\mathbf{K}} \cdot (\mathbf{T}' \otimes \mathbf{T}' \otimes \mathbf{T}')$$

with three higher-order tensors acting on the stress deviator

$$\mathbf{T}' := \mathbf{T} - {}^1\!/_3\, tr(\mathbf{T})\, \mathbf{I}\,.$$

The second term α_2 coincides with the v. MISES equivalent stress from above, and the latter two with a suggestion by REES (1983). The yield criterion is then a function of these three scalar functions. Conditions for the convexity of this yield surface are also given by BETTEN.

A slightly different way to construct anisotropic yield criteria is to first transform the stress deviator in a linear way, *i.e.*, by a tetrad $\boldsymbol{K}$ which induces the anisotropy

$$\underline{\mathbf{T}}' := \boldsymbol{K}[\mathbf{T}']$$

and to then use some isotropic equivalent stress as a function of $\underline{\mathbf{T}}'$. Such a transformation has been used by BARLAT/ LIAN (1989), BARLAT/ LEGE/ BREM (1991), KARAFILLIS/ BOYCE (1993), and extended by ARETZ/ BARLAT (2004), BRON/ BESSON (2004), CAZACU/ PLUNKETT/ BARLAT (2006), PLUNKETT/ LEBENSOHN/ CAZACU/ BARLAT (2006), BARLAT/ YOON/ CAZACU (2007), and many others.

---

[118] See also PARMA *et al.* (2018) and PIETRYGA/ VLADIMOROV/ REESE (2012)

Another way to implement anisotropy has been suggested by ORTIZ/ POPOV (1983). Here, the same HUBER-v. MISES equivalent stress (10.4) is used, which is compared to a yield stress that is not constant but instead depends on the direction of the stress tensor with respect to some tensorial internal variable, which can be identified by the back stress. If α is the angle between these two tensors in the deviatoric tensor space, then a FOURIER expansion is used for the yield stress

$$\sigma_Y(\alpha) = k_1 (1 + \sum_{n=2} k_n \cos n\alpha)$$

with scalars k_n, possibly depending on the stresses. It is shown that even for the simple twofold ansatz

$$\sigma_Y = k_1 (1 + k_2 \cos 2\alpha + k_3 \cos 3\alpha)$$

with only three scalar constants, a rather realistic description of distortional hardening can be achieved, at least for proportional processes. The condition for the convexity of this criterion is simply

$$5 |k_2| + 10 |k_3| \leq 1$$

after ORTIZ/ POPOV (1983). This approach has been further investigated by WEGENER/ SCHLEGEL (1996). A similar approach is that of FRANCOIS (2001).

Another important anisotropic yield criterion is the SCHMID law (1924), which is frequently used in crystal plasticity (see Chap. 10.5). Isotropic and anisotropic yield criteria for polycrystals based on the SCHMID law within the crystallites (grains) can be obtained by homogenisation methods, see, *e.g.*, LIAN/ CHEN (1991), DARRIEULAT/ PIOT (1996), DAWSON/ MACEWEN/ WU (2003), HABRAKEN (2004).

More representations of anisotropic functions based on invariants can be found in BOEHLER/ RACLIN (1982) and BRUHNS/ XIAO/ MEYERS (1999).

VEGTER/ VAN DEN BOOGAARD (2006) suggest a piecewise description of the anisotropic yield limit for a state of plane stress by means of BEZIER splines.

Another more geometrical approach to distortional hardening has been given by KURTYKA/ ŻYCZKOWSKI (1985, 1996). SHUTOV *et al.* (2010) suggest a model for distortional hardening which is based on a rheological model.

For *isotropic solids* with respect to an isotropic state, the symmetry transformation $\mathbf{A}$ is an arbitrary orthogonal tensor and $\sigma_{\mathbf{CS}}$ an isotropic tensor-function of $\mathbf{CS}$, which can always be represented as $\sigma^{iso}(I_{\mathbf{CS}}, II_{\mathbf{CS}}, III_{\mathbf{CS}})$.

The most simple example for such an isotropic law is obtained by taking the absolute value as equivalent stress

$$\sigma_{\mathbf{CS}}(\mathbf{CS}) \equiv |\mathbf{CS}|.$$

However, from a physics point of view this is not a realistic choice, as the shear stresses appear in the same way as the pressure, which can hardly be expected.

For many materials, especially for metals, the yield limit is almost independent of the hydrostatic pressure[119]. For such a **pressure-independent plasticity** the equivalent stress cannot depend on the trace of the CAUCHY stress tensor (which equals the trace of **CS**). In the isotropic case we are left with

$$\sigma_{CS}(\mathbf{CS}) \equiv \underline{\sigma}^{iso}(II_{\mathbf{T}}, III_{\mathbf{T}}) = \underline{\sigma}^{iso}(II_{\mathbf{CS}}, III_{\mathbf{CS}}) .$$

We can also imply the pressure-independence by using the stress deviator **T'** or

$$(\mathbf{CS})' := \mathbf{CS} - {}^{1}/_{3}\, tr(\mathbf{CS})\,\mathbf{I} = \mathbf{F}^{T}\,\mathbf{T}'\,\mathbf{F}^{-T} .$$

One commonly introduces principal invariants of the stress deviator as

$$J_1 := I_{\mathbf{T}'} = I_{(\mathbf{CS})'} = 0$$

(10.3) $$J_2 := -II_{\mathbf{T}'} = ½\, \mathbf{T}' \cdot \mathbf{T}'$$

$$= -II_{(\mathbf{CS})'} = ½\, tr((\mathbf{CS})'^{2}) = ½\, (\mathbf{CS})' \cdot (\mathbf{SC})' \geq 0$$

$$J_3 := III_{\mathbf{T}'} = III_{(\mathbf{CS})'}$$

and the equivalent stress as a function

$$\underline{\sigma}^{iso}(II_{\mathbf{T}}, III_{\mathbf{T}}) =: \sigma_J(J_2, J_3) .$$

An **example** for such a form is the equivalent stress suggested by DRUCKER[120] (1949)

$$\sigma_J(J_2, J_3) \equiv (J_2^{3} - c\, J_3^{2})^{1/6}$$

with a real constant c .

For some materials one finds a difference of the yield stress under tension and compression, although still being pressure-independent, which is called *strength differential effect*[121]. This effect has been measured for hcp materials and is due to the directional sensitivity of twinning. The effect can be described by the equivalent stress

$$\sigma_J(J_2, J_3) \equiv (J_2^{3/2} - c\, J_3)^{1/3}$$

suggested by FREUDENTAHL/ GOU (1969) and later by CAZACU/ BARLAT (2004). In the same paper, a generalisation of this criterion to anisotropy is given by the use of orthotropic invariants, like, *e.g.*,

$$a_1(\sigma_{xx} - \sigma_{yy})^2 + a_2(\sigma_{yy} - \sigma_{zz})^2 + a_3(\sigma_{zz} - \sigma_{xx})^2 + a_4\,\sigma_{yz}^{\,2} + a_5\,\sigma_{xz}^{\,2} + a_6\,\sigma_{xy}^{\,2}$$

---

[119] SPITZIG/ RICHMOND (1984) measured a pressure-dependence also for iron and aluminium.

[120] Daniel Charles Drucker (1918-2001)

[121] This effect is also described by OTA/ SHINDO/ FUKUOKA (1959) as an extension of HILL's (1948) criterion.

with six constants a_i inducing the anisotropy.

An example for a pressure-dependent isotropic equivalent stress is that by DRUCKER/ PRAGER[122] (1952)

$$\sigma_J(I_{\mathbf{T}}, J_2) \equiv {}^K\!/_3\, I_{\mathbf{T}} + J_2^{1/2}$$

with a real constant K, used for geomaterials, where inelastic volume changes occur.

A pressure-dependent generalisation of HILL´s and of the DRUCKER/ PRAGER criterion has been suggested by LIU/ HUANG/ STOUT (1997).

In particular, we obtain a **J_2-theory**, if no dependence on J_3 is given. This leads to the equivalent stress suggested by HUBER[123] (1904) and v. MISES (1913)

$$\sigma_M(J_2) := (3\, J_2)^{1/2} = \sqrt{3/2}\, |\mathbf{T}'| = ({}^3\!/_2\, \mathbf{T}' \cdot \mathbf{T}')^{1/2}.$$

It is proportional to the norm of the stress deviator. The factor of proportionality is determined in such a way that for a uniaxial stress state $\mathbf{T} \equiv \sigma_{11}\, \mathbf{e}_1 \otimes \mathbf{e}_1$ the identity

$$\sigma_M(J_2) \equiv |\sigma_{11}|$$

holds. The **HUBER-v. MISES yield criterion** is then

(10.4) $$\Phi_{\mathrm{M}p}(\mathbf{C}) = 3\, J_2 - \sigma_Y^2$$

with $J_2 = ½\, tr((\mathbf{CS})'^{\,2})$ and $\mathbf{S} = k_p(\mathbf{C})$, wherein σ_Y is the yield limit of the uniaxial tension test. The HUBER-v. MISES yield criterion can be equivalently expressed by the principal stresses σ_1, σ_2, and σ_3 as

$$\Phi_{\mathrm{M}p}(\mathbf{C}) = {}^1\!/_2\, [(\sigma_1 - \sigma_2)^2 + (\sigma_2 - \sigma_3)^2 + (\sigma_3 - \sigma_1)^2] - \sigma_Y^2.$$

Another important yield criterion is **TRESCA´s[124] yield criterion** (1865), which limits the maximum shear stress by a critical value τ_c. This can be expressed by the principal stresses as

$$\Phi_{\mathrm{Tr}p}(\mathbf{T}) = [(\sigma_1 - \sigma_2)^2 - 4\, \tau_c^2]\, [(\sigma_2 - \sigma_3)^2 - 4\, \tau_c^2]$$
$$[(\sigma_3 - \sigma_1)^2 - 4\, \tau_c^2] = 0.$$

Here it does not matter whether we use for σ_i CAUCHY´s principal stresses, those of CAUCHY´s deviator stresses, of **CS**, or of **(CS)'**. After REUSS[125] (1933) this yield limit can also be expressed by the invariants of the CAUCHY stresses

[122] William Prager (1903-1980)
[123] Maksymilian Tytus Huber (1872-1950)
[124] Henri Edouard Tresca (1814-1885)
[125] Endre Reuss (1900-1968)

$$\Phi_{\mathrm{Trp}}(\mathbf{T}) = 4\,J_2^{\,3} - 27\,J_3^{\,2} - 36\,\tau_c^{\,2}\,J_2^{\,2} + 96\,\tau_c^{\,4}\,J_2 - 64\,\tau_c^{\,6}.$$

TRESCA´s yield limit is constituted by piecewise smooth and differentiable surfaces. Only on their intersections is this criterion not differentiable.

A generalisation of the two above criteria has been given by HERSHEY (1954) with the equivalent stress

$$\sigma_H(J_2) := |\sigma_1 - \sigma_2|^n + |\sigma_2 - \sigma_3|^n + |\sigma_3 - \sigma_1|^n$$

with some real exponent n. For $n \equiv 2$ we again obtain the v. MISES equivalent stress, while for increasing n it tends to the TRESCA criterion. Such non-linear and non-quadratic equivalent stresses have been frequently used for both isotropic and anisotropic yield criteria, like HOSFORD (1979, 1993), HILL (1979), MONTHEILLET/ JONAS/ BENFERRAH (1991), CHU (1995).

A great variety of other yield criteria suggested in the literature can be found in SKRZYPEK § 3.2 (1993), BANABIC *et al.* (2000) p. 123 ff, ŻYCZKOWSKI (2001), and YU (2002, 2006).

Visualisations of the yield surfaces are not easy, as they are defined in a 6 or even 9-dimensional space. Therefore, only projections into subspaces are graphically representable. One frequently used visualisation is shown in the next figure, where we project the tripode of the principal stress axes into a plane, the *principal deviator stress plane* (also called octahedral plane, HAIGH-WESTERGAARD plane, Π-plane, or MELDAHL plane).

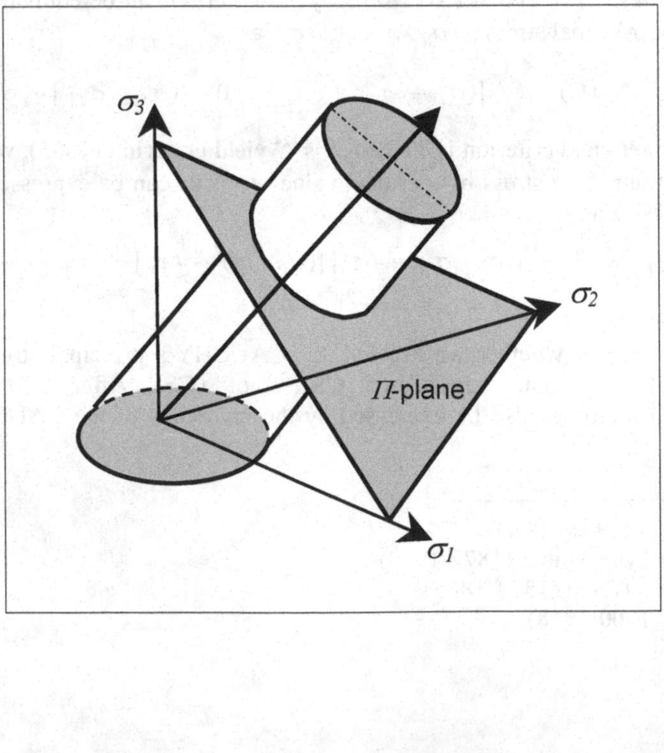

In this projection, all stress states with equal principal deviator stresses are projected into the same point, whatever their pressure is. In this plane, the yield limit of HUBER-v. MISES becomes a circle, and that of TRESCA a regular hexagon. The yield stress can either be calibrated such that the TRESCA polygon inscribes the v. MISES circle (for tensile tests), or vice versa (for torsion or shear tests).

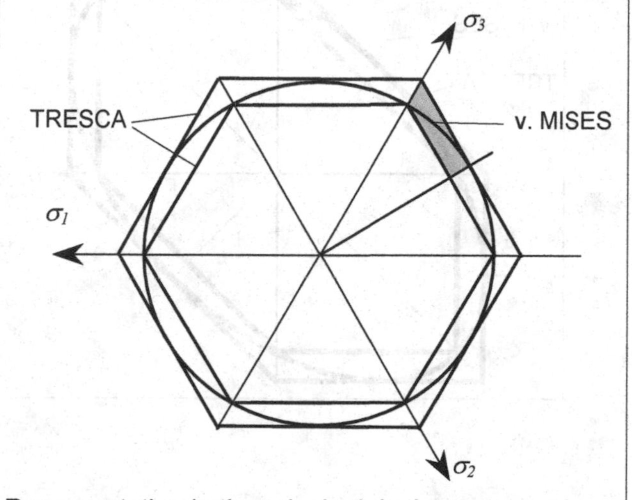

Representation in the principal deviator stress plane

For pressure-independent isotropic yield criteria it is generally sufficient to determine the yield limit in a *60°* segment in this figure. If one additionally assumes that the yield stress is equal in a tensile and a compression test, then a *30°* segment (grey range in the figure) is also sufficient. By symmetry transformations alone it can then be extended onto the whole stress space.

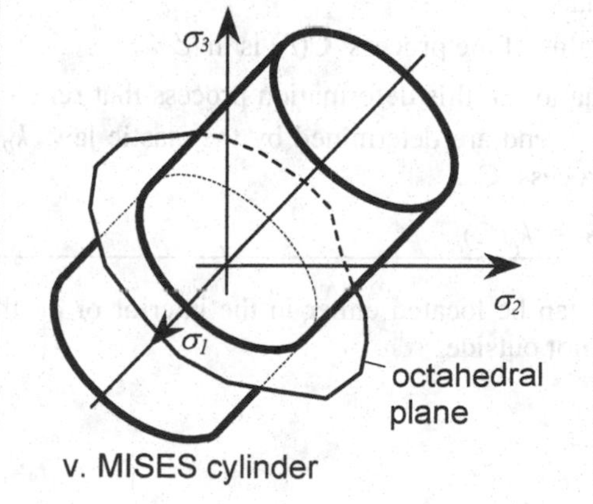

If we instead consider the intersection of the yield limit with the plane spanned by two principal deviator stresses, then we obtain an ellipse for the HUBER-v. MISES limit, and a hexagon for TRESCA´s yield limit.

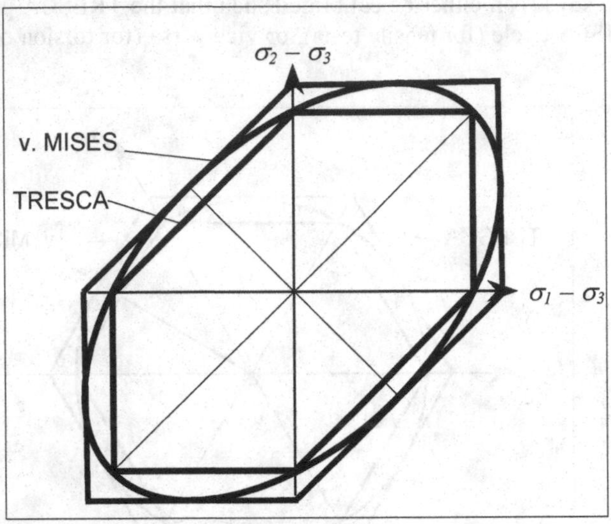

Elastic and Plastic Processes

For the constitution of our material class we now make the following assumption.

Assumption 10.2.
At the end of each deformation process $\mathbf{C}(\tau)\big|_0^t$, there exists an elastic range $\{\mathscr{E}la_p, k_p\}$ such that

- the terminate value of the process $\mathbf{C}(t)$ is in $\mathscr{E}la_p$
- for any continuation of this deformation process that remains entirely in $\mathscr{E}la_p$, the stresses at its end are determined by the elastic law k_p through the final value of that process $\mathbf{C}$

$$\mathbf{S} = k_p(\mathbf{C}) \qquad \forall\, \mathbf{C} \in \mathscr{E}la_p .$$

The deformation can be located either in the interior or on the boundary of the elastic ranges, but not outside.

If $\mathbf{C}(\tau)\big|_{t}^{t_1}$ is a continuation process of $\mathbf{C}(\tau)\big|_{0}^{t}$, which lies entirely in the current elastic range $\mathcal{E}la_p$, then $\mathcal{E}la_p$ remains constant, and the stresses at each instant are determined by the elastics law k_p. We will call such processes **elastic**. Variables and functions which carry the suffix p, depend on the current elastic range and, thus, do not alter during elastic processes.

In contrast to this, we call processes **inelastic** or **plastic**, which start at some instant t_1 in one elastic range $\mathcal{E}la_1$, and are at some later instant t_2 eventually outside of $\mathcal{E}la_1$: $\mathbf{C}(t_2) \notin \mathcal{E}la_1$. In this case, after Assumption 10.2., the material has necessarily changed the elastic range, and $\mathbf{C}(t_2)$ is in some other elastic range $\mathcal{E}la_2$ at t_2. As subprocesses of such plastic processes can be elastic until they pass the boundary of $\mathcal{E}la_p$ for the first time, we will restrict the notion of inelastic or plastic processes to those, which do not contain elastic subprocesses, and therefore continuously change the elastic ranges at all times, or yield. Accordingly, each process can be decomposed into a sequence of segments which are elastic and plastic. Note that we divide deformation processes into elastic and plastic, but not the deformations.

States[126] of **yielding** are characterised by two facts.

1) The deformation is currently on the yield limit, and thus fulfils its **yield criterion**

$$\Phi_p(\mathbf{C}(t)) = 0 .$$

2) It is about to leave the current elastic range, *i.e.*, the **loading condition**

$$\Phi_p^{\bullet} = \frac{d\Phi_p}{d\mathbf{C}} \cdot \mathbf{C}(t)^{\bullet} > 0$$

if Φ_p is differentiable in this state, or equivalently

$$\mathbf{N}_p \cdot \mathbf{C}(t)^{\bullet} > 0$$

is fulfilled. If the gradient is non-unique (like it is at edges and corners of the yield surface), then the loading condition must hold for at least one of these gradients.

Note that we formulated this condition in the strain space. Therefore, it does not necessarily imply an increment of the external load or of the stresses (hardening), but can also be fulfilled during softening[127].

These two conditions can be formally condensed into one single condition, like

$$\Phi_p + sgn(\Phi_p^{\bullet}) = 1 .$$

---

[126] In the sequel we will distinguish between *elastic states* and *plastic states* or *states of yielding*, although this does not fit into the state space concept as outlined in Chapter 9.

[127] See NAGHDI/ TRAPP (1975).

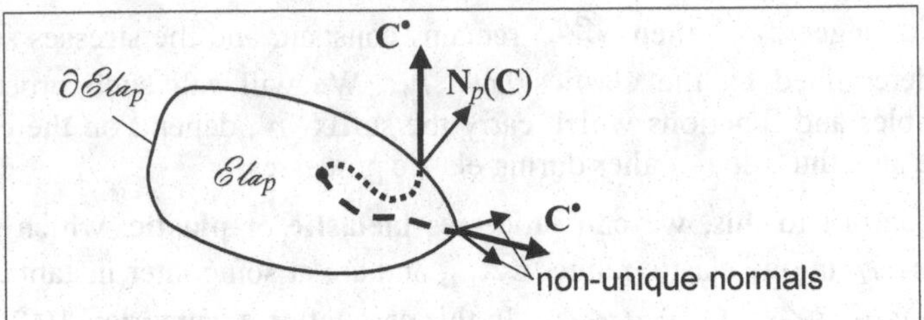

States of loading for two **C**-processes

If currently $\Phi_p = 0$ but $\Phi_p^{\bullet} < 0$, then this is a state of **unloading** and the process is elastic at this instant. The same holds for $\Phi_p^{\bullet} = 0$ on the yield surface. Such a state is called of **neutral loading**.

During plastic processes we will assume that the change of the elastic ranges takes place continuously, which means that for sufficiently short plastic processes, the elastic ranges (in the strain space) at the beginning and at the end are arbitrarily close to one-another.

As the current elastic range determines the stresses for the current configuration, then according to the principle of determinism the elastic range must be determined by the past **C**-process, *i.e.*, there must be a functional which assigns

$$e\colon \mathbf{C}(\tau)\Big|_{\tau=0}^{t} \mapsto \{\mathcal{E}\ell a p_p\,,\, k_p\}$$

with the condition $\mathbf{C}(t) \in \mathcal{E}\ell a p_p$.

If this process is followed by an elastic process until $t_2 > t$, then the elastic range remains constant between t and t_2

$$e\,\{\mathbf{C}(\tau)\Big|_{\tau=0}^{t}\} \;=\; e\,\{\mathbf{C}(\tau)\Big|_{\tau=0}^{t_2}\} \;=\; e\,\{\mathbf{C}(\tau)\Big|_{\tau=0}^{t_3}\} \;\;\forall\, t_3 \in [t\,,t_2]\,,$$

and the stress functional can be expressed by the function (10.1)

$$\mathbf{S}(t_2) \;=\; \mathfrak{K}\{\mathbf{C}(\tau)\Big|_{\tau=0}^{t_2}\} \;=\; k_p(\mathbf{C}(t_2))\,.$$

This is the first level of concretisation of the general stress functional $\mathfrak{K}$ for elastoplastic materials.

Isomorphy of the Elastic Ranges

For many materials it is a microphysically and experimentally well-substantiated fact that during yielding the elastic behaviour hardly alters even under very large deformations[128]. This reduces the effort for the identification tremendously, as otherwise one would have to identify the elastic constants at each step of the deformation anew. We now give this assumption a precise form.

Assumption 10.3. *The elastic laws of all elastic ranges of an elastic-plastic material are isomorphic.*

Accordingly, if $\{\mathscr{Ela}_1, k_1\}$ and $\{\mathscr{Ela}_2, k_2\}$ are two elastic ranges, then there is a $\mathbf{P}_{12} \in \mathscr{Inv}^+$, so that

- for the mass densities in the reference placements ρ_{01} of k_1 and ρ_{02} of k_2

$$\rho_{01} = \rho_{02}\, det(\mathbf{P}_{12})$$

  holds, and

- for the elastic laws we have the equality

$$k_2(\mathbf{C}) = \mathbf{P}_{12}\, k_1(\mathbf{P}_{12}^T\, \mathbf{C}\, \mathbf{P}_{12})\, \mathbf{P}_{12}^T \qquad \forall\, \mathbf{C} \in \mathscr{Psym}$$

  after (6.20).

As originally k_1 was defined only on $\mathscr{Ela}_1$ and k_2 only on $\mathscr{Ela}_2$, we first extended both laws appropriately on $\mathscr{Psym}$. This simplifies the formulation of the theory without restriction of the generality, as all deformations beyond the particular elastic range are principally beyond any physical identification.

As we have chosen a joint reference placement for all elastic laws (this was, however, not compulsory), we already have $\rho_{01} \equiv \rho_{02}$ and therefore $\mathbf{P}_{12} \in \mathscr{Unim}^+$, so that the first isomorphy condition is always fulfilled.

If all elastic laws belonging to different elastic ranges, are mutually isomorphic, then they all are isomorphic to a single, freely chosen, but fixed **elastic reference law** k_0. While the current elastic law k_p varies with time during yielding, this reference law k_0 is always constant. We thus have the isomorphy condition in the following form.

Theorem 10.4. *Let k_0 be the elastic reference law for an elasto-plastic material. Then for each elastic range $\{\mathscr{Ela}_p, k_p\}$ there is a $\mathbf{P} \in \mathscr{Unim}^+$ such that*

(10.5) $$\mathbf{S} = k_p(\mathbf{C}) = \mathbf{P}\, k_0(\mathbf{P}^T\, \mathbf{C}\, \mathbf{P})\, \mathbf{P}^T \qquad \forall\, \mathbf{C} \in \mathscr{Ela}_p\,.$$

Often one chooses k_0 as the elastic law of the initial elastic range. This choice, however, is neither compulsory nor natural, as initiality is only a vague notion in plasticity.

If the range is hyperelastic, then we have more specifically the isomorphy-condition

[128] See ŠILHAVÝ/ KRATOCHVÍL (1977).

$$\mathbf{S} = 2\rho \frac{dw_p}{d\mathbf{C}} = 2\rho\, \mathbf{P} \frac{dw_0(\mathbf{P}^T\mathbf{C}\mathbf{P})}{d\mathbf{C}} \mathbf{P}^T$$

with a **hyperelastic reference energy** w_0, so that after an appropriate normalisation

(10.6) $$w_p(\mathbf{C}) = w_0(\mathbf{P}^T\, \mathbf{C}\, \mathbf{P})$$

follows after (7.4).

For many elasto-plastic materials, the elastic deformations are so small that a linear elastic law like (6.39) is sufficiently exact

$$k_p(\mathbf{C}) = \tfrac{1}{2}\, \boldsymbol{K}_p[\mathbf{C} - \mathbf{C}_{up}]$$

with

$\boldsymbol{K}_p$: anisotropic elastic tetrad

$\mathbf{C}_{up}$: stress-free configuration.

The isomorphy condition of the linear elastic law with respect to an elastic reference law (10.5) gives

(10.7) $$\begin{aligned} k_p(\mathbf{C}) &= \tfrac{1}{2}\, \boldsymbol{K}_p[\mathbf{C} - \mathbf{C}_{up}] \\ &= \mathbf{P}\, k_0(\mathbf{P}^T\, \mathbf{C}\, \mathbf{P})\, \mathbf{P}^T = \mathbf{P}\, \tfrac{1}{2}\, \boldsymbol{K}_0[\mathbf{P}^T\, \mathbf{C}\, \mathbf{P} - \mathbf{C}_{u0}]\, \mathbf{P}^T \\ &= \tfrac{1}{2}\, \mathbf{P}\, \boldsymbol{K}_0[\mathbf{C}_e - \mathbf{C}_{u0}]\, \mathbf{P}^T \end{aligned}$$

with $\mathbf{C}_e := \mathbf{P}^T\, \mathbf{C}\, \mathbf{P}$ and the current elastic stiffness tetrad

$$\boldsymbol{K}_p = \mathbf{P} * \boldsymbol{K}_0\,,$$

which is assumed to have the usual symmetries (left and right subsymmetry), and the stress-free configuration

$$\mathbf{C}_{u0} = \mathbf{P}^T * \mathbf{C}_{up} = \mathbf{P}^T\, \mathbf{C}_{up}\, \mathbf{P}\,,$$

which can always be set to $\mathbf{I}$ by an appropriate choice of the reference placement (in contrast to $\mathbf{C}_{up}$).

With the isomorphy-condition (10.5), we can determine the stresses during all processes, both elastic and plastic, by a single elastic law k_0, if the current isomorphism $\mathbf{P}$ is known. Thus, $\mathbf{P}$ describes the influence of plastic processes on the current elastic stress function k_p. Note that $\mathbf{P}$ has *not* been introduced as a kinematical quantity such as deformation. We will call it instead **plastic transformation**[129]. As the choice of the reference law k_0 is arbitrary, so is the value of $\mathbf{P}$.

The plastic transformation $\mathbf{P}$ is introduced as a material isomorphism between the elastic laws k_0 and k_p. Therefore, we can directly apply the facts of Theorem 6.8.

[129] See WANG/ BLOOM (1974) and HALPHEN/ NGUYEN (1975).

Theorem 10.5. *Let* $\{\mathscr{E}la_0\ ,\ k_0\}$ *and* $\{\mathscr{E}la_p\ ,\ k_p\}$ *be two elastic ranges of an elasto-plastic material.*

1) If $\mathbf{P}$ *is the plastic transformation from* $\mathscr{E}la_0$ *to* $\mathscr{E}la_p$*, and* $\mathscr{G}_0$ *the symmetry group of* k_0*, then*

$$\mathscr{G}_p := \mathbf{P}\ \mathscr{G}_0\ \mathbf{P}^{-1}$$

is the symmetry group of k_p*.*

2) If $\mathbf{P}$ *is the plastic transformation from* $\mathscr{E}la_0$ *to* $\mathscr{E}la_p$*, then so is* $\mathbf{A}_p\,\mathbf{P}\,\mathbf{A}_0$ *for all* $\mathbf{A}_0 \in \mathscr{G}_0$ *and all* $\mathbf{A}_p \in \mathscr{G}_p$*.*

3) If $\mathbf{P}$ *and* $\underline{\mathbf{P}}$ *are plastic transformations from* $\mathscr{E}la_0$ *to* $\mathscr{E}la_p$*, then*

$$\mathbf{P}\,\underline{\mathbf{P}}^{-1} \in \mathscr{G}_p \qquad \text{and} \qquad \mathbf{P}^{-1}\,\underline{\mathbf{P}} \in \mathscr{G}_0$$

We thus see by *2)* that the plastic transformation is unique only up to both-sided symmetry transformations.

By *1)* we conclude that if

$$k_0(\mathbf{C}) = \mathbf{A}_0\,k_0(\mathbf{A}_0^T\,\mathbf{C}\,\mathbf{A}_0)\,\mathbf{A}_0^T \qquad \forall\,\mathbf{A}_0 \in \mathscr{G}_0$$

is a symmetry transformation of k_0, then

$$k_p(\mathbf{C}) = \mathbf{P}\,k_0\,(\mathbf{P}^T\,\mathbf{C}\,\mathbf{P})\,\mathbf{P}^T$$

$$= \mathbf{P}\,\mathbf{A}_0\,\mathbf{P}^{-1}\,k_p(\mathbf{P}^{-T}\,\mathbf{A}_0^T\,\mathbf{P}^T\,\mathbf{C}\,\mathbf{P}\,\mathbf{A}_0\,\mathbf{P}^{-1})\,\mathbf{P}^{-T}\,\mathbf{A}_0^T\,\mathbf{P}^T$$

$$\forall\,\mathbf{P}\,\mathbf{A}_0\,\mathbf{P}^{-1} \in \mathscr{G}_p$$

is a symmetry transformation of k_p.

If the reference placement with $\mathbf{C} \equiv \mathbf{I}$ is an undistorted state of k_0, then this is in general not one of k_p, because $\mathbf{A}_0 \in \mathscr{O}rth^+$ does not imply $\mathbf{P}\,\mathbf{A}_0\,\mathbf{P}^{-1} \in \mathscr{O}rth^+$. Only after a change of the reference placement with $\mathbf{P}^{-1}$ do we obtain the form of the elastic law

$$\underline{k_p}(\underline{\mathbf{C}}) = \mathbf{P}^{-1}\,k_p(\mathbf{P}^{-T}\,\underline{\mathbf{C}}\,\mathbf{P}^{-1})\,\mathbf{P}^{-T} = k_0(\underline{\mathbf{C}})$$

with $\underline{\mathbf{C}} := \mathbf{P}^T\,\mathbf{C}\,\mathbf{P}$ for $\mathscr{E}la_p$, which is also related to an undistorted state (reference placement), namely to $\mathbf{P}^T\,\mathbf{C}\,\mathbf{P}$. $\underline{k_p}$ has then the same symmetry group as k_0. This would mean, however, to introduce an individual reference placement for each elastic range (which MANDEL called *isoclinic*, see Chap. 10.5), whereas all the elastic laws k_p are related to a joint reference placement, which can be freely chosen.

If we consider *isotropic* elastic laws, then it is advantageous to choose a reference placement which is an isotropic state for the elastic reference law k_0 . Then k_0 is an isotropic tensor-function

$$k_0\,(\mathbf{Q}\,\mathbf{C}\,\mathbf{Q}^T) = \mathbf{Q}\,k_0\,(\mathbf{C})\,\mathbf{Q}^T \qquad \forall\,\mathbf{Q} \in \mathscr{G}_0 \equiv \mathscr{O}rth^+.$$

If $\mathbf{P}$ is the plastic transformation of an arbitrary elastic range $\{\mathcal{E}la_p, k_p\}$ of this isotropic material, then after *2)* of Theorem 10.5., $\mathbf{P}\,\mathbf{Q}$ is also a plastic transformation for all $\mathbf{Q} \in \mathcal{O}rth^+$ for the same range. If we now apply the polar decomposition to $\mathbf{P}$ and set $\mathbf{Q}^T$ equal to the orthogonal part of $\mathbf{P}$, then $\mathbf{P}\,\mathbf{Q}$ is symmetric. Hence, *in the isotropic case (and only in this case), we can generally assume that the plastic transformation is symmetric* (MANDEL 1971). Nevertheless, even in this case it can be reasonable to use a non-symmetric $\mathbf{P}$.

The stress power density is with (10.5)

$$
\begin{aligned}
l\rho &= \mathbf{T} \cdot \mathbf{L} = \mathbf{T} \cdot \mathbf{D} \\
&= \tfrac{1}{2}\, \mathbf{S} \cdot \mathbf{C}^{\bullet} \\
&= \tfrac{1}{2}\, k_p(\mathbf{C}) \cdot \mathbf{C}^{\bullet} \\
&= \tfrac{1}{2}\, \mathbf{P}\, k_0(\mathbf{P}^T\, \mathbf{C}\, \mathbf{P})\, \mathbf{P}^T \cdot \mathbf{C}^{\bullet} \\
&= \tfrac{1}{2}\, k_0(\mathbf{P}^T\, \mathbf{C}\, \mathbf{P}) \cdot \mathbf{P}^T\, \mathbf{C}^{\bullet}\, \mathbf{P} \\
&= \tfrac{1}{2}\, k_0(\mathbf{C}_e) \cdot (\mathbf{C}_e^{\bullet} - \mathbf{P}^{T\bullet}\, \mathbf{C}\, \mathbf{P} - \mathbf{P}^T\, \mathbf{C}\, \mathbf{P}^{\bullet}) \qquad (10.8) \\
&= \tfrac{1}{2}\, k_0(\mathbf{C}_e) \cdot \mathbf{C}_e^{\bullet} - k_0(\mathbf{C}_e) \cdot \mathbf{P}^T\, \mathbf{C}\, \mathbf{P}^{\bullet}\, \mathbf{P}^{-1}\, \mathbf{P} \\
&= \tfrac{1}{2}\, k_0(\mathbf{C}_e) \cdot \mathbf{C}_e^{\bullet} - \mathbf{P}\, k_0(\mathbf{C}_e)\, \mathbf{P}^T \cdot \mathbf{C}\, \mathbf{P}^{\bullet}\, \mathbf{P}^{-1} \\
&= \tfrac{1}{2}\, k_0(\mathbf{C}_e) \cdot \mathbf{C}_e^{\bullet} - \mathbf{C}\, \mathbf{S} \cdot \mathbf{P}^{\bullet}\, \mathbf{P}^{-1} \\
&= \tfrac{1}{2}\, k_0(\mathbf{C}_e) \cdot \mathbf{C}_e^{\bullet} + \mathbf{S}^p \cdot \mathbf{P}^{\bullet}
\end{aligned}
$$

with

$$
\mathbf{C}_e = \mathbf{P}^T\, \mathbf{C}\, \mathbf{P} \qquad (10.9)
$$

and

$$
\begin{aligned}
\mathbf{C}_e^{\bullet} &= (\mathbf{P}^T\, \mathbf{C}\, \mathbf{P})^{\bullet} = \mathbf{P}^T\, \mathbf{C}^{\bullet}\, \mathbf{P} + 2\, sym(\mathbf{P}^T\, \mathbf{C}\, \mathbf{P}^{\bullet}) \\
&= \mathbf{P}^T\, \mathbf{C}^{\bullet}\, \mathbf{P} + 2\, sym(\mathbf{C}_e\, \mathbf{P}^{-1}\, \mathbf{P}^{\bullet})\,,
\end{aligned}
$$

and with the non-symmetric **plastic stress tensor**

$$
\mathbf{S}^p := -\,\mathbf{C}\, \mathbf{S}\, \mathbf{P}^{-T}. \qquad (10.10)
$$

This name does not imply a decomposition of the stresses into elastic and plastic parts, but only refers to the fact that $\mathbf{S}^p$ is work-conjugate to the plastic transformation $\mathbf{P}$.

In the case of hyperelastic ranges, the first term is the differential of the reference strain energy

$$
\tfrac{1}{2}\, k_0(\mathbf{C}_e) \cdot \mathbf{C}_e^{\bullet} = \rho\, \frac{dw_0}{d\mathbf{C}_e} \cdot \mathbf{C}_e^{\bullet} = \rho\, w_0(\mathbf{C}_e)^{\bullet}.
$$

The rest is dissipated during yielding, which cannot be recovered by unloading. It will be left in the material as internal energy, partly converted into heat. During elastic processes, $\mathbf{P}^{\bullet} \equiv \mathbf{0}$ and, thus, the second term vanishes.

Flow and Hardening Rules

By the isomorphy-condition (10.5) we are able to bring the current elastic law k_p into a form which is the same for all elastic ranges. In order to do the same for the yield criterion, we make the following ansatz

(10.11) $$\Phi_p(\mathbf{C}) = \varphi(\mathbf{P}, \mathbf{C}, \mathbf{Z}) \mid \varphi : \mathscr{Unim}^+ \times \mathscr{Psym} \times \underline{\mathscr{Lin}} \to \mathscr{R}$$

wherein $\mathbf{Z}$ stands for **hardening variables** of the current elastic range being tensors of arbitrary order or even a vector of such tensors forming the linear space $\underline{\mathscr{Lin}}$, the specification of which depends on the hardening model (see below). During elastic processes, $\mathbf{P}$ and $\mathbf{Z}$ remain constant. These are the only internal variables of this theory. As we can determine the stresses by $\mathbf{C}$ and $\mathbf{P}$ after the isomorphy-condition (10.5), we did not include $\mathbf{S}$ in the list of arguments of the hardening rule. Note that the function Φ_p is associated individually with each elastic range, while the state function φ is assumed to hold in this form for *all* elastic ranges. The rates of the two forms are related by

$$\Phi_p(\mathbf{C})^\bullet = \frac{d\Phi_p(\mathbf{C})}{d\mathbf{C}} \cdot \mathbf{C}^\bullet = \frac{\partial \varphi(\mathbf{P},\mathbf{C},\mathbf{Z})}{\partial \mathbf{C}} \cdot \mathbf{C}^\bullet$$

being positive during yielding after the loading condition.

In $\dfrac{\partial \varphi}{\partial \mathbf{C}}$ the internal variables $\mathbf{P}$ and $\mathbf{Z}$ are kept constant for the (partial) differentiation. Thus, this expression is not a complete time-differential of φ.

As an **example** for hardening, we extend the yield criterion (10.2) by a **back stress** $\mathbf{S}_B \in \mathscr{Sym}$, which describes **kinematic hardening** and enters into the equivalent stress σ_{CS} as

$$\sigma_{CS}(\mathbf{C}\,(\mathbf{S} - \mathbf{S}_B)),$$

and a variable yield stress σ_Y, describing **isotropic hardening**. In the case of the HUBER-v. MISES criterion, $\mathbf{S}_B$ causes a shift of the yield cylinder in the stress space, and σ_Y its amplification for $\sigma_Y^\bullet > 0$. If the elastic law is invertible, one can likewise transform these stress-like hardening variables into deformation-like variables, such as

$$\mathbf{S}_B = k_p(\mathbf{C}_z) = \mathbf{P}\, k_0(\mathbf{P}^T\, \mathbf{C}_z\, \mathbf{P})\, \mathbf{P}^T$$

thus defining a **back strain** $\mathbf{C}_z \in \mathscr{Psym}$. Also the yield stress can be easily transformed into a strain-like variable by simply dividing it by an appropriate shear modulus G. If one wants to imply these two kinds of hardening, then we would have the identification

$$\mathbf{Z} \equiv \{\sigma_Y/G, \mathbf{C}_z\} \in \underline{\mathcal{Lin}} \equiv \mathcal{R}^+ \times \mathcal{Psym}\,.$$

Next we will consider the evolution of the plastic transformation during yielding. First of all, we conclude from the assumed continuity of the change of the elastic ranges, a continuous evolution of $\mathbf{P}$. We understand $\mathbf{P}$ as continuous and piecewise continuously differentiable function of time, being constant during elastic processes. Hence, an evolution functional

$$\not p: \mathbf{C}(\tau)\Big|_{\tau=0}^{t} \mapsto \mathbf{P}(t)$$

should exist, which assigns for each deformation process, starting from some initial state, the resulting plastic transformation at its end. Such a functional is called **flow rule**. $\not p$ must meet the condition that $\mathbf{P}$ remains constant under elastic processes, *i.e.*,

$$\mathbf{P}^\bullet = \not p\,\{\mathbf{C}(\tau)\Big|_{\tau=0}^{t}\}^\bullet = \mathbf{0}$$

for

$$\varphi(\mathbf{P}, \mathbf{C}, \mathbf{Z}) < 0 \qquad \text{(not on the yield surface)}$$

and/or

$$\frac{\partial\varphi}{\partial\mathbf{C}} \cdot \mathbf{C}^\bullet \le 0 \qquad \text{(no loading).}$$

For this evolution equation, rate-forms are commonly used

(10.12) $$\mathbf{P}^\bullet = p(\mathbf{P}, \mathbf{C}, \mathbf{Z}, \mathbf{C}^\bullet) \mid p : \mathcal{Unim}^+ \times \mathcal{Psym} \times \underline{\mathcal{Lin}} \times \mathcal{Sym} \to \mathcal{Lin}\,,$$

the same as for the hardening-rule

$$\mathbf{Z}^\bullet = z(\mathbf{P}, \mathbf{C}, \mathbf{Z}, \mathbf{C}^\bullet) \mid z : \mathcal{Unim}^+ \times \mathcal{Psym} \times \underline{\mathcal{Lin}} \times \mathcal{Sym} \to \underline{\mathcal{Lin}}\,.$$

As we can determine the current stresses after (10.5) by $\mathbf{C}$ and $\mathbf{P}$, we did not include them in the list of arguments of p and z. Since $\mathbf{P}$ and $\mathbf{Z}$ have to be constant during elastic processes, we expect

$$p(\mathbf{P}, \mathbf{C}, \mathbf{Z}, \mathbf{C}^\bullet) = \mathbf{0}$$
$$z(\mathbf{P}, \mathbf{C}, \mathbf{Z}, \mathbf{C}^\bullet) = \mathbf{0}$$

for

$$\varphi(\mathbf{P}, \mathbf{C}, \mathbf{Z}) < 0 \quad \text{and/or} \quad \frac{\partial\varphi}{\partial\mathbf{C}} \cdot \mathbf{C}^\bullet \le 0\,.$$

Thus, the flow and hardening rule p and z must have switchers to turn them on and off in plastic and elastic process segments, respectively.

For the elasto-plastic material under consideration we assume **rate-independence**. Thus, the rules p and z must have this property. Necessary and sufficient condition for this property is that p and z are positive-homogeneous functions of degree *1* in $\mathbf{C}^\bullet$, *i.e.*, for all positive reals α

$$p(\mathbf{P}, \mathbf{C}, \mathbf{Z}, \alpha\, \mathbf{C}^\bullet) = \alpha\, p(\mathbf{P}, \mathbf{C}, \mathbf{Z}, \mathbf{C}^\bullet)$$

and

$$z(\mathbf{P}, \mathbf{C}, \mathbf{Z}, \alpha\, \mathbf{C}^\bullet) = \alpha\, z(\mathbf{P}, \mathbf{C}, \mathbf{Z}, \mathbf{C}^\bullet)$$

must hold at any state.

These two evolution equations p and z cannot be chosen independently of the yield-criterion, but must be *consistent* with it. This means that the state must be at the yield limit during yielding. This can be achieved without restriction by prescribing only the direction of yielding, but not its magnitude, by the ansatz

$$\mathbf{P}^\bullet = \lambda\, p^\circ(\mathbf{P}, \mathbf{C}, \mathbf{Z}, \mathbf{C}^\circ)$$

$$\text{(10.13)} \qquad \mathbf{Z}^\bullet = \lambda\, z^\circ(\mathbf{P}, \mathbf{C}, \mathbf{Z}, \mathbf{C}^\circ)$$

where we use the notation $\mathbf{C}^\circ := \mathbf{C}^\bullet / |\mathbf{C}^\bullet|$ for the direction of a (non-zero) rate tensor, and a non-negative multiplicator λ, called **plastic consistency parameter**, which we will determine later by the consistency condition. Again, the functions p° and z° called **directional flow** and **hardening rule**, respectively, must contain a switcher which allows only for non-zero values, if both the yield criterion and the loading condition are simultaneously fulfilled. If this is not the case, no yielding occurs and λ can be put to zero. In such a form, the two rules determine the (positive) directions of flow

$$\mathbf{P}^\circ := p^\circ(\mathbf{P}, \mathbf{C}, \mathbf{Z}, \mathbf{C}^\circ) \quad \Rightarrow \quad \mathbf{P}^\bullet = \lambda\, \mathbf{P}^\circ$$

and hardening

$$\mathbf{Z}^\circ := z^\circ(\mathbf{P}, \mathbf{C}, \mathbf{Z}, \mathbf{C}^\circ) \quad \Rightarrow \quad \mathbf{Z}^\bullet = \lambda\, \mathbf{Z}^\circ,$$

the same as the ratio of the magnitude of yield to that of hardening, but not their absolute value.

In order to make this ansatz for the general hardening rule more concrete, we have to identify the hardening variables. We first consider the (simple) case of isotropic hardening, and then kinematic hardening.

Isotropic Hardening

Here $\mathbf{Z} \equiv \{\sigma_Y / G\}$ is a scalar. A simple linear ansatz for an isotropic hardening rule is the *strain hardening*

$$\sigma_Y^\bullet = A\, |\mathbf{Y}_p|$$

---

[131] http://krawietz.homepage.t-online.de/parallel.pdf; see also KRAWIETZ/ MATHIAK (1989).

with a real material constant A and the symmetric incremental plastic variable

$$\mathbf{Y}_p := \mathbf{C}_e^{\bullet} - \mathbf{P}^T \mathbf{C}^{\bullet} \mathbf{P} = (\mathbf{P}^T \mathbf{C} \mathbf{P})^{\bullet} - \mathbf{P}^T \mathbf{C}^{\bullet} \mathbf{P}$$
$$= \mathbf{P}^T \mathbf{C} \mathbf{P}^{\bullet} + \mathbf{P}^{T\bullet} \mathbf{C} \mathbf{P} = 2\, sym(\mathbf{P}^T \mathbf{C} \mathbf{P}^{\bullet})$$
$$= 2\, sym(\mathbf{P}^T \mathbf{C}\, p(\mathbf{P}, \mathbf{C}, \mathbf{Z}, \mathbf{C}^{\bullet}))\,,$$

which is only non-zero for states of yielding. Note that $\mathbf{Y}_p$ is not the time derivative of a state variable. For a positive constant A, σ_Y can only grow during plastic processes, and the material hardens.

As for real materials, hardening does not grow infinitely, but often converges to a saturation value, so the following ansatz is more realistic

$$\sigma_Y^{\bullet} = A\,(\sigma_\infty - \sigma_Y)\,|\,\mathbf{Y}_p\,|$$

with another positive material constant σ_∞.

An alternative to strain-hardening is *dissipation-hardening*

$$\sigma_Y^{\bullet} = c\,\mathbf{S}^p \cdot \mathbf{P}^{\bullet} = -c\,\mathbf{CS} \cdot \mathbf{P}^{\bullet}\,\mathbf{P}^{-1}$$

with a positive material constant c, or with saturation

$$\sigma_Y^{\bullet} = c\,(\sigma_\infty - \sigma_Y)\,\mathbf{S}^p \cdot \mathbf{P}^{\bullet}\,.$$

Kinematic Hardening

is – as we have already seen – determined by a back stress $\mathbf{S}_B \in \mathcal{Sym}$, which shall have the properties of the material stress. In particular, it shall be symmetric and invariant under EUCLIDean transformations. A possible choice for a hardening rule for the back stress is

$$\mathbf{S}_B^{\bullet} = \boldsymbol{K}_0[\mathbf{Y}_p] \qquad \in \mathcal{Sym}$$

with a stiffness-like tetrad $\boldsymbol{K}_0$ representing the initial material symmetries of the particular material class. The form of such tetrads is well known from linear elasticity. In any case, it should be such that the result is always symmetric and deviatoric. If one wants to limit the hardening, then this can be achieved by a saturation term like

$$\mathbf{S}_B^{\bullet} = (1 - K\,|\,\mathbf{S}_B\,|)\,\boldsymbol{K}_0[\mathbf{Y}_p] \qquad \in \mathcal{Sym}$$

with a real constant K, or in a tensorial form by

$$\mathbf{S}_B^{\bullet} = \boldsymbol{K}_0[\mathbf{Y}_p] - K\,\mathbf{S}_B\,. \qquad \in \mathcal{Sym}\,.$$

KRAWIETZ[131] suggest a similar kinematic hardening rule in the form

$$\mathbf{T}_{eB}^{2PK\,\bullet} = c\,(\mathbf{X}_p - k\,|\,\mathbf{X}_p\,|\,\mathbf{T}_{eB}^{2PK})$$

for a transformed 2nd PIOLA-KIRCHHOFF back stress tensor

$$\mathbf{T}_{eB}{}^{2\mathrm{PK}} := \mathbf{P}^{-1}\,\mathbf{T}_B{}^{2\mathrm{PK}}\,\mathbf{P}^{-T}$$

defined in analogy to the transformed 2nd PIOLA-KIRCHHOFF stress tensor

$$\mathbf{T}_e{}^{2\mathrm{PK}} := \mathbf{P}^{-1}\,\mathbf{T}^{2\mathrm{PK}}\,\mathbf{P}^{-T}$$

and an internal variable

$$\mathbf{X}_p := \boldsymbol{K}_0[\mathbf{T}_e{}^{2\mathrm{PK}}{}' - \mathbf{T}_{eB}{}^{2\mathrm{PK}}] \qquad \in \mathcal{S}ym\,.$$

Again, the tetrad $\boldsymbol{K}_0$ represents the initial material symmetries and makes sure that $\mathbf{X}_p$ is a symmetric deviator. k and c are scalar hardening variables or constants.

Of course, there are infinitely many other possible choices. Suggestions for kinematic hardening rules in the sense of PRAGER (1955) are

(10.14) $$\mathbf{S}_B{}^{\bullet} = K_1\,\mathbf{C}^{-1}\,\mathbf{Y}_p\,\mathbf{C}^{-1} \qquad \in \mathcal{S}ym$$

(10.15) $$(\mathbf{S}_B/\rho)^{\bullet} = K_2\,\mathbf{C}^{-1}\,\mathbf{Y}_p\,\mathbf{C}^{-1} \qquad \in \mathcal{S}ym$$

(10.16) $$(\mathbf{C}\;\mathbf{S}_B\mathbf{C})^{\bullet} = K_3\,\mathbf{Y}_p \qquad \in \mathcal{S}ym$$

or in the sense of ZIEGLER (1959)

(10.17) $$\mathbf{S}_B{}^{\bullet} = \lambda\,K_4\,(\mathbf{S}' - \mathbf{S}_B) \in \mathcal{S}ym \qquad \text{with } \mathbf{S}' := \mathbf{S} - {}^1\!/_3\, tr(\mathbf{C}\mathbf{S})\,\mathbf{C}^{-1}$$

(10.18) $$(\mathbf{C}\mathbf{S}_B)^{\bullet} = \lambda\,K_5\,\{(\mathbf{C}\mathbf{S})' - \mathbf{C}\mathbf{S}_B\} \qquad \in \mathcal{L}in$$

with material constants K_i . If we express the stress rate in (10.14) by the CAUCHY back stress $\mathbf{T}_B$, then we obtain the OLDROYD rate (3.27)

$$\mathbf{F}\,\mathbf{S}_B{}^{\bullet}\,\mathbf{F}^T = \mathbf{T}_B{}^{\bullet} - \mathbf{L}\,\mathbf{T}_B - \mathbf{T}_B\,\mathbf{L}^T = K_1\,\mathbf{F}^{-T}\,\mathbf{Y}_p\,\mathbf{F}^{-1}$$

the same as for (10.17)

$$\mathbf{T}_B{}^{\bullet} - \mathbf{L}\,\mathbf{T}_B - \mathbf{T}_B\,\mathbf{L}^T = \lambda\,K_4\,(\mathbf{T}' - \mathbf{T}_B)\,.$$

For the stress rate in (10.15) we obtain the TRUESDELL rate (3.28)

$$\rho\,\mathbf{F}\,(\mathbf{S}_B/\rho)^{\bullet}\,\mathbf{F}^T = J^{-1}\,\mathbf{F}\,\mathbf{T}^{2\mathrm{PK}}{}_B{}^{\bullet}\,\mathbf{F}^T$$
$$= \mathbf{T}_B{}^{\bullet} - \mathbf{L}\,\mathbf{T}_B - \mathbf{T}_B\,\mathbf{L}^T + \mathbf{T}_B\,div\,\mathbf{v} = K_2\,\mathbf{F}^{-T}\,\mathbf{Y}_p\,\mathbf{F}^{-1},$$

for (10.16) the COTTER-RIVLIN rate (3.29)

$$\mathbf{F}^{-T}\,(\mathbf{C}\,\mathbf{S}_B\mathbf{C})^{\bullet}\,\mathbf{F}^{-1} = \mathbf{T}_B{}^{\bullet} + \mathbf{L}^T\,\mathbf{T}_B + \mathbf{T}_B\,\mathbf{L} = K_3\,\mathbf{F}^{-T}\,\mathbf{Y}_p\,\mathbf{F}^{-1},$$

and for (10.18) the mixed OLDROYD rate (3.30)

$$\mathbf{T}_B{}^{\bullet} + \mathbf{L}^T\,\mathbf{T}_B - \mathbf{T}_B\,\mathbf{L}^T = \lambda\,K_5\,(\mathbf{T}' - \mathbf{T}_B)\,.$$

Note that in the latter cases, $\mathbf{S}_B$ can vary even for elastic processes. Only $(\mathbf{S}_B/\rho)$ or $(\mathbf{C}\,\mathbf{S}_B\,\mathbf{C})$ or $\mathbf{C}\mathbf{S}_B$, respectively, remain constant in such cases.

In this context also the ZAREMBA-JAUMANN rate (3.31) or the stress rate after GREEN/NAGHDI – GREEN/McINNIS – DIENES (3.32) are frequently used.

However, TRUESDELL/ NOLL (1965, p. 97) made the following comment: "The fluxes (...) are but two of the infinitely many possible invariant time fluxes that may be used. Clearly the properties of a material are *independent of the choice of flux*, which, like the choice of a measure of strain, is *absolutely immaterial*. Before the invariance to be required of constitutive equations was fully understood, there was some discussion of this point among the major theorists, but it has ceased. Thus we leave intentionally incited the blossoming literature on invariant time fluxes subjected to various arbitrary requirements."

Quite popular are **plastic potentials** originally suggested by v. MISES (1928) in the form of real-valued functions like

$$\pi : \mathcal{Lin} \times \mathcal{Lin} \times \mathcal{R}^+ \times \mathcal{Unim}^+ \rightarrow \mathcal{R}$$

$$(\mathbf{S}^p, \mathbf{C}\mathbf{S}_B, \sigma_Y, \mathbf{P}) \mapsto \pi(\mathbf{S}^p, \mathbf{C}\mathbf{S}_B, \sigma_Y, \mathbf{P}) \, .$$

Such potentials then give the flow rule

$$\mathbf{P}^\bullet = \lambda \frac{\partial \pi\left(\mathbf{S}^p, \mathbf{C}\mathbf{S}_B, \sigma_Y, \mathbf{P}\right)}{\partial \mathbf{S}^p}$$

and the kinematic hardening rule

$$\mathbf{C}_z^\bullet = \lambda \frac{\partial \pi\left(\mathbf{S}^p, \mathbf{C}\mathbf{S}_B, \sigma_Y, \mathbf{P}\right)}{\partial \mathbf{S}_B}$$

and the isotropic hardening rule

$$\sigma_Y^\bullet / G = \lambda \frac{\partial \pi\left(\mathbf{S}^p, \mathbf{C}\mathbf{S}_B, \sigma_Y, \mathbf{P}\right)}{\partial \sigma_Y}$$

for yield states.

In particular, we obtain an **associated plasticity theory**, if we use the yield criterion for the plastic potential after transforming the variables into the strain space.

Consistency Condition

As already mentioned, the yield criterion, the flow rule, and the hardening rule cannot be chosen independently, as during yielding the state of the material must permanently remain on the current yield limit. To investigate the implications of this restriction, we start from rate-independent yield and hardening rules (10.13). During yielding, the yield criterion must be permanently fulfilled

$$\varphi(\mathbf{C}, \mathbf{P}, \mathbf{Z}) = 0$$

and thus the **consistency condition**

$$0 = \varphi(\mathbf{C}, \mathbf{P}, \mathbf{Z})^{\bullet}$$

(10.19)
$$= \frac{\partial \varphi}{\partial \mathbf{C}} \cdot \mathbf{C}^{\bullet} + \frac{\partial \varphi}{\partial \mathbf{P}} \cdot \mathbf{P}^{\bullet} + \frac{\partial \varphi}{\partial \mathbf{Z}} \cdot \mathbf{Z}^{\bullet}$$

$$= \frac{\partial \varphi}{\partial \mathbf{C}} \cdot \mathbf{C}^{\bullet} + \frac{\partial \varphi}{\partial \mathbf{P}} \cdot (\lambda\, \mathbf{P}^{\circ}) + \frac{\partial \varphi}{\partial \mathbf{Z}} \cdot (\lambda\, \mathbf{Z}^{\circ}),$$

which is solved for λ by

$$\lambda(\mathbf{P}, \mathbf{C}, \mathbf{Z}, \mathbf{C}^{\bullet}) = \frac{-\dfrac{\partial \varphi}{\partial \mathbf{C}} \cdot \mathbf{C}^{\bullet}}{\dfrac{\partial \varphi}{\partial \mathbf{P}} \cdot \mathbf{P}^{\circ} + \dfrac{\partial \varphi}{\partial \mathbf{Z}} \cdot \mathbf{Z}^{\circ}}$$

Both, numerator and denominator of this ratio are always negative during yielding, and thus λ is positive. With it we obtain the **consistent yield** and **hardening rule**

(10.20)
$$\boxed{\begin{aligned} \mathbf{P}^{\bullet} &= \mathbf{P}^{\circ} \otimes \mathbf{A}[\mathbf{C}^{\bullet}] \\ \mathbf{Z}^{\bullet} &= \mathbf{Z}^{\circ} \otimes \mathbf{A}[\mathbf{C}^{\bullet}] \end{aligned}}$$

with the symmetric tensor

$$\mathbf{A}(\mathbf{P}, \mathbf{C}, \mathbf{Z}, \mathbf{C}^{\circ}) := \frac{-\dfrac{\partial \varphi}{\partial \mathbf{C}}}{\dfrac{\partial \varphi}{\partial \mathbf{P}} \cdot \mathbf{P}^{\circ} + \dfrac{\partial \varphi}{\partial \mathbf{Z}} \cdot \mathbf{Z}^{\circ}}.$$

Herein, all functions have to be evaluated at $(\mathbf{P}, \mathbf{C}, \mathbf{Z}, \mathbf{C}^{\circ})$ and all partial derivatives at $(\mathbf{P}, \mathbf{C}, \mathbf{Z})$. Because of the switchers in p° and z°, both rules are non-linear in $\mathbf{C}^{\bullet}$.

In all states (elastic *and* plastic), the **KUHN-TUCKER condition**

$$\lambda\, \varphi = 0 \qquad \text{with} \qquad \lambda \geq 0 \qquad \text{and} \qquad \varphi \leq 0$$

holds, as at any time, one of the two factors is zero.

For some purposes, one prefers an incremental form instead of the finite stress law. In it, the **material tangential stiffness tetrad** of the material is obtained by taking the differential of the stress law

$$\mathbf{S}^{\bullet} = \{k_p(\mathbf{C})\}^{\bullet} = \{\mathbf{P}\, k_0(\mathbf{P}^T \mathbf{C}\, \mathbf{P})\, \mathbf{P}^T\}^{\bullet}$$

$$= \mathbf{P}^{\bullet}\, k_0(\mathbf{P}^T \mathbf{C}\, \mathbf{P})\, \mathbf{P}^T + \mathbf{P}\, k_0(\mathbf{P}^T \mathbf{C}\, \mathbf{P})\, \mathbf{P}^{T\bullet} + \mathbf{P}\, \{\frac{dk_0}{d\mathbf{C}}\, [\mathbf{P}^T \mathbf{C}\, \mathbf{P}]^{\bullet}\}\, \mathbf{P}^T$$

$$= \mathbf{P}^{\bullet}\, k_0(\mathbf{P}^T \mathbf{C}\, \mathbf{P})\, \mathbf{P}^T + \mathbf{P}\, k_0(\mathbf{P}^T \mathbf{C}\, \mathbf{P})\, \mathbf{P}^{T\bullet}$$
$$+ \mathbf{P}\, \{\mathcal{K}_0[\mathbf{P}^T \mathbf{C}^{\bullet}\, \mathbf{P} + 2\, \mathbf{P}^T \mathbf{C}\, \mathbf{P}^{\bullet}]\}\, \mathbf{P}^T$$

$$(10.21) \qquad = (\mathbf{P}^\circ \otimes \mathbf{A}[\mathbf{C}^\bullet])\, k_0(\mathbf{P}^T \mathbf{C}\, \mathbf{P})\, \mathbf{P}^T + \mathbf{P}\, k_0(\mathbf{P}^T \mathbf{C}\, \mathbf{P})\, (\mathbf{P}^{\circ T} \otimes \mathbf{A}[\mathbf{C}^\bullet])$$

$$+ \mathbf{P}\, \{\boldsymbol{K}_0[\mathbf{P}^T \mathbf{C}^\bullet \mathbf{P} + 2\, \mathbf{P}^T \mathbf{C}\, (\mathbf{P}^\circ \otimes \mathbf{A}[\mathbf{C}^\bullet])]\}\, \mathbf{P}^T$$

$$= \mathbf{P}\, \boldsymbol{K}_0[\mathbf{P}^T \mathbf{C}^\bullet \mathbf{P}]\, \mathbf{P}^T + \{\mathbf{P}^\circ\, k_0(\mathbf{P}^T \mathbf{C}\, \mathbf{P})\, \mathbf{P}^T + \mathbf{P}\, k_0(\mathbf{P}^T \mathbf{C}\, \mathbf{P})\, \mathbf{P}^{\circ T}$$

$$+ 2\, \mathbf{P}\, \boldsymbol{K}_0[2\, \mathbf{P}^T \mathbf{C}\, \mathbf{P}^\circ]\, \mathbf{P}^T\} \otimes \mathbf{A}[\mathbf{C}^\bullet]$$

$$= \boldsymbol{K}_p[\mathbf{C}^\bullet] + \boldsymbol{V}_p[\mathbf{C}^\bullet]$$

with the elastic reference stiffness tetrad $\boldsymbol{K}_0 \;:=\; \dfrac{dk_0}{d\mathbf{C}}$ evaluated at $\mathbf{P}^T \mathbf{C}\, \mathbf{P}$ in analogy to (6.35), and the current elastic stiffness tetrad $\boldsymbol{K}_p = \mathbf{P} * \boldsymbol{K}_0$, which gives the elastic prediction

$$\boldsymbol{K}_p[\mathbf{C}^\bullet] \;:=\; \mathbf{P}\, \frac{dk_0}{d\mathbf{C}}\, [\mathbf{P}^T \mathbf{C}^\bullet \mathbf{P}]\, \mathbf{P}^T = (\mathbf{P} * \boldsymbol{K}_0)\, [\mathbf{C}^\bullet]\,.$$

The plastic corrector is $\boldsymbol{V}_p[\mathbf{C}^\bullet]$ with the tetrad

$$\boldsymbol{V}_p \;:=\; \{\mathbf{P}^\circ\, k_0(\mathbf{P}^T \mathbf{C}\, \mathbf{P})\, \mathbf{P}^T + \mathbf{P}\, k_0(\mathbf{P}^T \mathbf{C}\, \mathbf{P})\, \mathbf{P}^{\circ T} + 2\, \mathbf{P}\, \boldsymbol{K}_0[\mathbf{P}^T \mathbf{C}\, \mathbf{P}^\circ]\, \mathbf{P}^T\} \otimes \mathbf{A}$$

$$= \{\mathbf{P}^\circ\, \mathbf{P}^{-1}\, \mathbf{S} + \mathbf{S}\, \mathbf{P}^{-T}\, \mathbf{P}^{\circ T} + 2\, \boldsymbol{K}_p\, [\mathbf{C}\, \mathbf{P}^\circ\, \mathbf{P}^{-1}]\} \otimes \mathbf{A}\,,$$

which is only non-zero during yielding because of the switchers in p°. $\boldsymbol{K}_p$ possesses the two subsymmetries, and varies with $\mathbf{P}$. In general, it can also depend on $\mathbf{C}$. If the elastic reference law is assumed to be (physically) linear, then $\boldsymbol{K}_0$ is constant for all elastic ranges, but not $\boldsymbol{K}_p$.

Neither is $\boldsymbol{V}_p$ symmetric, but can be assumed to have both subsymmetries, as we in fact assumed before. Because of the switcher in p° and also in $\boldsymbol{V}_p$, $\mathbf{S}^\bullet$ is generally non-linear in $\mathbf{C}^\bullet$, a fact that is characteristic for elastic-plastic materials, which complicates the numerical integration of the evolution functions.

The decomposition of the stress rate can be brought into the form

$$\mathbf{S}^\bullet = \mathbf{P}\, \{\boldsymbol{K}_0\, [\mathbf{C}_e^\bullet]\}\, \mathbf{P}^T + 2\, sym(\mathbf{P}^\bullet\, \mathbf{P}^{-1}\, \mathbf{S})$$

with an elastic and a plastic increment. The plastic part is linear in $\mathbf{P}^\bullet$.

Symmetries in Plasticity

As the chosen dependent and independent variables in plasticity depend on the reference placement, we have to consider their transformation behaviour. So, let $\mathbf{K} := Grad(\kappa_0\, \underline{\kappa_0}^{-1}) \in \mathcal{Inv}^+$ be the local change of the reference placement, then we get the following transformations:

$\underline{\mathbf{F}} = \mathbf{F}\,\mathbf{K}$	deformation gradient
$\underline{\mathbf{C}} = \mathbf{K}^T \mathbf{C}\, \mathbf{K} = \mathbf{K}^T * \mathbf{C}$	right CAUCHY-GREEN tensor
$\mathbf{T} = \underline{\mathbf{T}}$	CAUCHY´s stress tensor

$$\underline{\mathbf{S}} = \mathbf{K}^{-1}\,\mathbf{S}\,\mathbf{K}^{-T} = \mathbf{K}^{-1} * \mathbf{S} \quad \text{material stress tensor}$$

$$\underline{\mathbf{P}} = \mathbf{P} \quad \text{plastic transformation}$$

$$\underline{\mathbf{C}}\,\underline{\mathbf{S}} = (\mathbf{K}^{T} * \mathbf{C})\,(\mathbf{K}^{-1} * \mathbf{S}) = \mathbf{K}^{T}\,\mathbf{C}\,\mathbf{S}\,\mathbf{K}^{-T}$$

$$\underline{\mathbf{S}}^{p} = -\,\mathbf{K}^{T}\,\mathbf{CS}\,\mathbf{K}^{-T}\,\mathbf{P}^{-T} \quad \text{plastic stress tensor}$$

We will further on assume that the hardening variables transform like deformations

$$\underline{\mathbf{Z}} = \mathbf{K}^{T} * \mathbf{Z} \quad \text{hardening variables}$$

and thus

$$\mathbf{Z}^{*} = \mathbf{K} * \underline{\mathbf{Z}}^{*} \quad \text{dual hardening variables.}$$

The symmetry transformations act analogously. $\mathbf{A}$ is in the symmetry group of the elasto-plastic material, if the constitutive equations fulfil the following conditions:

- elastic ref. law $\quad k_0(\mathbf{C}) = \mathbf{A}\,k_0(\mathbf{A}^{T}\,\mathbf{C}\,\mathbf{A})\,\mathbf{A}^{T}$
- yield criterion $\quad \varphi(\mathbf{P}\,,\mathbf{C}\,,\mathbf{Z}) = \varphi(\mathbf{P}\,,\mathbf{A}^{T} * \mathbf{C}\,,\mathbf{A}^{T} * \mathbf{Z})$
- flow rule $\quad p^{\circ}(\mathbf{P},\mathbf{C}\,,\mathbf{Z}\,,\mathbf{C}^{\circ}) = p^{\circ}(\mathbf{P}\,,\mathbf{A}^{T} * \mathbf{C}\,,\mathbf{A}^{T} * \mathbf{Z}\,,\mathbf{A}^{T} * \mathbf{C}^{\circ})$
- hardening rule $\quad z^{\circ}(\mathbf{P},\mathbf{C}\,,\mathbf{Z}\,,\mathbf{C}^{\circ}) = \mathbf{A}^{T} * z^{\circ}(\mathbf{P}\,,\mathbf{A}^{T} * \mathbf{C}\,,\mathbf{A}^{T} * \mathbf{Z}\,,\mathbf{A}^{T} * \mathbf{C}^{\circ})\,.$

If we formulate the yield criterion in $\mathbf{CS}$, then its symmetry transformation is

$$\Phi_{\sigma p}(\mathbf{CS}) = \Phi_{\sigma p}(\mathbf{A}^{T}\,\mathbf{CS}\,\mathbf{A}^{-T})\,.$$

Formally, each constitutive equation of the present elasto-plastic model may posses its own symmetry group. If, however, the material is known to be isotropic, orthotropic, cubic or else, then we would expect all constitutive equations to have the corresponding symmetry group.

As an **example** we consider a material with isotropic and kinematic hardening with the associated plastic potential

$$\begin{aligned}
\pi(\mathbf{S}^{p}, \mathbf{C}\,\mathbf{S}_B\,, \sigma_Y, \mathbf{P}) &\equiv \sigma_{\mathrm{CS}}(-\,\mathbf{S}^{p}\,\mathbf{P}^{T} - \mathbf{C}\,\mathbf{S}_B) - \sigma_Y^{2} \\
&= \sigma_{\mathrm{CS}}\{(\mathbf{S}_z - \mathbf{S}^{p})\mathbf{P}^{T}\} - \sigma_Y^{2} = \sigma_{\mathrm{CS}}\{\mathbf{C}\,(\mathbf{S} - \mathbf{S}_B)\} - \sigma_Y^{2} \\
&= \sigma_{\mathrm{CS}}\{\mathbf{C}\,k_p(\mathbf{C}) - \mathbf{C}\,k_p(\mathbf{C}_z)\} - \sigma_Y^{2} \\
&= \sigma_{\mathrm{CS}}\{\mathbf{C}\,\mathbf{P}\,k_0(\mathbf{P}^{T}\,\mathbf{C}\,\mathbf{P})\,\mathbf{P}^{T} - \mathbf{C}\,\mathbf{P}\,k_0(\mathbf{P}^{T}\,\mathbf{C}_z\,\mathbf{P})\,\mathbf{P}^{T}\} - \sigma_Y^{2} \\
&= \varphi(\mathbf{C}\,,\mathbf{P}\,,\sigma_Y/\,G, \mathbf{C}_z)\,.
\end{aligned}$$

with $\mathbf{S}_z :=\, -\,\mathbf{C}\,\mathbf{S}_B\,\mathbf{P}^{-T}$ in analogy to (10.10). By using the linear elastic law (10.7) this gives in particular

$$\begin{aligned}
&= \sigma_{\mathrm{CS}}(\mathbf{C}\,\tfrac{1}{2}\,\boldsymbol{K}_p[\mathbf{C} - \mathbf{C}_{up}] - \mathbf{C}\,\tfrac{1}{2}\,\boldsymbol{K}_p[\mathbf{C}_z - \mathbf{C}_{up}]) - \sigma_Y^{2} \\
&= \sigma_{\mathrm{CS}}(\tfrac{1}{2}\,\mathbf{C}\,\boldsymbol{K}_p[\mathbf{C} - \mathbf{C}_z]) - \sigma_Y^{2}\,.
\end{aligned}$$

As an isotropic linear elastic reference-law one could use

$$k_0(\mathbf{C}_e) = {}^{\kappa}\!/_2\, \{tr(\mathbf{C}_e) - 3\}\, \mathbf{I} + \mu\, (\mathbf{C}_e - \mathbf{I})$$

with two elastic constants κ and μ, which gives after (10.5)

$$\mathbf{S} = k_p(\mathbf{C}) = \{{}^{\kappa}\!/_2\, (tr(\mathbf{P}^T\, \mathbf{C}\, \mathbf{P}) - 3)\, \mathbf{I} + \mu\, (\mathbf{P}\, \mathbf{P}^T\, \mathbf{C} - \mathbf{I})\}\, \mathbf{P}\, \mathbf{P}^T$$

For the J_2-theory with back-stress $\mathbf{S}_B$, we have specifically

$$\sigma_{CS}(\mathbf{C}\, (\mathbf{S} - \mathbf{S}_B)) \equiv 3\, J_2$$

with

$$\begin{aligned} J_2 &= \tfrac{1}{2}\, (\mathbf{CS} - \mathbf{CS}_B)' \cdot (\mathbf{SC} - \mathbf{S}_B\mathbf{C})' \\ &= \tfrac{1}{2}\, ((\mathbf{CS})' - (\mathbf{CS}_B)') \cdot ((\mathbf{SC})' - (\mathbf{S}_B\mathbf{C})') \\ &= \tfrac{1}{2}\, (\mathbf{S}_z\, \mathbf{P}^T - \mathbf{S}^p\, \mathbf{P}^T)' \cdot (\mathbf{P}\, \mathbf{S}_z^{\,T} - \mathbf{P}\, \mathbf{S}^{pT})' \\ &= -\tfrac{1}{2}\, (\mathbf{S}_z\, \mathbf{P}^T - \mathbf{S}^p\, \mathbf{P}^T) \cdot \boldsymbol{D}[\mathbf{P}\, \mathbf{S}^{pT} - \mathbf{P}\, \mathbf{S}_z^{\,T}] \end{aligned}$$

after (10.3) with the deviatoriser $\boldsymbol{D}$ after (1.31). The chain rule gives

$$\frac{\partial J_2\left(\mathbf{S}^p, \mathbf{CS}_B, \mathbf{P}\right)}{\partial \mathbf{S}^p} = (\mathbf{SC} - \mathbf{S}_B\mathbf{C})'\, \mathbf{P}$$

and

$$\frac{\partial J_2\left(\mathbf{S}^p, \mathbf{CS}_B, \mathbf{P}\right)}{\partial \mathbf{S}_B} = -\, sym\{(\mathbf{CS} - \mathbf{CS}_B)'\, \mathbf{C}\}\ .$$

The associated flow rule is then

$$\mathbf{P}^\bullet = \lambda\, \frac{\partial \pi\left(\mathbf{S}^p, \mathbf{CS}_B, \sigma_Y, \mathbf{P}\right)}{\partial \mathbf{S}^p} = 3\lambda\, (\mathbf{SC} - \mathbf{S}_B\mathbf{C})'\, \mathbf{P}$$

and the associated hardening rule

$$\mathbf{C}_z^\bullet = \lambda\, \frac{\partial \pi\left(\mathbf{S}^p, \mathbf{CS}_B, \sigma_Y, \mathbf{P}\right)}{\partial \mathbf{S}_B} = -\, 3\lambda\, sym((\mathbf{CS} - \mathbf{CS}_B)'\, \mathbf{C})$$

which is always symmetric for a symmetric back-stress. By analogy, we set

$$\sigma_Y^\bullet / G = \lambda\, \frac{\partial \pi\left(\mathbf{S}^p, \mathbf{CS}_B, \sigma_Y, \mathbf{P}\right)}{\partial \sigma_Y} = -\, 2\lambda\, \sigma_Y\, .$$

Note that the flow rule for $\mathbf{P}^\bullet\, \mathbf{P}^{-1}$ is always deviatoric for $\mathbf{P} \in \mathcal{U}nim$, and symmetric for a vanishing back-stress.

The consistency condition (10.19) is now

$$0 = \varphi(\mathbf{C}, \mathbf{P}, \sigma_Y/G, \mathbf{C}_z)^\bullet$$

$$= \frac{\partial\varphi}{\partial\mathbf{C}} \cdot \mathbf{C}^\bullet + \frac{\partial\varphi}{\partial\mathbf{P}} \cdot \mathbf{P}^\bullet + \frac{\partial\varphi}{\partial\sigma_Y} \cdot \sigma_Y^\bullet + \frac{\partial\varphi}{\partial\mathbf{C}_z} \cdot \mathbf{C}_z^\bullet$$

$$= \frac{\partial\varphi}{\partial\mathbf{C}} \cdot \mathbf{C}^\bullet + \lambda \frac{\partial\varphi}{\partial\mathbf{P}} \cdot 3\,(\mathbf{SC} - \mathbf{S}_B\mathbf{C})'\,\mathbf{P}$$

$$- \lambda \frac{\partial\varphi}{\partial\sigma_Y} \cdot 2\,G\,\sigma_Y - \lambda \frac{\partial\varphi}{\partial\mathbf{C}_z} \cdot 3\,sym((\mathbf{CS} - \mathbf{CS}_B)'\,\mathbf{C})$$

so that the plastic consistency parameter is

$$\lambda = \frac{\partial\varphi}{\partial\mathbf{C}} \cdot \mathbf{C}^\bullet / \{-\frac{\partial\varphi}{\partial\mathbf{P}} \cdot 3\,(\mathbf{SC} - \mathbf{S}_B\mathbf{C})'\,\mathbf{P}$$

$$+ \frac{\partial\varphi}{\partial\sigma_Y} \cdot 2\,G\,\sigma_Y + \frac{\partial\varphi}{\partial\mathbf{C}_z} \cdot 3sym((\mathbf{CS} - \mathbf{CS}_B)'\,\mathbf{C})\} .$$

Conclusions

We consider *materials with isomorphic elastic ranges* or **elasto-plastic materials** going through the following steps of concretisation.

1) *A simple objective inelastic material*

The inelastic stress functional $\mathfrak{K}$ of (9.1) determines the stresses at the end of any **C**-process starting from some initial state after

$$\mathbf{S}(t) = \mathfrak{K}\{\mathbf{C}(\tau)\big|_{\tau=0}^{t}\} .$$

2) *A material with elastic ranges*

The functional e determines for each **C**-process starting from some initial state, the current elastic range

$$\{\mathcal{E}la_p, k_p\} = e\,\{\mathbf{C}(\tau)\big|_{\tau=0}^{t}\}.$$

The stresses can be calculated by the elastic law of the current elastic range by the configuration at the end of the process

$$\mathbf{S}(t) = k_p(\mathbf{C}(t)) .$$

3) *A material with isomorphic elastic ranges*

The current elastic law is isomorphic to a constant elastic reference law after (10.5)

$$\mathbf{S} = k_p(\mathbf{C}) = \mathbf{P}\,k_0(\mathbf{P}^T\,\mathbf{C}\,\mathbf{P})\,\mathbf{P}^T.$$

The integration of the coupled system of the algebro-differential equations (10.12)

$$\mathbf{P}^{\bullet} = p(\mathbf{P}, \mathbf{C}, \mathbf{Z}, \mathbf{C}^{\bullet})$$
$$\mathbf{Z}^{\bullet} = z(\mathbf{P}, \mathbf{C}, \mathbf{Z}, \mathbf{C}^{\bullet})$$

along a **C**-process starting from the initial state, gives us the current values of **P** and **Z**, which determine the current elastic range $\{\mathcal{E}la_p, k_p\}$. For this purpose, the initial values for **P** and **Z** at time $t_0 = 0$ must be known.

4) *Rate-independent elasto-plasticity*

By the constitutive functions φ (10.11), $p°$, and $z°$ (10.20) we can determine the consistent yield and hardening rules, which together with $k_0(\mathbf{C})$ (10.5) completes the model.

Hence, our elasto-plastic material model is determined by the following set of constitutive equations:

- elastic reference law $k_0(\mathbf{C})$
- yield criterion $\varphi(\mathbf{P}, \mathbf{C}, \mathbf{Z})$
- directional flow rule $p°(\mathbf{P}, \mathbf{C}, \mathbf{Z}, \mathbf{C}°)$
- directional hardening rule $z°(\mathbf{P}, \mathbf{C}, \mathbf{Z}, \mathbf{C}°)$

and initial values $\mathbf{P}(t_0)$ and $\mathbf{Z}(t_0)$. The deformation process can be prescribed arbitrarily. It determines for each material point the associated **C**-process. Along the latter we can integrate the accompanying processes in **P**, **Z**, and **S**. Numerically, this leads to a system of non-linear algebro-differential equations of first order in time, which can be solved by explicit or implicit methods with arbitrary exactness.

At each instant of time, we have the following scheme for the material model.

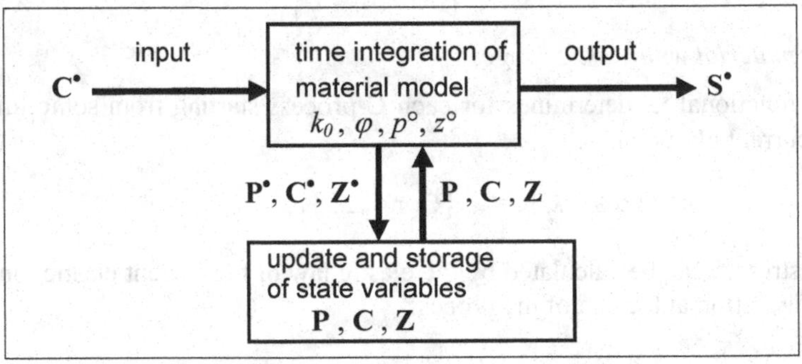

10.2 Thermoplasticity

A Comment on the Literature. The following theory was originally published in BERTRAM (2003). For further study we recommend ACHARYA/ SHAWKI (1996), HAUPT (1993), KRAWIETZ (1986), LUBARDA (2002), MAUGIN (1992), NEGAHBAN (2012), and SVENDSEN (1998).

Under plastic deformations, large amounts of energy are dissipated and primarily turned into heat[132]. We are therefore forced to include the thermodynamics of the process in consideration. For this purpose, we will next enlarge the concepts of the present theory to the thermodynamic variables.

The independent variables of this theory are given by the *thermo-kinematical process*

$$\{\mathbf{C}(\tau),\, \theta(\tau)\,,\, \mathbf{g}_0(\tau)\Big|_{\tau=0}^{t}\}\,,$$

which determines the *caloro-dynamical state* consisting of the stresses, the heat flux, the free energy, and the entropy

$$\{\mathbf{S}(t)\,,\, \mathbf{q}_0(t)\,,\, \psi(t)\,,\, \eta(t)\}\,.$$

The *PISM* is already identically fulfilled by consequently choosing material variables. The natural extension of the concept of an *elastic range* to thermodynamics is as follows.

Definition 10.6. A **thermo-elastic range** is a quintuple

$$\{\mathscr{Ela}_p\,, S_p\,, q_p\,, \psi_p\,, \eta_p\}$$

consisting of

1) a path-connected submanifold with boundary

$$\mathscr{Ela}_p \subset \mathscr{Psym} \times \mathscr{R}^+ \times \mathscr{V}$$

of the space of the thermo-kinematic variables, and

2) a set of thermo-elastic laws (as reduced forms)

$$S_p : \mathscr{Ela}_p \to \mathscr{Sym} \quad | \quad \{\mathbf{C}\,, \theta\,, \mathbf{g}_0\} \mapsto \mathbf{S}$$

$$q_p : \mathscr{Ela}_p \to \mathscr{V} \quad | \quad \{\mathbf{C}\,, \theta\,, \mathbf{g}_0\} \mapsto \mathbf{q}_0$$

[132] This has already been measured by FARREN/ TAYLOR (1925); see also RISTINMAA/ WALLIN/ OTTOSEN (2007).

$$\psi_p : \mathscr{E}\ell a_p \to \mathscr{R} \quad | \quad \{\mathbf{C}, \theta, \mathbf{g}_0\} \mapsto \psi$$

$$\eta_p : \mathscr{E}\ell a_p \to \mathscr{R} \quad | \quad \{\mathbf{C}, \theta, \mathbf{g}_0\} \mapsto \eta$$

that give for all thermo-kinematical processes $\{\mathbf{C}(\tau), \theta(\tau), \mathbf{g}_0(\tau)\Big|_{t_A}^{t}\}$, which are entirely in $\mathscr{E}\ell a_p$, the caloro-dynamic state $\{\mathbf{S}(t), \mathbf{q}_0(t), \psi(t), \eta(t)\}$ as

$$\mathbf{S}(t) = S_p(\mathbf{C}(t), \theta(t), \mathbf{g}_0(t))$$

$$\mathbf{q}_0(t) = q_p(\mathbf{C}(t), \theta(t), \mathbf{g}_0(t))$$

$$\psi(t) = \psi_p(\mathbf{C}(t), \theta(t), \mathbf{g}_0(t))$$

$$\eta(t) = \eta_p(\mathbf{C}(t), \theta(t), \mathbf{g}_0(t))$$

We assume that these functions are continuous and continuously differentiable on $\mathscr{E}\ell a_p$, and as such extensible on $\mathscr{P}sym \times \mathscr{R}^+ \times \mathscr{V}$.

Again, the **isomorphy** of the thermo-elastic ranges after Definition 6.11. plays an important role, and leads to the relations

$$S_p(\mathbf{C}, \theta, \mathbf{g}_0) = \mathbf{P}\, S_0(\mathbf{P}^T \mathbf{C}\, \mathbf{P}, \theta, \mathbf{P}^T \mathbf{g}_0)\, \mathbf{P}^T$$

$$q_p(\mathbf{C}, \theta, \mathbf{g}_0) = \mathbf{P}\, q_0(\mathbf{P}^T \mathbf{C}\, \mathbf{P}, \theta, \mathbf{P}^T \mathbf{g}_0)$$

$$\psi_p(\mathbf{C}, \theta, \mathbf{g}_0) = \psi_0(\mathbf{P}^T \mathbf{C}\, \mathbf{P}, \theta, \mathbf{P}^T \mathbf{g}_0) + \psi_c - \theta\, \eta_c \tag{10.22}$$

$$\eta_p(\mathbf{C}, \theta, \mathbf{g}_0) = \eta_0(\mathbf{P}^T \mathbf{C}\, \mathbf{P}, \theta, \mathbf{P}^T \mathbf{g}_0) + \eta_c$$

for all $(\mathbf{C}, \theta, \mathbf{g}_0) \in \mathscr{P}sym \times \mathscr{R}^+ \times \mathscr{V}$ with $\mathbf{P} \in \mathscr{U}nim^+$. The additive terms ψ_c and η_c do *not* depend on the current thermo-kinematical variables $\mathbf{C}$, θ, and $\mathbf{g}_0$[133].

The boundary $\partial\mathscr{E}\ell a_p$ of $\mathscr{E}\ell a_p$ is again called **yield limit** or **yield surface** of the thermo-elastic range. There is no material known for which the yield limit depends on the temperature gradient, so that $\mathscr{E}\ell a_p$ is trivial in its last component $\mathscr{V}$. We will further on suppress this last component of $\mathscr{E}\ell a_p$, so that $\mathscr{E}\ell a_p \subset \mathscr{P}sym \times \mathscr{R}^+$.

The **yield criterion** of a thermo-elastic range is then a mapping

$$\Phi_p : \mathscr{P}sym \times \mathscr{R}^+ \to \mathscr{R} \quad | \quad \{\mathbf{C}, \theta\} \mapsto \Phi_p(\mathbf{C}, \theta)$$

the kernel of which forms the yield surface

$$\Phi_p(\mathbf{C}, \theta) = 0 \quad \Leftrightarrow \quad \{\mathbf{C}, \theta\} \in \partial\mathscr{E}\ell a_p\,.$$

---

[133] See CASEY (1998).

For the distinction of states in the interior $\mathscr{Elap}^{o}$ and beyond the thermo-elastic range, we demand

$$\Phi_p(\mathbf{C}\,,\theta) < 0 \;\Leftrightarrow\; (\mathbf{C}\,,\theta) \in \mathscr{Elap}^{o}\,.$$

A simple **example** is

$$\Phi_p(\mathbf{C}\,,\theta) = \sigma_C(\mathbf{C}) - \sigma_Y(\theta)$$

with a strain function σ_C and a temperature-dependent yield stress σ_Y.

The **loading condition** now becomes

$$\Phi_p(\mathbf{C}\,,\theta)^{\bullet} = \frac{\partial \Phi_p}{\partial \mathbf{C}} \cdot \mathbf{C}^{\bullet} + \frac{\partial \Phi_p}{\partial \theta}\,\theta^{\bullet} > 0\,.$$

For the general yield criterion of all elastic ranges we use the ansatz with the vector of hardening variables $\mathbf{Z} \in \underline{\mathscr{Lin}}$

$$\Phi_p(\mathbf{C}\,,\theta) = \varphi(\mathbf{P}\,,\mathbf{C}\,,\theta\,,\mathbf{Z})$$

with

$$\varphi : \mathscr{Unim}^{+} \times \mathscr{Psym} \times \mathscr{R}^{+} \times \underline{\mathscr{Lin}} \to \mathscr{R}\,,$$

assumed to be differentiable in all arguments. The yield condition is then

$$\varphi(\mathbf{P}\,,\mathbf{C}\,,\theta\,,\mathbf{Z}) = 0$$

and the loading condition

$$\frac{\partial \varphi}{\partial \mathbf{C}} \cdot \mathbf{C}^{\bullet} + \frac{\partial \varphi}{\partial \theta}\,\theta^{\bullet} > 0\,,$$

which is not the complete time-derivative of φ.

By analogy of the mechanical ansatz (10.12) for the yield and hardening rules we set

$$\mathbf{P}^{\bullet} = p(\mathbf{P}\,,\mathbf{C}\,,\theta\,,\mathbf{g}_0\,,\mathbf{Z}\,,\mathbf{C}^{\bullet},\theta^{\bullet})$$
$$\mathbf{Z}^{\bullet} = z(\mathbf{P}\,,\mathbf{C}\,,\theta\,,\mathbf{g}_0\,,\mathbf{Z}\,,\mathbf{C}^{\bullet},\theta^{\bullet})$$

with two functions

$$p : \mathscr{Unim}^{+} \times \mathscr{Psym} \times \mathscr{R}^{+} \times \mathscr{V} \times \underline{\mathscr{Lin}} \times \mathscr{Sym} \times \mathscr{R} \to \mathscr{Lin}$$
$$z : \mathscr{Unim}^{+} \times \mathscr{Psym} \times \mathscr{R}^{+} \times \mathscr{V} \times \underline{\mathscr{Lin}} \times \mathscr{Sym} \times \mathscr{R} \to \underline{\mathscr{Lin}}\,.$$

We leave it open whether these two rules really depend on the temperature gradient.

We again assume that these two evolution equations are **rate-independent**. Necessary and sufficient condition for this property is that p and z are positive-homogeneous functions of degree 1 in $\mathbf{C}^\bullet$ and $\theta^\bullet$, *i.e.*, for all real numbers $\alpha > 0$ we have

$$p(\mathbf{P}, \mathbf{C}, \theta, \mathbf{g}_0, \mathbf{Z}, \alpha\mathbf{C}^\bullet, \alpha\,\theta^\bullet) = \alpha\, p(\mathbf{P}, \mathbf{C}, \theta, \mathbf{g}_0, \mathbf{Z}, \mathbf{C}^\bullet, \theta^\bullet)$$

$$z(\mathbf{P}, \mathbf{C}, \theta, \mathbf{g}_0, \mathbf{Z}, \alpha\mathbf{C}^\bullet, \alpha\,\theta^\bullet) = \alpha\, z(\mathbf{P}, \mathbf{C}, \theta, \mathbf{g}_0, \mathbf{Z}, \mathbf{C}^\bullet, \theta^\bullet)\,.$$

We can again fulfil the rate-independence identically by only determining the direction of yielding and hardening by an incremental evolution equation, but leaving their values by a non-negative multiplicator λ for the consistency condition (see below)

$$\mathbf{P}^\bullet = \lambda\, p^\circ(\mathbf{P}, \mathbf{C}, \theta, \mathbf{g}_0, \mathbf{Z}, \mathbf{C}^\circ, \theta^\circ)$$

$$\mathbf{Z}^\bullet = \lambda\, z^\circ(\mathbf{P}, \mathbf{C}, \theta, \mathbf{g}_0, \mathbf{Z}, \mathbf{C}^\circ, \theta^\circ)$$

with the normed increments

$$\mathbf{C}^\circ := \mathbf{C}^\bullet/\lambda_0 \qquad \text{and} \qquad \theta^\circ := \theta^\bullet/\lambda_0$$

by an arbitrary positive factor λ_0, such as

$$\lambda_0 := |\mathbf{C}^\bullet| + |\theta^\bullet|/\theta_0$$

with respect to an arbitrarily chosen reference temperature θ_0. Both functions p° and z° have a switcher, which sets the values to zero if both the yield criterion and the loading condition are not simultaneously fulfilled. We introduce the abbreviations for the yield direction

$$\mathbf{P}^\circ := \mathbf{P}^\bullet/\lambda = p^\circ(\mathbf{P}, \mathbf{C}, \theta, \mathbf{g}_0, \mathbf{Z}, \mathbf{C}^\circ, \theta^\circ)$$

and for the hardening direction

$$\mathbf{Z}^\circ := \mathbf{Z}^\bullet/\lambda = z^\circ(\mathbf{P}, \mathbf{C}, \theta, \mathbf{g}_0, \mathbf{Z}, \mathbf{C}^\circ, \theta^\circ)\,. \tag{10.23}$$

As the yield condition must permanently hold during yielding, we obtain the **consistency condition**

$$0 = \varphi(\mathbf{P}, \mathbf{C}, \theta, \mathbf{Z})^\bullet$$

$$= \frac{\partial\varphi}{\partial\mathbf{P}} \cdot \mathbf{P}^\bullet + \frac{\partial\varphi}{\partial\mathbf{C}} \cdot \mathbf{C}^\bullet + \frac{\partial\varphi}{\partial\theta} \cdot \theta^\bullet + \frac{\partial\varphi}{\partial\mathbf{Z}} \cdot \mathbf{Z}^\bullet$$

$$= \frac{\partial\varphi}{\partial\mathbf{C}} \cdot \mathbf{C}^\bullet + \frac{\partial\varphi}{\partial\theta} \cdot \theta^\bullet + \frac{\partial\varphi}{\partial\mathbf{P}} \cdot \lambda\,\mathbf{P}^\circ + \frac{\partial\varphi}{\partial\mathbf{Z}} \cdot \lambda\,\mathbf{Z}^\circ$$

which allows for determining the plastic multiplier as

$$\lambda(\mathbf{P}, \mathbf{C}, \theta, \mathbf{g}_0, \mathbf{Z}, \mathbf{C}^\bullet, \theta^\bullet) = \frac{-\dfrac{\partial\varphi}{\partial\mathbf{C}} \cdot \mathbf{C}^\bullet - \dfrac{\partial\varphi}{\partial\theta}\,\theta^\bullet}{\dfrac{\partial\varphi}{\partial\mathbf{P}} \cdot \mathbf{P}^\circ + \dfrac{\partial\varphi}{\partial\mathbf{Z}} \cdot \mathbf{Z}^\circ}\,.$$

Both its numerator and its denominator are negative for all loadings, and, hence, λ is positive. If the yield condition and the loading condition are not simultaneously fulfilled, then we set λ to zero. Again, the KUHN-TUCKER condition holds. We obtain the **consistent yield** and **hardening rules**

$$\boxed{\begin{aligned} \mathbf{P}^\bullet &= \mathbf{P}^\circ (\mathbf{A} \cdot \mathbf{C}^\bullet + \alpha\, \theta^\bullet) \\ \mathbf{Z}^\bullet &= \mathbf{Z}^\circ (\mathbf{A} \cdot \mathbf{C}^\bullet + \alpha\, \theta^\bullet) \end{aligned}} \tag{10.24}$$

with the symmetric tensor

$$\mathbf{A}(\mathbf{P}, \mathbf{C}, \theta, \mathbf{g}_0, \mathbf{Z}, \mathbf{C}^\circ, \theta^\circ) := \frac{-\dfrac{\partial \varphi}{\partial \mathbf{C}}}{\dfrac{\partial \varphi}{\partial \mathbf{P}} \cdot \mathbf{P}^\circ + \dfrac{\partial \varphi}{\partial \mathbf{Z}} \cdot \mathbf{Z}^\circ} \in \mathscr{Sym}$$

and the scalar

$$\alpha(\mathbf{P}, \mathbf{C}, \theta, \mathbf{g}_0, \mathbf{Z}, \mathbf{C}^\circ, \theta^\circ) := \frac{-\dfrac{\partial \varphi}{\partial \theta}}{\dfrac{\partial \varphi}{\partial \mathbf{P}} \cdot \mathbf{P}^\circ + \dfrac{\partial \varphi}{\partial \mathbf{Z}} \cdot \mathbf{Z}^\circ} \in \mathscr{R}.$$

All the functions therein have to be evaluated at $\{\mathbf{P}, \mathbf{C}, \theta, \mathbf{g}_0, \mathbf{Z}, \mathbf{C}^\circ, \theta^\circ\}$ and all partial derivatives to be taken at $\{\mathbf{P}, \mathbf{C}, \theta, \mathbf{Z}\}$.

The additive constants in the free energy and in the entropy ψ_c and η_c must remain constant during elastic processes because of the assumption of isomorphic thermo-elastic ranges, and thus cannot depend on $\mathbf{C}$, θ, or $\mathbf{g}_0$. They can only depend on those variables which are constant in the elastic ranges like the plastic transformation $\mathbf{P}$ and the hardening variables $\mathbf{Z}$, *i.e.*, $\psi_c(\mathbf{P}, \mathbf{Z})$, $\eta_c(\mathbf{P}, \mathbf{Z})$. We will further on assume that these dependencies are differentiable. Consequently, after (10.22) we must assume

$$\begin{aligned} \psi_p(\mathbf{C}, \theta, \mathbf{g}_0) &= \psi_0(\mathbf{P}^T \mathbf{C}\, \mathbf{P}, \theta, \mathbf{P}^T \mathbf{g}_0) + \psi_c(\mathbf{P}, \mathbf{Z}) - \theta\, \eta_c(\mathbf{P}, \mathbf{Z}) \\ \eta_p(\mathbf{C}, \theta, \mathbf{g}_0) &= \eta_0(\mathbf{P}^T \mathbf{C}\, \mathbf{P}, \theta, \mathbf{P}^T \mathbf{g}_0) + \eta_c(\mathbf{P}, \mathbf{Z}). \end{aligned}$$

In the literature an additive split of the free energy into *elastic* and *plastic* parts is often assumed. In the present context it is a consequence of the isomorphy-condition.

The material time-derivative of the free energy is

$$\begin{aligned} \psi^\bullet &= \psi_0(\mathbf{P}^T \mathbf{C}\, \mathbf{P}, \theta, \mathbf{P}^T \mathbf{g}_0)^\bullet + \psi_c(\mathbf{P}, \mathbf{Z})^\bullet - \theta\, \eta_c(\mathbf{P}, \mathbf{Z})^\bullet - \theta^\bullet\, \eta_c(\mathbf{P}, \mathbf{Z}) \\ &= \frac{\partial \psi_0}{\partial \mathbf{C}} \cdot (\mathbf{P}^T \mathbf{C}\, \mathbf{P})^\bullet + \frac{\partial \psi_0}{\partial \theta} \cdot \theta^\bullet + \frac{\partial \psi_0}{\partial \mathbf{g}_0} \cdot (\mathbf{P}^T \mathbf{g}_0)^\bullet \\ &\quad + \Big(\frac{\partial \psi_c}{\partial \mathbf{P}} - \theta \frac{\partial \eta_c}{\partial \mathbf{P}}\Big) \cdot \mathbf{P}^\bullet + \Big(\frac{\partial \psi_c}{\partial \mathbf{Z}} - \theta \frac{\partial \eta_c}{\partial \mathbf{Z}}\Big) \cdot \mathbf{Z}^\bullet - \theta^\bullet \eta_c \\ &= \mathbf{P} \frac{\partial \psi_0}{\partial \mathbf{C}}\, \mathbf{P}^T \cdot (\mathbf{C}^\bullet + 2\, \mathbf{C}\, \mathbf{P}^\bullet \mathbf{P}^{-1}) + \frac{\partial \psi_0}{\partial \theta} \cdot \theta^\bullet + \frac{\partial \psi_0}{\partial \mathbf{g}_0} \cdot (\mathbf{P}^T \mathbf{g}_0^\bullet + \mathbf{P}^{T\bullet} \mathbf{g}_0) \end{aligned}$$

$$+ \left(\frac{\partial \psi_c}{\partial \mathbf{P}} - \theta \frac{\partial \eta_c}{\partial \mathbf{P}}\right) \cdot \mathbf{P}^\bullet + \left(\frac{\partial \psi_c}{\partial \mathbf{Z}} - \theta \frac{\partial \eta_c}{\partial \mathbf{Z}}\right) \cdot \mathbf{Z}^\bullet - \theta^\bullet \eta_c$$

$$= \mathbf{P} \frac{\partial \psi_0}{\partial \mathbf{C}} \mathbf{P}^T \cdot \mathbf{C}^\bullet + \frac{\partial \psi_0}{\partial \theta} \cdot \theta^\bullet + \mathbf{P} \frac{\partial \psi_0}{\partial \mathbf{g}_0} \cdot \mathbf{g}_0^\bullet$$

$$+ \left(2\, \mathbf{C}\, \mathbf{P} \frac{\partial \psi_0}{\partial \mathbf{C}} + \mathbf{g}_0 \otimes \frac{\partial \psi_0}{\partial \mathbf{g}_0} + \frac{\partial \psi_c}{\partial \mathbf{P}} - \theta \frac{\partial \eta_c}{\partial \mathbf{P}}\right) \cdot \mathbf{P}^\circ (\mathbf{A} \cdot \mathbf{C}^\bullet + \alpha\, \theta^\bullet)$$

$$+ \left(\frac{\partial \psi_c}{\partial \mathbf{Z}} - \theta \frac{\partial \eta_c}{\partial \mathbf{Z}}\right) \cdot \mathbf{Z}^\circ (\mathbf{A} \cdot \mathbf{C}^\bullet + \alpha\, \theta^\bullet) - \theta^\bullet \eta_c .$$

All derivatives of the free energy have to be taken at $\{\mathbf{P}^T \mathbf{C}\, \mathbf{P},\ \theta,\ \mathbf{P}^T \mathbf{g}_0\}$.

The stress power density (10.8) is

$$\begin{aligned} l\, \rho &= \mathbf{T} \cdot \mathbf{L} \\ &= \tfrac{1}{2}\ \mathbf{S} \cdot \mathbf{C}^\bullet \\ &= \tfrac{1}{2}\ S_p(\mathbf{C}, \theta, \mathbf{g}_0) \cdot \mathbf{C}^\bullet \\ &= \tfrac{1}{2}\ \mathbf{P}\, S_0(\mathbf{P}^T \mathbf{C}\, \mathbf{P}, \theta, \mathbf{P}^T \mathbf{g}_0)\ \mathbf{P}^T \cdot \mathbf{C}^\bullet. \end{aligned}$$

We put this into the CLAUSIUS-DUHEM inequality (3.37)

$$0 \geq -l + \frac{\mathbf{q}_0 \cdot \mathbf{g}_0}{\rho_0\, \theta} + \psi^\bullet + \theta^\bullet \eta$$

$$= -\rho^{-1}\ \tfrac{1}{2}\ \mathbf{P}\, S_0(\mathbf{P}^T \mathbf{C}\, \mathbf{P}, \theta, \mathbf{P}^T \mathbf{g})\, \mathbf{P}^T \cdot \mathbf{C}^\bullet$$

$$+ \frac{\mathbf{g}_0}{\rho_0\, \theta} \cdot \mathbf{P}\, q_0(\mathbf{P}^T \mathbf{C}\, \mathbf{P}, \theta, \mathbf{P}^T \mathbf{g}_0) + \mathbf{P} \frac{\partial \psi_0}{\partial \mathbf{C}} \mathbf{P}^T \cdot \mathbf{C}^\bullet$$

$$+ \frac{\partial \psi_0}{\partial \theta} \cdot \theta^\bullet + \mathbf{P} \frac{\partial \psi_0}{\partial \mathbf{g}_0} \cdot \mathbf{g}_0^\bullet + \theta^\bullet \cdot \eta_0(\mathbf{P}^T \mathbf{C}\, \mathbf{P}, \theta, \mathbf{P}^T \mathbf{g}_0)$$

$$+ \left(2\, \mathbf{C}\, \mathbf{P} \frac{\partial \psi_0}{\partial \mathbf{C}} + \mathbf{g}_0 \otimes \frac{\partial \psi_0}{\partial \mathbf{g}_0} + \frac{\partial \psi_c}{\partial \mathbf{P}} - \theta \frac{\partial \eta_c}{\partial \mathbf{P}}\right) \cdot \mathbf{P}^\circ (\mathbf{A} \cdot \mathbf{C}^\bullet + \alpha\, \theta^\bullet)$$

$$+ \left(\frac{\partial \psi_c}{\partial \mathbf{Z}} - \theta \frac{\partial \eta_c}{\partial \mathbf{Z}}\right) \cdot \mathbf{Z}^\circ (\mathbf{A} \cdot \mathbf{C}^\bullet + \alpha\, \theta^\bullet) .$$

Theorem 10.7. *The CLAUSIUS-DUHEM inequality is fulfilled for arbitrary thermo-kinematical processes, if and only if the following conditions hold:*
1) the free energy is a potential for the stresses and the entropy (for the elastic reference range)

$$S_0 = 2\rho \frac{\partial \psi_0}{\partial \mathbf{C}} \qquad \eta_0 = -\frac{\partial \psi_0}{\partial \theta}$$

2) the free energy does not depend on the temperature gradient

$$\frac{\partial \psi_0}{\partial \mathbf{g}_0} = \mathbf{o}$$

3) the heat-conduction-inequality

$$\mathbf{q}_0 \cdot \mathbf{g}_0 \leq 0$$

4) the residual inequality

(10.25)
$$\rho^{-1}\,\mathbf{S}^p \cdot \mathbf{P}^\bullet - \Big(\frac{\partial \psi_c}{\partial \mathbf{P}} - \theta \frac{\partial \eta_c}{\partial \mathbf{P}}\Big) \cdot \mathbf{P}^\bullet - \Big(\frac{\partial \psi_c}{\partial \mathbf{Z}} - \theta \frac{\partial \eta_c}{\partial \mathbf{Z}}\Big) \cdot \mathbf{Z}^\bullet \geq 0$$

Proof. If the yield criterion or the loading conditions are not fulfilled, then $\mathbf{P}^\bullet$ and $\mathbf{Z}^\bullet$ are zero, and in the CLAUSIUS-DUHEM inequality remains only

$$0 \geq -\rho^{-1}\,½\;\mathbf{P}\,S_0(\mathbf{P}^T\mathbf{C}\,\mathbf{P},\,\theta,\,\mathbf{P}^T\mathbf{g})\,\mathbf{P}^T \cdot \mathbf{C}^\bullet$$

$$+ \frac{\mathbf{g}_0}{\rho_0\,\theta} \cdot \mathbf{P}\,q_0(\mathbf{P}^T\mathbf{C}\,\mathbf{P},\,\theta,\,\mathbf{P}^T\mathbf{g}_0) + \mathbf{P}\,\frac{\partial \psi_0}{\partial \mathbf{C}}\,\mathbf{P}^T \cdot \mathbf{C}^\bullet$$

$$+ \frac{\partial \psi_0}{\partial \theta} \cdot \theta^\bullet + \mathbf{P}\,\frac{\partial \psi_0}{\partial \mathbf{g}_0} \cdot \mathbf{g}_0^\bullet$$

$$+ \theta^\bullet \cdot \eta_0(\mathbf{P}^T\mathbf{C}\,\mathbf{P},\,\theta,\,\mathbf{P}^T\mathbf{g}_0)\,.$$

This holds for arbitrary increments, if and only if conditions *1)* to *3)* hold. As the thermo-elastic equations are continuous and continuously differentiable in $\mathscr{E}\ell a_p$, conditions *1)* to *3)* must also hold on the yield-surface. By use of (10.8) there remains the residual inequality; *q. e. d.*

The first three conditions are already known from thermo-elasticity. Note that they must hold for the thermo-elastic reference laws, and are then automatically valid for all isomorphic laws, including the current ones (left sides of (10.22)).

In the residual inequality three terms appear. The first stands for the dissipation of the stress $\mathbf{S}^p$ at the yielding $\mathbf{P}^\bullet$. While one may doubt whether the free energy can really depend upon the plastic transformation, there are clear indications for the hardening to be a dissipative process. If the BAUSCHINGER[134] effect is large, then the first term itself can be negative (*yielding against the stresses*, see

[134] Johann Bauschinger (1834-1893)

PHILLIPS *et al.* 1972 , PHILLIPS 1974, IKEGAMI 1982, BROWN/ CASEY/ NIKKEL 2003). This is then corrected by the other terms.[135]

By (10.23) the residual inequality can be given the equivalent form

$$(2\,\mathbf{C}\,\mathbf{P}\,\frac{\partial\psi_0}{\partial\mathbf{C}}+\frac{\partial\psi_c}{\partial\mathbf{P}}-\theta\,\frac{\partial\eta_c}{\partial\mathbf{P}})\cdot p^\circ(\mathbf{P}\,,\mathbf{C}\,,\theta\,,\mathbf{g}_0\,,\mathbf{Z}\,,\mathbf{C}^\circ,\,\theta^\circ)$$

$$+\,(\frac{\partial\psi_c}{\partial\mathbf{Z}}-\theta\,\frac{\partial\eta_c}{\partial\mathbf{Z}})\cdot z^\circ(\mathbf{P}\,,\mathbf{C}\,,\theta\,,\mathbf{g}_0\,,\mathbf{Z}\,,\mathbf{C}^\circ,\,\theta^\circ)\,\leq\,0$$

which contains all the constitutive functions being restricted by it.

10.3 Viscoplasticity

For many materials the assumption of rate-independence is only a coarse description of the reality. If one measures precisely enough, then one will detect that almost all materials creep (under constant stresses) or relax (under constant strains). Both effects are only possible in rate-dependent materials.

In order to include rate-dependence in our description, different possibilities exist. We will briefly mention just two of them in the sequel without going into detail.

Viscosity Based on Overstress

Classical plasticity theories assume the existence of elastic ranges, and states beyond the yield limit (*i.e.*, outside of these ranges) do not exist. In viscoplasticity, however, one allows also for such states as a result of viscous (over-) stresses[136]. We can include such effects easily in our format by some small modifications.

We will further on assume the existence of elastic ranges, within which all processes are purely elastic. Only during yielding shall it be possible that the state of stress or of strain is beyond the yield limit. Hence, the elastic law k_p must be also defined beyond the yield limit of $\mathscr{Ela}_p$, practically on the whole set $\mathscr{Psym}$. The isomorphy-condition (10.5) is assumed to still hold, so that we can determine

[135] See also the discussion in LUBLINER (1986), KRATOCHVÍL/ DILLON (1970), CLEJA-ŢIGOIU/ SOÓS (1990), and CASEY (1998).

[136] Models based on overstress are described, *e.g.*, in KRAUSZ/ KRAUSZ (1996), and in particular in Chap. 6 by KREMPL (for small deformations).

the stresses at all times, while $\mathbf{C}$ can lie in the interior of $\mathscr{Ela}_p$, on its boundary $\partial\mathscr{Ela}_p$, or elsewhere in $\mathscr{Psym} \backslash \mathscr{Ela}_p$.

The flow rule is now assumed to be rate-dependent by the ansatz

$$\mathbf{P}^{\bullet} = p(\mathbf{P}, \mathbf{C}, \mathbf{Z}, \mathbf{C}^{\bullet}) = \nu(\sigma_O)\, \boldsymbol{D}\, [\mathbf{C}\,(\mathbf{S} - \mathbf{S}_B)]$$

with the inverse **viscosity tetrad** $\nu\,\boldsymbol{D}$, the symmetric back-stress $\mathbf{S}_B$, and the **overstress** (intensity)

$$\sigma_O := \sigma_{CSp}\,(\mathbf{C}(\mathbf{S}' - \mathbf{S}_B)) - \sigma_Y,$$

on which the real-valued function ν depends. It contains a switcher, which switches its value to zero, if σ_O is not positive. A simple example for this ansatz is for positive arguments a NORTON-type creep law

$$\nu(\sigma_O) = \mu\,(\sigma_O / \sigma_{O0})^N$$

and the isotropic tensor

$$\boldsymbol{D} = \boldsymbol{I} - {}^1\!/_3\, \mathbf{I} \otimes \mathbf{I}$$

with three material constants σ_{O0}, μ, and N.

Relaxation Type

A different approach[137] to viscoplasticity is obtained by enlarging the flow and hardening rules (10.13) by rate-dependent relaxation terms

(10.26)
$$\mathbf{P}^{\bullet} = \lambda\, p^{\circ}(\mathbf{P}, \mathbf{C}, \mathbf{Z}, \mathbf{C}^{\circ}) + r(\mathbf{P}, \mathbf{C}, \mathbf{Z})$$
$$\mathbf{Z}^{\bullet} = \lambda\, z^{\circ}(\mathbf{P}, \mathbf{C}, \mathbf{Z}, \mathbf{C}^{\circ}) + s(\mathbf{P}, \mathbf{C}, \mathbf{Z})$$

with two functions

$$r : \mathscr{Inv}^{+} \times \mathscr{Psym} \times \underline{\mathscr{Lin}} \rightarrow \mathscr{Lin}$$
$$s : \mathscr{Inv}^{+} \times \mathscr{Psym} \times \underline{\mathscr{Lin}} \rightarrow \underline{\mathscr{Lin}}\,.$$

These can contribute to yielding and hardening, even if the yield and loading conditions are not fulfilled, and even if currently no deformation occurs and $\mathbf{C}^{\bullet} \equiv \mathbf{0}$.

For thermodynamic generalisation, we should also include the temperature in the list of arguments

$$\mathbf{P}^{\bullet} = \lambda\, p^{\circ}(\mathbf{P}, \mathbf{C}, \theta, \mathbf{g}_O, \mathbf{Z}, \mathbf{C}^{\circ}, \theta^{\circ}) + r(\mathbf{P}, \mathbf{C}, \theta, \mathbf{Z})$$
$$\mathbf{Z}^{\bullet} = \lambda\, z^{\circ}(\mathbf{P}, \mathbf{C}, \theta, \mathbf{g}_O, \mathbf{Z}, \mathbf{C}^{\circ}, \theta^{\circ}) + s(\mathbf{P}, \mathbf{C}, \theta, \mathbf{Z})$$

---

[137] See MANDEL (1973), p. 736, HALPHEN (1975), HALPHEN/ NGUYEN (1975).

with two extended functions

$$r : \mathscr{Inv}^+ \times \mathscr{Psym} \times \mathscr{R}^+ \times \underline{\mathscr{Lin}} \to \mathscr{Lin}$$

$$s : \mathscr{Inv}^+ \times \mathscr{Psym} \times \mathscr{R}^+ \times \underline{\mathscr{Lin}} \to \underline{\mathscr{Lin}} .$$

By the consistency condition we obtain for this ansatz

$$\begin{aligned} 0 &= \varphi(\mathbf{P}, \mathbf{C}, \theta, \mathbf{Z})^{\bullet} \\ &= \frac{\partial \varphi}{\partial \mathbf{C}} \cdot \mathbf{C}^{\bullet} + \frac{\partial \varphi}{\partial \theta} \theta^{\bullet} + \frac{\partial \varphi}{\partial \mathbf{P}} \cdot \lambda \mathbf{P}^{\circ} + \frac{\partial \varphi}{\partial \mathbf{Z}} \cdot \lambda \mathbf{Z}^{\circ} \\ &+ \frac{\partial \varphi}{\partial \mathbf{P}} \cdot r(\mathbf{P}, \mathbf{C}, \theta, \mathbf{Z}) + \frac{\partial \varphi}{\partial \mathbf{Z}} \cdot s(\mathbf{P}, \mathbf{C}, \theta, \mathbf{Z}) , \end{aligned}$$

so that in the solution for λ the two relaxation terms also appear. Still $\lambda = 0$, if not both the yield and the loading conditions are fulfilled simultaneously. If we substitute these rules into the CLAUSIUS-DUHEM inequality (3.37), we obtain the inequality

$$\begin{aligned} 0 \geq & \left(2\,\mathbf{C}\,\mathbf{P}\,\frac{\partial \psi_0}{\partial \mathbf{C}} + \frac{\partial \psi_c}{\partial \mathbf{P}} - \theta\,\frac{\partial \eta_c}{\partial \mathbf{P}}\right) \cdot r(\mathbf{P}, \mathbf{C}, \theta, \mathbf{Z}) \\ & + \left(\frac{\partial \psi_c}{\partial \mathbf{Z}} - \theta\,\frac{\partial \eta_c}{\partial \mathbf{Z}}\right) \cdot s(\mathbf{P}, \mathbf{C}, \theta, \mathbf{Z}) \end{aligned}$$

in addition to the conditions of (10.25). This follows obviously for a constant thermo-kinematical state ($\mathbf{C}^{\bullet} \equiv \mathbf{0}$, $\theta^{\bullet} \equiv 0$).

10.4 Plasticity Theories with Intermediate Placements

Most of the theories of *finite plasticity* in the literature are based on the notion of an *unloaded intermediate placement*. This is, besides the initial or reference placement and the current placement, a third one, which shall be obtained at each instant by a local unloading process. Such a procedure can also be embedded in the present format, as we will show next.

Multiplicative Decomposition

If the material is at an arbitrary instant in a deformation $\mathbf{F}(t)$, then we can think of elastically unloading it by a consecutive deformation, which we call $\mathbf{F}_e^{-1}$. The remaining stress-free deformation is then $\mathbf{F}_p := \mathbf{F}_e^{-1}\,\mathbf{F}(t)$. It is interpreted as a local **intermediate (stress-free) placement.** As this unloading is assumed to be elastic, the path of the unloading and the time it takes are irrelevant, as long as it remains in the current elastic range. With these notions we obtain the **multiplicative decomposition**[138]

(10.27) $$\mathbf{F} = \mathbf{F}_e\, \mathbf{F}_p$$

of the deformation gradient into an **elastic** and a **plastic part**. In general, none of these parts is a gradient of a motion of the body, as stress-free placements for the whole body need not exist. This thought experiment can be made at each instant, so that any **F**-process is associated with an accompanying $\mathbf{F}_e$–process and an $\mathbf{F}_p$–process.

We will now try to describe these notions within our format. First of all, we choose the reference placement, which has been arbitrary until now, such that the elastic reference law is stress-free in it

$$k_0(\mathbf{I}) = \mathbf{0}\,,$$

i.e., for the identification $\mathbf{P}^T\, \mathbf{C}_{up}\, \mathbf{P} \equiv \mathbf{I}$. This should be possible without loss of generality, at least locally. By the isomorphy condition (10.5), we obtain for an arbitrary elastic range $\{\mathscr{E}la_p\,,\, k_p\}$ for this configuration

$$\mathbf{S} = k_p(\mathbf{C}_{up} \equiv \mathbf{P}^{-T}\, \mathbf{P}^{-1}) = \mathbf{P}\, k_0(\mathbf{P}^T\, \mathbf{C}_{up}\, \mathbf{P})\, \mathbf{P}^T = \mathbf{P}\, k_0(\mathbf{I})\, \mathbf{P}^T = \mathbf{0}\,.$$

This means that such $\mathbf{C}_{up}$ describes for each elastic range a stress-free configuration (for all kinds of stress tensors). Note that we do not assume that this configuration is really contained in the particular elastic range $\mathscr{E}la_p$. It may as well lie outside of it, as it is only fictitious. If we interpret it as resulting from some placement $\mathbf{F}_p$, then the latter must solve the equation $\mathbf{C}_{up} = \mathbf{F}_p^{\,T}\, \mathbf{F}_p$. This is done by

$$\mathbf{F}_p \equiv \mathbf{O}\, \mathbf{P}^{-1} \in \mathscr{U}nim^+ \quad \Leftrightarrow \quad \mathbf{P} = \mathbf{F}_p^{\,-1}\, \mathbf{O} \in \mathscr{U}nim^+$$

with some arbitrary rotation $\mathbf{O} \in \mathscr{O}rth^+$. Clearly, if some placement is stress-free for a rate-independent material, then after the *PISM* it remains so under any rotation. Consequently

$$\mathbf{F}_e := \mathbf{F}\, \mathbf{F}_p^{\,-1} = \mathbf{F}\, \mathbf{P}\, \mathbf{O}^T \in \mathscr{I}nv^+$$

is the elastic part of the deformation gradient.

We can now determine the CAUCHY stresses in a particular elastic range $\{\mathscr{E}la_p\,,\, k_p\}$ by the isomorphy-condition (10.5)

$$\begin{aligned} \mathbf{T} &= \mathbf{F}\, \mathbf{S}\, \mathbf{F}^T \\ &= \mathbf{F}\, k_p(\mathbf{C})\, \mathbf{F}^T \end{aligned}$$

---

138 We leave it to the historians to find out, if this decomposition was first suggested by ECKART, KRÖNER, BILBY/ BOULLOGH, BACKMAN, BESSELING, FOX, GREEN/ NAGHDI, GREEN/ TOBOLSKI, LEE/ LIU, MANDEL, SEDOV, STOJANOVITCH/ DJURITCH/ VUJOSHEVITCH, STROH, TEODOSIU, WILLIS, or someone else, see BRUHNS (2015).

$$= \mathbf{F}\,\mathbf{P}\,k_0(\mathbf{P}^T\,\mathbf{C}\,\mathbf{P})\,\mathbf{P}^T\,\mathbf{F}^T$$

$$= \mathbf{F}_e\,\mathbf{O}\,k_0(\mathbf{O}^T\,\mathbf{F}_e^{\,T}\,\mathbf{F}_e\,\mathbf{O})\,\mathbf{O}^T\,\mathbf{F}_e^{\,T}$$

$$= \mathbf{F}_e\,\mathbf{O}\,k_0(\mathbf{O}^T\,\mathbf{C}_e\,\mathbf{O})\,\mathbf{O}^T\,\mathbf{F}_e^{\,T}$$

with

$$\mathbf{C}_e \,:=\, \mathbf{F}_e^{\,T}\,\mathbf{F}_e,$$

which coincides with the previous definition of $\mathbf{C}_e$ in (10.9) only for the case $\mathbf{O} \equiv \mathbf{I}$.

In general, the stresses cannot be determined by this equation, while $\mathbf{O}$ is not specified. There is only one case for which the choice of $\mathbf{O}$ has no influence on the stresses, and this is when the elastic reference law is isotropic. In all other cases an additional assumption on the intermediate placement has to be made.

If the elastic behaviour is isotropic and, thus, k_0 is an isotropic tensor-function, then

$$\mathbf{T} = \mathbf{F}_e\,\mathbf{O}\,k_0(\mathbf{O}^T\,\mathbf{C}_e\,\mathbf{O})\,\mathbf{O}^T\,\mathbf{F}_e^{\,T} = \mathbf{F}_e\,k_0(\mathbf{C}_e)\,\mathbf{F}_e^{\,T}.$$

If we, moreover, use hyperelastic ranges and a hyperelastic reference potential

$$k_0(\mathbf{C}_e) = 2\rho\,\frac{dw_0}{d\mathbf{C}_e}$$

then

$$\mathbf{T} = 2\rho\,\mathbf{F}_e\,\frac{dw_0}{d\mathbf{C}_e}\,\mathbf{F}_e^{\,T}.$$

As $\mathbf{P}$ is unimodular, then so is $\mathbf{F}_p$, and therefore

$$J = det(\mathbf{F}) = det(\mathbf{F}_e)$$

and

$$\mathbf{T} = \frac{2\rho_0}{det(\mathbf{F}_e)}\,\mathbf{F}_e\,\frac{dw_0}{d\mathbf{C}_e}\,\mathbf{F}_e^{\,T}.$$

This is equation (18) of the historical paper of LEE[139] (1969). LEE chose $\mathbf{F}_e$ as symmetric (Eq. 22), which can always be achieved by an appropriate choice of $\mathbf{O}$. Namely, by the polar decomposition of $\mathbf{F}_e$ we can choose $\mathbf{O}^T$ equal to its orthogonal part, and $\mathbf{F}_e$ turns out symmetric. Then $\mathbf{O}$ and with it also $\mathbf{F}_p$ and the intermediate placement become time-dependent, even during elastic processes.

If we instead choose $\mathbf{O} \equiv \mathbf{I}$ and calculate the 2nd PIOLA-KIRCHHOFF tensor with respect to the intermediate placement for the general anisotropic case

---

[139] Erastus Henry Lee (1916-2006)

$$\mathbf{T}_e^{\mathrm{2PK}} := det(\mathbf{F}_e)\, \mathbf{F}_e^{-1}\, \mathbf{T}\, \mathbf{F}_e^{-T} = J\, k_0(\mathbf{C}_e) = 2\rho_0 \frac{dw_0}{d\mathbf{C}_e}$$

or **MANDEL's stress tensor** with respect to the intermediate placement

$$\mathbf{T}_e^{\mathrm{M}} := \mathbf{C}_e\, \mathbf{T}_e^{\mathrm{2PK}} = det(\mathbf{F}_e)\, \mathbf{F}_e^{T}\, \mathbf{T}\, \mathbf{F}_e^{-T} = J\, \mathbf{C}_e\, k_0(\mathbf{C}_e)$$

$$= \frac{\rho_0}{\rho}\, \mathbf{P}^T\, \mathbf{C}\mathbf{S}\, \mathbf{P}^{-T},$$

then this corresponds to the suggestion of TEODOSIU (1970) and MANDEL (1971) of an **isoclinic placement**[140]. This placement is also time-dependent and has the property, that at any times the axes of anisotropy coincide in it. It is fixed during elastic processes, in contrast to LEE's intermediate placement, and the hyperelastic law w_0 and, in particular, its symmetry group are identical for all elastic ranges (like our reference law).

In all of these theories, the isomorphy condition is also fulfilled, even if it is not so clearly stated as in Assumption 10.3.

It shall be noted that the multiplicative decomposition can also be postulated in the reverse order as

$$\mathbf{F} = \underline{\mathbf{F}}_p\, \mathbf{F}_e \qquad \text{with } \underline{\mathbf{F}}_p := \mathbf{F}_e\, \mathbf{F}_p\, \mathbf{F}_e^{-1} \in \mathscr{Unim}^+.$$

Thus, the order of the multiplicative decomposition is by no means natural. If one would consider the **F**-process as a sequence of elastic and plastic segments, then

$$\mathbf{F} = \mathbf{F}_{e_n} \dots \mathbf{F}_{p_2}\, \mathbf{F}_{e_1}\, \mathbf{F}_{p_1},$$

would better correspond to the real process. However, all the $\mathbf{F}_{p_i}$ and $\mathbf{F}_{e_i}$ would not be appropriate internal variables, as they include much more information than actually needed to determine the state of the material[141].

Additive Decomposition

A different approach to finite plasticity was suggested by BACKMAN (1964) who determined the GREEN strains with respect to the intermediate placement ("elastic strains"). These can be expressed by the total strains and those of the intermediate placement with respect to the reference placement ("plastic strains") in a non-trivial way. The two leading linear terms are the basis of the suggestion of GREEN/ NAGHDI (1965, 1971). Here, a **plastic strain tensor** $\mathbf{E}_p \in \mathscr{Sym}$ is introduced as a primitive concept, and the elastic strain tensor is then its the difference with the total GREEN strain tensor

---

[140] V. d. GIESSEN (1989) and BESSELING/v. d. GIESSEN (1994) call it the *natural reference state.*

[141] See CLIFTON (1972), NEMAT-NASSER (1979), LUBARDA (1999), DAVISON (1995).

$$\mathbf{E}_e := \mathbf{E}^G - \mathbf{E}_p \in \mathscr{S}ym\,.$$

This leads to the **additive decomposition** of the strains

$$\mathbf{E}^G = \mathbf{E}_e + \mathbf{E}_p \tag{10.28}$$

and of the strain rates

$$\mathbf{E}^{G\bullet} = \mathbf{E}_e^{\bullet} + \mathbf{E}_p^{\bullet}.$$

This decomposition commutes, in contrast to the multiplicative one.

It is furthermore assumed that the stresses are given by a function of both parts

$$\mathbf{T}^{2PK} = g(\mathbf{E}_e\,, \mathbf{E}_p)\,.$$

We will next try to identify these concepts with the previous ones. We choose again $\mathbf{F}_p := \mathbf{P}^{-1}$ and

$$\mathbf{E}_p := \tfrac{1}{2}\,(\mathbf{F}_p^{\,T}\,\mathbf{F}_p - \mathbf{I}) = \tfrac{1}{2}\,(\mathbf{P}^{-T}\mathbf{P}^{-1} - \mathbf{I})\,.$$

Then we obtain

$$\begin{aligned}\mathbf{E}_e &= \mathbf{E}^G - \mathbf{E}_p\\ &= \tfrac{1}{2}\,(\mathbf{F}^T\,\mathbf{F} - \mathbf{I}) - \tfrac{1}{2}(\mathbf{F}_p^{\,T}\,\mathbf{F}_p - \mathbf{I})\\ &= \tfrac{1}{2}\,(\mathbf{F}_p^{\,T}\,\mathbf{C}_e\,\mathbf{F}_p - \mathbf{F}_p^{\,T}\,\mathbf{F}_p)\\ &= \mathbf{F}_p^{\,T}\,\tfrac{1}{2}\,(\mathbf{C}_e - \mathbf{I})\,\mathbf{F}_p\,.\end{aligned}$$

While $\mathbf{E}_p$ looks like a GREEN´s strain tensor, this does not apply to $\mathbf{E}_e$. If we push the sum forward into the intermediate placement, we get

$$\mathbf{F}_p^{\,-T}\,\mathbf{E}^G\,\mathbf{F}_p^{\,-1} = \tfrac{1}{2}\,(\mathbf{C}_e - \mathbf{I}) + \tfrac{1}{2}\,(\mathbf{I} - \mathbf{F}_p^{\,-T}\,\mathbf{F}_p^{\,-1})$$

wherein the last term is an ALMANSI-type tensor with respect to the intermediate placement. Then

$$\mathbf{C}_e = 2\,\mathbf{F}_p^{\,-T}\,\mathbf{E}_e\,\mathbf{F}_p^{\,-1} + \mathbf{I}$$

and by (10.26)

$$\mathbf{C} = 2\,\mathbf{E}^G + \mathbf{I} = 2\,\mathbf{E}_e + 2\,\mathbf{E}_p + \mathbf{I}\,.$$

For the 2nd PIOLA-KIRCHHOFF tensor in the reference placement we get

$$\begin{aligned}\mathbf{T}^{2PK} &= J\,k_p(\mathbf{C}) = J\,\mathbf{P}\,k_0(\mathbf{P}^T\,\mathbf{C}\,\mathbf{P})\,\mathbf{P}^T\\ &= J\,\mathbf{F}_p^{\,-1}\,k_0(\mathbf{F}_p^{\,-T}\,(2\mathbf{E}_e + 2\mathbf{E}_p + \mathbf{I})\,\mathbf{F}_p^{\,-1})\,\mathbf{F}_p^{\,-T},\end{aligned}$$

which can be understood as a function

$$\mathbf{T}^{2PK} = g(\mathbf{F}_p\,, \mathbf{E}_e)\,.$$

Once again we see that, in general, a symmetric plastic variable is not sufficient.

It is interesting to note that in none of these cases did we make use of the fact that the intermediate placement is stress-free. It is not unique and neither does it become so by this assumption. Nor is the used identification $\mathbf{F}_p = \mathbf{O}\,\mathbf{P}^{-1}$ unique.

In ITSKOV (2004) a comparison between these two types of decomposition is made, specified for a simple shear deformation.

Next let us consider the observer-dependence of these concepts. It is usual, but by no means compulsory, to introduce the reference placement as invariant and the current placement as objective under changes of observers. This excludes the possibility to choose the placement at some instant (like the initial one) as a reference placement for all observers. As a result of such an introduction, the deformation gradient transforms after (6.16) $\mathbf{F}^*(\mathbf{x}_0^*, t) = \mathbf{Q}(t)\,\mathbf{F}(\mathbf{x}_0, t)$. For the intermediate placement we can also choose in two ways. If we introduce them as invariant fields, then

$$\mathbf{F}_p^* = \mathbf{F}_p \qquad \text{and} \qquad \mathbf{F}_e^* = \mathbf{Q}\,\mathbf{F}_e\,.$$

If we introduce them as objective fields, then

$$\mathbf{F}_p^* = \mathbf{Q}\,\mathbf{F}_p \qquad \text{and} \qquad \mathbf{F}_e^* = \mathbf{Q}\,\mathbf{F}_e\,\mathbf{Q}^T.$$

These two choices are surely not the only ones. SIDOROFF (1970) and GREEN/ NAGHDI (1971) Eq. 35 postulated "objectivity" for the multiplicative decomposition with respect to two arbitrary and time-dependent rotations $\mathbf{Q}$, $\underline{\mathbf{Q}}$ $\in \mathcal{O}rth$ as

$$\mathbf{F}_p^* = \underline{\mathbf{Q}}\,\mathbf{F}_p \qquad \text{and} \qquad \mathbf{F}_e^* = \mathbf{Q}\,\mathbf{F}_e\,\underline{\mathbf{Q}}^T.$$

This postulate, however, turns out to be too restrictive, as such a rotational invariance leads to only isotropic elastic laws.[142]

The requirement of objectivity has led to a long controversy between the schools of multiplicative and additive decompositions[143]. We just want to state that both concepts can be brought into an objective or invariant form. And there are different possibilities to do so, as we already saw. This ambiguity is due to the fact that none of these decomposition-concepts are based on clear definitions, but were invented rather intuitively, and still contain conceptual deficiencies (see below).

The present theory of plasticity is based on both mathematically and physically clear definitions and does not leave space for such controversies. The variables $\mathbf{S}$, $\mathbf{C}$, $\mathbf{P}$, and $\mathbf{Z}$ have been introduced as invariant under change of observer, and, consequently, so are the constitutive functions k_p , k_0 , p , z , φ , etc. The entire model is invariant and identically fulfils the *PMO*, *PISM*, and *PFI*.

While we assumed the existence of one single flow rule of the form (10.12) for the complete rate of the plastic transformation (*i.e.*, for its symmetric *and* its skew

[142] A similar fallacy is given in a more general context in RIVLIN (2002); see also TSAKMAKIS (2004), and BERTRAM/ SVENDSEN (2004).
[143] See NAGHDI (1990), p. 326 ff.

part), in the literature a decomposition is often favoured, firstly of the velocity gradient

$$\mathbf{L} = \mathbf{F}^\bullet \mathbf{F}^{-1} = (\mathbf{F}_e \mathbf{F}_p)^\bullet \mathbf{F}_p^{-1} \mathbf{F}_e^{-1} = \mathbf{L}_e + \mathbf{L}_p \qquad \in \mathcal{Lin}$$

into an elastic

$$\mathbf{L}_e := \mathbf{F}_e^\bullet \mathbf{F}_e^{-1} \qquad \in \mathcal{Lin}$$

and a plastic part

$$\mathbf{L}_p := \mathbf{F}_e \mathbf{F}_p^\bullet \mathbf{F}_p^{-1} \mathbf{F}_e^{-1} \qquad \in \mathcal{Lin}.$$

Neither of them is in general a gradient field. Each part can be decomposed into symmetric and skew parts

$$\mathbf{L}_e = \mathbf{D}_e + \mathbf{W}_e \qquad \mathbf{D}_e \in \mathcal{Sym}, \mathbf{W}_e \in \mathcal{Skw}$$

$$\mathbf{L}_p = \mathbf{D}_p + \mathbf{W}_p \qquad \mathbf{D}_p \in \mathcal{Sym}, \mathbf{W}_p \in \mathcal{Skw}.$$

For the symmetric and the skew parts separate evolution equations are suggested. For $\mathbf{D}_p$, the *rate of plastic distortion* (MANDEL 1982), a plastic potential in the CAUCHY stresses is postulated

$$\mathbf{D}_p = \lambda_1 \frac{d\pi}{d\mathbf{T}} \qquad \in \mathcal{Sym}.$$

For $\mathbf{W}_p$, the **plastic spin**, the existence of which is even denied by some authors (which is reasonable only in the isotropic case by an appropriate choice of $\mathbf{O}$), DAFALIAS (1983, 1984,1985, 1990, 1998)[144] suggested a form

$$\mathbf{W}_p = \lambda_2\, \omega(\mathbf{T}, \mathbf{T}_B) \qquad \in \mathcal{Skw}$$

with a material function ω of the CAUCHY stresses $\mathbf{T}$ and the CAUCHY back-stress

$$\mathbf{T}_B = \mathbf{F}\,\mathbf{S}_B\,\mathbf{F}^T \qquad \in \mathcal{Sym}.$$

For the symmetric part $\mathbf{D}_e$ of the elastic rate $\mathbf{L}_e$ we can use the elastic law, while its skew part $\mathbf{W}_e$ can be eliminated by an appropriate choice of $\mathbf{O}$ (but not simultaneously with $\mathbf{W}_p$).

With the above identifications we obtain

$$\mathbf{L}_p = \mathbf{F}\,\mathbf{P}\,\mathbf{O}^T (\mathbf{O}\,\mathbf{P}^{-1})^\bullet\, \mathbf{F}^{-1}$$

$$= \mathbf{F}\,\mathbf{P}\,\mathbf{O}^T \mathbf{O}^\bullet\, \mathbf{P}^{-1} \mathbf{F}^{-1} - \mathbf{F}\,\mathbf{P}^\bullet\, \mathbf{P}^{-1} \mathbf{F}^{-1}$$

and

[144] See also AIFANTIS (1987), KHAN/ HUANG (1995) p. 261, LUBARDA/ SHIH (1994), v. d. GIESSEN (1991), NGUYEN/ RANIECKY (1999), TEODOSIU (1989), TONG/ TAO/ JIANG (2004).

$$\mathbf{L}_e = \mathbf{F}^\bullet \mathbf{F}^{-1} - \mathbf{L}_p .$$

The first term of $\mathbf{L}_p$ is the *push forward* of the skew spin $\mathbf{O}^T \mathbf{O}^\bullet \in \mathcal{S}kew$, which vanishes for $\mathbf{O} \equiv \mathbf{I}$, while the other term is the *push forward* of $\mathbf{P}^\bullet \mathbf{P}^{-1}$ from the reference placement into the current placement, given by the flow rule p, which can also give non-symmetric values.

The first term in the residual inequality (10.25) is then

$$\mathbf{S}^p \cdot \mathbf{P}^\bullet = -\mathbf{CS} \cdot \mathbf{P}^\bullet \mathbf{P}^{-1} = -J^{-1}\, \mathbf{T}_e^{\mathrm{M}} \cdot \mathbf{P}^\bullet \mathbf{P}^{-1}$$

$$= \mathbf{T} \cdot \mathbf{L}_p = \mathbf{T} \cdot \mathbf{D}_p .$$

The plastic spin does not directly contribute to the dissipation, but, of course, indirectly it does have influence on the further energetic behaviour of the material.

Conclusions

As we already mentioned before, one of the fundamental problems in finite plasticity theory is the introduction of appropriate internal variables and, in particular, that of some plastic or inelastic strain measure. Although the term *plastic deformation* seems to be rather customary in engineering, it turns out in the context of large deformations that it is extremely difficult to give it a precise definition. However, if this is not accurately done, it leaves space for controversial interpretations. Or, as W. NOLL (1974) writes, "*one cannot derive clear-cut conclusions from ill-formulated hypotheses*".

On the other hand, one can argue that the introduction of *plastic deformation* is neither necessary nor appropriate for plasticity. In fact, in the present theory there is no concept like *plastic* or *elastic deformation.* We only distinguished between elastic *states* and *states* of yielding. GILMAN (1960 p. 99) stated already: "It seems very unfortunate to me that the theory of plasticity was ever cast into the mould of stress-strain relations because 'strain' in the plastic case has no physical meaning that is related to the material of the body in question. It is rather like trying to deduce some properties of a liquid from the shape of the container that holds it." Moreover, the notion of *elastic* or *plastic deformations* or *strains* is misleading, because none of them is related to motion in a compatible way. Such expressions provoke the impression that such plastic or elastic 'strains' were kinematical concepts, which is principally false[145].

The concept of *unloaded states* or *intermediate configurations* in connection with the multiplicative decomposition of the deformation gradient has some well known shortcomings[146].

---

[145] See CLEJA-ŢIGOIU/ SOÓS 1990, p. 147.

[146] See CASEY/ NAGHDI (1980, 1992), NAGHDI (1990), CLEJA-ŢIGOIU/ SOÓS (1990), VIDOLI/ SCIARRA (2002), XIAO/ BRUHNS/ MEYERS (2006).

1. Rotations of such unloaded placements remain undetermined or arbitrary[147].
2. The unloaded state becomes fictitious in the presence of a strong BAUSCHINGER effect, which shifts the unloaded state outside of the current elastic range. Such behaviour has been experimentally observed[148].
3. Problems with objectivity and symmetry-transformations, which are somehow linked to 1[149].
4. Lack of a unified format to describe both isotropic and anisotropic behaviour. If we look at KHAN/ HUANG (1995) or at LUBARDA (2001), where both aspects are considered, we get the impression that such a unified theory is still not available there.

Some theories are limited to isotropic behaviour[150]. These are in particular the ones which make use of a *symmetric* plastic variable like LEE (1969), GREEN/ NAGHDI (1965, 1978), DASHNER (1986b), CASEY/ NAGHDI (1992), XIAO *et al.* (2000), and many others[151], although this restriction is quite often not explicitly stated or even recognised.

Other theories are made for anisotropic materials or crystal plasticity (see the next chapter), like those of MANDEL (1966, 1971, 1973, 1974), RICE (1971), KRAWIETZ (1986), NAGHDI/ SRINIVASA (1994), RUBIN (1994, 1996), but use concepts which are in turn not directly applicable in the isotropic case.

The second defect was overcome by a suggestion of CASEY/ NAGHDI (1992) and CASEY (1998), who introduced the concept of the *state of maximal unloading* in the deformation space to define plastic deformations[152]. However, this definition results also in a symmetric plastic variable, which is not sufficient for anisotropic materials.

While also most of the other shortcomings can be somehow overcome, the first cannot, in principle. Rotations are important, especially for anisotropic materials. But the concept of unloading does not determine such rotations. Because of this

---

[147] See SIDOROFF (1973), HAUPT (1985, 2000).

[148] See IVEY (1961), PHILLIPS *et al.* (1972), PHILLIPS (1974), IKEGAMI (1982).

[149] See GREEN/ NAGHDI (1971), ŠILHAVÝ (1977), DASHNER (1986).

[150] See the discussion in CLEJA-ŢIGOIU/ SOÓS (1990).

[151] In MIEHE (1998) a symmetric plastic variable is also introduced (*plastic metric*), complemented by a symmetric structural tensor for the inclusion of anisotropy. See also MENZEL/ STEINMANN (2003) and VLADIMIROV *et al.* (2010).

[152] See also WEISSMAN/ SACKMAN (2011).

fundamental deficiency the multiplicative decomposition does *not* give a "sound conceptual basis" for finite plasticity, as some authors wishfully claim[153].

MANDEL´s remedy to determine the rotations was the notion of an *isoclinic configuration, i.e.*, a time-dependent local reference placement with respect to which the elastic laws of all elastic ranges coincide[154]. This idea already comes quite close to the present suggestion of *isomorphic elastic ranges*. A similar idea stands behind the *state of relaxed strains* of ECKART (1948), the *equilibrium natural state* of HOLSAPPLE (1973), the *(local geometric) natural reference state* of BESSELING (1968), and BESSELING/ v. d. GIESSEN (1994), the *local current relaxed configuration* of CLEJA-ŢIGOIU (1990), the *natural configuration* of RAJAGOPAL/ SRINIVASA (1998), and *plastic indifference* of MIELKE (2003).

If we consequently follow MANDEL´s direction, we will arrive at the concept of *material isomorphisms* between elastic ranges, which simply states that the elastic properties, if properly identified, are not affected by yielding. This concept is strong enough, so that the notions of *unloaded states* or *intermediate configurations* or *plastic strain* are not needed anymore and, thus, have been entirely removed from the present theory. It leads to the introduction of a (non-symmetric) 2nd-order tensor as the fundamental plastic variable, called *plastic transformation*. Its mathematical properties can be easily determined. One of them is that it is only unique up to symmetry transformations. However, this non-uniqueness is natural, as it has no physically detectable influence on the resulting structure, as we already saw.

The isomorphy of the elastic ranges is fulfilled by almost all of the competing theories of finite plasticity, although not always stated in a clear and precise form[155]. However, it must remain clear that it is a strong constitutive assumption, which some materials fulfil, while others do not.

It should be noted here that the isomorphy condition also essentially holds for the theories of BESSELING (1968), LEE/ LIU (1967), LEE (1969), FOX (1968), RICE (1971), MANDEL (1971, 1973, 1974), LEE/ GERMAIN (1974), HALPHEN (1975), ANAND (1985), DASHNER (1986b), CLEJA-ŢIGOIU/ SOOS (1990), HACKL (1997), XIAO *et al.* (2000), NEMAT-NASSER (2004),

[153] See SANSOUR *et al.* (2006), HEIDARI et a. (2009), VLADIMIROV *et al.* (2010).

[154] MANDEL (1972, 1973, 1974); see also TEODOSIU (1970). For the multiplicative decomposition within crystal plasticity see also REINA/ CONTI (2014).

[155] As an exception, CLEJA-ŢIGOIU/ SOÓS (1990) do state such an assumption under the label of *temporal invariance*. See also LEE/ GERMAIN (1974) p. 122, 126, RICE (1971) p. 450, and MIELKE (2003).

and many others, in contrast to KRAWIETZ (1981, 1986), who suggested a more general concept.[156]

In BÖHLKE/ BERTRAM/ KREMPL (2003) a class of materials is investigated, for which the texture of the polycrystal has considerable influence on the elastic properties (*induced elastic anisotropy*). In such cases the elastic ranges are *not* isomorphic. However, the isomorphy still is a helpful concept and has been generalised there to describe such behaviour.

Once we have established the framework for elasto-plastic materials with isomorphic elastic ranges, the main open problem is to specify the hardening parameters and the hardening rule. Isotropic and kinematic hardening are just two choices, which are by no means sufficient to describe the experimentally well known effects of distortional hardening. And the examples of hardening rules given before, are hardly sufficient to describe the real behaviour of any specific material undergoing large deformations like, *e.g.*, in metal forming. To solve this problem will perhaps be the main challenge for researchers in plasticity during the next decades.

[156] See also SCHEIDLER/ WRIGHT (2001) and SVENDSEN (2001).

10.5 Crystal Plasticity

A Comment on the Literature. Apart from classical treatises of crystal plasticity like SCHMID/ BOAS (1935) and TAYLOR (1938), there are already several text-books and overview articles available with introductions to crystal plasticity, like ASARO (1983), FOLLANSBEE/ DAEHN (1991), HAVNER (1992), HOSFORD (1993), YANG/ LEE (1993), BASSANI (1994), KHAN/ HUANG (1995), KOCKS/ TOMÉ/ WENK (1998), LOVE/ ROLETT/ LUBARDA (2002), NEMAT-NASSER (2004), RAABE *et al.* (2004), GURTIN/ FRIED/ ANAND (2013).

Since the beginning of the 20th century, the micro-physical mechanisms that lead to permanent deformations in single crystals, were the subject of intensive investigations. The impact came from the application of x-rays to crystals (v. LAUE[157] since 1912), which were discovered in 1895 by RÖNTGEN[158]. Since the 1920s it has been proven that such deformations result from sliding of atoms along certain slip system, during which the crystal structure is preserved (TAYLOR/ ELAM, SCHMID, BURGERS, POLANYI, and others). These *slip planes* and *slip directions* coincide with those of the densest atomic packing. In the 1930s it was discovered that not entire atomic layers slip, but only dislocations in them, which is now well known as *dislocation glide.* Since then also other mechanisms of permanent deformations have been discovered, like diffusion (in particular at high temperatures), dislocation climb, phase changes, or twinning, etc., which are not necessarily linked to crystallographic slip systems. Nevertheless, slip system mechanisms are dominant under many practically relevant conditions.

Therefore, we will briefly describe the **slip system theory**. It is based on the following hypotheses.

(H1) The crystal consists – in spite of dislocations – of a regular crystal *lattice*, which remains existent even under arbitrarily large deformations.

(H2) Again we have *elastic ranges*, in which the stresses are related to the deformations of the lattice alone.

(H3) There exist crystallographic *slip systems*, constituted by slip planes and slip directions, in which inelastic deformations take place as simple shears.

(H4) Slip in a slip system can only be activated, if the shear stress in it reaches a critical value, called *critical resolved shear stress* (SCHMID´s law[159] 1924).

---

[157] Max von Laue (1879-1960)
[158] Wilhelm Conrad Röntgen (1845-1923)
[159] Erich Schmid (1896-1983)

(H5) The initial critical resolved shear stress can be considered as a material constant. However, it is subjected to hardening due to slips in the same slip system (self-hardening) as well as in the other slip systems (latent or cross hardening).

We will next translate these hypotheses into the context of the present theory of elasto-plastic materials.

After assumption *H2* we expect an anisotropic elastic law k_p in the elastic ranges, the symmetry group of which is determined by the lattice structure of the crystals, like rhombic, tetragonal, cubic, hexagonal, etc. (see Chap. 6.5).

As the stresses depend on the deformations of the lattices after *H2,* which is preserved even under very large deformations after *H1,* we again assume the isomorphy of the elastic law with respect to an elastic reference law (10.5). If the elastic ranges are small enough to allow for a linearisation of the elastic law (10.7), then it is natural to choose a crystallographic vector basis or a **lattice basis** $\{\mathbf{c}_{pi}\}$ and its dual $\{\mathbf{c}_p^{\,i}\}$ in the reference placement for the representations of

$$\boldsymbol{K}_p = \mathbf{P} * \boldsymbol{K}_0 = K^{ijkl}\ \mathbf{c}_{pi} \otimes \mathbf{c}_{pj} \otimes \mathbf{c}_{pk} \otimes \mathbf{c}_{pl}$$

and

$$\mathbf{C}_{up} = C_{ij}\ \mathbf{c}_p^{\,i} \otimes \mathbf{c}_p^{\,j}$$

and a second basis $\{\mathbf{c}_{0i} = \mathbf{P}^{-1}\,\mathbf{c}_{pi}\}$ with dual $\{\mathbf{c}_0^{\,i} = \mathbf{P}^T\,\mathbf{c}_p^{\,i}\}$ in the reference range. Note that it is generally not possible to have ONBs for both bases, unless the plastic transformation is orthogonal by chance. With respect to these, the elastic tensors have the same components

$$\boldsymbol{K}_0 = K^{ijkl}\ \mathbf{c}_{0i} \otimes \mathbf{c}_{0j} \otimes \mathbf{c}_{0k} \otimes \mathbf{c}_{0l}$$

and

$$\mathbf{C}_{u0} = \mathbf{P}^T * \mathbf{C}_{up} = C_{ij}\ \mathbf{c}_0^{\,i} \otimes \mathbf{c}_0^{\,j}.$$

Thus, the basis $\{\mathbf{c}_{0i}\}$ plays the same role for the elastic reference law k_0 as $\{\mathbf{c}_{pi}\}$ for the current elastic law k_p. In this sense, the elastic law or its anisotropy axes are *crystallographic.* $\{\mathbf{c}_{0i}\}$ and $\{\mathbf{c}_0^{\,i}\}$ are time-independent, while $\{\mathbf{c}_{pi}\}$ and $\{\mathbf{c}_p^{\,i}\}$ are time-dependent during yielding. For their time-derivatives we obtain

$$\mathbf{c}_{pi}^{\bullet} = \mathbf{P}^{\bullet}\,\mathbf{c}_{0i} = \mathbf{P}^{\bullet}\,\mathbf{P}^{-1}\,\mathbf{c}_{pi}\,.$$

Their push forward into the space are the spatial or EULERean lattice vectors

$$\mathbf{c}_{Ei}(t) \ := \ \mathbf{F}(t)\,\mathbf{c}_{pi}(t)\,.$$

If we choose for the elastic reference range the initial range at a time $t \equiv 0$ (which is not compulsory), then its initial values are

$$\mathbf{c}_{Ei}(0) \ := \ \mathbf{F}(0)\,\mathbf{c}_{0i}\,.$$

In the reference placement, only the vectors $\mathbf{c}_{0i}(0)$ are constant with respect to time (the reference lattice). In contrast to them, the current lattice vectors $\mathbf{c}_{pi}(t)$

are not material vectors, as they are even in the reference placement time-dependent during yielding.

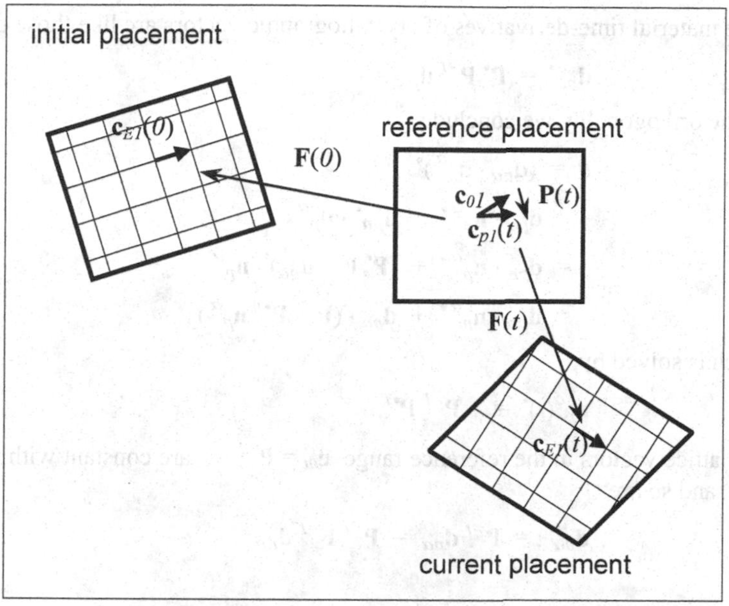

After *H3* we define a **slip system** in the current placement with index α as a pair $\{\mathbf{d}_{E\alpha}, \mathbf{n}_E{}^{\alpha}\}$ with

$\mathbf{d}_{E\alpha}$ the vector of the **slip direction** in the current placement,

$\mathbf{n}_E{}^{\alpha}$ the normal vector in the current placement being orthogonal to the **slip plane.**

The suffix E stands for EULERean representation. For many slip systems the sense of direction of these vectors is not important, *i.e.*, $\{\mathbf{d}_{E\alpha}, \mathbf{n}_E{}^{\alpha}\}$ describes the same slip system as $\{-\mathbf{d}_{E\alpha}, \mathbf{n}_E{}^{\alpha}\}$ or $\{-\mathbf{d}_{E\alpha}, -\mathbf{n}_E{}^{\alpha}\}$. As the slip direction must principally lie in the slip plane, we assume the orthogonality

$$\mathbf{d}_{E\alpha} \cdot \mathbf{n}_E{}^{\alpha} = 0 \,.$$

In order to obtain invariant representations, we pull the spatial or EULERean slip system vectors back to the reference placement. If these vectors are fixed to the lattice or crystallographic, then they transform like lattice vectors

$$\mathbf{d}_{p\alpha} := \mathbf{F}^{-1}\, \mathbf{d}_{E\alpha} \,,$$

i.e., like tangent vectors, and the normal vectors

$$\mathbf{n}_p{}^{\alpha} := \mathbf{F}^{T}\, \mathbf{n}_E{}^{\alpha}$$

like gradient vectors. These transformations preserve the orthogonality

$$\mathbf{d}_{p\alpha} \cdot \mathbf{n}_p{}^{\alpha} = 0 .$$

However, they are not normalised with respect to the EUCLIDean metric.

The material time-derivatives of crystallographic vectors are like those of $\mathbf{c}_{pi}$

$$\mathbf{d}_{p\alpha}{}^{\bullet} = \mathbf{P}^{\bullet}\, \mathbf{P}^{-1}\, \mathbf{d}_{p\alpha} .$$

By the orthogonality we conclude

$$\begin{aligned} 0 &= (\mathbf{d}_{p\alpha} \cdot \mathbf{n}_p{}^{\alpha})^{\bullet} \\ &= \mathbf{d}_{p\alpha} \cdot \mathbf{n}_p{}^{\alpha\bullet} + \mathbf{d}_{p\alpha}{}^{\bullet} \cdot \mathbf{n}_p{}^{\alpha} \\ &= \mathbf{d}_{p\alpha} \cdot \mathbf{n}_p{}^{\alpha\bullet} + (\mathbf{P}^{\bullet}\, \mathbf{P}^{-1}\, \mathbf{d}_{p\alpha}) \cdot \mathbf{n}_p{}^{\alpha} \\ &= \mathbf{d}_{p\alpha} \cdot \mathbf{n}_p{}^{\alpha\bullet} + \mathbf{d}_{p\alpha} \cdot (\mathbf{P}^{-T}\, \mathbf{P}^{\bullet T}\, \mathbf{n}_p{}^{\alpha}) , \end{aligned}$$

which is solved by

$$\mathbf{n}_p{}^{\alpha}(t)^{\bullet} = -\,\mathbf{P}^{-T}\, \mathbf{P}^{\bullet T}\, \mathbf{n}_p{}^{\alpha} .$$

The lattice vectors in the reference range $\mathbf{c}_{0i} = \mathbf{P}^{-1}\, \mathbf{c}_{pi}$ are constant with respect to time, and so are

$$\mathbf{d}_{0\alpha} := \mathbf{P}^{-1}\, \mathbf{d}_{p\alpha} = \mathbf{P}^{-1}\, \mathbf{F}^{-1}\, \mathbf{d}_{E\alpha}$$

and

$$\mathbf{n}_0{}^{\alpha} := \mathbf{P}^{T}\, \mathbf{n}_p{}^{\alpha} = \mathbf{P}^{T}\, \mathbf{F}^{T}\, \mathbf{n}_E{}^{\alpha} .$$

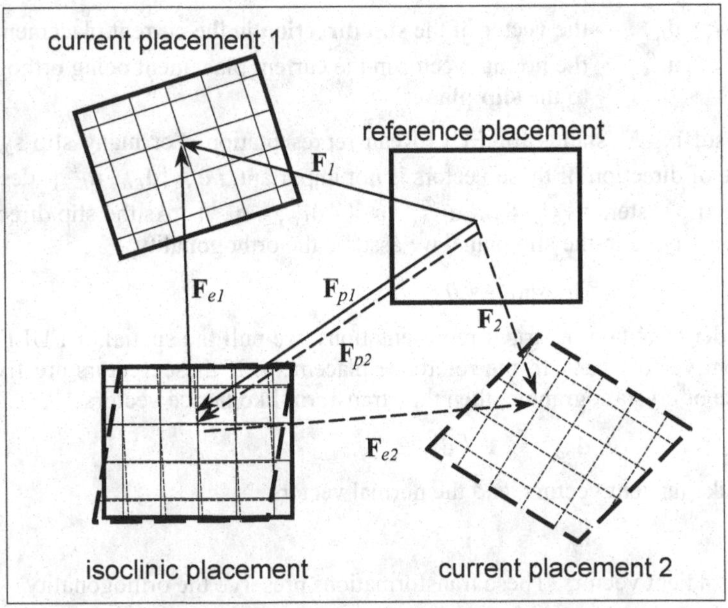

We can normalise these vectors to unit length. For the above time derivatives, we obtain

$$\mathbf{d}_{p\alpha}(t)^\bullet = \mathbf{P}(t)^\bullet \, \mathbf{d}_{0\alpha}$$

and

$$\mathbf{n}_p{}^\alpha(t)^\bullet = (\mathbf{P}(t)^{-T})^\bullet \, \mathbf{n}_0{}^\alpha = -\,\mathbf{P}(t)^{-T} \, \mathbf{P}(t)^{T\bullet} \, \mathbf{P}(t)^{-T} \, \mathbf{n}_0{}^\alpha .$$

The lattice structure gives a clear interpretation of MANDEL's isoclinic placement. We consider the crystalline body in two differently deformed placements. In contrast to the reference placement, the isoclinic placements are different for both states. At a material point, only the lattice directions coincide.

current placement		reference placement		isoclinic ref.
$\mathbf{c}_{Ei}(t)$	$\mathbf{F}$ ←	$\mathbf{c}_{pi}(t)$	$\mathbf{P}$ ←	$\mathbf{c}_{0i}$
$\mathbf{d}_{E\alpha}(t)$		$\mathbf{d}_{p\alpha}(t)$		$\mathbf{d}_{0\alpha}$
$\mathbf{c}_E{}^i(t)$	$\mathbf{F}^T$ →	$\mathbf{c}_p{}^i(t)$	$\mathbf{P}^T$ →	$\mathbf{c}_0{}^i$
$\mathbf{n}_E{}^\alpha(t)$		$\mathbf{n}_p{}^\alpha(t)$		$\mathbf{n}_0{}^\alpha$

As an **example** we consider a *face-centred cubic crystal* (fcc). The 12 primary octahedral slip systems are given with respect to the lattice vectors $\{\mathbf{c}_{Ei}\}$ in the current placement, which form an ONB in an undistorted state, by

$$\mathbf{n}_E{}^{1,2,3} = {}^1\!/_{\sqrt{3}}\,(\mathbf{c}_E{}^1 + \mathbf{c}_E{}^2 + \mathbf{c}_E{}^3)$$

and

$$\mathbf{d}_{E1} = {}^1\!/_{\sqrt{2}}\,(\mathbf{c}_{E2} - \mathbf{c}_{E3})$$

$$\mathbf{d}_{E2} = {}^1\!/_{\sqrt{2}}\,(\mathbf{c}_{E3} - \mathbf{c}_{E1})$$

$$\mathbf{d}_{E3} = {}^1\!/_{\sqrt{2}}\,(\mathbf{c}_{E1} - \mathbf{c}_{E2})$$

and all those, which are crystallographically equivalent. They can be generated by permutations of the signs in $\{\pm\,\mathbf{c}_{Ei}\}$. Besides, we have 6 secondary cubic slip systems with

$$\mathbf{n}_E{}^i = \mathbf{c}_E{}^i$$

and

$$\mathbf{d}_{Ei} = {}^1\!/_{\sqrt{2}}\,(\mathbf{c}_{Ei+1} - \mathbf{c}_{Ei+2}) .$$

For the body-centred cubic crystal (bcc) there are 6 slip planes with 2 slip directions each, *i.e.*, 12 primary slip systems.

The **resolved shear stress** in a slip system $\{\mathbf{d}_{E\alpha}, \mathbf{n}_E{}^{\alpha}\}$ can be calculated in the current placement by CAUCHY's stresses as the component in the direction of $\mathbf{d}_{E\alpha}$ of the stress vector in the plane $\mathbf{n}_E{}^{\alpha}$

$$\tau_s{}^{\alpha} := (\mathbf{T}\,\mathbf{n}_E{}^{\alpha}) \cdot \mathbf{d}_{E\alpha} = \mathbf{T} \cdot (\mathbf{d}_{E\alpha} \otimes \mathbf{n}_E{}^{\alpha})$$

or in the reference placement by the material stresses $\mathbf{S}$

$$\begin{aligned}
\tau_s{}^{\alpha} &= (\mathbf{F}\,\mathbf{S}\,\mathbf{F}^T) \cdot (\mathbf{F}\,\mathbf{d}_p{}^{\alpha}) \otimes (\mathbf{F}^{-T}\,\mathbf{n}_p{}^{\alpha}) \\
&= \mathbf{C}\,\mathbf{S} \cdot (\mathbf{d}_{p\alpha} \otimes \mathbf{n}_p{}^{\alpha}) \\
&= \mathbf{C}\,\mathbf{S} \cdot (\mathbf{d}_{p\alpha} \otimes \mathbf{P}^{-T}\,\mathbf{n}_0{}^{\alpha}) \qquad (10.29) \\
&= -\mathbf{S}^p \cdot (\mathbf{d}_{p\alpha} \otimes \mathbf{n}_0{}^{\alpha}) \qquad \text{with } \mathbf{S}^p := -\mathbf{C}\,\mathbf{S}\,\mathbf{P}^{-T} \\
&= \mathbf{C}\,\mathbf{S}\,\mathbf{P}^{-T} \cdot (\mathbf{P}\,\mathbf{d}_{0\alpha} \otimes \mathbf{n}_0{}^{\alpha}) \\
&= (\mathbf{P}^T\,\mathbf{C}\,\mathbf{S}\,\mathbf{P}^{-T}) \cdot (\mathbf{d}_{0\alpha} \otimes \mathbf{n}_0{}^{\alpha}) \\
&= (\mathbf{C}_e\,\mathbf{P}^{-1}\,\mathbf{S}\,\mathbf{P}^{-T}) \cdot (\mathbf{d}_{0\alpha} \otimes \mathbf{n}_0{}^{\alpha}) \qquad \text{with } \mathbf{C}_e := \mathbf{P}^T\,\mathbf{C}\,\mathbf{P}\,.
\end{aligned}$$

The rank-one tensor $\mathbf{d}_{0\alpha} \otimes \mathbf{n}_0{}^{\alpha}$ is called the (material) **SCHMID tensor** of the slip system. By the elastic law this function can be likewise transformed into the strain space

$$\begin{aligned}
\tau_s{}^{\alpha} &= \mathbf{C}\,k_p(\mathbf{C}) \cdot (\mathbf{d}_{p\alpha} \otimes \mathbf{n}_p{}^{\alpha}) \\
&= \mathbf{C}_e\,k_0(\mathbf{C}_e) \cdot (\mathbf{d}_{0\alpha} \otimes \mathbf{n}_0{}^{\alpha})
\end{aligned}$$

and by (10.7)

$$= (\mathbf{C}_e\,{}^1\!/\!_2\,\mathcal{K}_0[\mathbf{C}_e - \mathbf{C}_{u0}]) \cdot (\mathbf{d}_{0\alpha} \otimes \mathbf{n}_0{}^{\alpha})\,.$$

The time-derivative of the resolved shear stress is with (10.9)

$$\begin{aligned}
\tau_s{}^{\alpha\bullet} &= (\mathbf{C}_e^{\bullet}\,{}^1\!/\!_2\,\mathcal{K}_0[\mathbf{C}_e - \mathbf{C}_{u0}]) \cdot (\mathbf{d}_{0\alpha} \otimes \mathbf{n}_0{}^{\alpha}) \\
&\quad + (\mathbf{C}_e\,{}^1\!/\!_2\,\mathcal{K}_0[\mathbf{C}_e^{\bullet}]) \cdot (\mathbf{d}_{0\alpha} \otimes \mathbf{n}_0{}^{\alpha}) \\
&= (\{\mathbf{P}^T\,\mathbf{C}^{\bullet}\,\mathbf{P} + 2\,sym(\mathbf{C}_e\,\mathbf{P}^{-1}\,\mathbf{P}^{\bullet})\}\,{}^1\!/\!_2\,\mathcal{K}_0[\mathbf{C}_e - \mathbf{C}_{u0}]) \\
&\qquad \cdot (\mathbf{d}_{0\alpha} \otimes \mathbf{n}_0{}^{\alpha}) \\
&\quad + (\mathbf{C}_e\,{}^1\!/\!_2\,\mathcal{K}_0[\mathbf{P}^T\,\mathbf{C}^{\bullet}\,\mathbf{P} + 2\,\mathbf{C}_e\,\mathbf{P}^{-1}\,\mathbf{P}^{\bullet}]) \cdot (\mathbf{d}_{0\alpha} \otimes \mathbf{n}_0{}^{\alpha}) \\
&= \mathbf{A}_{\alpha} \cdot \mathbf{C}^{\bullet} + \mathbf{B}_{\alpha} \cdot \mathbf{P}^{-1}\,\mathbf{P}^{\bullet} \qquad (10.30)
\end{aligned}$$

with

$$\begin{aligned}
\mathbf{A}_{\alpha} &:= \mathbf{P}\,(\mathbf{d}_{0\alpha} \otimes \mathbf{n}_0{}^{\alpha}\,{}^1\!/\!_2\,\mathcal{K}_0[\mathbf{C}_e - \mathbf{C}_{u0}] + {}^1\!/\!_2\,\mathcal{K}_0[\mathbf{C}_e\,\mathbf{d}_{0\alpha} \otimes \mathbf{n}_0{}^{\alpha}])\,\mathbf{P}^T \\
\mathbf{B}_{\alpha} &:= \mathbf{C}_e\,\{sym(\mathcal{K}_0[\mathbf{C}_e - \mathbf{C}_{u0}]\,\mathbf{n}_0{}^{\alpha} \otimes \mathbf{d}_{0\alpha}) + \mathcal{K}_0[\mathbf{C}_e\,\mathbf{d}_{0\alpha} \otimes \mathbf{n}_0{}^{\alpha}]\}\,.
\end{aligned}$$

The tensor $\mathbf{A}_{\alpha}(\mathbf{P}, \mathbf{C}_e)$ can generally be symmetrised, while $\mathbf{B}_{\alpha}(\mathbf{C}_e)$ can not.

After *H4* the critical resolved shear stress of all slip systems constitutes the yield limit. It is reached, if at least the resolved shear stress of one slip system has obtained a critical value $\tau_c{}^\alpha$, called the **critical resolved shear stress**

$$|\,\tau_s{}^\alpha\,| = \tau_c{}^\alpha.$$

This yield limit is called **SCHMID's law** (1924). The corresponding yield criterion is

(10.31)
$$\varphi(\mathbf{S}^p) = \max_\alpha \{\,|\,\tau_s{}^\alpha(\,\mathbf{S}^p\,)\,| - \tau_c{}^\alpha\,\}.$$

In most cases, $\tau_c{}^\alpha$ can be expected to be positive, and we will generally assume this. However, by a strong BAUSCHINGER effect, it can also eventually happen that it becomes negative.

SCHMID's yield limit is – like TRESCA's – only piecewise differentiable. If the resolved shear stresses of more then one slip system have reached the critical value at the same time, then this state lies on the cross section of some hyperplanes. In such a corner, the criterion is not differentiable.

We consider first the case of only one active slip system (*single slip*). If we apply the flow rule associated with SCHMID's law, we obtain

$$\mathbf{P}^\bullet = \lambda^\alpha \frac{\partial \varphi}{\partial \mathbf{S}_p} = \lambda^\alpha \, sgn(\tau_s{}^\alpha) \frac{\partial \tau_s{}^\alpha}{\partial \mathbf{S}_p} = -\,\lambda^\alpha \, sgn(\tau_s{}^\alpha)\, \mathbf{d}_{p\alpha} \otimes \mathbf{n}_0{}^\alpha$$

$$\Leftrightarrow \qquad \mathbf{P}^\bullet\, \mathbf{P}^{-1} = -\,\lambda^\alpha \, sgn(\tau_s{}^\alpha)\, \mathbf{d}_{p\alpha} \otimes \mathbf{n}_p{}^\alpha$$

$$\Leftrightarrow \qquad \mathbf{P}^{-1}\, \mathbf{P}^\bullet = -\,\lambda^\alpha \, sgn(\tau_s{}^\alpha)\, \mathbf{d}_{0\alpha} \otimes \mathbf{n}_0{}^\alpha$$

$$\Leftrightarrow \qquad \mathbf{F}\,\mathbf{P}^\bullet\, \mathbf{P}^{-1}\,\mathbf{F}^{-1} = -\,\lambda^\alpha \, sgn(\tau_s{}^\alpha)\, \mathbf{d}_{E\alpha} \otimes \mathbf{n}_E{}^\alpha$$

with a plastic consistency parameter λ^α, which is positive for an active slip system. This is of the form of a simple shear (2.27) with a shear rate $\gamma^{\bullet\alpha} \equiv \lambda^\alpha$. This differential equation can be integrated over time and gives with the initial values $\mathbf{P}(0) = \mathbf{I}$, $\gamma^\alpha(0) = 0$

$$\mathbf{P}(t) = \mathbf{I} - \gamma^\alpha(t)\, sgn(\tau_s{}^\alpha)\, \mathbf{d}_{p\alpha} \otimes \mathbf{n}_p{}^\alpha \qquad \text{with } \gamma^{\bullet\alpha} = \gamma^{\alpha\bullet}.$$

During single slip we have $\mathbf{d}_{p\alpha}(t) = \mathbf{d}_{p\alpha}(0) = \mathbf{d}_{0\alpha}$, and $\mathbf{n}_p{}^\alpha(t) = \mathbf{n}_p{}^\alpha(0) = \mathbf{n}_0{}^\alpha$, *i.e.*, constant in time. This form is well substantiated by crystal physics. After *H3*, the plastic deformations take place as slips or shears in those slip systems, where SCHMID's law and the loading condition are fulfilled.

As $\mathbf{P}^\bullet\,\mathbf{P}^{-1}$ is traceless or deviatoric, the plastic transformation $\mathbf{P}$ in crystal plasticity is unimodular for appropriate initial values.

If more then one slip system is activated (multi-slip), then we superpose the slip rates to the general **slip rule** (KOITER 1953)

$$\mathbf{P}^{\bullet}\,\mathbf{P}^{-1} = -\sum_{\alpha} \gamma^{\bullet\alpha}(t)\, sgn(\tau_s^{\ \alpha})\, \mathbf{d}_{p\alpha}(t) \otimes \mathbf{n}_p^{\ \alpha}(t)$$

$$\Leftrightarrow \qquad \mathbf{P}^{\bullet} = -\sum_{\alpha} \gamma^{\bullet\alpha}(t)\, sgn(\tau_s^{\ \alpha})\, \mathbf{d}_{p\alpha}(t) \otimes \mathbf{n}_0^{\ \alpha}$$

$$(10.32) \qquad \Leftrightarrow \qquad \mathbf{P}^{-1}\,\mathbf{P}^{\bullet} = -\,\mathbf{F}_p^{\ \bullet}\,\mathbf{F}_p^{\ -1} = -\sum_{\alpha} \gamma^{\bullet\alpha}(t)\, sgn(\tau_s^{\ \alpha})\, \mathbf{d}_{0\alpha} \otimes \mathbf{n}_0^{\ \alpha}$$

(sum over all active slip systems) with the material SCHMID tensor $\mathbf{d}_{0\alpha} \otimes \mathbf{n}_0^{\ \alpha}$ being constant in time. Again, the current value $\mathbf{P}(t)$ of the plastic transformation is path-dependent, as usual in plasticity. After multi-slip, the $\gamma^{\bullet\alpha}$ cannot be anymore interpreted as time-derivatives of shears (this is the reason for not writing it as $\gamma^{\alpha\bullet}$). Neither does the associativity to SCHMID's law hold for the sum. Note that the $\gamma^{\bullet\alpha}$ are introduced as positive reals for active slip systems, and we set them to zero for inactive ones.

The above flow rule (10.32) is a non-symmetric flow rule, which determines both the symmetric and the skew part (plastic spin) of $\mathbf{P}^{\bullet}$ at the same time. We can clearly see that the plastic spin plays an important role for anisotropic materials, which can by no means be neglected[160]. Moreover, there is evidently no reason to use different or separate flow rules for the symmetric and the skew parts of $\mathbf{P}^{\bullet}\,\mathbf{P}^{-1}$.

The plastic dissipation power for crystals is after (10.8) with (10.29)

$$\rho\, l_p = -\,\mathbf{CS} \cdot \mathbf{P}^{\bullet}\,\mathbf{P}^{-1} = \mathbf{S}^p \cdot \mathbf{P}^{\bullet}$$
$$= \mathbf{CS} \cdot \sum_{\alpha} \gamma^{\bullet\alpha}\, sgn(\tau_s^{\ \alpha})\, \mathbf{d}_{p\alpha} \otimes \mathbf{n}_p^{\ \alpha} = \sum_{\alpha} |\,\tau_s^{\ \alpha}|\, \gamma^{\bullet\alpha}$$

which is positive in states of yielding and zero in elastic states.

Besides the flow rule, another material law is needed, which describes the hardening, *i.e.*, the evolution of the critical resolved shear stresses of all slip systems. After *H5* we distinguish for the hardening in the slip systems

- **self-hardening** as a result of slips in the same slip system
- **cross** or **latent hardening** as a result of slips in the other slip systems.

A simple and rather popular ansatz for these two kinds of hardening is the linear hardening rule (MANDEL 1966, HILL 1966)

$$(10.33) \qquad \tau_c^{\ \alpha\bullet} = \sum_{\beta} h_{\alpha\beta}\, \gamma^{\bullet\beta}$$

where the sum runs over all (active) slip systems, and $h_{\alpha\beta}$ stands for

- the self-hardening coefficient if $\alpha = \beta$ and for
- the latent hardening coefficient if $\alpha \neq \beta$.

---

160 See KRATOCHVIL (1971, 1973).

It has been well known since the pioneering times of TAYLOR, that in many cases the latent hardening is even greater than the self-hardening (see TAYLOR/ ELAM 1925), like $h_{\alpha\beta} / h_{\alpha\alpha} = 1.4$.

While after the above linear hardening law, the hardening can infinitely grow, the following (latent) hardening law after VOCE[161] (1955) has a saturation property

$$\tau_c{}^\alpha = \tau_\infty - (\tau_\infty - \tau_0)\, exp(-\gamma^\alpha/\gamma_0)$$

with constants τ_∞, τ_0, and γ_0.

The identification of the hardening behaviour by experiments or microphysical results is extremely difficult. In most cases one uses just very coarse estimations for these coefficients. The reason for this problem is that hardening mechanisms are due to rather complex microphysical mechanisms, which are difficult to quantify. It is not even experimentally confirmed that the hardening matrix is symmetric. Much information upon the hardening of crystals is given by KETTUNEN/ KUOKKALA (2002) and KOCKS/ MECKING (2003). More sophisticated hardening rules for single crystals can be found in HAVNER (1985), KHAN/ CHENG (1996), BASSANI (1990), and LUBARDA (2002).

We consider the current critical resolved shear stresses in N slip systems as the hardening variables and list them up in a vector

$$\mathbf{Z} := \{\tau_c{}^1, \tau_c{}^2, \ldots, \tau_c{}^N\} \in \mathscr{R}^N .$$

We will now try to determine the parameters $\gamma^{\bullet\alpha}$ for all active slip systems by the consistency condition

$$|\,\tau_s{}^\alpha(\mathbf{S}^p)\,| - \tau_c{}^\alpha = 0$$

or

$$|\,\tau_s{}^\alpha|^\bullet - \tau_c{}^{\alpha\bullet} = 0 = sgn(\tau_s{}^\alpha)\, \tau_s{}^{\alpha\bullet} - \tau_c{}^{\alpha\bullet} .$$

With (10.30), (10.32) and the hardening rule (10.33) this becomes

$$0 = sgn(\tau_s{}^\alpha)\, \mathbf{A}_\alpha \cdot \mathbf{C}^\bullet$$

$$- sgn(\tau_s{}^\alpha)\, \mathbf{B}_\alpha \cdot \sum_\beta \gamma^{\bullet\beta} sgn(\tau_s{}^\beta)\, \mathbf{d}_{0\beta} \otimes \mathbf{n}_0{}^\beta - \sum_\beta h_{\alpha\beta}\gamma^{\bullet\beta}$$

which forms for a given $\mathbf{C}^\bullet$ a system of inhomogeneous linear equations

$$\sum_{\alpha=1}^{N} g_{\alpha\beta}\gamma^{\bullet\beta} = a_\alpha$$

---

161 VOCE originally suggested this law for the one-dimensional phenomenological case, and not in the context of crystal plasticity.

with the coefficients

$$g_{\alpha\beta} := sgn(\tau_s^{\ \beta})\, \mathbf{B}_\alpha \cdot \mathbf{d}_{0\beta} \otimes \mathbf{n}_0^{\ \beta} + sgn(\tau_s^{\ \alpha})\, h_{\alpha\beta}$$

$$= sgn(\tau_s^{\ \beta})\, \mathbf{C}_e\, \{sym(\boldsymbol{K}_0[\mathbf{C}_e - \mathbf{C}_{u0}]\, \mathbf{n}_0^{\ \alpha} \otimes \mathbf{d}_{0\alpha})$$

$$+ \boldsymbol{K}_0[\mathbf{C}_e\, \mathbf{d}_{0\alpha} \otimes \mathbf{n}_0^{\ \alpha}]\} \cdot \mathbf{d}_{0\beta} \otimes \mathbf{n}_0^{\ \beta} + sgn(\tau_s^{\ \alpha})\, h_{\alpha\beta}$$

and

$$a_\alpha := \mathbf{A}_\alpha \cdot \mathbf{C}^\bullet$$

$$= \mathbf{P}\, (\mathbf{d}_{0\alpha} \otimes \mathbf{n}_0^{\ \alpha}\, \tfrac{1}{2}\, \boldsymbol{K}_0[\mathbf{C}_e - \mathbf{C}_{u0}] + \tfrac{1}{2}\, \boldsymbol{K}_0[\mathbf{C}_e\, \mathbf{d}_{0\alpha} \otimes \mathbf{n}_0^{\ \alpha}])\, \mathbf{P}^T \cdot \mathbf{C}^\bullet$$

for the slip rates $\gamma^{\bullet\beta}$. The first part of $g_{\alpha\beta}$ is a bilinear form of the SCHMID tensors of the slip systems, which depends on $\mathbf{C}_e$. Of course, only non-negative solutions are wanted. The matrix of the coefficients of this bilinear form is in general non-symmetric, even if the hardening matrix is symmetric. Only if we additionally assume small elastic deformations ($\mathbf{C}_e \approx \mathbf{C}_{u0}$), can its symmetry be achieved by an appropriate choice of the elastic reference law (STEINMANN 1996). What is even worse, is the fact that this matrix can become singular, so that a solution does not exist.

Clearly, as $\mathbf{P} \in \mathcal{U}nim^+$ has 8 degrees of freedom, but more than 8 slip systems can be activated, we cannot expect the consistency conditions to be sufficient[162] to uniquely determine all $\gamma^{\bullet\alpha}$. The selection of the active slip systems for these undetermined cases is the **TAYLOR problem**[163]. TAYLOR (1938) and later BISHOP/HILL (1951) suggested to choose that particular set of slip systems which minimises the plastic dissipation. However, this criterion is rather time consuming in practical applications, as all different combinations of slip systems have to be checked first. And neither is it unique in all cases. In the literature a number of other and more practical suggestions are given to determine a generalised inverse[164] for the present singular system of equations for the $\gamma^{\bullet\alpha}$.

In SCHURIG/ BERTRAM (2003)[165], SCHMID´s law is approximated by the smooth yield criterion

$$\varphi(\mathbf{S}^p, \mathbf{Z}) \equiv \frac{1}{n+1}\Big(\sum_\alpha \Big|\frac{\tau_s^\alpha}{\tau_c^\alpha}\Big|^{n+1}\Big) - 1$$

[162] See KRATOCHVIL (1973).

[163] Geoffry Ingram Taylor (1886-1975); see VAN HOUTTE (1991).

[164] See ANAND/ KOTHARI 1996, MIEHE/ SCHRÖDER/ SCHOTTE (1999), SCHRÖDER/ MIEHE (1997), KNOCKAERT/ CHASTEL/ MASSONI (2000), SCHMIDT-BALDASSARI (2003).

[165] See also GAMBIN (1991), ARMINJON (1991), DARRIEULAT/ PIOT (1996), MOLLICA/ SRINIVASA (2002), BORSCH/ SCHURIG (2008).

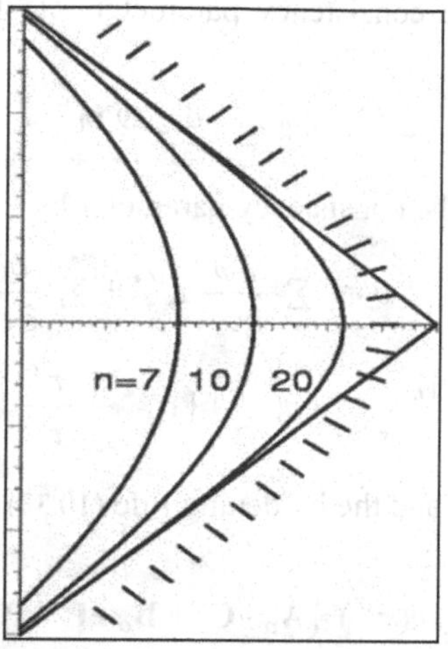

with a positive and sufficiently large constant n, where the resolved shear stress τ_s^{α} depends on $\mathbf{S}^p$ after (10.29). Obviously, this yield criterion is convex and smooth and prevents the resolved shear stresses to become larger than the critical ones, but with increasing n they can come arbitrarily close to them. In this sense, it converges towards SCHMID´s law from the interior. Its normal converges to the one of SCHMID´s law, wherever it exists. In the singular points, where the latter does not exists, it converges to some mean direction contained in the cone of the subdifferential of SCHMID´s law. In the figure the approximation of SCHMID´s law in a typical corner is shown for three different values of n, as well as the direction of the normal for $n \equiv 20$ to show the smooth transition in the corner.

As a flow rule we use the associated one

$$\mathbf{P}^{\bullet} = \lambda \frac{\partial \varphi}{\partial \mathbf{S}_p} = \lambda \sum_{\alpha} \frac{\partial \varphi}{\partial \tau_s^{\alpha}} \frac{d\tau_s^{\alpha}}{d\mathbf{S}_p}$$

$$= -\lambda \sum_{\alpha} sgn(\tau_s^{\alpha}) \frac{1}{\tau_c^{\alpha}} \left| \frac{\tau_s^{\alpha}}{\tau_c^{\alpha}} \right|^n \mathbf{d}_{p\alpha} \otimes \mathbf{n}_0^{\alpha}$$

$$= -\sum_{\alpha} sgn(\tau_s^{\alpha})\, \gamma^{\bullet\alpha}\, \mathbf{d}_{p\alpha} \otimes \mathbf{n}_0^{\alpha}$$

by (10.29) with

$$\gamma^{\bullet\alpha} = \lambda \frac{1}{\tau_c^{\alpha}} \left| \frac{\tau_s^{\alpha}}{\tau_c^{\alpha}} \right|^n$$

with a joint plastic consistency parameter λ for all slip systems, or likewise

$$\mathbf{P}^{-1}\,\mathbf{P}^{\bullet} = -\sum_{\alpha} \gamma^{\bullet\alpha}\, sgn(\tau_s^{\ \alpha})\, \mathbf{d}_{0\alpha} \otimes \mathbf{n}_0^{\ \alpha}.$$

We can determine the consistency parameter by the consistency condition

$$\varphi(\mathbf{S}^p, \mathbf{Z})^{\bullet} = 0 = \sum_{\alpha} \frac{\partial \varphi}{\partial \tau_s^{\alpha}}\, \tau_s^{\ \alpha\bullet} + \sum_{\alpha} \frac{\partial \varphi}{\partial \tau_c^{\alpha}}\, \tau_c^{\ \alpha\bullet}$$

$$= \sum_{\alpha} \left| \frac{\tau_s^{\alpha}}{\tau_c^{\alpha}} \right|^n \frac{1}{\tau_c^{\alpha}}\, sgn(\tau_s^{\ \alpha})\, \{\tau_s^{\ \alpha\bullet} - \frac{\tau_s^{\alpha}}{\tau_c^{\alpha}}\, \tau_c^{\ \alpha\bullet}\}$$

or by use of (10.30) and the hardening rule (10.33)

$$= \sum_{\alpha} \left| \frac{\tau_s^{\alpha}}{\tau_c^{\alpha}} \right|^n \frac{1}{\tau_c^{\alpha}}\, sgn(\tau_s^{\ \alpha})\, \{\mathbf{A}_{\alpha} \cdot \mathbf{C}^{\bullet} + \mathbf{B}_{\alpha} \cdot \mathbf{P}^{-1}\,\mathbf{P}^{\bullet} - \frac{\tau_s^{\alpha}}{\tau_c^{\alpha}} \sum_{\beta} h_{\alpha\beta}\, \gamma^{\bullet\beta}\}$$

$$= \sum_{\alpha} \left| \frac{\tau_s^{\alpha}}{\tau_c^{\alpha}} \right|^n \frac{1}{\tau_c^{\alpha}}\, sgn(\tau_s^{\ \alpha})\, \{\mathbf{A}_{\alpha} \cdot \mathbf{C}^{\bullet} - \mathbf{B}_{\alpha} \cdot \sum_{\beta} \gamma^{\bullet\beta}\, sgn(\tau_s^{\beta})\, \mathbf{d}_{0\beta} \otimes \mathbf{n}_0^{\beta}$$

$$- \frac{\tau_s^{\alpha}}{\tau_c^{\alpha}} \sum_{\beta} h_{\alpha\beta}\, \gamma^{\bullet\beta}\}$$

$$= \sum_{\alpha} \left| \frac{\tau_s^{\alpha}}{\tau_c^{\alpha}} \right|^n \frac{1}{\tau_c^{\alpha}}\, sgn(\tau_s^{\ \alpha})$$

$$\{\mathbf{A}_{\alpha} \cdot \mathbf{C}^{\bullet} - \sum_{\beta} \gamma^{\bullet\beta} \big(sgn(\tau_s^{\beta})\, \mathbf{B}_{\alpha} \cdot \mathbf{d}_{0\beta} \otimes \mathbf{n}_0^{\beta} + \frac{\tau_s^{\alpha}}{\tau_c^{\alpha}}\, h_{\alpha\beta}\big)\}$$

$$= \sum_{\alpha} \left| \frac{\tau_s^{\alpha}}{\tau_c^{\alpha}} \right|^n \frac{1}{\tau_c^{\alpha}}\, sgn(\tau_s^{\ \alpha})$$

$$\{\mathbf{A}_{\alpha} \cdot \mathbf{C}^{\bullet} - \lambda \sum_{\beta} \frac{1}{\tau_c^{\beta}} \left| \frac{\tau_s^{\beta}}{\tau_c^{\beta}} \right|^n \big(sgn(\tau_s^{\beta})\, \mathbf{B}_{\alpha} \cdot \mathbf{d}_{0\beta} \otimes \mathbf{n}_0^{\beta} + \frac{\tau_s^{\alpha}}{\tau_c^{\alpha}}\, h_{\alpha\beta}\big)\}\,,$$

which has the unique solution

$$\lambda = \{\sum_{\alpha} \left| \frac{\tau_s^{\alpha}}{\tau_c^{\alpha}} \right|^n \frac{1}{\tau_c^{\alpha}}\, sgn(\tau_s^{\ \alpha})\, \mathbf{A}_{\alpha} \cdot \mathbf{C}^{\bullet}\} \,/$$

$$\{\sum_{\alpha} \left| \frac{\tau_s^{\alpha}}{\tau_c^{\alpha}} \right|^n \frac{1}{\tau_c^{\alpha}}\, sgn(\tau_s^{\ \alpha}) \sum_{\beta} \frac{1}{\tau_c^{\beta}} \left| \frac{\tau_s^{\beta}}{\tau_c^{\beta}} \right|^n \big(sgn(\tau_s^{\beta})\, \mathbf{B}_{\alpha} \cdot \mathbf{d}_{0\beta} \otimes \mathbf{n}_0^{\beta}$$

$$+ \frac{\tau_s^{\alpha}}{\tau_c^{\alpha}} h_{\alpha\beta})\}.$$

In the limit for increasing n, this gives the consistent flow rule

$$\mathbf{P}^{-1}\,\mathbf{P}^{\bullet} = -\lambda \sum_{\alpha} sgn(\tau_s^{\alpha}) \frac{1}{\tau_c^{\alpha}} \mathbf{d}_{0\alpha} \otimes \mathbf{n}_0^{\alpha}$$

and the consistent hardening rule

$$\tau_c^{\alpha\bullet} = \lambda \sum_{\beta} \frac{1}{\tau_c^{\alpha}} h_{\alpha\beta}$$

with

$$\lambda = \{\sum_{\alpha} \frac{1}{\tau_c^{\alpha}} sgn(\tau_s^{\alpha})\,\mathbf{A}_{\alpha} \cdot \mathbf{C}^{\bullet}\} \,/$$

$$\{\sum_{\alpha} \frac{1}{\tau_c^{\alpha}} sgn(\tau_s^{\alpha}) \sum_{\beta} \frac{1}{\tau_c^{\beta}} \left(sgn(\tau_s^{\beta})\,\mathbf{B}_{\alpha} \cdot \mathbf{d}_{0\beta} \otimes \mathbf{n}_0^{\beta} + h_{\alpha\beta}\right)\} \;,$$

where the sums include all active slip systems.

Another choice to avoid the TAYLOR problem is the use of rate-dependent slip mechanisms. Practically, most if not all materials show a rate-dependence, in particular in the high temperature range. This leads to creep laws as suggested by ASARO (1983), HUTCHINSON (1976), CAILLETAUD (2001) and others. A frequently used example for such a law is the power law in the form

$$\gamma^{\bullet\alpha} = \mu |\,\tau_s^{\alpha} / \tau_c^{\alpha}\,|^{N}$$

with a positive constant μ (inverse to a viscosity) and a NORTON exponent N. If the latter is large enough, the creep rate is almost zero for stresses below the critical one. If, on the other hand, the stresses approach the critical resolved shear stress, the material will yield, so that the stresses can hardly go beyond the yield limit. However, in principle, such a material model does not have elastic ranges anymore. This can be avoided be introducing a threshold value in the above power law as suggested by MERIC/ POUBANNE/ CAILLETAUD (1991).

Conclusions

The crystal plastic model is constituted by the following constitutive equations:

- the elastic reference law (10.7)

$$k_0(\mathbf{C}) = \tfrac{1}{2}\,\mathbf{K}_0[\mathbf{C} - \mathbf{C}_{u0}]$$

- the isomorphy-condition (10.5)

$$k_p(\mathbf{C}) = \mathbf{P}\, k_0(\mathbf{P}^T\,\mathbf{C}\,\mathbf{P})\mathbf{P}^T$$

- the slip systems: $\{\mathbf{d}_{p\alpha}, \mathbf{n}_p{}^{\alpha}\}$
- the resolved shear stresses in the slip systems with index α (10.29)

$$\tau_s{}^{\alpha} := \mathbf{CS} \cdot \mathbf{d}_{p\alpha} \otimes \mathbf{n}_p{}^{\alpha}$$

- the initial critical resolved shear stresses $\tau_c{}^{\alpha}$
- the flow rule (10.32)

$$\mathbf{P}^{-1}\,\mathbf{P}^{\bullet} = -\sum_{\alpha} \gamma^{\bullet\alpha}\, sgn(\tau_s{}^{\alpha})\, \mathbf{d}_{0\alpha} \otimes \mathbf{n}_0{}^{\alpha}$$

- a prescription for the selection of active slip systems
- and a hardening-rule like (10.33)

$$\tau_c{}^{\alpha\bullet} = \sum_{\beta} h_{\alpha\beta}\, |\gamma^{\bullet\beta}| \,.$$

Thus, in crystal plasticity the structure of the material model is already known in detail, in contrast to phenomenological plasticity. Therefore, crystal plasticity is of great importance for understanding plasticity. In BERTRAM/ KRASKA (1995a, b) calculations have been made for tensile tests, which show anisotropic effects due to the crystal structure.

10.6 Material Plasticity

In the foregoing chapter on plasticity, the main assumption was that of isomorphic elastic ranges. The underlying notion is that for certain materials like single crystals, the anisotropy of the elastic behaviour within the elastic ranges results from a structure of the material which is preserved under plastic deformations, but is immaterial in the sense that it can evolve independently of the material. As a consequence, one has to distinguish between lattice vectors or, more general, symmetry axes on the one hand, and material line elements on the other.

However, there are also elastoplastic materials with a microstructure for which this does not hold. Examples for this second class of microstructures are reinforced materials where the fibres of the reinforcement deform together with the matrix in which they are embedded. If the fibres are not randomly distributed, but instead arranged in certain directions, then the elastic behaviour within the elastic ranges can be expected to be anisotropic. Under plastic deformations, both the fibres and the matrix material will jointly deform. So the anisotropy is a material property since it is fixed to the material (in contrast to a lattice). Therefore, such behaviour is called ***material plasticity***[166].

[166] See FOREST/ PARISOT (2000), who introduced this expression for twinning phenomena.

A possible way to modify the above theory towards a *material plasticity* would be to choose two *different* plastic transformations like $\mathbf{P}_K$, $\mathbf{P}_C \in \mathcal{U}nim^+$ such that

$$k_p(\mathbf{C}) = \tfrac{1}{2}(\mathbf{P}_K * \boldsymbol{K}_0)\,[\mathbf{C} - (\mathbf{P}_C^{-T} * \mathbf{C}_{u0})]$$

holds instead of the plastic transformation (10.7). Clearly, with this the isomorphy does not hold anymore. In this case, the elastic stiffness can evolve differently from the unloaded configuration. However, it preserves all properties like symmetries, definiteness, etc. This track has been followed by WEBER *et al.* (2020)[167].

The yield criterion would then be a function of these variables and of the hardening variables $\mathbf{Z}$

$$\varphi(\mathbf{P}_K, \mathbf{P}_C, \mathbf{C}, \mathbf{Z})$$

and two flow rules are required which are assumed to have the usual rate-independent form

$$\mathbf{P}_K^{\bullet} = \lambda\, p^{\circ}(\mathbf{P}_K, \mathbf{P}_C, \mathbf{C}, \mathbf{Z}, \mathbf{C}^{\circ}) =: \lambda\, \mathbf{P}_K^{\circ}$$

$$\mathbf{P}_C^{\bullet} = \lambda\, p^{\circ}(\mathbf{P}_K, \mathbf{P}_C, \mathbf{C}, \mathbf{Z}, \mathbf{C}^{\circ}) =: \lambda\, \mathbf{P}_C^{\circ}$$

with a joint plastic parameter λ which is determined by the consistancy condition

$$\begin{aligned} 0 &= \varphi(\mathbf{P}_K, \mathbf{P}_C, \mathbf{C}, \mathbf{Z})^{\bullet} \\ &= \frac{\partial\varphi}{\partial\mathbf{C}} \cdot \mathbf{C}^{\bullet} + \frac{\partial\varphi}{\partial\mathbf{P}_K} \cdot \mathbf{P}_K^{\bullet} + \frac{\partial\varphi}{\partial\mathbf{P}_C} \cdot \mathbf{P}_C^{\bullet} + \frac{\partial\varphi}{\partial\mathbf{Z}} \cdot \mathbf{Z}^{\bullet} \\ &= \frac{\partial\varphi}{\partial\mathbf{C}} \cdot \mathbf{C}^{\bullet} + \frac{\partial\varphi}{\partial\mathbf{P}_K} \cdot (\lambda\, \mathbf{P}_K^{\circ}) + \frac{\partial\varphi}{\partial\mathbf{P}_C} \cdot (\lambda\, \mathbf{P}_C^{\circ}) + \frac{\partial\varphi}{\partial\mathbf{Z}} \cdot (\lambda\, \mathbf{Z}^{\circ}), \end{aligned}$$

which is solved for λ by

$$\lambda(\mathbf{P}_K, \mathbf{P}_C, \mathbf{C}, \mathbf{Z}, \mathbf{C}^{\bullet}) = \frac{-\dfrac{\partial\varphi}{\partial\mathbf{C}} \cdot \mathbf{C}^{\bullet}}{\dfrac{\partial\varphi}{\partial\mathbf{P}_K} \cdot \mathbf{P}_K^{\circ} + \dfrac{\partial\varphi}{\partial\mathbf{P}_C} \cdot \mathbf{P}_C^{\circ} + \dfrac{\partial\varphi}{\partial\mathbf{Z}} \cdot \mathbf{Z}^{\circ}}.$$

[167] A different approach can be found in STAHN *et al.* (2020) where a fibre reinforced material is modelled as a composition of transversly isotropic components.

11 Rheology

Elasticity and plasticity describe rate-independent behaviour. However, many materials like polymers or metals in their high-temperature regimes show rate-dependent effects like creep (deformation changes under constant stresses) and relaxation (stress changes under constant strains). Therefore, it would be desirable to have a viscoelastic format for finite deformations in addition to the elastic and the elastoplastic ones of the previous parts of this book.

Within the context of small deformations, (linear) viscoelasticity or rheology is a well-established branch of material modelling[168]. One uses rheological models, *i.e.* networks of elastic springs and viscous dampers to construct infinitely many linear viscoelastic models in an algorithmically clear way. This can be done uniaxially or in three dimensions. In the isotropic case, one simply decomposes the stresses and strains into their deviatoric and spherical parts and applies the method to both of them separately. Finally, they are again superimposed to give the complete three-dimensional stresses and strains. In the anisotropic case, a similar method has been suggested by BERTRAM/ OLSCHEWSKI (1993).

The extension of this method to finite deformations, however, is neither unique nor straightforward, and has led to some controversy which is still going on[169]. In particular, the connection in series causes problems, while a parallel connection of such elements turns out to be less controversial. Therefore, we will begin our treatment of such models with parallel connections of elements, and discuss the connections in series afterwards. First of all we describe the two kinds of rheological elements in such networks in detail.

11.1 Rheological Elements

The **HOOKE element** symbolized by a spring is described by a hyperelastic law like (see Sect. 7.3)

[168] see BERTRAM/ GLÜGE (2015) Chapt. 1.3 for further references.

[169] See also REESE/ GOVINDJEE (1998), LION (2000), SVENDSEN (2001), BRÖCKER/ MATZENMILLER (2014), ALTMEYER *et al.* (2016), DONNER/ IHLEMANN (2016), KIESSLING *et al.* (2016), DONNER *et al.* (2017).

A. Bertram, *Elasticity and Plasticity of Large Deformations*,
https://doi.org/10.1007/978-3-030-72328-6_11

(11.1) $$\mathbf{T}^{2PK} = E(\mathbf{E}^G) = \rho_0 \frac{dw(\mathbf{E}^G)}{d\mathbf{E}^G}$$

with an elastic energy $w(\mathbf{E}^G)$, or likewise as an integrable rate form

$$\mathbf{T}^{2PK\bullet} = \frac{dE(\mathbf{E}^G)}{d\mathbf{E}^G}[\mathbf{E}^{G\bullet}] = \rho_0 \frac{d^2 w(\mathbf{E}^G)}{d\mathbf{E}^{G2}}[\mathbf{E}^{G\bullet}] .$$

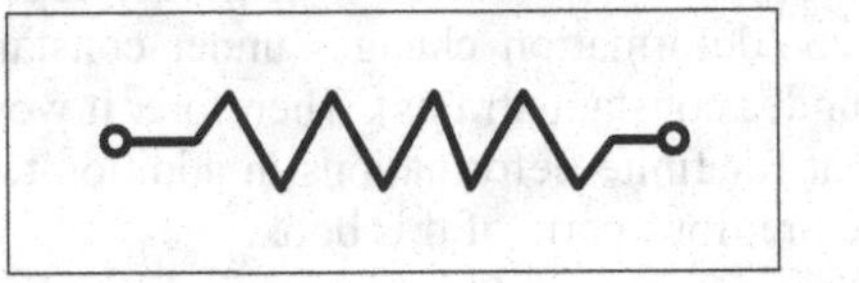

HOOKE element

We call the behaviour of the HOOKE element *conservative*, since the complete stress work is stored and can be reactivated upon unloading, as constitutive in hyperelasticity.

The elastic law can be linear or nonlinear, isotropic or anisotropic. It is here assumed to be invertible

$$\mathbf{E}^G = E^{-1}(\mathbf{T}^{2PK})$$

or as a rate form

$$\mathbf{E}^{G\bullet} = \frac{dE^{-1}(\mathbf{T}^{2PK})}{d\mathbf{T}^{2PK}}[\mathbf{T}^{2PK\bullet}].$$

At each state the tetrads are mutually inverse, *i.e.*,

$$\frac{dE^{-1}(\mathbf{T}^{2PK})}{d\mathbf{T}^{2PK}} = \Big(\frac{dE(\mathbf{E}^G)}{d\mathbf{E}^G}\Big)^{-1} \quad \text{at } \mathbf{T}^{2PK} = E(\mathbf{E}^G).$$

Both are symmetric because of integrability (see Chapt. 7.1).

The second element is called the **NEWTON element** and symbolized by a damper. It introduces viscosity into the models.

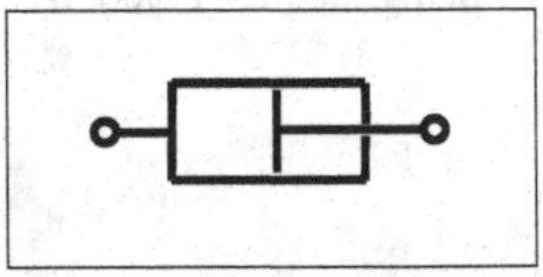

NEWTON element

The NEWTON element is governed by a viscous law

$$\mathbf{T} = d(\mathbf{D})$$

if it is isotropic, or as a reduced form more generally

(11.2) $$\mathbf{T}^{2PK} = D(\mathbf{E}^G, \mathbf{E}^{G\bullet}).$$

The latter function can be linear or non-linear, and isotropic or anisotropic. It is here assumed to also be invertible in the rates, such that it allows for the following reduced form

$$\mathbf{E}^{G\bullet} = D^{-1}(\mathbf{E}^G, \mathbf{T}^{2PK}) .$$

Both elements shall comply with the CLAUSIUS-PLANCK inequality (3.38) in the form

$$\frac{1}{\rho_0} \mathbf{T}^{2PK} \cdot \mathbf{E}^{G\bullet} \geq w^\bullet .$$

If this is fulfilled for all configuration processes, we call the elements **passive** after KRAWIETZ (1986).

For the HOOKE element, the entire stress work is stored as the elastic energy (hyperelasticity), so that equality holds in the CLAUSIUS-PLANCK inequality for conservative elements.

For the NEWTON element, no energy at all is stored in a recoverable way, so that its energy $w \equiv 0$, and the CLAUSIUS-PLANCK inequality tells us that the total stress power is dissipated

$$\mathbf{T}^{2PK} \cdot \mathbf{E}^{G\bullet} = D(\mathbf{E}^G, \mathbf{E}^{G\bullet}) \cdot \mathbf{E}^{G\bullet} \geq 0 \qquad \forall\, \mathbf{E}^G, \mathbf{E}^{G\bullet} \in \mathcal{Sym} .$$

So the behaviour is *dissipative.*

Example. We reconsider the linear viscous fluid from Chapt. 4.3 in the form

$$\mathbf{T} = \nu(J^\bullet)\,\mathbf{I} + 2\mu\,\mathbf{D}$$

$$= J^{-1}\,\mathbf{F}\,\mathbf{T}^{2PK}\,\mathbf{F}^T = \nu(J^\bullet)\,\mathbf{I} + 2\mu\,½\,\mathbf{F}^{-T}\,\mathbf{C}^\bullet\,\mathbf{F}^{-1}$$

or

$$\mathbf{T}^{2PK} = \nu(J^\bullet)\,J\,\mathbf{C}^{-1} + \mu\,\mathbf{C}^{-1}\mathbf{C}^\bullet\mathbf{C}^{-1} =: D(\mathbf{C}, \mathbf{C}^\bullet)$$

with a volumetric viscosity as a linear function $\nu(J^\bullet)$ and a shear viscosity μ . Instead of the right CAUCHY-GREEN tensor $\mathbf{C}$, one can, of course, also use GREEN´s tensor $\mathbf{E}^G = ½\,(\mathbf{C} - \mathbf{I})$ to obtain (11.2).

We now come to connect such elements in rheological models. Such connections can be either in series or in parallel.

11.1.1 Parallel Connections

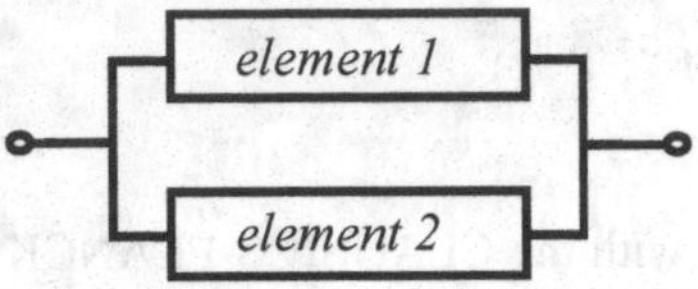

For parallel connections of springs and dampers we assume that ***the deformations in the two elements are equal***. This means that the deformation gradients are identical at all times

$$\mathbf{F}_1 = \mathbf{F}_2$$

and, consequently, all other local deformation measures as well, in particular also the GREEN tensors

(11.3) $$\mathbf{E}^G{}_1 = \mathbf{E}^G{}_2 .$$

So there is no need for indexing strains in parallel connections.

Secondly, ***the CAUCHY stresses are assumed to be additive***

$$\mathbf{T} = \mathbf{T}_1 + \mathbf{T}_2 .$$

In connection with the first assumption, the additivity will also hold for all other stress measures like, *e.g.*, the second PIOLA-KIRCHHOFF stress tensor

(11.4) $$\mathbf{T}^{2PK} = \mathbf{T}^{2PK}{}_1 + \mathbf{T}^{2PK}{}_2 .$$

Obviously, the order of parallel connections has no influence on its behaviour.

Since we prefer material variables for material modelling because of their invariance properties, we have to choose a reference placement. However, its choice as well as the choice of the particular stress and strain measures is immaterial in this theory.

An extension of this concept to multiple parallel connections is straightforward. Even for an infinite number of elements this should be possible, as long as the stresses converge in the sum. In KRAWIETZ (1986), Chapt. 10.2 examples of such multiple connections are discussed in some detail.

One has to make sure that any combination in parallel of thermodynamically consistent elements is again thermodynamically consistent or passive.

Theorem 11.1. *If all elements of a parallel connection are passive, then so is the connection itself.*

Proof. We show this for only two elements. The extension to a multiple connection is then straight. The stress power is

$$\mathbf{T}^{2PK} \cdot \mathbf{E}^{G\bullet} = \mathbf{T}^{2PK}{}_1 \cdot \mathbf{E}^{G\bullet} + \mathbf{T}^{2PK}{}_2 \cdot \mathbf{E}^{G\bullet}$$
$$= \mathbf{T}^{2PK}{}_1 \cdot \mathbf{E}^{G}{}_1{}^{\bullet} + \mathbf{T}^{2PK}{}_2 \cdot \mathbf{E}^{G}{}_2{}^{\bullet} .$$

If both terms are non-negative for all processes, then so is their sum, *q.e.d.*

We will next consider a simple, but nevertheless important example of such parallel connections.

KELVIN model

If we connect a spring E and a damper D in parallel, we obtain the KELVIN model. Here the strains in the spring and in the damper are equal after (11.3)

$$\mathbf{E}^G := \mathbf{E}^G{}_E = \mathbf{E}^G{}_D$$

and the stresses are additive after (11.4)

$$\mathbf{T}^{2PK} = \mathbf{T}^{2PK}{}_E + \mathbf{T}^{2PK}{}_D$$

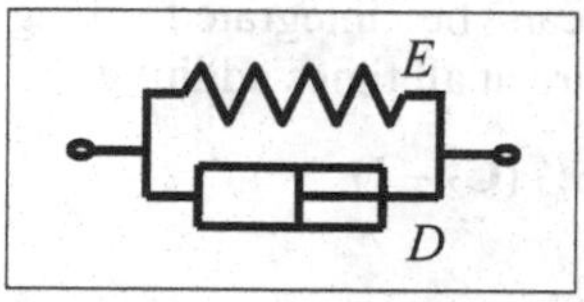

KELVIN model

with the element laws for the spring element

$$\mathbf{T}^{2PK}{}_E = E(\mathbf{E}^G{}_E) = \rho_0 \frac{dw(\mathbf{E}^G{}_E)}{d\mathbf{E}^G{}_E} = E(\mathbf{E}^G)$$

and for the damper element

$$\mathbf{T}^{2PK}{}_D = D(\mathbf{E}^G{}_D, \mathbf{E}^G{}_D{}^{\bullet}) = D(\mathbf{E}^G, \mathbf{E}^{G\bullet})$$

so that

$$\mathbf{T}^{2PK} = E(\mathbf{E}^G) + D(\mathbf{E}^G, \mathbf{E}^{G\bullet})$$

is the constitutive law for the KELVIN model.

If one takes instead of the HOOKE element a PRANDTL element, this connection will describe rigid viscoplasticity with overstresses, as we will show later in this chapter.

11.1.2 Connections in Series

The second type of connection is the one in series.

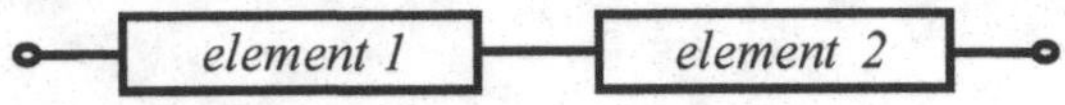

In this case, the basic assumptions are more controversial than in the case of a parallel connection.

Many authors like, *e.g.*, TROSTEL (1971) and PALMOW (1984) assume an additive decomposition of the strain *rates*[170]

$$\mathbf{E}^{G\bullet} = \mathbf{E}^{G}{}_1{}^{\bullet} + \mathbf{E}^{G}{}_2{}^{\bullet}$$

which can equivalently be expressed by the rates of the right CAUCHY-GREEN tensors

$$\mathbf{C}^{\bullet} = \mathbf{C}_1^{\bullet} + \mathbf{C}_2^{\bullet}.$$

By choosing appropriate initial values, this can be integrated along the deformation process so that also the deformations are at all times additive

$$\mathbf{E}^{G} = \mathbf{E}^{G}{}_1 + \mathbf{E}^{G}{}_2 = \tfrac{1}{2}\,(\mathbf{C}_1 - \mathbf{I}) + \tfrac{1}{2}\,(\mathbf{C}_2 - \mathbf{I}) \tag{11.5}$$

$$= \tfrac{1}{2}\,(\mathbf{C}_1 + \mathbf{C}_2) - \mathbf{I}\,.$$

Consequently, it essentially does not matter whether we assume the additivity of the strains or of the strain rates.

The second fundamental assumption for the serial connection is that ***the CAUCHY stresses in both elements are equal***

$$\mathbf{T} = \mathbf{T}_1 = \mathbf{T}_2\,.$$

If we pull these stresses back to the reference placement by the same total deformation gradient, then the second PIOLA-KIRCHHOFF stresses are also equal

$$\mathbf{T}^{2PK} = \mathbf{T}^{2PK}{}_1 = \mathbf{T}^{2PK}{}_2\,. \tag{11.6}$$

Obviously, the order of these connections does not matter.

---

[170] These two authors were surely neither the first nor the last to use this approach, see also ALTMEYER *et al.* (2016), DONNER/ IHLEMANN (2016) and DONNER *et al.* (2017). KRAWIETZ (1986) gives a critical description of this approach which he calls *unechte Hintereinanderschaltung* (false series connection).

We again assume that all elements in a rheological model are passive. We have to assure further that the connections are also thermodynamically consistent or passive.

Theorem 11.2. *If all elements of a series connection are passive, then so is the connection itself.*

Proof. For series connections we have

$$\mathbf{T}^{2PK} \cdot \mathbf{E}^{G\bullet} = \mathbf{T}^{2PK} \cdot \mathbf{E}^{G}{}_{1}{}^{\bullet} + \mathbf{T}^{2PK} \cdot \mathbf{E}^{G}{}_{2}{}^{\bullet} = \mathbf{T}^{2PK}{}_{1} \cdot \mathbf{E}^{G}{}_{1}{}^{\bullet} + \mathbf{T}^{2PK}{}_{2} \cdot \mathbf{E}^{G}{}_{2}{}^{\bullet} .$$

Again, if both terms are passive, then so is their sum, *q.e.d.*

We will next exemplify such a connection for

MAXWELL model

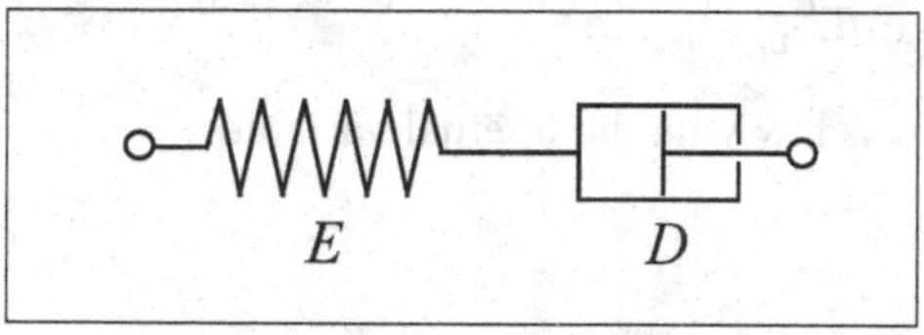

Here a spring E and a damper D are connected in series. We assume a hyperelastic law for the spring element after (11.1)

$$\mathbf{T}^{2PK}{}_{E} = E(\mathbf{E}^{G}{}_{E}) = \rho_0 \frac{dw(\mathbf{E}^{G}{}_{E})}{d\mathbf{E}^{G}{}_{E}}$$

such that its rate form is

$$\mathbf{T}^{2PK}{}_{E}{}^{\bullet} = \frac{dE(\mathbf{E}^{G}{}_{E})}{d\mathbf{E}^{G}{}_{E}} [\mathbf{E}^{G}{}_{E}{}^{\bullet}] = \rho_0 \frac{dw^2(\mathbf{E}^{G}{}_{E})}{d\mathbf{E}^{G}{}_{E}{}^{2}} [\mathbf{E}^{G}{}_{E}{}^{\bullet}] .$$

The inverse law is

$$\mathbf{E}^{G}{}_{E} = E^{-1}(\mathbf{T}^{2PK}{}_{E})$$

or as a rate form

$$\mathbf{E}^{G}{}_{E}{}^{\bullet} = \frac{dE^{-1}(\mathbf{T}^{2PK}{}_{E})}{d\mathbf{T}^{2PK}{}_{E}} [\mathbf{T}^{2PK}{}_{E}{}^{\bullet}] .$$

For the damper element we assume a viscous law after (11.2)

$$\mathbf{T}^{2PK}{}_{D} = D(\mathbf{E}^{G}{}_{D}, \mathbf{E}^{G}{}_{D}{}^{\bullet})$$

which is assumed to be dissipative and invertible in the time rate

$$\mathbf{E}^{G}{}_{D}{}^{\bullet} = D^{-1}(\mathbf{E}^{G}{}_{D}, \mathbf{T}^{2PK}{}_{D}) .$$

As the series connection of a spring and a damper in the MAXWELL model, the deformations or their rates are additive according to (11.5)

$$\mathbf{E}^{G\bullet} = \mathbf{E}^{G}{}_{E}{}^{\bullet} + \mathbf{E}^{G}{}_{D}{}^{\bullet} = \frac{dE^{-1}(\mathbf{T}^{2PK}{}_{E})}{d\mathbf{T}^{2PK}{}_{E}} [\mathbf{T}^{2PK}{}_{E}{}^{\bullet}] + D^{-1}(\mathbf{E}^{G}{}_{D}, \mathbf{T}^{2PK}{}_{D})$$

and the stresses are equal after (11.6)

$$\mathbf{T}^{2PK} = \mathbf{T}^{2PK}{}_{E} = \mathbf{T}^{2PK}{}_{D}$$

so that we can drop the indices here, and find

$$\mathbf{E}^{G\bullet} = \frac{dE^{-1}(\mathbf{T}^{2PK})}{d\mathbf{T}^{2PK}} [\mathbf{T}^{2PK\bullet}] + D^{-1}(\mathbf{E}^{G}{}_{D}, \mathbf{T}^{2PK})$$

or inversly

$$\mathbf{T}^{2PK\bullet} = \frac{dE(\mathbf{E}^{G}{}_{E})}{d\mathbf{E}^{G}{}_{E}} [\, \mathbf{E}^{G\bullet} - D^{-1}(\mathbf{E}^{G}{}_{D}, \mathbf{T}^{2PK})] .$$

This is a rate form of the stress law with the internal variable

$$\mathbf{E}^{G}{}_{E} = E^{-1}(\mathbf{T}^{2PK})$$

or

$$\mathbf{E}^{G}{}_{D} = \mathbf{E}^{G} - \mathbf{E}^{G}{}_{E} .$$

With the foregoing assumptions on parallel and series connections we are, in principle, able to calculate all the rheological models constructed by connections of springs and dampers, like those of the following table. In all cases, the resulting constitutive equations are coupled linear or non-linear ODEs which can be integrated along a given stress or strain path.

The following list[171] contains the simplest rheological models. For some of them, different arrangements lead to identical behaviour.

[171] taken from BERTRAM/ GLÜGE (2015)

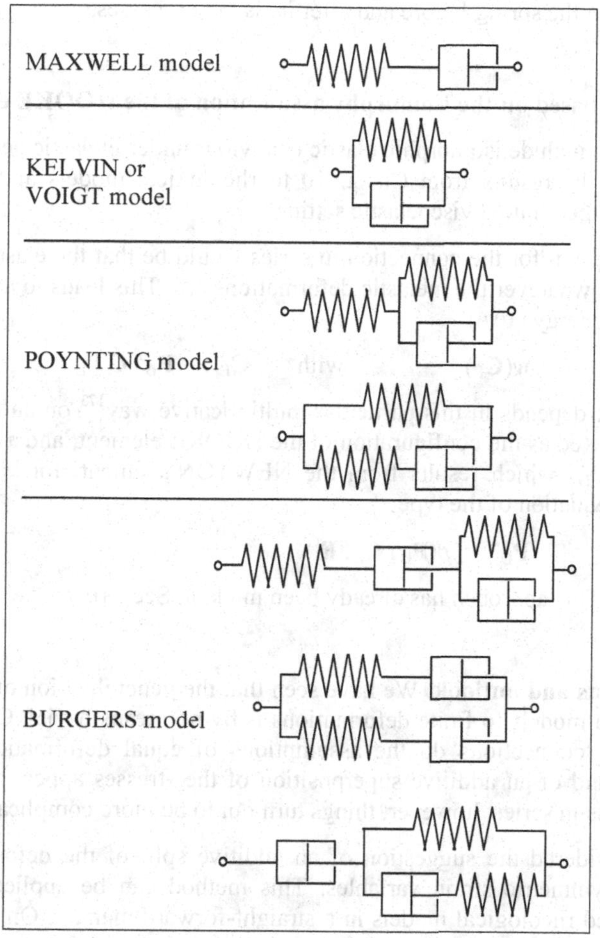

Critical remark. The approach of constructing complex rheological models by an additive composition of the contribution of the rheological elements has the severe defect that it is rather limited in its applications[172], since the elastic stress law of the spring(s) is independent of the viscous deformations of the damper(s). Even if we include the deformations of the damper $\mathbf{E}^G{}_D$ into the independent variables

$$\mathbf{T}^{2PK}{}_E = E(\mathbf{E}^G{}_E, \mathbf{E}^G{}_D)$$

[172] See also KRAWIETZ (1986) for critical comments on this approach.

this would still be rather particular, as we have seen already in the context of the additive elastoplasticity in Sect. 10.4. There we have shown that the introduction of an additive symmetric plastic variable does not allow for isomorphic elastic behaviour in the spring before and after inelastic processes.

Approach based on the isomorphy assumption of the HOOKE elements

In order to include isomorphic elastic behaviour under inelastic deformations, we will apply the results from Chapt. 10 to rheological models and show how to implement them into a viscoelastic setting.

The assumption for the connection in series would be that the elastic laws remain isomorphic whatever the inelastic deformations are. This leads to an expression of the elastic energy (10.6)

$$w(\mathbf{C}_E) \qquad \text{with} \qquad \mathbf{C}_E := \mathbf{P}_D{}^T\, \mathbf{C}\, \mathbf{P}_D \tag{11.7}$$

such that it depends in this particular multiplicative way[173] on an elastic variable $\mathbf{C}_E$ interpreted as the configuration of the HOOKE element, and a non-symmetric variable $\mathbf{P}_D$ which results from the NEWTON element, for which a viscous evolution equation of the type

$$\mathbf{P}_D{}^{\bullet} = p(\mathbf{P}_D, \mathbf{C}_E, \mathbf{E}^{G\bullet})$$

is needed. This approach has already been made in Sect. 10.7.

Conclusions and outlook. We have seen that the generalization of the method of rheological models to finite deformations is by no means trivial. Only in the case of parallel connections do the assumptions of equal deformations in the two elements and of an additive superposition of the stresses appear reasonable. For connections in series, however, things turn out to be more complicated.

We considered the suggestion of an additive split of the deformations in the form of symmetric strain variables. This method can be applied also to more complicated rheological models in a straight-forward manner. On the other hand, it suffers from severe restrictions of its applicability since it cannot even reproduce the isomorphy of the elastic law.

An alternative procedure is to assume isomorphy of the HOOKE elements during inelastic deformations. This leads to a multiplicative split of the (non-symmetric) deformation gradient into elastic and inelastic parts. Hence an evolution equation for a non-symmetric inelastic variable is needed.

The situation reminds us of the discussion on finite plasticity of the preceding chapter. There we also discussed an additive split of a symmetric deformation tensor into elastic and inelastic parts after GREEN/ NAGHDI (1965)

---

[173] See KRAWIETZ (1986) who calls this approach *echte Hintereinanderschaltung.*

and showed its limited applicability. In rheology, the situation is almost the same. The only important difference is that in the case of plasticity, the evolution equation for the inelastic internal variable would be rate-independent, probably with a threshold, and rate-dependent in the case of viscoelasticity, see Sect. 10.7.

The assumption of elastic isomorphy is also not general enough to cover all phenomena. If the inelastic deformations influence the elastic behaviour in a non-isomorphic way, one needs a more general ansatz than (11.7). This could be done by assuming an additional tensorial internal variable $\mathbf{M}$ which accounts for this influence. In this case, we can make the ansatz $w(\mathbf{C}_E, \mathbf{M})$ for the energy. Such an internal variable has to be further specified physically, and an evolution equation for it is needed.

11.2 Models for Viscoplasticity

All the models that we have considered before in this chapter constitute viscoelastic behaviour. We will now extend such models to also include plastic behaviour. In doing so, we will derive examples for viscoplasticity for which a general framework has already been given in Sect. 10.3 Before we start doing this, we introduce a simple model representing elastoplasticity. It is called the

PRANDTL model

The PRANDTL model describes elastoplasticity and can be used to visualize the plasticity concept of the previous chapter on plasticity. The symbol of the PRANDTL model is a block with dry friction called **COULOMB**[174] **element** or **ST. VENANT element** and a HOOKE element connected in series. We do not take the COULOMB element alone, since it does not allow for the *Principle of Determinisms* in its rigid regime.

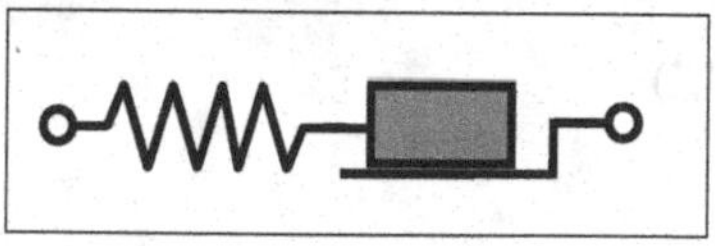

PRANDTL model

If we approach the behaviour of this model in the same way as we did before, then we can again apply (10.21)

$$\mathbf{T}^{2PK\bullet} = (\boldsymbol{H}_p + J\,\boldsymbol{V}_p)\,[\mathbf{C}^\bullet] \tag{11.8}$$

[174] Charles Augustin Coulomb (1736-1806)

as the incremental form in elastoplasticity with two stiffness tetrads. The tetrad $\boldsymbol{V}_p$ is only non-zero if yielding occurs. In other cases, only the (symmetric) elasticity tetrad $\boldsymbol{H}_p$ is activated. In the general cases, both tetrads can depend on the current configuration and the state variables. We assume that in all cases the complete rate-form is invertible into

$$\mathbf{C}^{\bullet} = (\boldsymbol{H}_p + J\,\boldsymbol{V}_p)^{-1}\,[\mathbf{T}^{2\text{PK}\,\bullet}]\,.$$

This form can be interpreted as an additive superposition of two rates, one due to the elasticity of the spring, the other due to the movement of the COULOMB element accounting for the plastic deformation.

This incremental form of the constitutive equation describes elastoplasticity, *i.e.*, the rate-independent behaviour of materials with elastic ranges. It can be isotropic or anisotropic, allowing for any kind of hardening. The thermodynamic consistency has been checked in the previous chapter, as well as the evolution equations (flow and hardening rule) for the internal variables on which the tetrads depend.

The PRANDTL model can be further combined with dampers called NEWTON elements to simulate viscoplastic behaviour. An example for such behaviour is the

Overstress Model

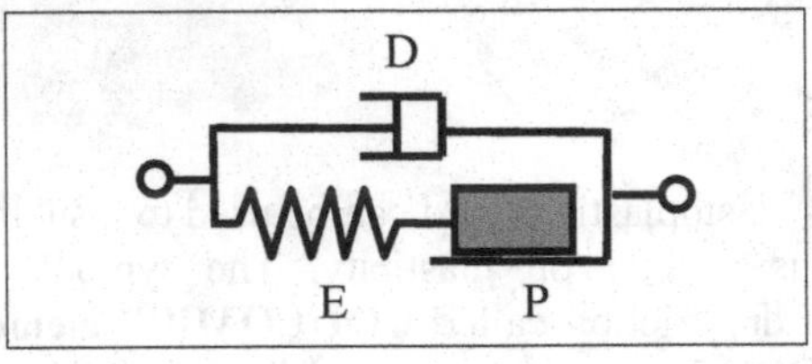

overstress model

Here we connect the PRANDTL model in parallel with a NEWTON element, *i.e.*, a damper with the viscous law

$$\mathbf{T}^{2\text{PK}}{}_{\text{D}} = D(\mathbf{C}\,,\mathbf{C}^{\bullet})\,.$$

so that

$$\mathbf{T}^{2\text{PK}}{}_{\text{D}}{}^{\bullet} = \frac{\partial D(\mathbf{C},\mathbf{C}^{\bullet})}{\partial \mathbf{C}}\,[\mathbf{C}^{\bullet}] + \frac{\partial D(\mathbf{C},\mathbf{C}^{\bullet})}{\partial \mathbf{C}^{\bullet}}\,[\mathbf{C}^{\bullet\bullet}]\,.$$

The stresses in the PRANDTL model are given by (11.8) as before.

In a parallel connection, the deformations in the two branches of the model are equal, while the stresses in the NEWTON element $\mathbf{T}^{2\text{PK}}{}_{\text{D}}$ and in the PRANDTL model $\mathbf{T}^{2\text{PK}}{}_{\text{P}}$ can be superimposed

$$\mathbf{T}^{2\text{PK}} = \mathbf{T}^{2\text{PK}}{}_{\text{D}} + \mathbf{T}^{2\text{PK}}{}_{\text{P}}\,.$$

We substitute this into (11.8)

$$\mathbf{C}^\bullet = (\boldsymbol{H}_p + J\,\boldsymbol{V}_p)^{-1}\,[\mathbf{T}^{2\mathrm{PK}\,\bullet} - \mathbf{T}^{2\mathrm{PK}}{}_{\mathrm{D}}{}^\bullet] \tag{11.9}$$

$$= (\boldsymbol{H}_p + J\,\boldsymbol{V}_p)^{-1}\,(\mathbf{T}^{2\mathrm{PK}\,\bullet} - \frac{\partial D(\mathbf{C},\mathbf{C}^\bullet)}{\partial \mathbf{C}}\,[\mathbf{C}^\bullet] - \frac{\partial D(\mathbf{C},\mathbf{C}^\bullet)}{\partial \mathbf{C}^\bullet}\,[\mathbf{C}^{\bullet\bullet}])$$

and bring this equation into the rate form

$$\mathbf{T}^{2\mathrm{PK}\,\bullet} = \frac{\partial D(\mathbf{C},\mathbf{C}^\bullet)}{\partial \mathbf{C}}\,[\mathbf{C}^\bullet] + \frac{\partial D(\mathbf{C},\mathbf{C}^\bullet)}{\partial \mathbf{C}^\bullet}\,[\mathbf{C}^{\bullet\bullet}] + (\boldsymbol{H}_p + J\,\boldsymbol{V}_p)\,[\mathbf{C}^\bullet]\,.$$

This is an ordinary differential equation for the stress increment, which can be integrated along a given deformation process to determine the resulting stresses.

If the stress in the PRANDTL element is at the yield stress, the block can move, *i.e.*, yielding occurs. In this case, the stresses consist of the yield stress plus the stress from the damper, which is called *overstress*. So the total stresses exceed the yield stresses.

If one wants to combine this overstress model with a spring E in series to achieve a more realistic model, one can do this by again applying (11.7) to the spring and interpreting the internal variable $\mathbf{P}_{\mathrm{D}}$ as resulting from the inelastic deformation from the overstress part of the model. In this case, however, a constitutive equation of type (11.9) would not be sufficient, since a non-symmetric variable is needed.

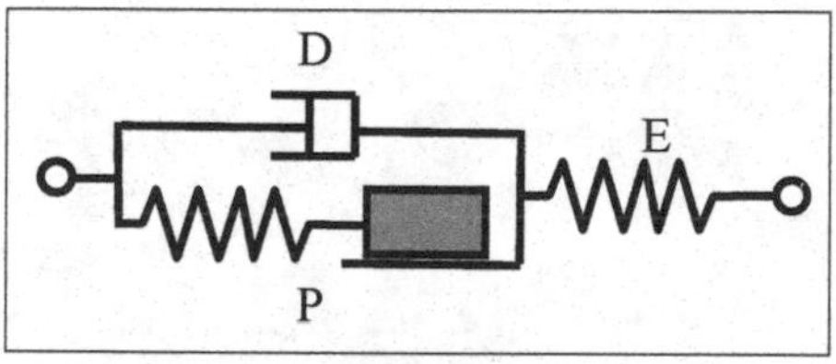

This model gives an example for the overstress format of Sect. 10.3.

12 Essay on Gradient Materials

In the preceding chapters we dealt exclusively with simple materials. However, there is some evidence that within this class certain effects cannot be described, in principle, like, *e.g.*, size effects.

In fact, since the fundamental kinematical variable of simple materials, namely the deformation gradient $\mathbf{F}$, is dimensionless, there is no space for an internal length scale in simple materials. The situation changes if we include $Grad\,Grad\,\boldsymbol{\chi} = Grad\,\mathbf{F}$, which has the dimension $[\text{length}]^{-1}$, or, more generally, the K-th order gradient $Grad^K\,\boldsymbol{\chi}$ with dimension $[\text{length}]^{1-K}$. Such material models are called **gradient materials.**

Moreover, higher gradients in the balance laws open up the possibility of applying non-classical boundary conditions such as line forces acting on the edges and, in the case of third-order gradient theories, concentrated forces acting on the corners of a body. This is a challenging new capability of gradient materials.

Such extensions to gradient materials had already been suggested in the nineteenth century by PIOLA (1848), CAUCHY (1851), and ST.-VENANT (1869a). The first elaborate theories with the inclusion of gradients have been published by TOUPIN (1962) and MINDLIN (1964, 1965). These early gradient theories considered mainly linear elastic models and did not become very recognized at the time. Later on, one tried to include gradient effects into elasto-plastic models with the intention of describing plastic gradient effects as observed in the dislocation dynamics of single crystals. This branch of continuum mechanics has recently become more and more influential, manifested in broad research and publication activities with applications to crystal plasticity, damage mechanics, microstructures, computational mechanics, etc.[175]

Most of these results are restricted to small deformations where formats up to the Nth-order can be constructed in a straight-forward manner. In contrast, the extension of gradient theories to finite deformations is challenging, both from a theoretical and a practical point of view.

A naive approach would be to generalize the material functional (4.1) by including higher deformation gradients up the Kth-order into the list of independent variables of the general material functional as

$$\mathbf{T}(\mathbf{x}_0, t) = \mathfrak{F}_n\{\boldsymbol{\chi}(\mathbf{x}_0, \tau), Grad\,\boldsymbol{\chi}(\mathbf{x}_0, \tau), \ldots, Grad^K\boldsymbol{\chi}(\mathbf{x}_0, \tau)\big|_{\tau=0}^{t}\}.$$

[175] see BERTRAM/ FOREST (2020)

A. Bertram, *Elasticity and Plasticity of Large Deformations*,
https://doi.org/10.1007/978-3-030-72328-6_12

However, it turns out that such an extension beyond the simple materials is not compatible with the second law of thermodynamics. Even in the hyperelastic case this can be easily shown[176]. In fact, if we assume an elastic energy as a function

$$w(\mathbf{F}\,,\, Grad\,\mathbf{F})$$

then the CLAUSIUS-PLANCK inequality (3.38) requires that

$$l = \frac{1}{\rho}\ \mathbf{T}\cdot\mathbf{L} = \frac{1}{\rho}\ \mathbf{T}\,\mathbf{F}^{-T}\cdot\mathbf{F}^{\bullet}$$

$$\geq w^{\bullet} = \frac{\partial w}{\partial \mathbf{F}}\cdot\mathbf{F}^{\bullet} + \frac{\partial w}{\partial Grad\,\mathbf{F}}\cdot Grad\,\mathbf{F}^{\bullet}$$

holds for all values of $\mathbf{F}^{\bullet}$ and $Grad\,\mathbf{F}^{\bullet}$. It is obvious that this is only the case if the energy does *not* depend on $Grad\,\mathbf{F}$. The remedy for this shortcoming is to introduce higher stress tensors in the stress power on the left-hand side of this equation, as we will show in the sequel.

Some fundamental questions would then immediately arise, such as:

- If we have to introduce higher stress tensors, then what is their transformation behaviour, what are their symmetry properties, etc.?
- And do we need additional balance equations to determine them?
- What are the generalized boundary conditions that are needed for these additional fields?
- What would a finite elasticity and plasticity that includes these variables look like?
- How can we extend the concepts of material isomorphy and symmetry to this class of materials?
- How can we assure thermodynamic consistency?

Obviously, these questions are rather fundamental. In order to give them satisfactory answers, one has to go back to the roots of continuum mechanics and give them a much broader framework than that spanned in the previous chapters of this book.

In the following Essay, this challenging task will be embraced. Again, we do not restrict our concern to small deformations, but to finite ones. Naturally, this can only be done in a rather sketchy way, since it is beyond the format of this book to present a detailed theory of gradient materials. For brevity we will only consider

---

[176] see GURTIN (1965)

second gradient materials, but no higher ones. If the reader is interested in the details of this theory, one is refered to BERTRAM (2019)[177].

Finite gradient theories have already been suggested by PIOLA (1848), LE ROUX (1911, 1913), CASAL (1961), TOUPIN (1962), GREEN/ RIVLIN (1964), DUVAUT (1964), CROSS (1973), CHEVERTON/ BEATTY (1975), POLIZZOTTO (2009), SVENDSEN/ NEFF/ MENZEL (2009), SIEVERT (2011), CLEJA-TIGOIU (2002, 2010, 2012), MIEHE (2014), and others in the sequel.

Notations

We will try to use the same tensor notations in this Essay as already used in the rest of this book. However, in some cases this turns out to be too restrictive. So, we will eventually introduce some new products locally between tensors of different order.

One of them is rather useful and shall be introduced already here. It describes the application of an invertible dyad $\mathbf{T}$ to a higher-order tensor $\boldsymbol{A}$ denoted by

(12.1) $$\mathbf{T} \circ \boldsymbol{A} \;:=\; A_{ij\ldots k}\,(\mathbf{T}^{-T}\,\mathbf{e}_i) \otimes (\mathbf{T}\,\mathbf{e}_j) \otimes \ldots \otimes (\mathbf{T}\,\mathbf{e}_k)\,.$$

This shall not be confused with the RAYLEIGH product (1.35). So the first entry of $\boldsymbol{A}$ is transformed differently from all the others. It will be used for the pull-back/ push-forward operation of higher gradients. In such cases, the first entry has a covariant character, while the rest is contravariant.

We find the relation with the RAYLEIGH product

$$\mathbf{T} \circ \boldsymbol{A} \;=\; \mathbf{T} * (\mathbf{T}^{-1}\,\mathbf{T}^{-T}\,\boldsymbol{A})$$

$$=\mathbf{T}^{-T}\mathbf{T}^{-1}\,(\mathbf{T} * \boldsymbol{A})\,.$$

The second-order identity tensor also gives the identity mapping

$$\mathbf{I} \circ \boldsymbol{A} \;=\; \boldsymbol{A}$$

and the inversion is done by

$$\mathbf{T} \circ (\mathbf{T}^{-1} \circ \boldsymbol{A}) \;=\; \boldsymbol{A}\,.$$

Furthermore, the product is associative

$$\mathbf{S} \circ (\mathbf{T} \circ \boldsymbol{A}) \;=\; (\mathbf{S}\,\mathbf{T}) \circ \boldsymbol{A}$$

for all invertible dyads $\mathbf{S}$ and $\mathbf{T}$ and linear in the higher-order tensors $\boldsymbol{A}$.

For the case of $\mathbf{T}$ being orthogonal, this transformation coincides with the RAYLEIGH product.

---

[177] Make sure to download the latest version of this Compendium. The present Essay is partly identical to the Compendium.

12.1 Balance Laws

In Chapt. 3 of this book, we derived the balance laws for simple materials from EULER´s laws of motion and the assumptions of the stress analysis of EULER and CAUCHY, namely the distributions of forces in the interior and on the surface of the body. For gradient materials these distributions become more complicated and can hardly be guessed at.

Most authors in the field like HELLINGER (1914) and GERMAIN (1972, 1973a, 1973b) choose an extended *Principle of virtual work* as a starting point for the balance laws. But also for this procedure one needs a detailed *a priori* knowledge of the forces and torques.

This is why we will choose here a different axiomatic approach based on the *Principle of Objectivity*. This has been used already by the COSSERATs (1909) in their famous approach to polar media. In the present, different context, we will start from an expression of the power which contains the velocity and its first and second gradients, and assume its objectivity. We will see in the sequel that this assumption renders all balance laws in local and global forms.

We will write again $\mathscr{B}_t$ for the region that the body currently occupies in space, and $\mathscr{A}_t$ for its surface. This is subdivided into a finite set of surface segments, bounded by edges, which are denoted by $\mathscr{L}_t$. In order to avoid sums and indices, the expression $\int\limits_{\mathscr{L}_t}$ stands for the line integrals over *all* edges of *all* surface segments. So, every edge line enters such integral expressions twice, since it belongs to two adjacent surface regions.

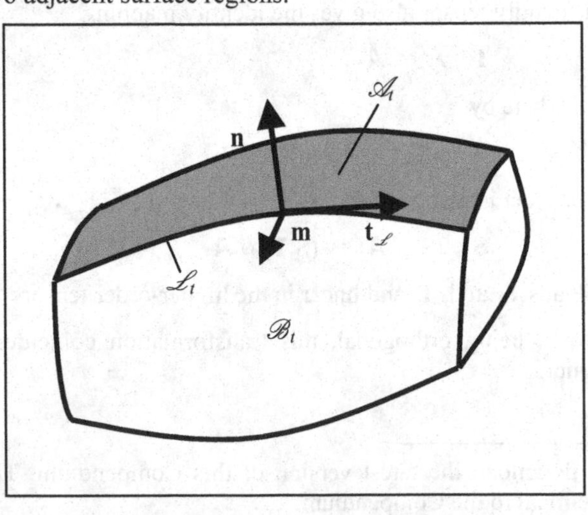

Further we need a vector basis on each point of the edge belonging to some surface element. Such local ONB is given by

$\mathbf{n}$	outer normal to the surface segment
$\mathbf{t}_{\mathscr{L}}$	FRENET's tangent to the edge line
$\mathbf{m} := \mathbf{t}_{\mathscr{L}} \times \mathbf{n}$	bi-normal

In order to apply partial integrations of gradients on the surface[178], we decompose the gradient of a differentiable function φ at surface points into its normal and its tangential parts

$$grad\ \varphi = grad_n\, \varphi + grad_t\, \varphi$$

which corresponds to the natural split of the spatial nabla operator

$$\boldsymbol{\nabla} = \boldsymbol{\nabla}_n + \boldsymbol{\nabla}_t$$

with its normal part

$$\boldsymbol{\nabla}_n := (\mathbf{n} \cdot \boldsymbol{\nabla})\, \mathbf{n} = (\mathbf{n} \otimes \mathbf{n})\, \boldsymbol{\nabla} = \frac{\partial}{\partial x_n}\, \mathbf{n}$$

with the the normal coordinate x_n in the direction of the outer normal $\mathbf{n}$, and the tangential part as the complement

$$\boldsymbol{\nabla}_t := (\mathbf{I} - \mathbf{n} \otimes \mathbf{n})\, \boldsymbol{\nabla}.$$

The trace of these gradients is the divergence, which is also decomposed

$$div\ \varphi = div_n\, \varphi + div_t\, \varphi\,.$$

Then the surface divergence theorem holds in the form[179]

$$\text{(12.2)} \qquad \int_{\mathscr{A}_t} div_t\, \mathbf{v}\; dA = \int_{\mathscr{A}_t} (div_t\, \mathbf{n})\; \mathbf{v} \cdot \mathbf{n}\; dA + \int_{\mathscr{L}_t} \mathbf{v} \cdot \mathbf{m}\; dL$$

The starting point for our approach to second order gradient materials is an expression for the (global) **power** which we assume in the form

---

[178] See TOUPIN (1962), MINDLIN (1965), BLEUSTEIN (1967), DILLON/ KRATOCHVIL (1970), TROSTEL (1985), PODIO-GUIDUGLI/ VIANELLO (2013), POLIZZOTTO (2013), JAVILI/ DELL'ISOLA/ STEINMANN (2013), NEFF/ MÜNCH / GHIBA/ MADEO (2016), MADEO/ GHIBA/ NEFF/ MÜNCH (2016).

[179] See BRAND (1947) p. 222, DELL´ISOLA/ SEPPECHER (1995, 1997), JAVILI/ DELL'ISOLA/ STEINMANN (2013), and CORDERO/ FOREST/ BUSSO (2016).

$$L := \int_{\mathscr{B}_t} (\mathbf{t} \cdot \mathbf{v} + \mathbf{T} \cdot grad\, \mathbf{v} + \boldsymbol{\mathsf{T}} \cdot grad\, grad\, \mathbf{v})\, dV$$

with the velocity $\mathbf{v}$, the first and second velocity gradient, and three conjugate dynamical variables of order one (vector field) $\mathbf{t}$, of order two (dyad or tensor field) $\mathbf{T}$, and of order three (triad) $\boldsymbol{\mathsf{T}}$. Since $grad\, grad\, \mathbf{v}$ has the right sub-symmetry, we will also assume it for $\boldsymbol{\mathsf{T}}$ ($\mathsf{T}_{ijk} = \mathsf{T}_{ikj}$) thus eliminating all power-neutral parts from the third term from the outset.

We will next extend *Assumption 4.4* on the objectivity with the following.

Assumption 12.1. Principle of material objectivity (PMO)
The power is objective under EUCLIDean transformations.

For scalars like the mechanical power, the definitions of objectivity and invariance coincide. So, the power is also assumed to be invariant under changes of observer.

In the sequel we will show that all laws of motion can be obtained by this assumption.

The EUCLIDean transformation (3.15) determines the transformation for the velocity

$$\begin{aligned} \mathbf{v}^* &= \mathbf{Q}\, \mathbf{v} + \mathbf{Q}^\bullet\, \mathbf{Q}^T\, (\mathbf{x}^* - \mathbf{c}) + \mathbf{c}^\bullet \\ &= \mathbf{Q}\, \mathbf{v} + \boldsymbol{\omega} \times (\mathbf{x}^* - \mathbf{c}) + \mathbf{c}^\bullet \end{aligned}$$

with a (time-dependent) orthogonal tensor $\mathbf{Q}$, the skew tensor $\mathbf{Q}^\bullet\, \mathbf{Q}^T$ and its axial vector $\boldsymbol{\omega}$. This gives for the gradient

$$grad^*\, \mathbf{v}^* = \mathbf{Q} * grad\, \mathbf{v} + \mathbf{Q}^\bullet\, \mathbf{Q}^T$$

and for the second gradient

$$grad^*\, grad^*\, \mathbf{v}^* = \mathbf{Q} * grad\, grad\, \mathbf{v}$$

which is, thus, objective.

The objectivity requirement of *Assumption 12.1* demands that for any two observers, the following holds:

$$\begin{aligned} &\mathbf{t} \cdot \mathbf{v} + \mathbf{T} \cdot grad\, \mathbf{v} + \boldsymbol{\mathsf{T}} \cdot grad\, grad\, \mathbf{v} \\ &= \mathbf{t}^* \cdot \mathbf{v}^* + \mathbf{T}^* \cdot grad^*\, \mathbf{v}^* + \boldsymbol{\mathsf{T}}^* \cdot grad^*\, grad^*\, \mathbf{v}^* \\ &= \mathbf{t}^* \cdot [\mathbf{Q}\, \mathbf{v} + \mathbf{Q}^\bullet\, \mathbf{Q}^T\, (\mathbf{x}^* - \mathbf{c}) + \mathbf{c}^\bullet] \\ &\quad + \mathbf{T}^* \cdot [\mathbf{Q} * grad\, \mathbf{v} + \mathbf{Q}^\bullet\, \mathbf{Q}^T] + \boldsymbol{\mathsf{T}}^* \cdot [\mathbf{Q} * grad\, grad\, \mathbf{v}] \end{aligned}$$

wherein we denoted all variables with respect to the second observer by an asterisk. We will now draw some conclusions from this invariance expression.

The objectivity of the power is only fulfilled if three necessary and sufficient conditions hold.

1) For all observers the dynamical quantities are objective

$$\begin{aligned} \mathbf{t} &= \mathbf{Q}^T \mathbf{t}^* &\Leftrightarrow&\quad \mathbf{t}^* = \mathbf{Q}\,\mathbf{t} \\ \mathbf{T} &= \mathbf{Q}^T * \mathbf{T}^* &\Leftrightarrow&\quad \mathbf{T}^* = \mathbf{Q} * \mathbf{T} \qquad (12.3) \\ \boldsymbol{T} &= \mathbf{Q}^T * \boldsymbol{T}^* &\Leftrightarrow&\quad \boldsymbol{T}^* = \mathbf{Q} * \boldsymbol{T}\,. \end{aligned}$$

2) Because of the arbitrariness of $\mathbf{c}^\bullet$, we conclude

$$\mathbf{t}^* = \mathbf{o} \qquad (12.4)$$

for all observers, which does not contradict its objectivity.

3) Because of the arbitrariness of the skew part $\mathbf{Q}^\bullet\,\mathbf{Q}^T$, $\mathbf{T}^*$ must be symmetric for all observers

$$\mathbf{T}^* = \mathbf{T}^{*T}\,. \qquad (12.5)$$

By integration by parts[180] using the GAUSS-OSTROGRADSKY formula and (12.2), the global power can be brought into the following form

$$\int_{\mathscr{B}_t} (\mathbf{t} \cdot \mathbf{v} + \mathbf{T} \cdot grad\,\mathbf{v} + \boldsymbol{T} \cdot grad\,grad\,\mathbf{v})\, dV$$

$$= \int_{\mathscr{B}_t} (\mathbf{t} - div\,\mathbf{T} + div^2\,\boldsymbol{T}) \cdot \mathbf{v}\, dV$$

$$+ \int_{\mathscr{A}_t} \{[(\mathbf{T} - div_n\,\boldsymbol{T} - 2\,div_t\,\boldsymbol{T})\,\mathbf{n}] \cdot \mathbf{v}$$

$$+ \boldsymbol{T} \cdot [\mathbf{v} \otimes (div_t\,\mathbf{n}\;\mathbf{n} \otimes \mathbf{n} - grad_t\,\mathbf{n})] + (\boldsymbol{T}\,\mathbf{n}) \cdot grad_n\,\mathbf{v}\}\; dA$$

$$+ \int_{\mathscr{L}_t} \boldsymbol{T} \cdot (\mathbf{v} \otimes \mathbf{m} \otimes \mathbf{n})\, dL\,.$$

This expression is objective if

$$\int_{\mathscr{B}_t} (\mathbf{t} - div\,\mathbf{T} + div^2\,\boldsymbol{T}) \cdot \mathbf{v}\, dV$$

$$+ \int_{\mathscr{A}_t} \{[(\mathbf{T} - div_n\,\boldsymbol{T} - 2\,div_t\,\boldsymbol{T})\,\mathbf{n}] \cdot \mathbf{v}$$

[180] For details see BERTRAM (2019).

$$+\ \boldsymbol{T}\cdot[\mathbf{v}\otimes(div_t\,\mathbf{n}\,\mathbf{n}\otimes\mathbf{n}-grad_t\,\mathbf{n})]\ +(\boldsymbol{T}\,\mathbf{n})\cdot grad_n\,\mathbf{v}\}\ dA$$

$$+\int_{\mathscr{L}_t}\ \boldsymbol{T}\cdot(\mathbf{v}\otimes\ \mathbf{m}\otimes\mathbf{n})\ dL$$

$$=\int_{\mathscr{B}_t}\ (\mathbf{t}^*-div\,\mathbf{T}^*+div^2\,\boldsymbol{T}^*)\cdot\mathbf{v}^*\ dV$$

$$+\int_{\mathscr{A}_t}\ \{[(\mathbf{T}^*-div_n\,\boldsymbol{T}^*-2\,div_t\,\boldsymbol{T}^*)\,\mathbf{n}^*]\cdot\mathbf{v}^*$$

$$+\boldsymbol{T}^*\cdot[\mathbf{v}^*\otimes(div_t\,\mathbf{n}^*\,\mathbf{n}^*\otimes\mathbf{n}^*-grad_t\,\mathbf{n}^*)]$$

$$+(\boldsymbol{T}^*\,\mathbf{n}^*)\cdot grad_n\,\mathbf{v}^*\}\ dA$$

$$+\int_{\mathscr{L}_t}\ \boldsymbol{T}^*\cdot(\mathbf{v}^*\otimes\mathbf{m}^*\otimes\mathbf{n}^*)\ dL$$

$$=\int_{\mathscr{B}_t}\ (\mathbf{t}^*-div\,\mathbf{T}^*+div^2\,\boldsymbol{T}^*)\cdot[\mathbf{Q}\,\mathbf{v}+\mathbf{Q}^\bullet\,\mathbf{Q}^T(\mathbf{x}^*-\mathbf{c})+\mathbf{c}^\bullet]\ dV$$

$$+\int_{\mathscr{A}_t}\ \{[(\mathbf{T}^*-div_n\,\boldsymbol{T}^*-2\,div_t\,\boldsymbol{T}^*)\,\mathbf{n}^*]\cdot[\mathbf{Q}\,\mathbf{v}+\mathbf{Q}^\bullet\,\mathbf{Q}^T(\mathbf{x}^*-\mathbf{c})+\mathbf{c}^\bullet]$$

$$+\boldsymbol{T}^*\cdot[\mathbf{Q}\,\mathbf{v}+\mathbf{Q}^\bullet\,\mathbf{Q}^T(\mathbf{x}^*-\mathbf{c})+\mathbf{c}^\bullet]\otimes(div_t\,\mathbf{n}^*\,\mathbf{n}^*\otimes\mathbf{n}^*-grad_t\,\mathbf{n}^*)]$$

$$+(\boldsymbol{T}^*\,\mathbf{n}^*)\cdot[\mathbf{Q}*grad_n\,\mathbf{v}+grad_n(\boldsymbol{\omega}\times(\mathbf{x}^*-\mathbf{c}))]\}\ dA$$

$$+\int_{\mathscr{L}_t}\ (\boldsymbol{T}^*\cdot[\mathbf{Q}\,\mathbf{v}+\mathbf{Q}^\bullet\,\mathbf{Q}^T(\mathbf{x}^*-\mathbf{c})+\mathbf{c}^\bullet]\otimes\mathbf{m}^*\otimes\mathbf{n}^*)\ dL\,.$$

By the objectivity of the dynamical quantities (12.3), only the following residual equation remains

$$0=\int_{\mathscr{B}_t}\ (\mathbf{t}^*-div\,\mathbf{T}^*+div^2\,\boldsymbol{T}^*)\cdot[\mathbf{Q}^\bullet\,\mathbf{Q}^T(\mathbf{x}^*-\mathbf{c})+\mathbf{c}^\bullet]\ dV$$

$$+\int_{\mathscr{A}_t}\ \{[(\mathbf{T}^*-div_n\,\boldsymbol{T}^*-2\,div_t\,\boldsymbol{T}^*)\,\mathbf{n}^*]\ \cdot[\mathbf{Q}^\bullet\,\mathbf{Q}^T(\mathbf{x}^*-\mathbf{c})+\mathbf{c}^\bullet]$$

$$+\boldsymbol{T}^*\cdot[\mathbf{Q}^\bullet\,\mathbf{Q}^T(\mathbf{x}^*-\mathbf{c})+\mathbf{c}^\bullet]\otimes(div_t\,\mathbf{n}^*\,\mathbf{n}^*\otimes\mathbf{n}^*-grad_t\,\mathbf{n}^*)]$$

$$+ (\boldsymbol{T}^* \, \mathbf{n}^*) \cdot [grad_n(\boldsymbol{\omega} \times (\mathbf{x}^* - \mathbf{c}))]\} \, dA$$

$$+ \int_{\mathscr{L}_t} (\boldsymbol{T}^* \cdot [\mathbf{Q}^\bullet \, \mathbf{Q}^T (\mathbf{x}^* - \mathbf{c}) + \mathbf{c}^\bullet] \otimes \mathbf{m}^* \otimes \mathbf{n}^*) \, dL \, .$$

Because of the arbitrariness of $\mathbf{c}^\bullet$, we conclude the **balance of forces**

$$\mathbf{o} = \int_{\mathscr{B}_t} (\mathbf{t} - div \, \mathbf{T} + div^2 \, \boldsymbol{T}) \, dV$$

$$+ \int_{\mathscr{A}_t} \{(\mathbf{T} - div_n \, \boldsymbol{T} - 2 \, div_t \, \boldsymbol{T}) \, \mathbf{n} + \boldsymbol{T} : (div_t \, \mathbf{n} \, \mathbf{n} \otimes \mathbf{n} - grad_t \, \mathbf{n})\} \, dA$$

$$+ \int_{\mathscr{L}_t} \boldsymbol{T} : \mathbf{m} \otimes \mathbf{n} \, dL$$

with a double contraction $:$. We dropped the asterisks, since this form of the equation holds in the same form for all observers.

It is customary to introduce the **body force** in the first integrand

$$\rho \, \mathbf{b} \, := \, \rho \, \mathbf{a} + \mathbf{t} - div \, \mathbf{T} + div^2 \boldsymbol{T}$$

and the **specific surface tension**

$$\mathbf{t}_{\mathscr{A}} \, := \, (\mathbf{T} - div_n \, \boldsymbol{T} - 2 \, div_t \, \boldsymbol{T}) \, \mathbf{n} + \boldsymbol{T} : (div_t \, \mathbf{n} \, \mathbf{n} \otimes \mathbf{n} - grad_t \, \mathbf{n})$$

and the **specific line force** on the edge

$$\mathbf{t}_{\mathscr{L}} \, := \, \boldsymbol{T} : (\mathbf{m} \otimes \mathbf{n})$$

so that the **balance of linear momentum** becomes

$$\int_{\mathscr{B}_t} \mathbf{a} \, dm = \int_{\mathscr{B}_t} \mathbf{b} \, dm + \int_{\mathscr{A}_t} \mathbf{t}_{\mathscr{A}} \, dA + \int_{\mathscr{L}_t} \mathbf{t}_{\mathscr{L}} \, dL \, .$$

Secondly, the remaining parts are w.r.t. an appropriate point of reference that lets $\mathbf{c}$ vanish

$$0 = \int_{\mathscr{B}_t} (\mathbf{t}^* - div \, \mathbf{T}^* + div^2 \, \boldsymbol{T}^*) \cdot (\mathbf{Q}^\bullet \, \mathbf{Q}^T \, \mathbf{x}^*) \, dV$$

$$+ \int_{\mathscr{A}_t} \{[(\mathbf{T}^* - div_n \, \boldsymbol{T}^* - 2 \, div_t \, \boldsymbol{T}^*) \, \mathbf{n}^*] \cdot (\mathbf{Q}^\bullet \, \mathbf{Q}^T \, \mathbf{x}^*)$$

$$+ \boldsymbol{T}^* \cdot [\mathbf{Q}^\bullet \, \mathbf{Q}^T \, \mathbf{x}^*) \otimes (div_t \, \mathbf{n}^* \, \mathbf{n}^* \otimes \mathbf{n}^* - grad_t \, \mathbf{n}^*)]$$

$$+ (\boldsymbol{T}^* \mathbf{n}^*) \cdot grad_n(\boldsymbol{\omega} \times \mathbf{x}^*)\} \, dA$$

$$+ \int_{\mathscr{L}_t} (\boldsymbol{T}^* \cdot (\mathbf{Q}^\bullet \mathbf{Q}^T \mathbf{x}^*) \otimes \mathbf{m}^* \otimes \mathbf{n}^*) \, dL$$

$$= \int_{\mathscr{B}_t} (\mathbf{b}^* - \mathbf{a}^*) \cdot (\boldsymbol{\omega} \times \mathbf{x}) \, dm$$

$$+ \int_{\mathscr{A}_t} \mathbf{t}_{\mathscr{A}}^* \cdot (\boldsymbol{\omega} \times \mathbf{x}^*) + 2 \, axi_n (\boldsymbol{T}^* \mathbf{n}^*) \cdot \boldsymbol{\omega} \, dA$$

$$+ \int_{\mathscr{L}_t} \mathbf{t}_{\mathscr{L}}^* \cdot (\boldsymbol{\omega} \times \mathbf{x}^*) \, dL$$

$$= \boldsymbol{\omega} \cdot \{ \int_{\mathscr{B}_t} \mathbf{x}^* \times (\mathbf{b}^* - \mathbf{a}^*) \, dm + \int_{\mathscr{A}_t} \mathbf{x}^* \times \mathbf{t}_{\mathscr{A}} + 2 \, axi_n (\boldsymbol{T}^* \mathbf{n}^*) \, dA + \int_{\mathscr{L}_t} \mathbf{x}^* \times \mathbf{t}_{\mathscr{L}} \, dL\}$$

with the normal axial vector $axi_n(\boldsymbol{T}^* \mathbf{n}^*)$ of a second order tensor $\boldsymbol{T}^* \mathbf{n}^*$ defined by

$$2 \, axi_n(\boldsymbol{T}^* \mathbf{n}^*) \cdot \boldsymbol{\omega} \; := \; (\boldsymbol{T}^* \mathbf{n}^*) \cdot grad_n(\boldsymbol{\omega} \times \mathbf{x}^*)$$

which gives

$$2 \, axi_n(\boldsymbol{T}^* \mathbf{n}^*) \; = \; - (\boldsymbol{T}^* \mathbf{n}^*)_{23} \, \mathbf{e}_1 + (\boldsymbol{T}^* \mathbf{n}^*)_{13} \, \mathbf{e}_2$$

w.r.t. an orthonormal vector basis of the form $\{\mathbf{e}_1 , \mathbf{e}_2 , \mathbf{n}\}$.

We can again drop the asterisks, since this equation holds in the same form for all observers. Because of the arbitrariness of $\mathbf{Q}^\bullet \mathbf{Q}^T$ or its axial vector $\boldsymbol{\omega}$, we conclude the **balance of moment of momentum**

$$\int_{\mathscr{B}_t} \mathbf{x} \times \mathbf{a} \, dm \; = \; \int_{\mathscr{B}_t} \mathbf{x} \times \mathbf{b} \, dm$$

$$+ \int_{\mathscr{A}_t} [\mathbf{x} \times \mathbf{t}_{\mathscr{A}} + 2 \, axi_n (\boldsymbol{T} \mathbf{n})] \, dA + \int_{\mathscr{L}_t} \mathbf{x} \times \mathbf{t}_{\mathscr{L}} \, dL$$

With these results, it is not difficult to show that the PMO is equivalent to each of the following statements.

Theorem 12.1
The generalized EULER laws of motion

- **balance of linear momentum**

$$\int_{\mathcal{B}_t} \mathbf{b}\, dm + \int_{\mathcal{A}_t} \mathbf{t}_{\mathcal{A}}\, dA + \int_{\mathcal{L}_t} \mathbf{t}_{\mathcal{L}}\, dL = \int_{\mathcal{B}_t} \mathbf{a}\, dm$$

- **balance of moment of momentum**

$$\int_{\mathcal{B}_t} \mathbf{x} \times \mathbf{b}\, dm + \int_{\mathcal{A}_t} \{\mathbf{x} \times \mathbf{t}_{\mathcal{A}} + 2\, axi_n(\boldsymbol{T}\mathbf{n})\}\, dA + \int_{\mathcal{L}_t} \mathbf{x} \times \mathbf{t}_{\mathcal{L}}\, dL$$

$$= \int_{\mathcal{B}_t} \mathbf{x} \times \mathbf{a}\, dm$$

hold for the body for one observer (and hence for all).

Theorem 12.2
The extended CAUCHY laws

(12.6) $$div\,(\mathbf{T} - div\,\boldsymbol{T}) + \rho\,\mathbf{b} = \rho\,\mathbf{a}$$

$$\mathbf{T} = \mathbf{T}^T$$

hold everywhere in the body.

Theorem 12.3
The ***Principle of virtual power*** *holds in the form*

$$\int_{\mathcal{B}_t} (\mathbf{b} - \mathbf{a}) \cdot \delta\mathbf{v}\, dm + \int_{\mathcal{A}_t} [\mathbf{t}_{\mathcal{A}} \cdot \delta\mathbf{v} + (\boldsymbol{T}\mathbf{n}) \cdot grad_n\, \delta\mathbf{v}]\, dA$$

$$+ \int_{\mathcal{L}_t} \mathbf{t}_{\mathcal{L}} \cdot \delta\mathbf{v}\, dL$$

$$= \int_{\mathcal{B}_t} (\mathbf{T} \cdot \delta\mathbf{D} + \boldsymbol{T} \cdot grad\, grad\, \delta\mathbf{v})\, dV$$

with $\delta\mathbf{D} := sym\ grad\ \delta\mathbf{v}$ *for all vector fields* $\delta\mathbf{v}$ *for one observer (and hence for all).*

From the above theorems we also see the dynamic or **NEUMANN boundary conditions** for the body. We can prescribe

- the vector field of the tractions $\mathbf{t}_{\mathscr{A}}$ on $\mathscr{A}_t$ working on $\delta\mathbf{v}$
- the line forces $\mathbf{t}_{\mathscr{L}}$ on edges on $\mathscr{L}_t$ working on $\delta\mathbf{v}$
- and the tensor field of the double tractions $\boldsymbol{T}\,\mathbf{n}$ in the normal direction on $\mathscr{A}_t$ working on $grad_n\,\delta\mathbf{v}$.

The **DIRICHLET boundary conditions** are then the prescription of

- the displacement field $\mathbf{u}$ on the surface
- its normal gradient $grad_n\,\mathbf{u}$ on the surface of the body
- and the displacements of the edges (in a compatible way).

A more detailed discussion of the boundary conditions for gradient materials can be found in KRAWIETZ (2021) where an interesting analogy to a crust shell is demonstrated.

The same procedure can be applied if one wants to include higher gradients than just the second one as we did here. If one includes the third velocity gradient in the power one can apply an integration by parts from line elements to vertex points. This allows for concentrated point forces acting on the vertices (corners) of a body[181].

It is interesting to note that the classical FOURIER theory of heat conduction is also a gradient theory, since not only the temperature but also its gradient are included in the list of independent variables. The conjugate variables of them are the entropy and the heat flux, respectively. For them boundary conditions can prescribe the temperature and the heat flux. The latter is proportional to the temperature gradient in the classical theory, see Sect. 3.5.

[181] see BERTRAM (2019), REIHER/ BERTRAM (2018, 2020)

12.2 Material Theory of Second-Order Continua

The preceding part of this Essay is dedicated to those items of gradient materials that hold for all of them, regardless of the particular material behaviour. This shall be next specified for two important material classes, namely elasticity and elasto-plasticity. Before we start with this, however, we will first make some general remarks concerning all second-order gradient materials.

In contrast to simple materials which we exclusively considered in the previous parts of this book, we will here need more than one constitutive equation for the stresses. In fact, for an N-th order theory we have to determine N hyperstress tensors of all orders between 2 and $N+1$. When we limit our concern to second gradient materials, we will thus need constitutive equations for a second-order CAUCHY-like stress tensor $\mathbf{T}$ and a third-order hyperstress tensor $\boldsymbol{T}$.

The most fundamental assumption is again the **Principle of Determinism.** We can use it in the same form as in *Assumption 4.1* if we understand stresses as both $\mathbf{T}$ and $\boldsymbol{T}$. Also the **Principle of Local Action** 4.2 remains valid if we properly extend it by new variables.

Assumption 12.2. Principle of Local Action for second-gradient materials
The stresses $\mathbf{T}$ *and* $\boldsymbol{T}$ *at a material point depend on the motion and the first and the second deformation gradient at this point.*

We cast this assumption into the form of two functionals for the stresses

$$\mathbf{T}(\mathbf{x}_0\,,t) \;=\; \mathfrak{F}_1\{\boldsymbol{\chi}(\mathbf{x}_0\,,\tau)\,,\,\mathbf{F}(\mathbf{x}_0\,,\tau)\,,\,Grad\,\mathbf{F}(\mathbf{x}_0\,,\tau)\big|_{\tau=0}^{t}\}$$

$$\boldsymbol{T}(\mathbf{x}_0\,,t) \;=\; \mathfrak{F}_2\{\boldsymbol{\chi}(\mathbf{x}_0\,,\tau)\,,\,\mathbf{F}(\mathbf{x}_0\,,\tau)\,,\,Grad\,\mathbf{F}(\mathbf{x}_0\,,\tau)\big|_{\tau=0}^{t}\}$$

after which the two stress tensors depend on the process of the motion and its first and second gradient at this material point.

The remaining invariance principle PISM is rather strong. Therefore, we restate it here in its complete form.

Assumption 12.3.
Principle of invariance under superimposed rigid body motions (PISM)
If $\mathbf{T}(\mathbf{x}_0\,,t)$ and $\boldsymbol{T}(\mathbf{x}_0\,,t)$ *are the stresses after a motion*

$$\boldsymbol{\chi}(\mathbf{x}\,,\tau)\big|_{\tau=0}^{t}$$

then

$$\mathbf{Q}(t) * \mathbf{T}(\mathbf{x}_0\,,t) \quad \text{and} \quad \mathbf{Q}(t) * \boldsymbol{T}(\mathbf{x}_0\,,t)$$

are the stresses after superimposing a rigid body modification upon the original motion

$$\{\mathbf{Q}(\tau)\,\boldsymbol{\chi}(\mathbf{x}\,,\,\tau) + \mathbf{c}(\tau)\Big|_{\tau=0}^{t}\,\}$$

with arbitrary differentiable time functions $\mathbf{Q}(\tau) \in \mathcal{O}rth^{+}$ *and* $\mathbf{c}(\tau) \in \mathcal{V}$.

This principle can be identically fulfilled by introducing appropriate invariant varibles both for the strains and for the stresses. This procedure results in *reduced forms* as we had them already for simple materials.

And again, neither the choice of such material variables nor the resulting reduced forms are unique. In fact, there are infinitely many of such choices. But this choice is immaterial, *i.e.*, they all can describe identical material behaviour. So we can make this choice guided by practical aspects.

We can transform the variables within the stress power into material variables like

(12.7) $$L = \int_{\mathcal{B}_t} (\mathbf{T} \cdot \mathbf{D} + \boldsymbol{\mathsf{T}} \cdot grad^2\, \mathbf{v})\, dV$$

$$= \int_{\mathcal{B}_0} J\,(\tfrac{1}{2}\,\mathbf{T}^{2PK} \cdot \mathbf{C}^{\bullet} + \boldsymbol{\mathsf{S}} \cdot \boldsymbol{\mathsf{C}}^{\bullet})\, dV_0$$

with the second-order right CAUCHY-GREEN tensor $\mathbf{C} := \mathbf{F}^T\,\mathbf{F}$ and the third-order **configuration tensor**

(2.10) $$\boldsymbol{\mathsf{C}} := \mathbf{F}^{-1}\, Grad\,\mathbf{F}$$

and its work conjugate material hyperstress tensor

(2.26) $$\boldsymbol{\mathsf{S}} := \mathbf{F}^{-1} \circ \boldsymbol{\mathsf{T}}$$

with the product $\circ$ defined in (12.1). It is not difficult to see that both triads $\boldsymbol{\mathsf{C}}$ and $\boldsymbol{\mathsf{S}}$ are invariant under both changes of observers and superimposed rigid body motions. Of course, this is only one out of infinitely many choices for invariant third-order configuration and stress tensors.

The triad $\boldsymbol{\mathsf{C}}$ has been used by CHAMBON/ CAILLERIE/ TAMAGNINI (2001), FOREST/ SIEVERT (2003), CLEJA-TIGOIU (2013), STEINMANN (2015), and others. It is sometimes called the *connection* or *curvature*, although this might lead to confusion with nabla or the well-known RIEMANN´s curvature tensor (1.47). We therefore call it here the **configuration tensor**.

The configuration tensor has the right subsymmetry by definition

$$C_{ijk} = C_{ikj}$$

so that it has only 18 independent components. The same right subsymmetry is also assumed for $\boldsymbol{\mathsf{S}}$.

In the present framework of second-order gradient materials, the pair $(\mathbf{C}, \boldsymbol{\mathsf{C}})$ constitutes the local **configuration space** $\mathcal{C}onf$, which is 6+18 = 24 dimensional because of the imposed (sub)symmetries. The elements of $\mathcal{C}onf$ are invariant under both changes of observer and rigid body modifications, like all LAGRANGEan or material variables.

However, they do depend on the choice of the reference placement. If this coincides with the current placement then $\mathbf{C} \equiv \mathbf{I}$ and $\boldsymbol{\mathsf{C}} \equiv \boldsymbol{\mathsf{O}}$ the latter being the third-order zero tensor.

12.3 Hyperelastic Gradient Materials

We will next extend the definition of hyperelasticity from simple materials to second-order materials.

Definition 12.1. We call a material a **second-order hyperelastic material** if there exists a specific **elastic energy** as a function

$$w(\boldsymbol{\chi}, Grad\, \boldsymbol{\chi}, Grad\, Grad\, \boldsymbol{\chi})$$

such that the specific stress power after (12.7) equals

$$l = w^{\bullet}$$

This form of the energy, however, does not satisfy the *Principle of Invariance under Superimposed Rigid Body Motions* (PISM). This powerfull principle is identically fulfilled by the following reduced form

(12.8) $$w(\mathbf{C}, \boldsymbol{\mathsf{C}})$$

by using only invariant kinematical variables.

By the chain rule and (12.7), this yields for the stress power

$$l = 1/\rho_0 \left(\tfrac{1}{2}\, \mathbf{T}^{2PK} \cdot \mathbf{C}^{\bullet} + \boldsymbol{\mathsf{S}} \cdot \boldsymbol{\mathsf{C}}^{\bullet}\right)$$

$$= \partial w(\mathbf{C}, \boldsymbol{\mathsf{C}}) / \partial \mathbf{C} \cdot \mathbf{C}^{\bullet} + \partial w(\mathbf{C}, \boldsymbol{\mathsf{C}}) / \partial \boldsymbol{\mathsf{C}} \cdot \boldsymbol{\mathsf{C}}^{\bullet}$$

and by comparison we obtain the potential relations

(12.9) $$\mathbf{T}^{2PK} = 2\rho_0\, \partial w(\mathbf{C}, \boldsymbol{\mathsf{C}}) / \partial \mathbf{C}$$

$$\boldsymbol{\mathsf{S}} = \rho_0\, \partial w(\mathbf{C}, \boldsymbol{\mathsf{C}}) / \partial \boldsymbol{\mathsf{C}}.$$

By means of these potentials our material class identically fulfils the CLAUSIUS-PLANCK inequality (3.38), which turns into an equation in the absence of mechanical dissipation.

12.3.1 Change of the Reference Placement

All of the chosen material variables depend on the choice of the reference placement. However, this choice is immaterial in the following sense. If we use one particular reference placement for the constitutive model, we can uniquely transform the constitutive equations with respect to any other reference placement. After this transformation they will look quite different, although they describe exactly the same behaviour. This transformation under change of reference placement will further play a crucial role in the context of material isomorphisms and symmetry transformations, as we have seen already in Chapt. 6.

Changes of the reference placements have already been considered in Sect. 6.3. If $\mathbf{K} := Grad(\kappa_0\ \underline{\kappa}_0^{-1}) \in \mathcal{Inv}^+$ is again the local gradient of the change of the reference placement, the transformation of $\mathbf{C}$ is given by (6.17)

$$\underline{\mathbf{C}} = \mathbf{K}^T\, \mathbf{C}\, \mathbf{K} = \mathbf{K}^T * \mathbf{C}\,.$$

With the abbreviation

$$\boldsymbol{K} := \mathbf{K}^{-1}\ \underline{Grad}\,\mathbf{K}$$

we find the transformations of the configuration triad as

$$\underline{\boldsymbol{C}} = \boldsymbol{K} + \mathbf{K}^T \circ \boldsymbol{C}$$

with the product defined in (12.1), while the dynamical variables transform like

$$\underline{\mathbf{T}}^{2PK} = \mathbf{K}^{-1} * (J_K\, \mathbf{T}^{2PK})$$

$$\underline{\boldsymbol{S}} = \mathbf{K}^{-1} \circ (J_K\, \boldsymbol{S})\,.$$

We are now able to precisely define what we understand by identical material behaviour of two hyperelastic laws.

Definition 12.2. Two elastic material points with elastic energies w_1 and w_2 are called **isomorphic** if one can find reference placements for each of them with respect to which the elastic energy functions are identical.

This definition constitutes an equivalence relation on all elastic energies of the type (12.8), the equivalence classes of which are the *second-order hyperelastic materials*. By applying the above transformations of our variables under changes of reference placements, we can prove the following representation.

Theorem 12.4. *Two elastic energies* w_1 *and* w_2 *are isomorphic if and only if there are two tensors, namely a second-order invertible tensor* $\mathbf{P}$ *and a third-order tensor with right subsymmetry* $\boldsymbol{P}$, *and a scalar* w_c *such that*

(12.10) $$w_2(\mathbf{C}, \boldsymbol{C}) = w_1(\mathbf{P}^T * \mathbf{C}, \mathbf{P}^T \circ \boldsymbol{C} + \boldsymbol{P}) + w_c$$

holds for all configurations $(\mathbf{C}, \boldsymbol{C}) \in \mathscr{Conf}$.

The constant w_c does not play an important role in elasticity and can be dropped for simplicity. In plasticity, however, it becomes important, as we will see later.

12.3.2 Elastic Symmetry

The application of the isomorphism to only one point constitutes an automorphism or a *symmetry transformation.*

Definition 12.3. A **symmetry transformation** of an elastic energy w is a couple $(\mathbf{A}, \boldsymbol{A})$ of a dyad $\mathbf{A}$ and a triad $\boldsymbol{A}$ such that

$$w(\mathbf{C}, \boldsymbol{C}) = w(\mathbf{A}^T * \mathbf{C}, \mathbf{A}^T \circ \boldsymbol{C} + \boldsymbol{A})$$

holds for all configurations $(\mathbf{C}, \boldsymbol{C}) \in \mathscr{Conf}$.

Using the potentials (12.9) we obtain for the stress laws

(12.11) $$k(\mathbf{C}, \boldsymbol{C}) = \mathbf{A} * k(\mathbf{A}^T * \mathbf{C}, \mathbf{A}^T \circ \boldsymbol{C} + \boldsymbol{A})$$
$$K(\mathbf{C}, \boldsymbol{C}) = \mathbf{A} \circ K(\mathbf{A}^T * \mathbf{C}, \mathbf{A}^T \circ \boldsymbol{C} + \boldsymbol{A}).$$

The set of all symmetry transformations for a given energy constitutes the **symmetry group** $\mathscr{G}$ of the material. In fact, they give rise for a group structure in the algebraic sense.

The composition of two symmetry transformations is given by

$$(\mathbf{A}, \boldsymbol{A})\,(\mathbf{B}, \boldsymbol{B}) = (\mathbf{A}\,\mathbf{B}, \mathbf{B}^T \circ \boldsymbol{A} + \boldsymbol{B})$$

for all $(\mathbf{A}, \boldsymbol{A}), (\mathbf{B}, \boldsymbol{B}) \in \mathscr{G}$, which is not commutative.

The identity of the group is $(\mathbf{I}, \boldsymbol{O}) \in \mathscr{G}$ with the third-order zero tensor $\boldsymbol{O}$, and the inverse of some $(\mathbf{A}, \boldsymbol{A}) \in \mathscr{G}$ is $(\mathbf{A}^{-1}, \mathbf{A}^{-T} \circ \boldsymbol{A})$.

While the role of the second-order part $\mathbf{A}$ of a symmetry transformation is already familiar to us from the simple materials, less is known about the role of the third-order part $\boldsymbol{A}$ until today.

If a material contains with all proper symmetry transformations $(\mathbf{Q}, \boldsymbol{A})$ also the corresponding improper ones $(-\mathbf{Q}, \boldsymbol{A})$, it is called **centro-symmetric**. The centro-symmetry is often tacitly included, although it is a restriction with strong consequences for gradient materials, in contrast to simply materials where it turns out to be redundant.

We will call a material **isotropic**, if the symmetry group contains all orthogonal tensors (*general orthogonal group*) in the first entry. If it contains only the proper orthogonal tensors (*special orthogonal group*), it is called **hemitropic**. So hemitropy combined with centro-symmetry gives isotropy.

In all of these cases, we obtain after (12.11) for symmetry transformation of the form $(\mathbf{Q}, \boldsymbol{\mathsf{O}})$ with an orthogonal $\mathbf{Q}$

$$w(\mathbf{C}, \boldsymbol{\mathsf{C}}) = w(\mathbf{Q} * \mathbf{C}, \mathbf{Q} * \boldsymbol{\mathsf{C}})$$

$$\mathbf{Q} * k(\mathbf{C}, \boldsymbol{\mathsf{C}}) = k(\mathbf{Q} * \mathbf{C}, \mathbf{Q} * \boldsymbol{\mathsf{C}})$$

$$\mathbf{Q} * K(\mathbf{C}, \boldsymbol{\mathsf{C}}) = K(\mathbf{Q} * \mathbf{C}, \mathbf{Q} * \boldsymbol{\mathsf{C}})$$

$\forall\, (\mathbf{C}, \boldsymbol{\mathsf{C}}) \in \mathcal{C}\!onf$. Thus, for an isotropic/hemitropic material the elastic laws are isotropic/hemitropic tensor functions.

12.3.3 Finite Linear Elasticity

For some applications it is justified to linearise the stress laws (12.9) around some unloaded configuration (physically linear theory) to find extensions of the ST. VENANT-KIRCHHOFF law (6.38).

We choose the reference placement as stress-free (unloaded), and describe the deformation by couples of GREEN´s strain tensor and the configuration tensor $(\mathbf{E}^G, \boldsymbol{\mathsf{C}})$. In the reference placements both have value zero. The strains are assumed to be small which means that

$$|\,\mathbf{E}^G\,| \ll 1 \qquad \text{and} \qquad L\,|\,\boldsymbol{\mathsf{C}}\,| \ll 1$$

with a scaling parameter L of dimension *length*.

In tensor notation a quadratic energy has the following form

$$(12.12) \qquad \rho_0\, w(\mathbf{E}^G, \boldsymbol{\mathsf{C}}) =$$

$$\tfrac{1}{2}\, \overset{\langle 4\rangle}{\boldsymbol{\mathsf{E}}}_{22} \cdot (\mathbf{E}^G \otimes \mathbf{E}^G) + \overset{\langle 5\rangle}{\boldsymbol{\mathsf{E}}}_{23} \cdot (\mathbf{E}^G \otimes \boldsymbol{\mathsf{C}}) + \tfrac{1}{2}\, \overset{\langle 6\rangle}{\boldsymbol{\mathsf{E}}}_{33} \cdot (\boldsymbol{\mathsf{C}} \otimes \boldsymbol{\mathsf{C}})$$

with three higher-order elasticity tensors $\overset{\langle 4\rangle}{\boldsymbol{\mathsf{E}}}_{22}\,, \overset{\langle 5\rangle}{\boldsymbol{\mathsf{E}}}_{23}\,, \overset{\langle 6\rangle}{\boldsymbol{\mathsf{E}}}_{33}$.

These elasticities can be submitted to the following symmetry conditions:

$\overset{\langle 4 \rangle}{\boldsymbol{E}}_{22}$:

- left subsymmetry $\{ij\ kl\} = \{ji\ kl\}$
- right subsymmetry $\{ij\ kl\} = \{ij\ lk\}$
- and the major symmetry $\{ij\ kl\} = \{kl\ ij\}$

with 21 independent constants as customary from classical elasticity

$\overset{\langle 5 \rangle}{\boldsymbol{E}}_{23}$:

- left subsymmetry $\{ij\ klm\} = \{ji\ klm\}$
- right subsymmetry $\{ij\ klm\} = \{ij\ kml\}$

with 108 independent parameters

$\overset{\langle 6 \rangle}{\boldsymbol{E}}_{33}$:

- left subsymmetries $\{ijk\ lmn\} = \{ikj\ lmn\}$
- right subsymmetries $\{ijk\ lmn\} = \{ijk\ lnm\}$
- and major symmetry $\{ijk\ lmn\} = \{lmn\ ijk\}$

with 171 independent parameters.

This gives in total 300 independent constants in the general case, which can be further reduced if we account for material symmetries.

In the general anisotropic case, the elastic energy (12.12) acts as a potential for the stresses after (12.9)

$$\mathbf{T}^{2PK} = k(\mathbf{E}^G, \boldsymbol{C}) = \overset{\langle 4 \rangle}{\boldsymbol{E}}_{22}\,[\mathbf{E}^G] + \overset{\langle 5 \rangle}{\boldsymbol{E}}_{23}\,[\boldsymbol{C}]$$

$$\boldsymbol{S} = K(\mathbf{E}^G, \boldsymbol{C}) = [\mathbf{E}^G]\,\overset{\langle 5 \rangle}{\boldsymbol{E}}_{23} + \overset{\langle 6 \rangle}{\boldsymbol{E}}_{33}\,[\boldsymbol{C}].$$

Between the different elasticity tensors and the kinematic variables in brackets [] there are as many contractions as the order of this particular variable is.

These laws are straightforward extensions of the ST.-VENANT-KIRCHHOFF law to gradient elasticity. They are physically linear, but geometrically nonlinear, and they fulfil the EUCLIDean invariance requirement and are, thus, reduced forms. Note that the linear theory depends on the choice of the stress and configuration variables, in contrast to the preceding non-linear theory. However, for small deformations, the differences remain negligible.

If the material is *centro-symmetric*, then any combination of odd and even-order strain tensors must give zero energy, so that the coupling elasticities $\overset{\langle 5 \rangle}{\boldsymbol{E}}_{23}$ must vanish, which reduces the number of independent elastic constants to 192.

In the *hemitropic* case the elasticity tensors must be hemitropic tensors, *i.e.*,

$$\overset{\langle 4 \rangle}{\boldsymbol{E}}_{22} = \mathbf{Q} * \overset{\langle 4 \rangle}{\boldsymbol{E}}_{22}$$

$$\overset{\langle 5 \rangle}{\boldsymbol{E}}_{23} = \mathbf{Q} * \overset{\langle 5 \rangle}{\boldsymbol{E}}_{23}$$

$$\overset{\langle 6 \rangle}{\boldsymbol{E}}_{33} = \mathbf{Q} * \overset{\langle 6 \rangle}{\boldsymbol{E}}_{33}$$

must hold for all proper orthogonal tensors $\mathbf{Q}$. For even-order tensors, hemitropy and isotropy coincide, so that $\overset{\langle 4 \rangle}{\boldsymbol{E}}_{22}$ and $\overset{\langle 6 \rangle}{\boldsymbol{E}}_{33}$ must be isotropic tensors.

4th-order isotropic tensors are multiples of the three tensors

- $\mathbf{I} \otimes \mathbf{I}$
- the fourth-order identity tensor $\overset{\langle 4 \rangle}{\boldsymbol{I}} = \mathbf{e}_i \otimes \mathbf{e}_j \otimes \mathbf{e}_i \otimes \mathbf{e}_j$
- the fourth-order symmetrizer (1.29)

$$\tfrac{1}{4}\,(\mathbf{e}_i \otimes \mathbf{e}_j + \mathbf{e}_j \otimes \mathbf{e}_i) \otimes (\mathbf{e}_i \otimes \mathbf{e}_j + \mathbf{e}_j \otimes \mathbf{e}_i)\,.$$

However, if applied to the symmetric right CAUCHY-GREEN tensor $\mathbf{C}$, the latter two give the same result. So only one of them enters the square form of the energy.

5th-order hemitropic tensors are scalar multiples of products between the second order identity and the third-order permutation tensor $\overset{\langle 3 \rangle}{\boldsymbol{\varepsilon}}$ after WEYL (1939). Ten of them have been listed by, *e.g.*, CALDONAZZO (1932), KEARSLEY/ FONG (1975), and SILBER (1988)

$$\overset{\langle 5 \rangle}{\boldsymbol{H}}_1 = \varepsilon_{ijk}\, \mathbf{e}_i \otimes \mathbf{e}_l \otimes \mathbf{e}_l \otimes \mathbf{e}_j \otimes \mathbf{e}_k = \varepsilon_{ijk}\, \mathbf{e}_i \otimes \mathbf{I} \otimes \mathbf{e}_j \otimes \mathbf{e}_k$$

$$\overset{\langle 5 \rangle}{\boldsymbol{H}}_2 = \varepsilon_{ijk}\, \mathbf{e}_i \otimes \mathbf{e}_l \otimes \mathbf{e}_j \otimes \mathbf{e}_l \otimes \mathbf{e}_k$$

$$\overset{\langle 5 \rangle}{\boldsymbol{H}}_3 = \varepsilon_{ijk}\, \mathbf{e}_i \otimes \mathbf{e}_l \otimes \mathbf{e}_j \otimes \mathbf{e}_k \otimes \mathbf{e}_l$$

$$\overset{\langle 5 \rangle}{\boldsymbol{H}}_4 = \varepsilon_{ijk}\, \mathbf{e}_i \otimes \mathbf{e}_j \otimes \mathbf{e}_l \otimes \mathbf{e}_l \otimes \mathbf{e}_k = \varepsilon_{ijk}\, \mathbf{e}_i \otimes \mathbf{e}_j \otimes \mathbf{I} \otimes \mathbf{e}_k$$

$$\overset{\langle 5 \rangle}{\boldsymbol{H}}_5 = \varepsilon_{ijk}\, \mathbf{e}_i \otimes \mathbf{e}_j \otimes \mathbf{e}_l \otimes \mathbf{e}_k \otimes \mathbf{e}_l$$

$$\overset{\langle 5 \rangle}{\boldsymbol{H}}_6 = \varepsilon_{ijk}\, \mathbf{e}_i \otimes \mathbf{e}_j \otimes \mathbf{e}_k \otimes \mathbf{e}_l \otimes \mathbf{e}_l = \overset{\langle 3 \rangle}{\boldsymbol{\varepsilon}} \otimes \mathbf{I}$$

$$\overset{\langle 5 \rangle}{\boldsymbol{H}}_7 = \varepsilon_{ijk}\, \mathbf{e}_l \otimes \mathbf{e}_l \otimes \mathbf{e}_i \otimes \mathbf{e}_j \otimes \mathbf{e}_k = \mathbf{I} \otimes \overset{\langle 3 \rangle}{\boldsymbol{\varepsilon}}$$

$$\overset{\langle 5 \rangle}{\boldsymbol{H}}_8 = \varepsilon_{ijk}\, \mathbf{e}_l \otimes \mathbf{e}_i \otimes \mathbf{e}_l \otimes \mathbf{e}_j \otimes \mathbf{e}_k$$

$$\overset{\langle 5 \rangle}{\boldsymbol{H}}_9 = \varepsilon_{ijk}\, \mathbf{e}_l \otimes \mathbf{e}_i \otimes \mathbf{e}_j \otimes \mathbf{e}_l \otimes \mathbf{e}_k$$

$$\overset{\langle 5 \rangle}{\boldsymbol{H}}_{10} = \varepsilon_{ijk}\, \mathbf{e}_l \otimes \mathbf{e}_i \otimes \mathbf{e}_j \otimes \mathbf{e}_k \otimes \mathbf{e}_l \,.$$

All of them can be mutually generated by transpositions.

In the linear elastic law, we need such hemitropic pentadics as linear mappings between triads and dyads as in (12.12)

$$\overset{\langle 5 \rangle}{\boldsymbol{E}}_{23} \cdot (\mathbf{E}^{\mathrm{G}} \otimes \boldsymbol{C})$$

with symmetric dyads $\mathbf{E}^{\mathrm{G}}$ and triads $\boldsymbol{C}$ with right subsymmetry. Therefore, we can demand a symmetry in the first and in the last two entries. For this reason $\overset{\langle 5 \rangle}{\boldsymbol{H}}_1, \overset{\langle 5 \rangle}{\boldsymbol{H}}_4, \overset{\langle 5 \rangle}{\boldsymbol{H}}_5, \overset{\langle 5 \rangle}{\boldsymbol{H}}_6, \overset{\langle 5 \rangle}{\boldsymbol{H}}_7, \overset{\langle 5 \rangle}{\boldsymbol{H}}_8$ will not be needed.

Only scalar multiples of the following hemitropic pentadic

$$\overset{\langle 5 \rangle}{\boldsymbol{H}}_2 + \overset{\langle 5 \rangle}{\boldsymbol{H}}_3 + \overset{\langle 5 \rangle}{\boldsymbol{H}}_9 + \overset{\langle 5 \rangle}{\boldsymbol{H}}_{10}$$

$$= \varepsilon_{ijk}\, \mathbf{e}_i \otimes \mathbf{e}_l \otimes \mathbf{e}_j \otimes \mathbf{e}_l \otimes \mathbf{e}_k + \varepsilon_{ijk}\, \mathbf{e}_i \otimes \mathbf{e}_l \otimes \mathbf{e}_j \otimes \mathbf{e}_k \otimes \mathbf{e}_l$$

$$+ \varepsilon_{ijk}\, \mathbf{e}_l \otimes \mathbf{e}_i \otimes \mathbf{e}_j \otimes \mathbf{e}_l \otimes \mathbf{e}_k + \varepsilon_{ijk}\, \mathbf{e}_l \otimes \mathbf{e}_i \otimes \mathbf{e}_j \otimes \mathbf{e}_k \otimes \mathbf{e}_l$$

show all the required symmetries. However, this gives the same results as any of them

$$^{1}/_{4}\,(\overset{\langle 5 \rangle}{\boldsymbol{H}}_2 + \overset{\langle 5 \rangle}{\boldsymbol{H}}_3 + \overset{\langle 5 \rangle}{\boldsymbol{H}}_9 + \overset{\langle 5 \rangle}{\boldsymbol{H}}_{10}) \cdot (\mathbf{E}^{\mathrm{G}} \otimes \boldsymbol{C})$$

$$= \overset{\langle 5 \rangle}{\boldsymbol{H}}_2 \cdot \mathbf{E}^{\mathrm{G}} \otimes \boldsymbol{C} = \overset{\langle 5 \rangle}{\boldsymbol{H}}_3 \cdot (\mathbf{E}^{\mathrm{G}} \otimes \boldsymbol{C})$$

$$= \overset{\langle 5 \rangle}{\boldsymbol{H}}_9 \cdot \mathbf{E}^G \otimes \boldsymbol{C} = \overset{\langle 5 \rangle}{\boldsymbol{H}}_{10} \cdot (\mathbf{E}^G \otimes \boldsymbol{C})$$

$$= \mathbf{E}^G \cdot (\overset{\langle 3 \rangle}{\boldsymbol{\varepsilon}} : \boldsymbol{C}) = (\overset{\langle 3 \rangle}{\boldsymbol{\varepsilon}} \cdot \mathbf{E}^G) : \boldsymbol{C} .$$

for all symmetric dyads $\mathbf{E}^G$ and triads $\boldsymbol{C}$ with right subsymmetry, where : stands for a double contraction.

6th-order isotropic tensors are linear combinations of the following hexadics after KEARSLY/ FONG (1975)

$$\overset{\langle 6 \rangle}{\boldsymbol{H}}_1 = \mathbf{e}_i \otimes \mathbf{e}_i \otimes \mathbf{e}_k \otimes \mathbf{e}_k \otimes \mathbf{e}_p \otimes \mathbf{e}_p = \mathbf{I} \otimes \mathbf{I} \otimes \mathbf{I}$$

$$\overset{\langle 6 \rangle}{\boldsymbol{H}}_2 = \mathbf{e}_i \otimes \mathbf{e}_i \otimes \mathbf{e}_k \otimes \mathbf{e}_m \otimes \mathbf{e}_k \otimes \mathbf{e}_m = \mathbf{I} \otimes \mathbf{e}_k \otimes \mathbf{e}_m \otimes \mathbf{e}_k \otimes \mathbf{e}_m$$

$$\overset{\langle 6 \rangle}{\boldsymbol{H}}_3 = \mathbf{e}_i \otimes \mathbf{e}_i \otimes \mathbf{e}_k \otimes \mathbf{e}_m \otimes \mathbf{e}_m \otimes \mathbf{e}_k = \mathbf{I} \otimes \mathbf{e}_k \otimes \mathbf{I} \otimes \mathbf{e}_k$$

$$\overset{\langle 6 \rangle}{\boldsymbol{H}}_4 = \mathbf{e}_i \otimes \mathbf{e}_j \otimes \mathbf{e}_i \otimes \mathbf{e}_j \otimes \mathbf{e}_p \otimes \mathbf{e}_p = \mathbf{e}_i \otimes \mathbf{e}_j \otimes \mathbf{e}_i \otimes \mathbf{e}_j \otimes \mathbf{I}$$

$$\overset{\langle 6 \rangle}{\boldsymbol{H}}_5 = \mathbf{e}_i \otimes \mathbf{e}_j \otimes \mathbf{e}_i \otimes \mathbf{e}_m \otimes \mathbf{e}_j \otimes \mathbf{e}_m$$

$$\overset{\langle 6 \rangle}{\boldsymbol{H}}_6 = \mathbf{e}_i \otimes \mathbf{e}_j \otimes \mathbf{e}_i \otimes \mathbf{e}_m \otimes \mathbf{e}_m \otimes \mathbf{e}_j = \mathbf{e}_i \otimes \mathbf{e}_j \otimes \mathbf{e}_i \otimes \mathbf{I} \otimes \mathbf{e}_j$$

$$\overset{\langle 6 \rangle}{\boldsymbol{H}}_7 = \mathbf{e}_i \otimes \mathbf{e}_j \otimes \mathbf{e}_j \otimes \mathbf{e}_i \otimes \mathbf{e}_p \otimes \mathbf{e}_p = \mathbf{e}_i \otimes \mathbf{I} \otimes \mathbf{e}_i \otimes \mathbf{I}$$

$$\overset{\langle 6 \rangle}{\boldsymbol{H}}_8 = \mathbf{e}_i \otimes \mathbf{e}_j \otimes \mathbf{e}_k \otimes \mathbf{e}_i \otimes \mathbf{e}_j \otimes \mathbf{e}_k = \overset{\langle 6 \rangle}{\boldsymbol{I}}$$

$$\overset{\langle 6 \rangle}{\boldsymbol{H}}_9 = \mathbf{e}_i \otimes \mathbf{e}_j \otimes \mathbf{e}_k \otimes \mathbf{e}_i \otimes \mathbf{e}_k \otimes \mathbf{e}_j$$

$$\overset{\langle 6 \rangle}{\boldsymbol{H}}_{10} = \mathbf{e}_i \otimes \mathbf{e}_j \otimes \mathbf{e}_j \otimes \mathbf{e}_m \otimes \mathbf{e}_i \otimes \mathbf{e}_m$$

$$\overset{\langle 6 \rangle}{\boldsymbol{H}}_{11} = \mathbf{e}_i \otimes \mathbf{e}_j \otimes \mathbf{e}_k \otimes \mathbf{e}_j \otimes \mathbf{e}_i \otimes \mathbf{e}_k$$

$$\overset{\langle 6 \rangle}{\boldsymbol{H}}_{12} = \mathbf{e}_i \otimes \mathbf{e}_j \otimes \mathbf{e}_k \otimes \mathbf{e}_k \otimes \mathbf{e}_i \otimes \mathbf{e}_j$$

$$\overset{\langle 6 \rangle}{\boldsymbol{H}}_{13} = \mathbf{e}_i \otimes \mathbf{e}_j \otimes \mathbf{e}_j \otimes \mathbf{e}_m \otimes \mathbf{e}_m \otimes \mathbf{e}_i = \mathbf{e}_i \otimes \mathbf{I} \otimes \mathbf{I} \otimes \mathbf{e}_i$$

$$\overset{\langle 6 \rangle}{\boldsymbol{H}}_{14} = \mathbf{e}_i \otimes \mathbf{e}_j \otimes \mathbf{e}_k \otimes \mathbf{e}_j \otimes \mathbf{e}_k \otimes \mathbf{e}_i$$

$$\overset{\langle 6 \rangle}{\boldsymbol{H}}_{15} = \mathbf{e}_i \otimes \mathbf{e}_j \otimes \mathbf{e}_k \otimes \mathbf{e}_k \otimes \mathbf{e}_j \otimes \mathbf{e}_i$$

all of which are transpositions of $\mathbf{I} \otimes \mathbf{I} \otimes \mathbf{I}$.

We can reduce this list by imposing the following symmetries, namely the major symmetry, and the subsymmetries of the second and third entries and of the fifth and sixth ones because of the corresponding one of the curvature tensor $\boldsymbol{C}$. Only the following hexadics show these symmetries

$$\overset{\langle 6 \rangle}{\boldsymbol{H}}_{7} = \mathbf{e}_i \otimes \mathbf{e}_j \otimes \mathbf{e}_j \otimes \mathbf{e}_i \otimes \mathbf{e}_p \otimes \mathbf{e}_p = \mathbf{e}_i \otimes \mathbf{I} \otimes \mathbf{e}_i \otimes \mathbf{I}$$

$$\overset{\langle 6 \rangle}{\boldsymbol{H}}_{8} + \overset{\langle 6 \rangle}{\boldsymbol{H}}_{9} = \mathbf{e}_i \otimes \mathbf{e}_j \otimes \mathbf{e}_k \otimes \mathbf{e}_i \otimes \mathbf{e}_j \otimes \mathbf{e}_k + \mathbf{e}_i \otimes \mathbf{e}_j \otimes \mathbf{e}_k \otimes \mathbf{e}_i \otimes \mathbf{e}_k \otimes \mathbf{e}_j$$

$$\overset{\langle 6 \rangle}{\boldsymbol{H}}_{11} + \overset{\langle 6 \rangle}{\boldsymbol{H}}_{14} + \overset{\langle 6 \rangle}{\boldsymbol{H}}_{12} + \overset{\langle 6 \rangle}{\boldsymbol{H}}_{15}$$

$$= \mathbf{e}_i \otimes \mathbf{e}_j \otimes \mathbf{e}_k \otimes \mathbf{e}_j \otimes \mathbf{e}_i \otimes \mathbf{e}_k + \mathbf{e}_i \otimes \mathbf{e}_j \otimes \mathbf{e}_k \otimes \mathbf{e}_j \otimes \mathbf{e}_k \otimes \mathbf{e}_i$$

$$+ \mathbf{e}_i \otimes \mathbf{e}_j \otimes \mathbf{e}_k \otimes \mathbf{e}_k \otimes \mathbf{e}_i \otimes \mathbf{e}_j + \mathbf{e}_i \otimes \mathbf{e}_j \otimes \mathbf{e}_k \otimes \mathbf{e}_k \otimes \mathbf{e}_j \otimes \mathbf{e}_i$$

$$\overset{\langle 6 \rangle}{\boldsymbol{H}}_{1} + \overset{\langle 6 \rangle}{\boldsymbol{H}}_{4} + \overset{\langle 6 \rangle}{\boldsymbol{H}}_{13} + \overset{\langle 6 \rangle}{\boldsymbol{H}}_{10}$$

$$= \mathbf{e}_i \otimes \mathbf{e}_i \otimes \mathbf{e}_k \otimes \mathbf{e}_k \otimes \mathbf{e}_p \otimes \mathbf{e}_p + \mathbf{e}_i \otimes \mathbf{e}_j \otimes \mathbf{e}_i \otimes \mathbf{e}_j \otimes \mathbf{e}_p \otimes \mathbf{e}_p$$

$$+ \mathbf{e}_i \otimes \mathbf{e}_j \otimes \mathbf{e}_j \otimes \mathbf{e}_m \otimes \mathbf{e}_m \otimes \mathbf{e}_i + \mathbf{e}_i \otimes \mathbf{e}_j \otimes \mathbf{e}_j \otimes \mathbf{e}_m \otimes \mathbf{e}_i \otimes \mathbf{e}_m$$

$$\overset{\langle 6 \rangle}{\boldsymbol{H}}_{2} + \overset{\langle 6 \rangle}{\boldsymbol{H}}_{3} + \overset{\langle 6 \rangle}{\boldsymbol{H}}_{5} + \overset{\langle 6 \rangle}{\boldsymbol{H}}_{6}$$

$$= \mathbf{e}_i \otimes \mathbf{e}_i \otimes \mathbf{e}_k \otimes \mathbf{e}_m \otimes \mathbf{e}_k \otimes \mathbf{e}_m + \mathbf{e}_i \otimes \mathbf{e}_i \otimes \mathbf{e}_k \otimes \mathbf{e}_m \otimes \mathbf{e}_m \otimes \mathbf{e}_k$$

$$+ \mathbf{e}_i \otimes \mathbf{e}_j \otimes \mathbf{e}_i \otimes \mathbf{e}_m \otimes \mathbf{e}_j \otimes \mathbf{e}_m + \mathbf{e}_i \otimes \mathbf{e}_j \otimes \mathbf{e}_i \otimes \mathbf{e}_m \otimes \mathbf{e}_m \otimes \mathbf{e}_j \,.$$

Instead of $\overset{\langle 6 \rangle}{\boldsymbol{H}}_{8} + \overset{\langle 6 \rangle}{\boldsymbol{H}}_{9}$ the 6th-order identity would give the same result in the square-form of the energy.

The complete hemitropic elastic energy is

(12.13)
$$2\,\rho_0 w = a_1 tr^2\,\mathbf{E}^G + a_2 tr\,(\mathbf{E}^{G\,2}) + 2\,b_1\,\mathbf{E}^G \cdot (\overset{\langle 3\rangle}{\boldsymbol{\varepsilon}} : \mathbf{C})$$
$$+\, b_2\,\mathbf{C}\cdot\mathbf{C} + b_3\,\mathbf{C}\cdot\mathbf{C}^{[12]} + b_4(\mathbf{C}:\mathbf{I})\cdot(\mathbf{C}:\mathbf{I})$$
$$+\, b_5\,\mathbf{I}:\mathbf{C}\cdot\mathbf{C}:\mathbf{I} + b_6\,(\mathbf{I}:\mathbf{C})\cdot(\mathbf{I}:\mathbf{C})\,.$$

Here $\cdot$ denotes the scalar product and $:$ a double contraction. a_i play the role of the two classical LAMÉ constants, and b_i six additional scalar elastic constants. $\mathbf{C}^{[12]}$ stands for the transposition of $\mathbf{C}$ with respect to its first and second entry, and $\overset{\langle 3\rangle}{\boldsymbol{\varepsilon}}$ is the third-order permutation tensor.

This gives the following stresses after (12.9)

$$\mathbf{T}^{2PK} = a_1(\mathbf{E}^G:\mathbf{I})\,\mathbf{I} + a_2\,\mathbf{E}^G + b_6\,\overset{\langle 3\rangle}{\boldsymbol{\varepsilon}}:\mathbf{C}$$

$$\mathbf{S} = sym^{[23]}\big[b_1\,\mathbf{C}:\mathbf{I}\otimes\mathbf{I} + b_2\,\mathbf{I}\otimes\mathbf{C}:\mathbf{I}$$
$$+\, b_3\,\mathbf{I}\otimes\mathbf{I}:\mathbf{C} + b_4\,\mathbf{C} + b_5\,\mathbf{C}^{[12]} + b_6\,\overset{\langle 3\rangle}{\boldsymbol{\varepsilon}}\,\mathbf{E}^G\big]$$
$$= sym^{[23]}\big[b_2\,\mathbf{I}\otimes\mathbf{C}:\mathbf{I} + b_3\,\mathbf{I}\otimes\mathbf{I}:\mathbf{C} + b_6\,\overset{\langle 3\rangle}{\boldsymbol{\varepsilon}}\,\mathbf{E}^G\big]$$
$$+\, b_1\,\mathbf{C}:\mathbf{I}\otimes\mathbf{I} + b_4\,\mathbf{C} + b_5/2\,(\mathbf{C}^{[12]} + \mathbf{C}^{[13]})\,.$$

$sym^{[23]}$ stands for the symmetrization in the second and third entries. If we additionally assume *isotropy*, then the hemitropic part with b_6 vanishes and we obtain the isotropic version of MINDLIN/ ESHEL (1968) with only 7 independent parameters[182].

The linearity of the elastic laws will not be assumed in what follows, in order to preserve full generality.

12.4 Finite Gradient Elastoplasticity

The extension of elastoplasticity of Chapt. 10 to second-gradient materials[183] is straight-forward since all steps here are in complete analogy to those of simple materials. The only difference so far is the extension of the kinematic set which becomes here the extended configuration space $\mathscr{Conf}$, and the analogous extension of the dynamical set consisting of the two stress tensors. Both sets have the dimension 24 if we use all symmetries.

---

[182] see SUIKER/ CHANG (2000), DELL´ISOLA/ SCIARRA/ VIDOLI (2009), AUFFRAY/ LE QUANG/ HE (2013), BERTRAM/ FOREST (2014), and BEHESHTI (2017)

[183] This section is based on BERTRAM (2015).

The basic concept for plasticity is again that of elastic ranges.

Definition 12.4. A (hyper-) **elastic range** is a pair $\{\mathscr{E}_p, w_p\}$ consisting of

1.) a path-connected submanifold with boundary $\mathscr{E}_p \subset \mathscr{Conf}$ of the configuration space

2.) and an elastic energy function

$$w_p : \mathscr{E}_p \to \mathscr{R} \;\; | (\mathbf{C}, \boldsymbol{\mathsf{C}}) \mapsto w_p(\mathbf{C}, \boldsymbol{\mathsf{C}})$$

such that after any continuation process $\{\mathbf{C}(\tau), \boldsymbol{\mathsf{C}}(\tau)\} \,|_{t_0}^{t}$, which remains entirely in $\mathscr{E}_p$

$$\{\mathbf{C}(\tau), \boldsymbol{\mathsf{C}}(\tau)\} \in \mathscr{E}_p \qquad \forall\, \tau \in [t_0, t]$$

the stresses are determined after (12.9) by the final values of the process as

$$\mathbf{T}^{2PK}(t) = 2\rho_0\, \partial w_p(\mathbf{C}, \boldsymbol{\mathsf{C}})/\partial \mathbf{C} =: k_p(\mathbf{C}, \boldsymbol{\mathsf{C}})$$

$$\boldsymbol{\mathsf{S}}(t) = \rho_0\, \partial w_p(\mathbf{C}, \boldsymbol{\mathsf{C}})/\partial \boldsymbol{\mathsf{C}} =: K_p(\mathbf{C}, \boldsymbol{\mathsf{C}}) .$$

In order to simplify notations in what follows, we will extend the elastic laws beyond $\mathscr{E}_p$ to the whole configuration space $\mathscr{Conf}$ so that it is easier to compare the behaviour within two elastic ranges.

The most important assumption in Chapt. 10 was that of the isomorphy of the elastic ranges, which we will also use in the present context.

To make this statement more precise, let $\{\mathscr{E}_{p1}, w_1\}$ and $\{\mathscr{E}_{p2}, w_2\}$ be two elastic ranges. Then according to Theorem 12.4 the elastic laws of the two elastic ranges are isomorphic if there exist two tensors of order two and three $(\mathbf{P}_{12}, \boldsymbol{\mathsf{P}}_{12})$ and a scalar w_c such that

- for the mass densities ρ_{01} and ρ_{02} in the reference placements at this particular point and some neighbourhood the following holds

$$\rho_{01} = \rho_{02}\, det\, \mathbf{P}_{12} \tag{12.14}$$

- for the elastic energies we have the equality after (12.10)

$$w_2(\mathbf{C}, \boldsymbol{\mathsf{C}}) = w_1(\mathbf{P}^T * \mathbf{C}, \mathbf{P}^T \circ \boldsymbol{\mathsf{C}} + \boldsymbol{\mathsf{P}}) + w_c$$

such that the elastic stress laws are after (12.9)

$$k_2(\mathbf{C}, \boldsymbol{\mathsf{C}}) = (det^{-1}\mathbf{P}_{12})\, [\mathbf{P}_{12} * k_1(\mathbf{P}_{12}^T * \mathbf{C}, \mathbf{P}_{12}^T \circ \boldsymbol{\mathsf{C}} + \boldsymbol{\mathsf{P}}_{12})]$$

$$K_2(\mathbf{C}, \boldsymbol{\mathsf{C}}) = (det^{-1}\mathbf{P}_{12})\, [\mathbf{P}_{12} \circ K_1(\mathbf{P}_{12}^T * \mathbf{C}, \mathbf{P}_{12}^T \circ \boldsymbol{\mathsf{C}} + \boldsymbol{\mathsf{P}}_{12})]$$

$\forall\, (\mathbf{C}, \boldsymbol{\mathsf{C}}) \in \mathscr{Conf}$.

As we have chosen a joint reference placement for all elastic laws of one particular material point (this is, however, not compulsory), we already have

$\rho_{01} \equiv \rho_{02}$, and therefore $\mathbf{P}_{12}$ must be proper unimodular, so that the first isomorphy condition (12.14) is always fulfilled.

The constant w_c is linked to the two elastic ranges under consideration and constant during processes that do not leave the current elastic range, *i.e.*, during elastic process segments. For different elastic ranges we will have to expect different constants as well.

Again we can use some freely chosen **elastic reference energy** w_0 so that the isomorphy condition can be cast into the following form.

Theorem 12.5. *Let* w_0 *be the elastic reference energy for an elasto-plastic material. Then for each elastic range* $\{\mathscr{E}_p, w_p\}$ *there is a tensor couple* $(\mathbf{P}, \boldsymbol{\mathsf{P}})$ *and a scalar* w_{c0} *such that*

(12.15) $$w_p(\mathbf{C}, \boldsymbol{\mathsf{C}}) = w_0(\mathbf{P}^T * \mathbf{C}, \mathbf{P}^T \circ \boldsymbol{\mathsf{C}} + \boldsymbol{\mathsf{P}}) + w_{c0}$$

for all $(\mathbf{C}, \boldsymbol{\mathsf{C}}) \in \mathscr{Conf}$.

We consider the second-order unimodular $\mathbf{P}$ and the third-order $\boldsymbol{\mathsf{P}}$ as independent internal variables (*unconstrained gradient plasticity*). This is in contrast to theories of constrained gradient plasticity, where the higher-order plastic variables result from the second-order one by differentiation.

Practically one will choose (*i*) the elastic reference energy w_0, and (*ii*) introduce $(\mathbf{P}, \boldsymbol{\mathsf{P}})$ as internal variables for which evolution equations are needed, which we call *flow rules*. The two tensors $(\mathbf{P}, \boldsymbol{\mathsf{P}})$ are the plastic variables in this theory, eventually complemented by hardening variables.

While w_{c0} does not play any role in elasticity, in plasticity the situation is completely different. Because of the isomorphy condition, w_{c0} is constant for one elastic range. Hence it cannot depend on $\mathbf{C}$ and $\boldsymbol{\mathsf{C}}$, but it can depend on all other internal or plastic variables of the constitutive model.

The elastic laws are then given by the potentials

$$\begin{aligned}\mathbf{T}^{2PK} &= k_p(\mathbf{C}, \boldsymbol{\mathsf{C}}) = 2\rho_0\, \partial w_p(\mathbf{C}, \boldsymbol{\mathsf{C}})/\partial \mathbf{C} \\ &= 2\rho_0\, \partial w_0(\mathbf{P}^T * \mathbf{C}, \mathbf{P}^T \circ \boldsymbol{\mathsf{C}}) + \boldsymbol{\mathsf{P}})/\partial \mathbf{C} \\ &= \mathbf{P} * k_0(\mathbf{P}^T * \mathbf{C}, \mathbf{P}^T \circ \boldsymbol{\mathsf{C}} + \boldsymbol{\mathsf{P}})\end{aligned}$$

and

$$\begin{aligned}\boldsymbol{\mathsf{S}} &= K_p(\mathbf{C}, \boldsymbol{\mathsf{C}}) = \rho_0\, \partial w_p(\mathbf{C}, \boldsymbol{\mathsf{C}})/\partial \boldsymbol{\mathsf{C}} \\ &= \rho_0\, \partial w_0(\mathbf{P}^T * \mathbf{C}, \mathbf{P}^T \circ \boldsymbol{\mathsf{C}} + \boldsymbol{\mathsf{P}})/\partial \boldsymbol{\mathsf{C}} \\ &= \mathbf{P} \circ K_0(\mathbf{P}^T * \mathbf{C}, \mathbf{P}^T \circ \boldsymbol{\mathsf{C}} + \boldsymbol{\mathsf{P}})\end{aligned}$$

for all $(\mathbf{C}, \boldsymbol{\mathsf{C}}) \in \mathscr{Conf}$.

By extending the concept of a yield criterion from Sect. 10.1 to gradient materials, we assume the general form of the **yield criterion** as

$$\varphi(\mathbf{P}, \boldsymbol{P}, \mathbf{C}, \boldsymbol{C}, \mathbf{Z}_p)$$

with hardening variables $\mathbf{Z}_p$ of any order. With this extension we obtain for the **yield condition**

$$\varphi(\mathbf{P}, \boldsymbol{P}, \mathbf{C}, \boldsymbol{C}, \mathbf{Z}_p) = 0 \tag{12.16}$$

and for the **loading condition**

$$\partial\varphi / \partial\mathbf{C} \cdot \mathbf{C}^{\bullet} + \partial\varphi / \partial\boldsymbol{C} \cdot \boldsymbol{C}^{\bullet} > 0 \tag{12.17}$$

where the plastic variables are kept constant. Again, these two conditions are necessary and sufficient for yielding.

We can now decompose the stress power for our elastoplastic material into a conservative (elastic) and a dissipative (plastic) part. The specific stress power (12.7) is

$$\begin{aligned} l &= 1/\rho_0 \, (\tfrac{1}{2}\, \mathbf{T}^{2PK} \cdot \mathbf{C}^{\bullet} + \boldsymbol{S} \cdot \boldsymbol{C}^{\bullet}) \\ &= 1/\rho_0 \, [\tfrac{1}{2}\, k_p(\mathbf{C}, \boldsymbol{C}) \cdot \mathbf{C}^{\bullet} + K_p(\mathbf{C}, \boldsymbol{C}) \cdot \boldsymbol{C}^{\bullet}] \\ &= 1/\rho_0 \, [\tfrac{1}{2}\, \mathbf{P} * k_0 (\mathbf{P}^T * \mathbf{C}, \mathbf{P}^T \circ \boldsymbol{C} + \boldsymbol{P}) \cdot \mathbf{C}^{\bullet} \\ &\qquad + \mathbf{P} \circ K_0 (\mathbf{P}^T * \mathbf{C}, \mathbf{P}^{-1} \mathbf{P}^{-T} (\mathbf{P}^T * \boldsymbol{K}) + \boldsymbol{P}) \cdot \boldsymbol{C}^{\bullet}] \\ &= 1/\rho_0 \, [\tfrac{1}{2}\, k_0(\mathbf{C}_e, \boldsymbol{C}_e) \cdot (\mathbf{P}^T * \mathbf{C}^{\bullet}) + K_0 (\mathbf{C}_e, \boldsymbol{C}_e) \cdot (\mathbf{P}^T \circ \boldsymbol{C}^{\bullet})] \end{aligned} \tag{12.18}$$

with the abbreviations

$$\begin{aligned} \mathbf{C}_e &:= \mathbf{P}^T * \mathbf{C} = \mathbf{P}^T \, \mathbf{C} \, \mathbf{P} \\ \boldsymbol{C}_e &:= \mathbf{P}^T \circ \boldsymbol{C} + \boldsymbol{P} . \end{aligned}$$

This gives for the rates

$$\begin{aligned} \mathbf{C}_e^{\bullet} &= (\mathbf{P}^T \, \mathbf{C} \, \mathbf{P})^{\bullet} = \mathbf{P}^T \, \mathbf{C}^{\bullet} \, \mathbf{P} + 2\, sym(\mathbf{P}^T \, \mathbf{C} \, \mathbf{P}^{\bullet}) \\ &= \mathbf{P}^T * \mathbf{C}^{\bullet} + 2\, sym(\mathbf{C}_e \, \mathbf{P}^{-1} \, \mathbf{P}^{\bullet}) \end{aligned}$$

where *sym* stands for the symmetric part, and

$$\begin{aligned} \boldsymbol{C}_e^{\bullet} &= \{\mathbf{P}^T \circ \boldsymbol{C} + \boldsymbol{P}\}^{\bullet} \\ &= \mathbf{P}^T \circ \boldsymbol{C}^{\bullet} + \boldsymbol{P}^{\bullet} + C_{ijk} \, \{(\mathbf{P}^{-1\bullet} \, \mathbf{e}_i) \otimes (\mathbf{P}^T \, \mathbf{e}_j) \otimes (\mathbf{P}^T \, \mathbf{e}_k) \\ &\quad + (\mathbf{P}^{-1} \, \mathbf{e}_i) \otimes (\mathbf{P}^{T\bullet} \, \mathbf{e}_j) \otimes (\mathbf{P}^T \, \mathbf{e}_k) + (\mathbf{P}^{-1} \, \mathbf{e}_i) \otimes (\mathbf{P}^T \, \mathbf{e}_j) \otimes (\mathbf{P}^{T\bullet} \, \mathbf{e}_k)\} \\ &= \mathbf{P}^T \circ \boldsymbol{C}^{\bullet} + \boldsymbol{P}^{\bullet} - \mathbf{P}^{-1} \, \mathbf{P}^{\bullet} (\boldsymbol{C}_e - \boldsymbol{P}) \\ &\quad + 2\, subsym \, [(\boldsymbol{C}_e - \boldsymbol{P}) \, \mathbf{P}^{-1} \, \mathbf{P}^{\bullet}] \end{aligned}$$

with *subsym* indicating the symmetric part with respect to the right subsymmetry. We insert this into (12.18) to obtain

$$l = 1/\rho_0 \,\{ \tfrac{1}{2}\, k_0(\mathbf{C}_e, \boldsymbol{C}_e) \cdot [\mathbf{C}_e^\bullet - 2\, sym(\mathbf{C}_e\, \mathbf{P}^{-1}\, \mathbf{P}^\bullet)]$$

$$+ K_0(\mathbf{C}_e, \boldsymbol{K}_e) \cdot [\boldsymbol{C}_e^\bullet - \boldsymbol{P}^\bullet + \mathbf{P}^{-1}\, \mathbf{P}^\bullet\, (\boldsymbol{C}_e - \boldsymbol{P})$$

$$- 2\, subsym\, [(\boldsymbol{C}_e - \boldsymbol{P})\, \mathbf{P}^{-1}\, \mathbf{P}^\bullet]\}$$

and because of the symmetries of the stress tensors

$$= 1/\rho_0 \,\{ \tfrac{1}{2}\, k_0(\mathbf{C}_e, \boldsymbol{C}_e) \cdot [\mathbf{C}_e^\bullet - 2\, sym(\mathbf{C}_e\, \mathbf{P}^{-1}\, \mathbf{P}^\bullet)]$$

$$+ K_0(\mathbf{C}_e, \boldsymbol{K}_e) \cdot [\boldsymbol{K}_e^\bullet - \boldsymbol{P}^\bullet + \mathbf{P}^{-1}\, \mathbf{P}^\bullet\, (\boldsymbol{K}_e - \boldsymbol{P}) - 2\, (\boldsymbol{K}_e - \boldsymbol{P})\, \mathbf{P}^{-1} \cdot \mathbf{P}^\bullet\}$$

$$= 1/\rho_0 \,\{ \tfrac{1}{2}\, k_0(\mathbf{C}_e, \boldsymbol{K}_e) \cdot \mathbf{C}_e^\bullet + K_0(\mathbf{C}_e, \boldsymbol{K}_e) \cdot \boldsymbol{K}_e^\bullet$$

$$- \tfrac{1}{2}\, k_0(\mathbf{C}_e, \boldsymbol{K}_e) \cdot (2\, \mathbf{C}_e\, \mathbf{P}^{-1}\, \mathbf{P}^\bullet)$$

$$(12.19) \qquad - K_0(\mathbf{C}_e, \boldsymbol{K}_e) \cdot [\boldsymbol{P}^\bullet - \mathbf{P}^{-1}\, \mathbf{P}^\bullet\, (\boldsymbol{K}_e - \boldsymbol{P}) + 2\, (\boldsymbol{K}_e - \boldsymbol{P})\, \mathbf{P}^{-1}\, \mathbf{P}^\bullet]\}$$

$$= \partial_{\mathbf{C}}\, w_0(\mathbf{C}_e, \boldsymbol{K}_e) \cdot \mathbf{C}_e^\bullet + \partial_{\boldsymbol{K}}\, w_0\, (\mathbf{C}_e, \boldsymbol{K}_e) \cdot \boldsymbol{K}_e^\bullet$$

$$- \tfrac{1}{2}\, k_0(\mathbf{C}_e, \boldsymbol{K}_e) \cdot (2\, \mathbf{C}_e\, \mathbf{P}^{-1}\, \mathbf{P}^\bullet)$$

$$- K_0(\mathbf{C}_e, \boldsymbol{K}_e) \cdot [\boldsymbol{P}^\bullet - \mathbf{P}^{-1}\, \mathbf{P}^\bullet\, (\boldsymbol{K}_e - \boldsymbol{P}) + 2\, (\boldsymbol{K}_e - \boldsymbol{P})\, \mathbf{P}^{-1}\, \mathbf{P}^\bullet]$$

$$= w_0(\mathbf{C}_e, \boldsymbol{K}_e)^\bullet + \mathbf{T}^{2PK}{}_p \cdot \mathbf{P}^\bullet$$

$$+ \boldsymbol{S}_p \cdot [\boldsymbol{P}^\bullet - \mathbf{P}^{-1}\, \mathbf{P}^\bullet\, (\boldsymbol{K}_e - \boldsymbol{P}) + 2\, (\boldsymbol{K}_e - \boldsymbol{P})\, \mathbf{P}^{-1}\, \mathbf{P}^\bullet]$$

with the **plastic stress tensor** defined as

$$\mathbf{T}^{2PK}{}_p := -\mathbf{P}^{-T}\, \mathbf{C}_e\, k_0(\mathbf{C}_e, \boldsymbol{C}_e) = -\mathbf{P}^{-T}\, \mathbf{P}^T\, \mathbf{C}\, \mathbf{P}\, \mathbf{P}^{-1}\, \mathbf{T}^{2PK}\, \mathbf{P}^{-T}$$

$$= -\mathbf{C}\, \mathbf{T}^{2PK}\, \mathbf{P}^{-T}$$

and **plastic hyperstress tensor**

$$\boldsymbol{S}_p := -K_0(\mathbf{C}_e, \boldsymbol{C}_e) = -\mathbf{P}^T \circ \boldsymbol{S} .$$

According to (12.19), the stress power goes into a change of the elastic reference energy and a dissipative part that is only active during yielding, and works on the rates $\mathbf{P}^\bullet$ and $\boldsymbol{P}^\bullet$.

For these two plastic internal variables and for the hardening variable(s) $\mathbf{Z}_p$ we need evolution equations, namely two **flow rules** and a **hardening rule**, for which we make an ansatz as rate-independent ODEs as customary in plasticity

$$\mathbf{P}^\bullet = f(\mathbf{P}, \boldsymbol{P}, \mathbf{C}, \boldsymbol{C}, \mathbf{Z}_p, \mathbf{C}^\bullet, \boldsymbol{C}^\bullet)$$

$$\boldsymbol{P}^\bullet = F(\mathbf{P}, \boldsymbol{P}, \mathbf{C}, \boldsymbol{C}, \mathbf{Z}_p, \mathbf{C}^\bullet, \boldsymbol{C}^\bullet)$$

$$\mathbf{Z}_p^\bullet = h(\mathbf{P}, \boldsymbol{P}, \mathbf{C}, \boldsymbol{C}, \mathbf{Z}_p, \mathbf{C}^\bullet, \boldsymbol{C}^\bullet) .$$

The rate-independence can be assured in the usual way by the introduction of a joint **plastic consistency parameter** $\lambda \geq 0$

$$\mathbf{P}^\bullet = \lambda f^\circ(\mathbf{P}, \boldsymbol{\mathsf{P}}, \mathbf{C}, \boldsymbol{\mathsf{C}}, \mathbf{Z}_p, \mathbf{C}^\circ, \boldsymbol{\mathsf{C}}^\circ)$$

(12.20)
$$\boldsymbol{\mathsf{P}}^\bullet = \lambda F^\circ(\mathbf{P}, \boldsymbol{\mathsf{P}}, \mathbf{C}, \boldsymbol{\mathsf{C}}, \mathbf{Z}_p, \mathbf{C}^\circ, \boldsymbol{\mathsf{C}}^\circ)$$

$$\mathbf{Z}_p^\bullet = \lambda h^\circ(\mathbf{P}, \boldsymbol{\mathsf{P}}, \mathbf{C}, \boldsymbol{\mathsf{C}}, \mathbf{Z}_p, \mathbf{C}^\circ, \boldsymbol{\mathsf{C}}^\circ)$$

where we normed the increments of the kinematical variables

$$\mathbf{C}^\circ := \mathbf{C}^\bullet/\mu \qquad \text{and} \qquad \boldsymbol{\mathsf{C}}^\circ := \boldsymbol{\mathsf{C}}^\bullet/\mu$$

by a factor

$$\mu := \sqrt{(|\mathbf{C}^\bullet|^2 + L^2\, |\boldsymbol{\mathsf{C}}^\bullet|^2)}$$

which is positive during yielding. The positive constant L with the dimension of a length is necessary for dimensional reasons and controls the ratio of yielding due to $\mathbf{C}^\bullet$ and $\boldsymbol{\mathsf{C}}^\bullet$.

The three functions f°, F°, h° give the directions of the flow and hardening, while the amount is determined by the consistency parameter λ. It is zero during elastic segments and positive during yielding when it can be calculated by the consistency condition (12.16)

$$\begin{aligned}
0 &= \varphi(\mathbf{P}, \boldsymbol{\mathsf{P}}, \mathbf{C}, \boldsymbol{\mathsf{C}}, \mathbf{Z}_p)^\bullet \\
&= \partial\varphi/\partial\mathbf{P} \cdot \mathbf{P}^\bullet + \partial\varphi/\partial\boldsymbol{\mathsf{P}} \cdot \boldsymbol{\mathsf{P}}^\bullet + \partial\varphi/\partial\mathbf{C} \cdot \mathbf{C}^\bullet \\
&\quad + \partial\varphi/\partial\boldsymbol{\mathsf{C}} \cdot \boldsymbol{\mathsf{C}}^\bullet + \partial\varphi/\partial\mathbf{Z}_p \cdot \mathbf{Z}_p^\bullet \\
&= \partial\varphi/\partial\mathbf{P} \cdot \lambda f^\circ(\mathbf{P}, \boldsymbol{\mathsf{P}}, \mathbf{C}, \boldsymbol{\mathsf{C}}, \mathbf{Z}_p, \mathbf{C}^\circ, \boldsymbol{\mathsf{C}}^\circ) \\
&\quad + \partial\varphi/\partial\boldsymbol{\mathsf{P}} \cdot \lambda F^\circ(\mathbf{P}, \boldsymbol{\mathsf{P}}, \mathbf{C}, \boldsymbol{\mathsf{C}}, \mathbf{Z}_p, \mathbf{C}^\circ, \boldsymbol{\mathsf{C}}^\circ) \\
&\quad + \partial\varphi/\partial\mathbf{C} \cdot \mathbf{C}^\bullet + \partial\varphi/\partial\boldsymbol{\mathsf{C}} \cdot \boldsymbol{\mathsf{C}}^\bullet \\
&\quad + \partial\varphi/\partial\mathbf{Z}_p \cdot \lambda h^\circ(\mathbf{P}, \boldsymbol{\mathsf{P}}, \mathbf{C}, \boldsymbol{\mathsf{C}}, \mathbf{Z}_p, \mathbf{C}^\circ, \boldsymbol{\mathsf{C}}^\circ)
\end{aligned}$$

which gives for the solution the quotient

$$\begin{aligned}
\lambda &= -[\partial\varphi/\partial\mathbf{C} \cdot \mathbf{C}^\bullet + \partial\varphi/\partial\boldsymbol{\mathsf{C}} \cdot \boldsymbol{\mathsf{C}}^\bullet] / \\
&\quad [\partial\varphi/\partial\mathbf{P} \cdot f^\circ(\mathbf{P}, \boldsymbol{\mathsf{P}}, \mathbf{C}, \boldsymbol{\mathsf{C}}, \mathbf{Z}_p, \mathbf{C}^\circ, \boldsymbol{\mathsf{C}}^\circ) \\
&\quad + \partial\varphi/\partial\boldsymbol{\mathsf{P}} \cdot F^\circ(\mathbf{P}, \boldsymbol{\mathsf{P}}, \mathbf{C}, \boldsymbol{\mathsf{C}}_p, \mathbf{Z}, \mathbf{C}^\circ, \boldsymbol{\mathsf{C}}^\circ) \\
&\quad + \partial\varphi/\partial\mathbf{Z}_p \cdot h^\circ(\mathbf{P}, \boldsymbol{\mathsf{P}}, \mathbf{C}, \boldsymbol{\mathsf{C}}, \mathbf{Z}_p, \mathbf{C}^\circ, \boldsymbol{\mathsf{C}}^\circ)].
\end{aligned}$$

Both, numerator and denominator of this ratio are always negative during yielding as a consequence of the loading condition (12.17), and, thus, λ is positive in this case.

If we substitute this value of λ into (12.20), we obtain the **consistent flow** and **hardening rules**. In all cases (elastic *and* plastic), the KUHN-TUCKER condition

$$\lambda\,\varphi = 0 \qquad \text{with} \qquad \lambda \geq 0 \qquad \text{and} \qquad \varphi \leq 0$$

holds since at any time one of the two factors is zero.

In BERTRAM (2019) an example for such a gradient elastoplastic model is given, which extends the J_2 –theory of simple materials (anisotropic v. MISES yield criterion[184] and associated flow rule[185]).

Thermodynamic Consistency

We will finally show under which conditions our elasto-plastic gradient model satisfies the thermodynamic consistency. We will do this by using the CLAUSIUS-PLANCK inequality (3.38) which requires with (12.19) that

$$w^\bullet \leq l = w_0(\mathbf{C}_e, \boldsymbol{K}_e)^\bullet + \mathbf{T}^{2PK}{}_p \cdot \mathbf{P}^\bullet$$
$$+ \boldsymbol{S}_p \cdot [\boldsymbol{P}^\bullet - \mathbf{P}^{-1}\,\mathbf{P}^\bullet\,(\boldsymbol{K}_e - \boldsymbol{P}) + 2\,(\boldsymbol{K}_e - \boldsymbol{P})\,\mathbf{P}^{-1}\,\mathbf{P}^\bullet]$$

holds for all processes. If we identify w with

$$w_p(\mathbf{C}, \boldsymbol{C}) = w_0\,(\mathbf{P}^T * \mathbf{C}, \mathbf{P}^T \circ \boldsymbol{C} + \boldsymbol{P}) + w_{c0}\,(\mathbf{P}, \boldsymbol{P}, \mathbf{Z}_p)$$

after (12.15), then the rate of the energy w_0 drops out on both sides of the inequality, and only the ***residual dissipation inequality***

$$\mathbf{T}^{2PK}{}_p \cdot \mathbf{P}^\bullet + \boldsymbol{S}_p \cdot [\boldsymbol{P}^\bullet - \mathbf{P}^{-1}\,\mathbf{P}^\bullet\,(\boldsymbol{K}_e - \boldsymbol{P}) + 2\,(\boldsymbol{K}_e - \boldsymbol{P})\,\mathbf{P}^{-1}\,\mathbf{P}^\bullet]$$
$$\geq \partial w_{c0}/\partial \mathbf{P} \cdot \mathbf{P}^\bullet + \partial w_{c0}/\partial \boldsymbol{P} \cdot \boldsymbol{P}^\bullet + \partial w_{c0}/\partial \mathbf{Z}_p \cdot \mathbf{Z}_p^\bullet$$

remains. This can be further specified by the flow and hardening rules (12.20)

$$(\mathbf{T}^{2PK}{}_p - \partial w_{c0}/\partial \mathbf{P}) \cdot \lambda\,\mathbf{P}^\circ$$
$$+ \boldsymbol{S}_p \cdot [\lambda\,\boldsymbol{P}^\circ - \mathbf{P}^{-1}\,\lambda\,\mathbf{P}^\circ\,(\boldsymbol{K}_e - \boldsymbol{P}) + 2\,(\boldsymbol{K}_e - \boldsymbol{P})\,\mathbf{P}^{-1}\,\lambda\,\mathbf{P}^\circ]$$
$$- \partial w_{c0}/\partial \boldsymbol{P} \cdot \lambda\,\boldsymbol{P}^\circ - \partial w_{c0}/\partial \mathbf{Z}_p \cdot \lambda\,\mathbf{Z}_p^\circ \geq 0$$

with the normed increments

$$\mathbf{P}^\circ := f^\circ(\mathbf{P}, \boldsymbol{P}, \mathbf{C}, \boldsymbol{C}, \mathbf{Z}_p, \mathbf{C}^\circ, \boldsymbol{C}^\circ)$$
$$\boldsymbol{P}^\circ = F^\circ(\mathbf{P}, \boldsymbol{P}, \mathbf{C}, \boldsymbol{C}, \mathbf{Z}_p, \mathbf{C}^\circ, \boldsymbol{C}^\circ)$$
$$\mathbf{Z}_p^\circ = h^\circ(\mathbf{P}, \boldsymbol{P}, \mathbf{C}, \boldsymbol{C}, \mathbf{Z}_p, \mathbf{C}^\circ, \boldsymbol{C}^\circ)$$

or, for a non-negative plastic parameter λ

184 v. MISES (1928)

185 see BERTRAM/ FOREST (2014)

$$(\mathbf{T}^{2PK}{}_p - \partial w_{c0}/\partial \mathbf{P}) \cdot \mathbf{P}^\circ$$
$$+ \boldsymbol{S}_p \cdot [\boldsymbol{P}^\circ - \mathbf{P}^{-1}\, \mathbf{P}^\circ (\boldsymbol{K}_e - \boldsymbol{P}) + 2\, (\boldsymbol{K}_e - \boldsymbol{P})\, \mathbf{P}^{-1}\, \mathbf{P}^\circ]$$
$$- \partial w_{c0}/\partial \boldsymbol{P} \cdot \boldsymbol{P}^\circ - \partial w_{c0}/\partial \mathbf{Z}_p \cdot \mathbf{Z}_p^\circ \geq 0$$

which is a restriction upon the flow and hardening rules.

The full thermomechanical theory of gradient elastoplasticity can be found in BERTRAM (2016, 2017, 2019). An extension of it to third-order gradients is given in REIHER/ BERTRAM (2018, 2020) and BERTRAM (2917, 2019).

12.5 Gradient Fluids

While the focus of the preceding sections on gradient materials was on solids, we will now consider viscous fluids. ST.-VENANT (1869) already stated that the velocity profile of a turbulent channel or pipe flow cannot be described by a simple fluid, neither linear by a NAVIER-STOKES law (4.4) nor non-linear like the REINER fluid (4.3), as we nowadays call them. Instead one should take into account higher velocity gradients as well.

The same idea was set into reality a century later by the Berlin mechanics school of TROSTEL[186] and co-workers. Unfortunately these works became little known in the scientific community. Later on FRIED/ GURTIN (2006) suggested the same ansatz. TROSTEL´s intention was originally to describe fully developed turbulence. SILBER (1993) applied gradient models to describe blood flow, while FRIED/ GURTIN (2006) intended to describe nano flows. Recently, KRAWIETZ (2021) uses a similar ansatz and particularized it for the two-dimensional problem of a flow between two rotating cylinders (stirring process). This way, he can discuss in detail the role of the reaction stresses and of different boundary conditions.

We will here briefly recall this approach based on the works of TROSTEL (1985, 1988, 2010) and SILBER *et al.* (1986)[187]. For brevity we will only consider gradient fluids of order two, although a theory of order three also exists[188].

While the approach of NAVIER-STOKES (4.4) assumes a dependence of the CAUCHY stresses upon the symmetric part of the velocity gradient, we extend the kinematic set by the second velocity gradient, and the dynamical set by a third-order hyperstress tensor. Of interest here are the viscous parts, while the elastic

[186] Rudolf Trostel (1928 - 2016)
[187] For more details see BERTRAM (2019, 2020a) and KRAWIETZ (2021).
[188] See SILBER (1986).

part of a fluid, namely the dependence of the pressure upon the density is beyond the focus of this chapter.

So, for simplicity, we assume incompressibility. Internal constraints for gradient materials have been investigated by BERTRAM/ GLÜGE (2016). It leads to a decomposition of the stress tensors into reaction stresses and extra stresses. In what follows we will mainly focus on the extra stresses which must be given by a constitutive equation.

The basic assumption is that the CAUCHY extra stress tensor $\mathbf{T}_E$ and a third-order hyperstress tensor $\boldsymbol{T}_E$ depend on the velocity gradient $grad\ \mathbf{v}$ and $grad\ grad\ \mathbf{v}$ at the same point and the same instant of time. So we start with the assumption that two material laws exist such that

$$\mathbf{T}_E = f(grad\ \mathbf{v},\ grad\ grad\ \mathbf{v})$$

$$\boldsymbol{T}_E = F(grad\ \mathbf{v},\ grad\ grad\ \mathbf{v})\,.$$

By the *Principle of material objectivity* both stress tensors have to be objective. This ansatz fulfils the *Principle of equipresence* and the *Principle of local action*, but not yet the *Principle of invariance under superimposed rigid body modifications* that demands for all velocity fields

$$\mathbf{Q} * f(grad\ \mathbf{v},\ grad\ grad\ \mathbf{v}) = f(\mathbf{Q} * grad\ \mathbf{v} + \mathbf{Q}^{\bullet}\,\mathbf{Q}^T,\ \mathbf{Q} * grad\ grad\ \mathbf{v})$$

$$\mathbf{Q} * F(grad\ \mathbf{v},\ grad\ grad\ \mathbf{v}) = F(\mathbf{Q} * grad\ \mathbf{v} + \mathbf{Q}^{\bullet}\,\mathbf{Q}^T,\ \mathbf{Q} * grad\ grad\ \mathbf{v})$$

for all proper orthogonal time-dependent tensors $\mathbf{Q}$.

We have already seen that $\mathbf{Q}^{\bullet}\,\mathbf{Q}^T$ is skew. By its arbitrariness, we have to limit the dependence of the stresses upon its symmetric part, the rate of deformation tensor $\mathbf{D} = sym\ grad\ \mathbf{v}$. So we can reduce our considerations to functions of the form

$$\mathbf{T}_E = f(\mathbf{D},\ grad\ grad\ \mathbf{v})$$

$$\boldsymbol{T}_E = F(\mathbf{D},\ grad\ grad\ \mathbf{v})$$

for which we use the same symbol as before. The *Principle of invariance under superimposed rigid body modifications* demands for these laws

$$\mathbf{Q} * f(\mathbf{D},\ grad\ grad\ \mathbf{v}) = f(\mathbf{Q} * \mathbf{D},\ \mathbf{Q} * grad\ grad\ \mathbf{v})$$

$$\mathbf{Q} * F(\mathbf{D},\ grad\ grad\ \mathbf{v}) = F(\mathbf{Q} * \mathbf{D},\ \mathbf{Q} * grad\ grad\ \mathbf{v})$$

for all proper orthogonal tensors $\mathbf{Q}$. So the two stress laws must be hemitropic tensor functions. These functions can be linear or non-linear. In what follows, however, we will restrict our considerations to the linear case.

Linear Viscous Gradient Fluids

It is convenient to start with a **dissipation potential** as a positive semidefinite and hemitropic square form, which is similar to the elastic potential of (12.12)

$$\delta(\mathbf{D}, grad\, grad\, \mathbf{v})$$

$$= \tfrac{1}{2} \overset{\langle 4\rangle}{\boldsymbol{D}}_{22} \cdot (\mathbf{D} \otimes \mathbf{D}) + \overset{\langle 5\rangle}{\boldsymbol{D}}_{23} \cdot (\mathbf{D} \otimes grad\, grad\, \mathbf{v})$$

$$+ \tfrac{1}{2} \overset{\langle 6\rangle}{\boldsymbol{D}}_{33} \cdot (grad\, grad\, \mathbf{v} \otimes grad\, grad\, \mathbf{v})$$

with three hemitropic viscosity tensors $\overset{\langle 4\rangle}{\boldsymbol{D}}_{22}$, $\overset{\langle 5\rangle}{\boldsymbol{D}}_{23}$, and $\overset{\langle 6\rangle}{\boldsymbol{D}}_{33}$ of order 4, 5, and 6, respectively. This function acts as a potential for the extra stresses

$$\mathbf{T}_E = \partial\delta / \partial\mathbf{D} = \overset{\langle 4\rangle}{\boldsymbol{D}}_{22} [\mathbf{D}] + \overset{\langle 5\rangle}{\boldsymbol{D}}_{23} [grad\, grad\, \mathbf{v}]$$

$$\boldsymbol{T}_E = \partial\delta / \partial\, grad\, grad\, \mathbf{v} = [\mathbf{D}]\, \overset{\langle 5\rangle}{\boldsymbol{D}}_{23} + \overset{\langle 6\rangle}{\boldsymbol{D}}_{33} [grad\, grad\, \mathbf{v}]$$

where the brackets [] stand for a multiple contraction of the adjacent tensors according to the order of the argument in these brackets.

These hemitropic tensors can be submitted to the (sub)symmetries which **D** and *grad grad* **v** show. This restriction reduces the number of such hemitropic tensors drastically. In fact, in the present case only the following hemitropic tensors are needed as we have already shown in (12.13)

$$\overset{\langle 4\rangle}{\boldsymbol{D}}_{22} = a_1 \mathbf{I} \otimes \mathbf{I} + a_2 \overset{\langle 4\rangle}{\boldsymbol{I}}$$

$$\overset{\langle 6\rangle}{\boldsymbol{D}}_{33} = a_3 \mathbf{I} \otimes \mathbf{I} \otimes \mathbf{I}^{[46]} + a_4 \mathbf{I} \otimes \mathbf{I} \otimes \mathbf{I} + a_5 \mathbf{I} \otimes \mathbf{I} \otimes \mathbf{I}^{[13]\,[46]}$$

$$+ a_6 \overset{\langle 6\rangle}{\boldsymbol{I}} + a_7 \overset{\langle 6\rangle}{\boldsymbol{I}}{}^{[13]}$$

$$\overset{\langle 5\rangle}{\boldsymbol{D}}_{23} = a_8\, \varepsilon_{ijk}\, \mathbf{e}_i \otimes \mathbf{e}_l \otimes \mathbf{e}_j \otimes \mathbf{e}_l \otimes \mathbf{e}_k$$

with respect to an arbitrary ONB $\{\mathbf{e}_i\}$ with eight scalar factors a_i . This leads to the following extra stresses

$$\mathbf{T}_E = a_2\, \mathbf{D} + a_8\, sym[\, \overset{\langle 3\rangle}{\boldsymbol{\varepsilon}} : (grad\, grad\, \mathbf{v})]$$

$$\boldsymbol{T}_{\mathrm{E}} = sym^{[23]}\,[a_5\, grad\, grad\, \mathbf{v} : \mathbf{I} \otimes \mathbf{I} + a_6\, grad\, grad\, \mathbf{v} + a_7\, (grad\, grad \mathbf{v})^{[13]} + a_8\, \overset{\langle 3 \rangle}{\boldsymbol{\varepsilon}}\, \mathbf{D}]$$

as a complete generalization of the NAVIER-STOKES law to second gradient fluids.[189]

We assume **incompressibility** by the constraint equation[190] in an EULERian form as

$$div\, \mathbf{v} = \mathbf{I} \cdot\cdot\, \mathbf{D} = 0\,.$$

If the constraint holds in the entire fluid, the second-order constraint must also hold

$$(grad\, div\, \mathbf{v}) \cdot \mathbf{r} = (\mathbf{I} \otimes \mathbf{r}) \cdot (grad\, grad\, \mathbf{v}) = 0$$

with an arbitrary vector field $\mathbf{r}$. Then the second-order reaction stress is a hydrostatic pressure

$$\mathbf{T}_{\mathrm{R}} = -p\, \mathbf{I}$$

and the third-order reaction stress is

$$\boldsymbol{T}_{\mathrm{R}} = sym^{[23]}\, (\mathbf{I} \otimes \mathbf{r})$$

with an undetermined vector $\mathbf{r}$, so that none of them dissipates in any flow compatible with the constraint.[191]

The local balance of linear momentum (12.6) becomes in this case

$$\begin{aligned} \rho\, (\mathbf{a} - \mathbf{b}) &= div\, (\mathbf{T}_{\mathrm{E}} + \mathbf{T}_{\mathrm{R}} - div\, \boldsymbol{T}_{\mathrm{E}} - div\, \boldsymbol{T}_{\mathrm{R}}) \\ &= a_2/2\, \Delta\mathbf{v} - a_3\, curl\, \Delta\mathbf{v} - (a_4 + a_6)\, \Delta\Delta\mathbf{v} - grad\, p - grad\, div\, \mathbf{r}\,. \end{aligned}$$

This system of partial differential equations can be analytically solved for particular cases, like the COUETTE flow or the HAGEN/ POISEULLE flow.

Conclusions and Outlook

We have seen that the inclusion of higher gradients in the list of independent variables offers new possibilities for modelling the behaviour in the presence of length scale effects and of non-classical boundary conditions. Such a

---

189 The coupled (hemitropic) terms have not been considered by TROSTEL (1985), FRIED/ GURTIN (2006), and KRAWIETZ (2021).

190 For higher-order internal constraints see BERTRAM/ GLÜGE (2016) and BERTRAM (2017, 2019)

191 This third-order reaction stress has not been considered by TROSTEL (1985). In FRIED/ GURTIN (2006) it appears but has not been submitted to the subsymmetries.

theory can still be put in a familiar way into a thermodynamic frame with restrictions posed by the second law of thermodynamics. This opens the door for applications in many fields of continuum mechanics, and already a blossoming variety of them has grown and probably will continue to grow much more in the future.

Such gradient materials are part of the so-called **generalized continua**. Another group is formed by theories where deformation-like variables are introduced which are independent or at least different from the classical ones and are often called *micro-deformations*. Such models are called **micromorphic**. If one restricts these micro-variables to pure rotations, the theory is labelled **micropolar**.

All these theories have attracted tremendous interest in the recent past and will surely play an important role in the future.

References

Acharya, A.; Shawki, T. G.: *The Clausius-Duhem inequality and the structure of rate-independent plasticity*. Int. J. Plast. **22**,2, 229-283 (1996)
Aifantis, E.C.: *The physics of plastic deformation*. Int. J. Plasticity **3**,3, 211-247 (1987)
Akivis, M. A.; Goldberg, V. V.: *Tensor Calculus with Applications*. World Scientific Pub., Singapore (2003)
Altenbach, H.: *Kontinuumsmechanik*. Springer Vieweg (2012)
Altmeyer, G.; Panicaud, B.; Rouhaud, E.; Wang, M.; Roos, A.; Kerner, R.:*Viscoelasticity behavior for finite deformations, using a consistent hypoelastic model based on Rivlin materials*. Continuum Mech. Thermodyn. **28**, 1741–1758 (2016)
Anand, L.: *Constitutive equations for hot-working of metals*. Int. J. Plasticity **1**, 213-231 (1985)
Anand, L.; Kothari, M.: *A computational procedure for rate-independent crystal plasticity*. J. Mech. Phys. Solids **44**,4, 525-558 (1996)
Anandarajah, A.: *Computational Methods in Elasticity and Plasticity: Solids and Porous Media*. Springer (2010)
Antman, S. S.: *Nonlinear Problems of Elasticity*. Springer, New York (1995)
Aretz, H.; Barlat, F.: *General orthotropic yield functions based on linear stress deviator transformations*. NUMIFORM 2004. Eds.: S. Ghosh, J. M. Castro, J. K. Lee. Amer. Inst. Physics, 147-151 (2004)
Arminjon, M.: *A regular form of the Schmid law. Application to the ambiguity problem*. Textures and Microstructures **14-18**, 1121-1128 (1991)
Arramon, Y. P.; Mehrabadi, M. M.; Martin, D. W.; Cowin, S. C.: *A multi-dimensional anisotropic strength criterion based on Kelvin modes*. Int. J. Solids Structures **37**, 2915-2935 (2000)
Arruda, E. M.; Boyce, M. C.: *A three-dimensional constitutive model for the large stretch behavior of rubber elastic materials*. J. Mech. Phys. Solids **41**,2, 389-412 (1993)
Asaro, R. J.: *Micromechanics of crystals and polycrystals*. In: *Advances in Applied Mechanics*. Eds.: J. W. Hutchinson, T. Y. Wu, Academic Press **23**, 1-115 (1983)
Asaro, R. J.: *Crystal plasticity*. J. Appl. Mech. **50**, 921-934 (1983)
Atkin, R. J.; Fox, N.: *An Introduction to the Theory of Elasticity*. Longman, London (1980)
Attard, M. M.: *Finite strain-isotropic hyperelasticity*. Int. J. Solids Structures **40**,17, 4353-4378 (2003)
Auffray, N.; Le Quang, H.; He, Q. C. : *Matrix representations for 3D strain-gradient elasticity*. J. Mechanics Physics Solids **61**, 1202-1223 (2013)
Backman, M. E.: *Form for the relation between stress and finite elastic and plastic strains under impulsive loading*. J. Appl. Phys. **35**,8, 2524-2533 (1964)

A. Bertram, *Elasticity and Plasticity of Large Deformations*,
https://doi.org/10.1007/978-3-030-72328-6

Ball, J. M.: *Convexity conditions and existence theorems in nonlinear elasticity.* Arch. Rational Mech. Anal. **63**, 337-403 (1977)

Banabic, D.; Bunge, H.-J.; Pöhlandt, K.; Tekkaya, A. E.: *Formability of Metallic Materials.* Springer, Berlin (2000)

Barlat, F.; Lian, J.: *Plastic behavior and stretchability of sheet metals. Part I: A yield function for orthotropic sheets under plane stress conditions*. Int. J. Plasticity **5**, 51-66 (1989)

Barlat, F.; Lege, D. J.; Brem, J. C.: *A six-component yield function for anisotropic materials*. Int. J. Plasticity 7, 693-712 (1991)

Barlat, F.; Yoon, J. W.; Cazacu, O.: *On linear transformations of stress tensors for the description of plastic anisotropy.* Int. J. Plasticity **23**, 876-896 (2007)

Başar, Y.; Weichert, D.: *Nonlinear Continuum Mechanics of Solids*. Springer, Berlin (2000)

Bassani, J. L.: *Single crystal hardening.* Appl. Mech. Rev. **43**,5,2, 320-327 (1990)

Bassani, J. L.: *Plastic flow of crystals.* In: *Advances in Applied Mechanics.* Eds.: J. W. Hutchinson, T. Y. Wu, Academic Press **30**, 191-258 (1994)

Beatty, M. F.: *A class of universal relations in isotropic elasticity theory.* J. Elasticity **17**, 113-121 (1987)

Beatty, M. F.: *Topics in finite elasticity: Hyperelasticity of rubber, elastomers, and biological tissues – with examples.* Appl. Mech. Rev. **40**,12, 1699-1734 (1987)

Beatty, M. F.; Hayes, M. A.: *Deformations of an elastic, internally constrained material. Part 1: Homogeneous deformations.* J. Elasticity **29**, 1-84 (1992)

Beatty, M. F.: *Seven lectures on finite elasticity.* In: *Topics in Finite Elasticity.* Eds.: M. Hayes, G. Saccomandi. CISM course 424. Springer, Wien, 31-93 (2001)

Beatty, M. F.: *An average-stretch full-network model for rubber elasticity.* In: *The Rational Spirit in Modern Continuum Mechanics*, Eds.: C.-S. Man, R. L. Fosdick, Kluwer Academic Publishers, Dordrecht, 65-86 (2004)

Beheshti, A.: *Generalization of strain-gradient theory to finite elastic deformation for isotropic materials.* Continuum Mech. Thermodyn. **29**, 493-507 (2017)

Bell, J. F.: *The experimental foundations of solid mechanics.* Encyclopedia of Physics VIa/1, Springer-Verlag (1973)

Bell, J. F.: *Contemporary perspectives in finite strain plasticity.* Int. J. Plasticity **1**,1, 3-27 (1985)

Bell, J. F.: *The decrease of volume during loading to finite plastic strain.* Meccanica **31**, 461-472 (1996)

Bermúdez de Castro, A.: *Continuum Thermomechanics.* Birkhäuser, Basel (2005)

Bertram, A.; Haupt, P.: *A Note on Andreussi - Guidugli's theory of thermo-mechanical constraints in simple materials.* Bull. acad. polonaise sci., Serie sci. techn. **14**,1, 47-51 (1976)

Bertram, A.: *Material systems – a framework for the description of material behavior*. Arch. Rational Mech. Anal. **80**,2, 99-133 (1982)

Bertram, A.: *Axiomatische Einführung in die Kontinuumsmechanik.* BI Wissenschaftsverlag, Mannheim (1989)

Bertram, A.: *Description of finite inelastic deformations.* In: Proceedings of MECAMAT'92 *Multiaxial Plasticity* (1992) in Cachan, France. Eds.: A. Benallal, R. Billardon, D. Marquis, 821-835 (1992)

Bertram, A.: *What is the general constitutive equation?* In: *Beiträge zur Mechanik.* Eds.: C. Alexandru, G. Gödert, U. Görn, R. Parchem, J. Villwock, TU Berlin, 28-37 (1993)

Bertram, A.; Olschewski, J.: *Zur Formulierung linearer anelastischer Stoffgleichungen mit Hilfe einer Projektionsmethode.* Z. ang. Math. Mech. **73**,4-5, T401-3 (1993)

Bertram, A.; Kraska, M.: *Determination of finite plastic deformations in single crystals.* Arch. Mech. **47**, 2, 203-222 (1995a)

Bertram, A.; Kraska, M.: *Description of finite plastic deformations in single crystals by material isomorphisms.* Proceedings of IUTAM & ISIMM Symposium on *Anisotropy, Inhomogeneity and Nonlinearity in Solid Mechanics* (1994) in Nottingham, Kluwer Academic Publ., Eds.: D. F. Parker, A. H. England, Dordrecht, 77-90 (1995b)

Bertram, A.: *An alternative approach to finite plasticity based on material isomorphisms.* Int. J. Plasticity **52**,3, 353-374 (1998)

Bertram, A.: *On general frameworks for material modeling.* In: Conference Papers of the 4. Int. Conf. on *Constitutive Laws for Engineering Materials*. Eds.: R. C. Picu, E. Krempl, Troy, 46-49 (1999)

Bertram, A.; Svendsen, B.: *On material objectivity and reduced constitutive equations.* Arch. Mech. **53**,6, 653-675 (2001)

Bertram, A.: *Finite thermoplasticity based on isomorphisms.* Int. J. Plasticity **19**, 2027-2050 (2003)

Bertram, A.; Svendsen, B.: *Reply to Rivlin's Material symmetry revisited – or Much ado about nothing.* GAMM-Mitteilungen **27**,1, 88-93 (2004)

Bertram, A.; Böhlke, T.; Šilhavy, M.: *On the rank 1 convexity of stored energy functions of physically linear stress-strain relations.* J. Elasticity **86**, 235-243 (2007)

Bertram, A.; Forest, S.: *Mechanics based on an objective power functional.* Technische Mechanik **27**,1,1-17 (2007)

Bertram, A.; Tomas, J. (Eds.): *Micro-Macro-Interactions in Structured Media and Particle Systems.* Springer, Berlin (2008)

Bertram, A.: *On the History of Material Theory - A Critical Review.* In: *The History of Theoretical, Material and Computational Mechanics - Mathematics Meets Mechanics and Engineering*, Lecture Notes in Applied Mathematics and Mechanics Volume 1, Editor: Erwin Stein, Springer, pp 119-131 (2014)

Bertram, A.; Forest, S.: *The thermodynamics of gradient elastoplasticity.* Continuum Mech. Thermodyn. **26**, 269-286 (2014)

Bertram, A.; Glüge R.: *Solid Mechanics. Theory, Modeling, and Problems.* Springer-Verlag (2015) 330 pages

Bertram, A.: *Finite gradient elasticity and plasticity: a constitutive mechanical framework.* Continuum Mech. Thermodyn., **27**,6, 1039-1058 (2015)

Bertram, A.; Glüge, R.: *Gradient Materials with Internal Constraints.* Mathematics and Mechanics of Complex Systems **4**,1, 1-15 (2016)

Bertram, A.: *Compendium on Gradient Materials*. TU Berlin 253 pages. https://www.lkm.tuberlin.de/fileadmin/fg49/publikationen/bertram/Compendium_on_Gradient_ Materials_June_2019.pdf (2016, 2017, 2019)
Bertram, A.: *On viscous gradient fluids*. Continuum Mech. Thermodyn. **32**,5, 1385-1401 (2020)
Bertram, A.; Forest, S. (eds.): *Mechanics of Strain Gradient Materials*. Springer-Verlag (2020)
Besseling, J. F.: *A thermodynamic approach to rheology*. In: *Irreversible Aspects of Continuum Mechanics and Transfer of Physical Characteristics in Moving Fluids*. Eds.: H. Parkus, L. I. Sedov, Springer, Wien, 16-51 (1968)
Besseling, J. F.; Giessen, E. v. d.: *Mathematical Modelling of Inelastic Deformation*. Chapman & Hall, London (1994)
Besson, J.; Cailletaud; G.; Chaboche, J.-L.; Forest, S.: *Mécanique non linéaire des matériaux*. Hermes Science, Paris (2001)
Besson, J.; Cailletaud; G.; Chaboche, J.-L.; Forest, S. ; Blétry, M.: *Non-Linear Mechanics of Materials*. Springer, Dordrecht (2010)
Betten, J.: *Über die Konvexität von Fließkörpern isotroper und anisotroper Stoffe*. Acta Mech. **32**, 233-247 (1979)
Betten, J.: *Kontinuumsmechanik*. Springer, Berlin (1993), 2nd rev. ed. (2001)
Biegler, M. W.; Mehrabadi, M. M.: *An energy-based constitutive model for anisotropic solids subject to damage*. Mech. Materials **19**, 151-164 (1995)
Billington, E. W.: *Introduction to the Mechanics and Physics of Solids*. Adam Hilger, Bristol (1986)
Bishop, J. F. W.; Hill, R.: *A theory of the plastic distortion of a polycrystalline aggregate under combined stresses*. Phil. Mag. Ser. 7, **42**, 414-427 (1951)
Bishop, R. L.; Goldberg, S. I.: *Tensor Analysis on Manifolds*. Dover Pub., New York (1968)
Blatz, P. J.; Ko, W. L.: *Application of finite elasticity theory to deformation of rubbery materials*. Tans. Soc. Rheol. **6**, 223-251 (1962)
Bleustein, J. L.: *A note on the boundary conditions of Toupin's strain-gradient theory*. Int. J. Solids Structures **3**, 1053-1057 (1967)
Blume, J. A.: *On the form of the inverted stress-strain law for isotropic hyper-elastic solids*. Int. J. Non-Linear Mechnics **27**,3, 413-421 (1992)
Boehler, J. P.; Raclin, J.: *Écrouissage anisotrope des matériaux orthotropes prédéformés*. J. Méca. Théor. Appl., Numéro spécial 23-44 (1982)
Böhlke, T.; Bertram, A.; Krempl, E.: *Modeling of deformation induced anisotropy in free-end torsion*. Int. J. Plasticity **19**, 1867-1884 (2003)
Boer, R. de: *Vektor- und Tensorrechnung für Ingenieure*. Springer, Berlin (1982)
Bonet, J.; Wood, R. D.: *Nonlinear Continuum Mechanics for Finite Element Analysis*. Cambridge Univ. Press (1997)
Borisenko, A. I.; Tarapov, *I.E.*: *Vector and Tensor Analysis with Applications*. Prentice-Hall, Englewood Cliffs (1968)
Borja, R. I.: *Plasticity - Modeling and Computation*. Springer (2013)
Borsch, S.; Schurig, M.: *Regularisation of the Schmid law in crystal plasticity*. In Bertram, A.; Tomas, J., 77-91 (2008)
Borst, R. de; Giessen, E. v. d. (Eds.): *Material Instabilities in Solids*. John Wiley & Sons, Chichester (1998)

Bowen, R. M.; Wang, C.-C.: *Introduction to Vectors and Tensors*. 2 volumes. Plenum Press, New York (1976)

Brand, L.: *Vector and Tensor Analysis*. New York, John Wiley & Sons (1947)

Brillouin, L.: *Tensors in Mechanics and Elasticity*. Acad. Press, New York (1964)

Bröcker, C.; Matzenmiller, A.: *On the generalization of uniaxial thermo-viscoplasticity with damage to finite deformations based on enhanced rheological models*. Techn. Mech. **34**,3-4, 142-165 (2014)

Bron, F.; Besson, J.: *A yield function for anisotropic materials – Application to aluminum alloys*. Int. J. Plasticity **20**,4-5, 937-963 (2004)

Brown, A. A.; Casey, J.; Nikkel, D. J.: *Experiments conducted in the context of the strain-space formulation of plasticity*. Int. J. Plasticity **19**,11, 1965-2005 (2003)

Bruhns, O. T.; Xiao, H.; Meyers, A.: *On representations of yield functions for crystals, quasicrystals and transversely isotropic solids*. Eur. J. Mech. A/Solids **18**, 47-67 (1999)

Bruhns, O. T.: *The multiplicative decomposition of the deformation gradient in plasticity - origin and limitations*. In: H. Altenbach, T. Matsuda, D. Okumura (edts.) *From Creep Damage Mechanics to Homogenization Methods*. Springer-Verlag, 37-66 (2015)

Bruhns, O. T.: *History of Plasticity*. In: H. Altenbach, A. Öchsner (edts.) *Encyclopedia of Continuum Mechanics*. Springer, 1129-1190 (2020)

Cailletaud, G.: *Crystalline viscoplasticity applied to single crystals*. In: *Handbook of Materials Behavior Models*. Vol. 1. Ed.: J. Lemaitre, 308-317 (2001)

Caldonazzo, B.: *Osservazione sui tensori quintupli emisotropi*. Rend. d. Reale Acad. Naz. dei Lincei **15**, 840-843 (1932)

Carlson, D. E.; Shield, R. T. (eds.): *Finite Elasticity*. Martinus Nijhoff Pub., The Hague (1982)

Carroll, M. M.: *Finite strain solutions in compressible isotropic elasticity*. J. Elasticity **20**, 65-92 (1988)

Casal, P.: *La capillarité interne en mécanique*. CNRS **VI**,3, 31-37 (1961)

Casey, J., Naghdi, P. M.: *A remark on the use of the decomposition $F = F_e\, F_p$ in plasticity*. J. Appl. Mech. **47**, 672-675 (1980)

Casey, J.; Naghdi, P. M.: *A prescription for the identification of finite plastic strain*. Int. J. Engng. Sci. **30**,10, 1257-1278 (1992)

Casey, J.: *On elastic-thermo-plastic materials at finite deformations*. Int. J. Plasticity **14**,1-3, 173-191 (1998)

Cauchy, A.-L.: *Note sur l'équlibre et les mouvements vibratoires des corps solides*. C. R. Acad. Sci. Paris **32**, 323-326 (1851)

Cazacu, O.; Barlat, F.: *A criterion for description of anisotropy and yield differential effects in pressure-insensitive metals*. Int. J. Plasticity **20**, 2027-2045 (2004)

Cazacu, O.; Plunkett, B.; Barlat, F.: *Orthotropic yield criterion for hexagonal closed packed metals*. Int. J. Plasticity **22**, 1171-1194 (2006)

Chadwick, P.: *Continuum Mechanics – Concise Theory and Problems*. George Allen & Unwin, London (1976)

Chambon, R.; Caillerie, D.; Tamagnini, C.: *A finite deformation second gradient theory of plasticity*. Comptes Rendus de l'Académie des Sciences Paris - Series IIb - Mechanics **329**, 797-802 (2001)

Chaves, E. W. V.: *Notes on Continuum Mechanics*. Springer (2013)

Cheverton, K.J.; Beatty, M.F.: *On the mathematical theory of the behavior of some non-simple materials.* Archive Rat. Mech. Anal. **60**,1, 1-16 (1975)

Choquet-Bruhat, Y.; de Witt-Morette, C.; Dillard-Bleick, M.: *Analysis, Manifolds and Physics*. North Holland, Amsterdam (1977)

Chu, E.: *Generalization of Hill's 1979 anisotropic yield criteria.* J. Mat. Process. Techn. **50**, 207-215 (1995)

Ciarlet, P. G.: *Lectures on Three-Dimensional Elasticity.* Tata Inst. of Fundamental Research, Bombay (1983)

Ciarlet, P. G.: *Mathematical Elasticity*. Vol. I. North Holland, Amsterdam (1988)

Cleja-Ţigoiu, E.; Soós, S.: *Elastoviscoplastic models with relaxed configurations and internal state variables*. Appl. Mech. Rev. **43**,7, 131-151 (1990)

Cleja-Ţigoiu, S.: *Large elasto-plastic deformations of materials with relaxed configurations. I. Constitutive assumptions*. Int. J. Engng. Sci. **28**,3, 171-180 (1990)

Cleja-Ţigoiu, S.: *Couple stresses and non-Riemannian plastic connection in finite elasto-plasticity*. Z. angew. Math. Phys. 53, 996-1013 (2002)

Cleja-Ţigoiu, S.: *Elasto-plastic materials with lattice defects modeled by second-order deformations with non-zero curvature.* Int. J. Fract. **166**, 61-75 (2010)

Cleja-Ţigoiu, S.; Paşcan, R.: *Non-local elasto-viscoplastic models with dislocations in finite elasto-plasticity. Part II: Influence of dislocations in crystal plasticity.* Mathematics Mechanics Solids **18**,4, 373-396 (2012)

Cleja-Ţigoiu, S.: *Non-local elasto-viscoplastic models with dislocations in finite elasto-plasticity. Part I: Constitutive framework.* Mathematics Mechanics Solids **18,4**, 349-372 (2013)

Clifton, R. J.: *On the equivalence of F^e F^p and F^p F^e*. J. Appl. Mech. **39**, 287-289 (1972)

Coleman, B. D.; Noll, W.: *Material symmetry and thermostatic inequalities in finite elastic deformations.* Arch. Rational Mech. Anal. **15**, 87-111 (1964)

Cordero, N. M.; Forest, S.; Busso, E. P.: *Second strain gradient elasticity of nano-objects*. J. Mech. Phys. Solids. **97**, 92-124 (2016)

Cosserat, E.; Cosserat, F.: *Théorie des corps déformables*. Herman et fils, Paris (1909)

Criscione, J. C.; Humphrey, J. D.; Douglas, A. S.; Hunter, W. C.: *An invariant basis for natural strain which yields orthogonal stress response terms in isotropic hyperelasticity.* J. Mech. Phys. Solids **48**,12, 2445-2465 (2000)

Criscione, J. C.: *Rivlin's representation formula is ill-conceived for the determination of response functions via biaxial testing.* In: *The Rational Spirit in Modern Continuum Mechanics*, Eds.: C.-S. Man, R. L. Fosdick, Kluwer Academic Publishers, Dordrecht, 197-215 (2004)

Cross, J. J.: *Mixtures of fluids and isotropic solids*. Archives of Mechanics **25**,6, 1025-1039 (1973)

Dafalias, Y. F.: *Corotational rates for kinematic hardening at large plastic deformations.* J. Appl. Mech. **50**, 561-565 (1983)

Dafalias, Y. F.: *The plastic spin concept and a simple illustration of its role in finite plastic transformation.* Mech. Materials **3**, 223-233 (1984)
Dafalias, Y. F.: *The plastic spin.* J. Appl. Mech. **52**, 865-871 (1985)
Dafalias, Y. F.: *On the microscopic origin of the plastic spin.* Acta Mech. **82**, 31-48 (1990)
Dafalias, Y. F.: *Plastic spin: necessity or redundancy.* Int. J. Plasticity **14**,9, 909-931(1998)
Darrieulat, M.; Piot, D.: *A method of generating analytical yield surfaces of crystalline materials.* Int. J. Plasticity **12**,5, 575-610 (1996)
Dashner, P. A.: *Invariance considerations in large strain elasto-plasticity.* J. Appl. Mech. **53**, 55-60 (1986a)
Dashner, P. A.: *Plastic potential theory in large strain elastoplasticity.* Int. J. Solids Structures **22**,6, 593-623 (1986b)
Davison, L.: *Kinematics of finite elastoplastic deformation.* Mech. Materials **21**, 73-88 (1995)
Dawson, P. R.; MacEwen, S. R.; Wu, P.-D.: *Advances in sheet metal forming analyses: dealing with mechanical anisotropy from crystallographic texture.* Int. Mater. Rev. **28**,2, 86-122 (2003)
Dell'Isola, F.; Seppecher, P.: *The relationship between edge contact forces, double forces and interstitial working allowed by the principle of virtual power.* Comptes Rendus de l´Academie de Sciences - Series IIB 7, 43-48 (1995)
Dell'Isola, F.; Seppecher, P.: *Edge contact forces and quasi-balanced power.* Meccanica, **32**,1, 33-52 (1997)
Dell´Isola, F.; Sciarra, G.; Vidoli, S.: *Generalized Hooke´s law for isotropic second gradient materials.* Proc. R. Soc. A. **465**, 2177-2196 (2009)
Del Piero, G.: *On the elastic-plastic material element.* Arch. Rational Mech. Anal. **59**,2, 111-130 (1975)
Desmorat, R., Marull, R.: *Non-quadratic Kelvin modes based plasticity criteria for anisotropic materials.* Int. J. Plasticity **27**, 328-351 (2011)
Dillon, O.W.; Kratochvil, J.: *A strain gradient theory of plasticity.* Int. J. Solids Structures **6**, 1513-1533 (1970)
Dimitrienko, Y. I.: *Tensor Analysis and Nonlinear Tensor Functions.* Kluwer Acad. Pub., Dordrecht (2002)
Dimitrienko, Y. I.: *Nonlinear Continuum Mechanics and Large Inelastic Deformations.* Springer, Dordrecht (2011)
Doghri, I.: *Mechanics of Deformable Solids.* Springer, Berlin (2000)
Donner, H.; Ihlemann, J.: *A numerical framework for rheological models based on the decomposition of the deformation rate tensor.* PAMM - Proceedings in Applied Mathematics and Mechanics **16**,1, 319-320 (2016)
Donner, H.; Kanzenbach, L.; Naumann, C.; Ihlemann, J.: *Efficiency of rubber material modelling and characterisation.* In: Lion, A.; Johlitz, M. (edts.): Proceedings of the European Conference on Constitutive Models for Rubbers X. - London : CRC Press, 19-29 (2017)
Doyle, T. C.; Ericksen, J. L.: *Nonlinear Elasticity.* In: *Advances in Appl. Mech.* IV. Eds.: H. I. Dryden, T. v. Karman, Acad. Press, New York, 53-115 (1956)

Drozdov, A. D.: *Finite Elasticity and Viscoelasticity*. World Scientific Pub., Singapore (1996)

Dunne, F.; Petrinic, N.: *Introduction to Computational Plasticity*. Oxford Univ. Press (2005)

Dunwoody, J.; *On universal deformations with non-uniform temperatures in isotropic, incompressible elastic solids*. Math. Mech. Solids **8**,5, 507-513 (2003)

Duvaut, G.: *Application du principe de l'indifférence matérielle à un milieu élastique matériellement polarisé*. Compte Rendus Acad. Sci. Paris **258**, 3631-3634 (1964)

Eckart, C.: *The thermodynamics of irreversible processes. IV. The theory of elasticity and anelasticity*. Physical Review **73**,4, 373-382 (1948)

Ehlers, W.; Eipper, G.: *The simple tension problem at large volumetric strains computed from finite hyperelastic material laws*. Acta Mech. **130**, 17-27 (1998)

Ericksen, J. L.: *Deformations possible in every compressible, isotropic, perfectly elastic material*. J. Math. Phys. **34**, 126-128 (1955)

Ericksen, J. L.: *Tensor Fields*. In: C. Truesdell, R. A. Toupin: *The Classical Field Theories*. In: Handbuch der Physik. Vol. III/1. Ed.: S. Flügge, Springer, Berlin (1960)

Ericksen, J. L.: *Introduction to the Thermodynamics of Solids*. Chapman & Hall, London (1991)

Farren, W. S.; Taylor, G. I.: *The heat developed during plastic extension of metals*. Proc. Royal Soc. London, Ser. A, **107**, 422-451 (1925)

Feigenbaum, H. P.; Dafalias, Y. F.: *Directional distortional hardening in metal plasticity within thermodynamics*. Int. J. Solids Structures **44**, 7526-7542 (2007)

Finger, J.: *Über die gegenseitigen Beziehungen von gewissen in der Mechanik mit Vortheil anwendbaren Flächen zweiter Ordnung nebst Anwendungen auf Probleme der Astatik*. Sitzungsber. der Kaiserl. Akad. the Wissenschaften, Mathematisch-Naturwiss. Classe **101**, IIa, 1105-1142 (1892)

Finger, J.: *Über die allgemeinsten Beziehungen zwischen endlichen Deformationen und den zugehörigen Spannungen in aeolotropen and isotropen Substanzen*. Sitzungsber. der Kaiserl. Akad. the Wissenschaften, Mathematisch-Naturwiss. Classe **103**,10,IIa, 1073-1100 (1894)

Fitzgerald, J. E.: *A tensorial Hencky measure of strain and strain rate for finite deformations*. J. Appl. Phys. **51**,10, 5111-5115 (1980)

Fleming, W.: *Functions of Several Variables*. Springer, New York (1977)

Forest, S.; Parisot, R.: *Material crystal plasticity and deformation twinning*. Rend. Sem. Mat. Univ. Pol. Torino, **58**,1, 99-111 (2000)

Forest, S.; Sievert, R.: *Elastoviscoplastic constitutive frameworks for generalized continua*. Acta Mechanica **160**, 71-111 (2003)

Fosdick, R. L.; Serrin, J.: *On the impossibility of linear Cauchy and Piola-Kirchhoff constitutive theories for stress in solids*. J. Elasticity **9**,1, 83-89 (1979)

Fox, N.: *On the continuum theories of dislocations and plasticity*. Quart. J. Mech. Appl. Math. **21**, 67-75 (1968)

Fraeijs de Veubeke, B. M.: *A Course in Elasticity*. Springer, New York (1979)

François, D.; Pineau, A.; Zaoui, A.: *Mechanical Behaviour of Materials. Vol 1: Elasticity and Plasticity*. Kluwer Acad. Pub., Dordrecht (1998)

François, M.: *A plasticity model with yield surface distortion for non proportional loading.* Int. J. Plasticity **17**, 703-717 (2001)

Freed, A. D.: *Soft Solids.* Birkhäuser (2014)

Freudenthal, A. M.; Gou, P. F.: *Second order effects in the theory of plasticity.* Acta Mech. **8**, 34-52 (1969)

Frewer, M.: *More clarity on the concept of material frame-indifference in classical continuum mechanics.* Acta Mech. **202**, 213-246 (2009)

Fried, E., Gurtin, M.E.: *Tractions, balances, and boundary conditions for non-simple materials with application to liquid flow at small-length scales.* Arch. Ration. Mech. Anal. 182, 513-554, (2006)

Frischmuth, F.; Kosiński, W.; Perzyna, P.: *Remarks on mathematical theory of materials.* Arch. Mech. **38**,1-2, 59-69 (1986)

Fung, Y. C. B.: *Elasticity of soft tissues in simple elongation.* American J. Physiology **213**,6, 1532-1544 (1967)

Gambin, W.: *Plasticity of crystals with interacting slip systems.* Engn. Trans. **39**,3-4, 303-324 (1991)

Gambin, W.: *Crystal plasticity based on yield surfaces with rounded-off corners.* Z. Angew. Math. Mech. **71**,4, T265-T268 (1991)

Gent, A. N.: *A new constitutive relation for rubber.* Rubber Chem. Technol. **69**,1, 59-61 (1996)

Germain, P.: *Sur l'application de la méthode des puissances virtuelles en mécanique des milieux continus.* C.R. Acad. Sci. Paris Sér. A **274**,2, 1051-1055 (1972)

Germain, P.: *La méthode des puissances virtuelles en mécanique des milieux continus, première partie : théorie du second gradient.* J. Mécanique **12**, 235-274 (1973a)

Germain, P.: *The method of virtual power in continuum mechanics. Part 2: Microstructure.* J. Appl. Math. **25**, 556-575 (1973b)

Germain, P.; Nguyen, Q. S.; Suquet, P.: *Continuum Thermodynamics.* J. Appl. Mech. **50**, 1010-1020 (1983)

Giessen, E. v. d.: *Continuum models of large deformation plasticity – Part I.* Eur. J. Mech. A/Solids **8**,1, 15-34 (1989)

Giessen, E. v. d.: *Micromechanical and thermodynamic aspects of the plastic spin.* Int. J. Plasticity **7**, 365-386 (1991)

Gilman, J. J.: *Physical nature of plastic flow and fracture.* In: *Plasticity*, Proc. 2nd Symp. on Naval Structural Mechanics. Eds.: E. H. Lee, P. S. Symonds, Pergamon Press, Oxford, 43-99 (1960)

Goldenblat, I. I.: *On a problem in the mechanics of finite deformation of continuous media* (in Russian). C. R. Dokl. Acad. Sci. SSR **70**,6, 973-976 (1950)

Goldenblat, I. I.; Kopnov, V. A.: *Strength of glass-reinforced plastics in the complex stress state.* Mekhanika Polimerov, vol.1 (1965). English translation: *Polymer Mechanics*, vol. 1, Faraday Press (1966)

Gonzalez, O.; Stuart, A.M.: *A First Course in Continuum Mechanics*, Cambridge University Press (2008)

Green, A. E.; Zerna, W.: *Theoretical Elasticity.* Oxford at the Clarendon Press (1954, 1968, 1975, 2002)

Green, A. E.; Rivlin, R. S.: *Simple force and stress multipoles*. Archive Rat. Mech. Anal. **16**,5, 325-353 (1964)

Green, A. E.; Adkins, J. E.: *Large Elastic Deformations.* Oxford Univ. Press, Oxford (1960), 2nd ed. (1970)

Green, A. E.; Naghdi, P. M.: *A general theory of an elastic-plastic continuum.* Arch. Rational Mech. Anal. **18**,4, 251-281 (1965)

Green, A. E.; Naghdi, P. M.; Trapp, J. A.: *Thermodynamics of a continuum with internal constraints*. Int. J. Engn. Sci. **8**, 891-908 (1970)

Green, A. E.; Naghdi, P. M.: *Some remarks on elastic-plastic deformation at finite strain*. Int. J. Engn. Sci., **9**, 1219-1229 (1971)

Greve, R.: *Kontinuumsmechanik. Ein Grundkurs.* Springer, Berlin (2003)

Gurtin, M. E.: *Thermodynamics and the possibility of spacial interaction in elastic materials*. Archive Rat. Mech. Anal. **19**,5, 339-352 (1965)

Gurtin, M. E.; Podio Guidugli, P.: *The thermodynamics of constrained materials*. Arch. Rat. Mech. Anal. **51**,3, 192-208 (1973)

Gurtin, M. E.: *An Introduction to Continuum Mechanics.* Academic Press, New York (1981)

Gurtin, M. E.; Fried, E.; Anand, L.: *The Mechanics and Thermodynamics of Continua.* Cambridge University Press (2013)

Habraken, A. M.: *Modelling the plastic anisotropy of metals.* Arch. Comput. Meth. Engng. **11**,1, 3-96 (2004)

Hackl, K.: *Generalized standard media and variational principles in classical and finite strain elastoplasticity*. J. Mech. Phys. Solids **45**,5, 667-688 (1997)

Halmos, P. R.: *Finite-Dimensional Vector Spaces.* Springer, New York (1974, 1987)

Halphen, M. B.; Nguyen, Q. S.: *Sur les matériaux standards généralisés*. J. Méca. **14**,1, 39-63 (1975)

Halphen, M. B.: *Sur le champ des vitesses en thermoplasticité finie.* Int. J. Solids Structures **11**,9, 947-960 (1975)

Hamel, G.: *Theoretische Mechanik.* Springer, Berlin (1949)

Han, W.; Reddy, B. D.: *Plasticity – Mathematical Theory and Numerical Analysis*. Springer, Berlin (1999)

Hanyga, A.; Ogden, R. W.: *Mathematical Theory of Non-Linear Elasticity.* Ellis Horwood Pub., Chichester (1985)

Hashiguchi, K.: *Elastoplasticity Theory.* Springer, 2nd edition (2014)

Haupt, P.: *Viskoelastizität und Plastizität.* Springer, Berlin (1977)

Haupt, P.: *On the concept of an intermediate configuration and its application to a representation of viscoelastic-plastic material behavior*. Int. J. Plasticity **1**, 303-316 (1985)

Haupt, P.; Tsakmakis, C.: *On the application of dual variables in continuum mechanics*. J. Continuum Mech. Thermodyn. **1**, 165-196 (1989)

Haupt, P.: *Thermodynamics of solids*. In: *Non-Equilibrium Thermodynamics with Application to Solids*. W. Muschik (Edt.). Springer-Verlag Wien, CISM Course 336, (1993)

Haupt, P.: *Continuum Mechanics and Theory of Materials*. Springer, Berlin (2000), 2nd rev. ed. (2002)

Havner, K. S.: *A discrete model for the prediction of subsequent yield surfaces in polycrystalline plasticity*. Int. J. Solids Structures 7, 719-730 (1971)
Havner, K. S.: *Comparisons of crystal hardening laws in multiple slip*. Int. J. Plasticity **1**, 111-124 (1985)
Havner, K. S.: *Finite Plastic Deformation of Crystalline Solids*. Cambridge University Press (1992)
Hayes, M.; Saccomandi, G. (eds.): *Topics in Finite Elasticity*. CISM course 424. Springer, Wien (2001)
Heidari, M.; Vafai, A.; Desai, C.: *An Eulerian multiplicative constitutive model of finite elastoplasticity*. Eur. J. Mech. A/Solids **28**, 1088-1097 (2009)
Hellinger, E.: *Die allgemeinen Ansätze der Mechanik der Kontinua*. In: Encyklopädie der mathematischen Wissenschaften. Bd. IV-30, 601-694 (1914)
Hershey, A. V.: *The plasticity of an isotropic aggregate of anisotropic face-centered cubic crystals*. J. Appl. Mech. **21**, 241-249 (1954)
Hill, R.: *A theory of the yielding and plastic flow of anisotropic metals*. Proc. Roy. Soc. London **A 193**, 281-297 (1948)
Hill, R.: *The Mathematical Theory of Plasticity*. Clarendon Press, Oxford (1950, 1998)
Hill, R.: *Elastic properties of reinforced solids: some theoretical principles*. J. Mech. Phys. Solids **11**, 357-372 (1963)
Hill, R.: *Continuum micro-mechanics of elastoplastic polycrystals*. J. Mech. Phys. Solids **13**, 89-101 (1965)
Hill, R.: *Generalized constitutive relations for incremental deformation of metal crystals by multislip*. J. Mech. Phys. Solids **14**,2, 95-102 (1966)
Hill, R.: *On constitutive inequalities for simple materials – I*. J. Mech. Phys. Solids **16**, 229-242 (1968)
Hill, R.: *On constitutive macro-variables for heterogeneous solids at finite strain*. Proc. R. Soc. Lond. A. **326**, 131-147 (1972)
Hill, R.; Rice, J. R.: *Constitutive analysis of elastic-plastic crystals at arbitrary strain*. J. Mech. Phys. Solids **20**, 401-413 (1972)
Hill, R.: *Aspects of invariance in solid mechanics*. In: *Advances in Applied Mechanics*. Ed.: C.-S. Yih, Academic Press **18**, 1-75 (1978)
Hill, R.: *Theoretical plasticity of textured aggregates*. Math. Proc. Camb. Phil Soc. **85**, 179-191 (1979)
Hill, R.: *On macroscopic effects of heterogeneity in elastoplastic media at finite strain*. Math. Proc. Camb. Phil. Soc. 95, 481-494 (1984)
Hill, R.: *Constitutive modelling of orthotropic plasticity in sheet metals*. J. Mech. Phys. Solids **38**,3, 405-417 (1990)
Hill, R.: *A user-friendly theory of orthotropic plasticity in sheet metals*. Int. J. Mech. Sci. **35**,1,19-25 (1993)
Hill, J. M.; Arrigo, D. J.: *A note on Ericksen's problem for radially symmetric deformations*. Math. Mech. Solids **4**, 395-405 (1999)
Holsapple, K. A.: *On natural states and plastic strain in simple materials*. Z. Angew. Math. Mech. **53**, 9-16 (1973)
Holzapfel, G. A.: *Nonlinear Solid Mechanics. A Continuum Approach for Engineering*. John Wiley & Sons, Chichester (2000)

Honeycombe, R. W. K.: *The Plastic Deformation of Metals*. Edward Arnold, ASM (1968, 1984)

Horgan, C. O.; Saccomandi, G.: *Constitutive modelling of rubber-like and biological materials with limiting chain extensibility*. Math. Mech. Solids **7**, 353-371 (2002)

Hosford, W. F.: *On the yield loci of anisotropic cubic metals*. 7th North. Amer. Metalworking Conf., S. M. E., Dearborn MI, 191-197 (1979

Hosford, W. F.; Caddell, R. M.: *Metal Forming – Mechanics and Metallurgy*. Prentice-Hall, Englewood Cliffs (1983, 1993)

Hosford, W. F.: *The Mechanics of Crystals and Textured Polycrystals*. Oxford University Press, New York (1993)

Hoss, L.; Marczak, R. J.: *A new constitutive model for rubber-like materials*. Mecánica Computacional **29**, 2759-2773 (2010)

Hsu, T. C.: *A theory of the yield locus and flow rule of anisotropic materials*. J. Strain Anal. **1**,3, 204-215 (1966)

Huber, M. T.: *Właściwa praca odkształcenia jako miara wytężenia materyału (The specific strain work as a measure of material effort)*. Czasopismo Techniczne 22, 81-83 (1904)

Huilgol, R. R.: *On the structure of the group* g_1* *appearing in hyperelasticity*. Int. J. Non-Linear Mechanics **6**, 677-681 (1971)

Hutchinson, J. W.: *Bounds and self-consistent estimates for creep of polycrystalline materials*. Proc. R. Soc. Lond. A. **348**,101-127 (1976)

Hutter, K.; Baaser, H. (eds.): *Deformation and Failure in Metallic Materials*. Springer, Berlin (2003)

Ignatieff, Y. A.: *The Mathematical World of Walter Noll. A Scientific Biography*. Springer, Berlin (1996)

Ikegami, K.: *Experimental plasticity on the anisotropy of metals*. In: Colloques internationaux du CNRS **295** *Comportement mécanique de solides anisotropes*. 201-242 (1982)

Irgens, F.: *Continuum Mechanics*. Springer, Berlin (2008)

Itskov, M.: *On the application of the additive decomposition of generalized strain measures in large strain plasticity*. Mech. Res. Comm. **31**, 507-517 (2004)

Itskov, M.: *Tensor Algebra and Tensor Analysis for Engineers*. Springer, Berlin (2007)

Ivey, H. J.: *Plastic stress-strain relations and yield surfaces for aluminium alloys*. J. Mech. Engng. Sci. **3**,1, 15-31 (1961)

Jaumann, G.: *Geschlossenes System physikalischer und chemischer Differentialgesetze*. Sitzungsber. Akad. Wiss. Wien (IIa) **120**, 385-530 (1911)

Javili, A.; Dell'Isola, F.; Steinmann, P.: *Geometrically nonlinear higher-gradient elasticity with energetic boundaries*. J. Mechanics Physics Solids **61**,12, 2381-2401 (2013)

Kappus, R.: *Zur Elastizitätstheorie endlicher Verschiebungen*. Z. Angew. Math. Mech. **19**,5, 271-285, **19**,6,344-361 (1939)

Karafillis, A. P.; Boyce, M. C.: *A general anisotropic yield criterion using bounds and a transformation weighting tensor*. J. Mech. Phys. Solids **41**,12, 1859-1886 (1993)

Karni, Z.; Reiner, M.: *The general measure of deformation*. In: *Second-order Effects in Elasticity, Plasticity and Fluid Dynamics*. Eds.: M. Reiner, D. Abir, Jerusalem Academic Press, Pergamon Press, Oxford, 217-227 (1964)

Kearsley, E. A., Fong, J. T.: *Linearly independent sets of isotropic Cartesian tensors of ranks up to eigtht*. J. Research Nat. Bur. Stand. -B. Mathematical Sciences, 79B,1-2 (1975)

Kellogg, O. D.: *Foundations of Potential Theory*. F. Ungar Pub. Comp., New York (1929)

Kettunen, P. O.; Kuokkala, V.-T.: *Plastic Deformation and Strain Hardening*. Trans Tech Pub., Zürich (2003)

Khan, A. S.; Huang, S.: *Continuum Theory of Plasticity*. John Wiley & Sons, New York (1995)

Khan, A. S.; P. Cheng: *An anisotropic elastic-plastic constitutive model for single and polycrystalline metals*. I- *Theoretical Developments*. Int. J. Plasticity **12**,2, 147-162 (1996)

Kießling, R; Landgraf, R.; Scherzer, R.; Ihlemann, J.: *Introducing the concept of directly connected rheological elements by reviewing rheological models at large strains*. Int. J. Sol. Structures **97-98**, 650-667 (2016)

Knockaert, R.; Chastel, Y.; Massoni, E.: *Rate-independent crystalline and polycrystalline plasticity, application to FCC materials*. Int. J. Plasticity **16**, 179-198 (2000)

Knops, R. J.; Payne, L. E.: *Uniqueness Theorems in Linear Elasticity*. Springer, Berlin (1971)

Knops, R. J.; Wilkes, E. W.: *Theory of elastic stability*. In: Handbuch der Physik. Vol. VIa/3. Ed.: S. Flügge, Springer, Berlin (1973)

Kocks, U. F.; Tomé, C. N.; Wenk, H.-R.: *Texture and Anisotropy: Preferred Orientations in Polycrystals and their Effect on Material Properties*. Cambridge Univ. Press (1998)

Kocks, U. F.; Mecking, H.: *Physics and phenomenology of strain hardening: the FCC case*. Progr. Mat. Sci. **48**, 171-273 (2003)

Koiter, W. T.: *Stress-strain relations, uniqueness and variational theorems for elastic-plastic materials with a singular yield surface*. Quart. Appl. Math. **11**,3, 350-354 (1953)

Knowles, J. K.: *The finite anti-plane shear field near the tip of a crack for a class of incompressible elastic solids*. Int. J. Fracture **13**,5, 611-639 (1977)

Korobeynikov, S. N.: *Objective tensor rates and applications in formulation of hyperelastic relations*. J. Elast **93**, 105-140 (2008)

Kratochvíl, J.; Dillon, O. W.: *Thermodynamics of crystalline elastic-visco-plastic materials*. J. Appl. Phys. **41**,4, 1470-1479 (1970)

Kratochvíl, J.: *Finite-strain theory of crystalline elastic-inelastic materials*. J. Appl. Phys. **42**,3, 1104-1108 (1971)

Kratochvíl, J.: *On a finite strain theory of elastic-inelastic materials*. Acta Mech. **16**,1-2, 127-142 (1973)

Krausz, A. S.; Krausz, K. (eds.): *Unified Constitutive Laws of Plastic Deformation*. Acad. Press, San Diego (1996)

Krawietz, A.: *Passivität, Konvexität und Normalität bei elastisch-plastischem Material*. Ingenieur-Archiv **51**, 257-274 (1981)

Krawietz, A.: *Materialtheorie*. Springer, Berlin (1986)

Krawietz, A.; Mathiak, F.: *Constitutive behavior of sheet metal – theoretical and experimental investigations*. In: Proc. of COMPLAS II, Barcelona (1989)

Krawietz, A.: *Surface tension and reaction stresses of a linear incompressible second gradient fluid*. Continuum Mechanics Thermodynamics. DOI: 10.1007/s00161-020-00951-8 (2021)

Krempl, E.: *A small-strain viscoplasticity theory based on overstress*. In: *Unified Constitutive Laws of Plastic Deformation*. Eds.: A. S. Krausz, K. Krausz, Acad. Press, San Diego, 282-318 (1996)

Kurtyka, T.; Życzkowski, M.: *A geometric description of distortional plastic hardening of deviatoric materials*. Arch. Mech. **37**,4-5, 383-395 (1985)

Kurtyka, T.; Życzkowski, M.: *Evolution equations for distortional plastic hardening*. Int. J. Plasticity. **12**,2, 191-213 (1996)

Lainé, E.; Vallée, C.; Fortuné, D.: *Nonlinear isotropic constitutive laws: choice of the three invariants, convex potentials and constitutive inequalities*. Int. J. Engn. Sci. 37, 1927-1941 (1999)

Latorre, M.; Montáns, F. J.: *On the interpretation of the logarithmic strain tensor in an arbitrary system of representation*. Int. J. Solids Structures **51**, 1507-1515 (2014)

Lebedev, L. P.; Cloud, M. J.: *Tensor Analysis*. World Scientific Pub., Singapore (2003)

Lee, E. H.; Liu, D. T.: *Finite-strain elastic-plastic theory with application to plane-wave analysis*. J. Appl. Phys. **38**,1, 19-27 (1967)

Lee, E. H.: *Elastic-plastic deformation at finite strains*. J. Appl. Mech. **36**, 1-6 (1969)

Lee, E. H., Germain, P. : *Elastic-plastic theory at finite strain*. In: *Problems of Plasticity*. Int. Symp., Warsaw 1972, Ed.: A. Sawczuk, Nordhoff, Leyden, 117-133 (1974)

Leigh, D. C.: *Nonlinear Continuum Mechanics*. McGraw-Hill, New York (1968)

Lemaitre, J. (Ed.): *Handbook of Materials Behavior Models*. Vol. 1. Academic Press, San Diego (2001)

Le Roux, M. J.: *Étude géométrique de la torsion et de la flexion dans la déformation infinitésimale d'un milieu continu*. Annal. Scien. É.N.S. **3**,28, 523-579 (1911)

Le Roux, M. J.: Recherches sur la géometrie des déformations finies. Annal. Scien. É.N.S. **3**,30, 193-245 (1913)

Lian, J.; Chen, J.: *Isotropic polycrystal yield surfaces of b.c.c. and f.c.c. metals: crystallographic and continuum mechanics approaches*. Acta Metall. Mater. **39**,10, 2285-2294 (1991)

Lion, A.: Constitutive modelling in finite thermoviscoplasticity: a physical *approach based on nonlinear rheological models*. Int. J. Plasticity **16**, 469-494 (2000)

Liu, I-S.: *Continuum Mechanics*. Springer, Berlin (2002)

Liu, I-S.: *On Euclidean objectivity and the principle of material frame-indifference*. J. Continuum Mech. Thermodyn. **16**,1-2, 177-183 (2004)

Liu, C.; Huang, Y.; Stout, M. G.: *On the asymmetric yield surface of plastically orthotropic materials: A phenomenological study.* Acta Mater. **45**,6, 2397-2406 (1997)

Loomis, L. H.; Sternberg, S.: *Advanced Calculus.* Addison-Wesley, Reading (1968)

Lowe, T. C.; Rolett, A. D.; Follansbee, P. S.; Daehn, G. S. (eds.): *Modeling the Deformation of Crystalline Solids.* TMS, Warrendale (1991)

Lubarda, V. A.; Shih, C. F.: *Plastic spin and related issues in phenomenological plasticity.* J. Appl. Mech. **61**, 524-529 (1994)

Lubarda, V. A.: *Duality in constitutive formulation of finite-strain elasto-plasticity based on* $F = F_e F_p$ *and* $F = F^p F^e$ *decompositions.* Int. J. Plasticity **15**, 1277-1290 (1999)

Lubarda, V. A.: *Elastoplasticity Theory.* CRC Press, Boca Raton (2002)

Lubliner, J.: *Normality rules in large-deformation plasticity.* Mech. Materials **5**, 29-34 (1986)

Lubliner, J.: *Plasticity Theory.* Macmillan, New York (1990)

Lütkepohl, H.: *Handbook of Matrices.* John Wiley & Sons, Chichester (1996)

Lurie, A. I.: *Nonlinear Theory of Elasticity.* North Holland, Amsterdam (1990)

Macvean, D. B.: *Die Elementararbeit in einem Kontinuum und die Zuordnung von Spannungs- und Verzerrungstensoren.* Z. angew. Math. Phys. **19**, 157-184 (1968)

Madeo, A.; Ghiba, I.-D.; Neff, P.; Münch, I.: *A new view on boundary conditions in the Grioli-Koiter-Mindlin-Toupin indeterminate couple stress model.* Europ. J. Mecha. A/ Solids **59**, 294-322 (1916)

Mandel, J.: *Contribution théorique à l'étude de l'écrouissage et des lois de l'écoulement plastique.* In: *Proc. 11th Int. Congr. Appl. Mech.* München 1964, Ed.: H. Görtler, Springer, Berlin, 502-509 (1966)

Mandel, J.: *Plasticité classique et viscoplasticité.* CISM course No. 97, Springer, Wien (1971)

Mandel, J.: *Equations constitutive et directeurs dans les milieux plastiques et viscoplastique.* Int. J. Solids Structures **9**,6, 725-740 (1973)

Mandel, J.:. *Thermodynamics and plasticity.* In: Proc. Int. Symp. *Foundations of Continuum Thermodynamics.* Eds.: J. J. Delgado Domingos, M. N. R. Nina, J. H. Whitlaw, McMillan, London, 283-304 (1974)

Mandel, J.: *Définition d'un repère privilégié pour l'étude des transformations anélastiques du polycristal.* J. Méca. Théor. Appl. **1**,1, 7-23 (1982)

Marckmann, G.; **Verron**, E.: *Comparison of hyperelastic models for rubberlike materials.* Rubber Chemistry Techn., **79**,5, 835-858 (2006)

Marris, A. W.; Shiau, J. F.: *Universal deformations in isotropic incompressible hyperelastic materials when the deformation tensor has equal proper values.* Arch. Rat. Mech. Anal. **36**,2, 135-160 (1970)

Marris, A. W.: *Two new theorems on Ericksen's problem.* Arch. Rat. Mech. Anal. **79**,2, 131-173 (1982).

Marsden, J. E.; Hughes, T. J. R.: *Mathematical Foundations of Elasticity.* Prentice-Hall, Englewood Cliffs (1983)

Maugin, G. A.: *The Thermomechanics of Plasticity and Fracture.* Cambridge Univ. Press, Cambridge (1992)

Maugin, G. A.: *Thermomechanics of Nonlinear Irreversible Behaviors: An Introduction*, No.27 of WSP Series on Nonlinear Science, World Scientific Publ., Singapore (1999)

McConnell, A. J.: *Applications of Tensor Analysis.* Dover, New York (1957)

McLellan, A. G.: *The Classical Thermodynamics of Deformable Materials.* Cambridge Univ. Press (1980)

Menzel, A.; Steinmann, P.: *On the spatial formulation of anisotropic multiplicative elasto-plasticity.* Comput. Methods Appl. Mech. Engn. **192**, 3431-3470 (2003)

Méric, L.; Poubanne, P.; Cailletaud, G.: *Single crystal modeling for structural calculations: Part 1– Model presentation.* J. Engrg. Mat. Techn. **113**, 162-182 (1991)

Miehe, C.: *A constitutive frame of elastoplasticity at large strains based on the notion of a plastic metric.* Int. J. Solids Structures **35**,30, 3859-3897 (1998)

Miehe, C.; Schröder, J.; Schotte, J.: *Computational homogenization analysis in finite plasticity – Simulation of texture development in polycrystalline materials.* Comput. Methods Appl. Mech. Engrg. **171**, 387-418 (1999)

Miehe, C.: *Computational micro-to-macro transitions for discretized microstructures of heterogeneous materials at finite strains based on the minimization of averaged incremental energy.* Comput. Methods Appl. Mech. Engn. **192,** 559-591 (2003)

Miehe, C.: *Variational gradient plasticity at finite strains. Part I: Mixed potentials for the evolution and update problems of gradient-extended dissipative solids.* Comput. Methods Appl. Mech. Engn. **268**, 677-703 (2014)

Mielke, A.: *Energetic formulation of multiplicative elasto-plasticity using dissipation distances.* Continuum Mech. Thermodyn. **15**, 351-382 (2003)

Mindlin, R. D.: *Micro-structure in linear elasticity.* Archive Rat. Mech. Anal. **16**,1, 51-78 (1964)

Mindlin, R. D.: *Second gradient of strain and surface-tension in linear elasticity.* Int. J. Solids Structures **1**, 417-438 (1965)

Mindlin, R. D.; Eshel, N. N.: *On first strain-gradient theories in linear elasticity.* Int. J. Solids Structures **4**, 109-124 (1968)

Mises, R. v.: *Mechanik der festen Körper im plastisch-deformablen Zustand.* Nachr. Kgl. Ges. Wiss. Göttingen, Math. Phys. Klasse 582-592 (1913)

Mises, R. v.: *Mechanik der plastischen Formänderung von Kristallen.* Z. Angew. Math. Mech. **8**,3, 161-185 (1928)

Mollica, F.; Srinivasa, A. R.: *A general framework for generating convex yield surfaces for anisotropic metals.* Acta Mech. 154, 61-84 (2002)

Montheillet, F.; Jonas, J. J.; Benferrah, M.: *Development of anisotropy during the cold rolling of aluminium sheet.* Int. J. Mech. Sci. **33**,3,197-209 (1991)

Mooney, M.: *A theory of large elastic deformation.* J. Appl Phys., **11**, 582-592 (1940)

Müller, W. C.: *Universal solutions for simple thermodynamic bodies.* Arch. Rational Mech. Anal. **35**, 220- 225 (1970)

Müller, I.: *Thermodynamics.* Pitman, Boston (1985)

Müller, I.; Strehlow, P.: *Rubber and Rubber Balloons. Paradigms of Thermodynamics*. Lecture Notes in Physics **637**, Springer, Heidelberg (2004)
Murnaghan, F. D.: *Finite deformations of an elastic solid.* Amer. J. Math. **59**, 235-260 (1937)
Naghdi, P.: *Stress-strain relations in plasticity and thermoplasticity.* In: *Plasticity*, Proc. 2nd Symp. on Naval Structural Mechanics. Eds.: E. H. Lee, P. S. Symonds, Pergamon Press, Oxford, 121-169 (1960)
Naghdi, P. M.; Trapp, J. A.: *The significance of formulating plasticity theory with reference to loading surfaces in strain space.* Int. J. Engn. Sci. **13**, 785-797 (1975)
Naghdi, P.: *A critical review of the state of finite plasticity*. J. Appl. Math. Phys. **41**, 315-394 (1990)
Naghdi, P. M.; Srinivasa, A.: *Some general results in the theory of crystallographic slip*. Z. angew. Math. Phys. **45**, 687-732 (1994)
Narasimhan, M. N. L.: *Principles of Continuum Mechanics*. John Wiley & Sons, New York (1993)
Neff, P.; Münch, I.; Ghiba, I.-D.; Madeo, A.: *On some fundamental misunderstandings in the indeterminate couple stress model. A comment on recent papers of A. R. Hadjesfandiari and G. F. Dargush.* Int. J. Solids Structures **81**, 233-243 (2016)
Negahban, M.: *The Mechanical and Thermodynamical Theory of Plasticity.* CRC Press (2012)
Nemat-Nasser, S.: *Decomposition of strain measures and their rates in finite deformation elastoplasticity.* Int. J. Solids Structures **15**, 155-166 (1979)
Nemat-Nasser, S.; Hori, M.: *Micromechanics: Overall Properties of Heterogeneous Materials.* North-Holland, Amsterdam (1993)
Nemat-Nasser, S.: *Plasticity.* Cambridge University Press (2004)
Neumann, C.: *Zur Theorie der Elasticität.* Borchardt's J. für reine and angewandte Math. **57**, 281-318 (1860)
Neutsch, W.: *Koordinaten.* Spektrum Akademischer Verlag, Heidelberg (1995)
Nguyen, H. V.; Raniecky, B.: *Observable plastic spin and comparison with other approaches*. Arch. Mech. **51**,2, 207-221 (1999)
Nguyen, Q. S.: *Stability and Nonlinear Solid Mechanics*. John Wiley & Sons Chichester (2000)
Noll, W.: *On the continuity of the solid and fluid states*. J. Rational Mech. Anal. **4**, 3-81 (1955)
Noll, W.: *A mathematical theory of the mechanical behavior of continuous media.* Arch. Rational Mech. Anal. **2**, 197-226 (1958)
Noll, W.: *Euclidean geometry and Minkowskian chronometry*. Amer. Math. Monthly **71**, 129-144 (1964)
Noll, W.: *Proof of the maximality of the orthogonal group in the unimodular group.* Arch. Rational Mech. Anal. **18**, 100-102 (1965)
Noll, W.: *A new mathematical theory of simple materials*. Arch. Rational Mech. Anal. **48**, 1-50 (1972)
Noll, W.: *The Foundations of Mechanics and Thermodynamics. Selected Papers.* Springer, Berlin (1974)

Noll, W.: *Finite-Dimensional Spaces. Algebra, Geometry, and Analysis*. Martinus Nijhoff Publ., Dordrecht (1987)

Noll, W.: *A frame-free formulation of elasticity*. J. Elasticity **83**,3, 291-307 (2006)

Ogden, R. W.: *Large deformation isotropic elasticity – on the correlation of theory and experiment for incompressible rubber-like solids.* Proc. Roy. Soc. A **326**, 565-584 (1972)

Ogden, R. W.: *Non-Linear Elastic Deformations*. John Wiley & Sons, New York (1984a)

Ogden, R. W.: *On Eulerian and Lagrangean objectivity in continuum mechanics.* Arch. Mech. **36**,2, 207-218 (1984b)

Ogden, R. W.: *Nonlinear elasticity, anisotropy, material stability and residual stresses in soft tissue.* In: CISM Course 441 *Biomechanics of Soft Tissue in Cardiovascular Systems.* Eds.: G. A. Holzapfel, R. W. Ogden, Springer, Wien (2003)

Ortiz, M.; Popov, E. P.: *Distortional hardening rules for metal plasticity.* J. Eng. Mech. ASCE **109**,4, 1042-1057 (1983)

Ota, T.; Shindo, A.; Fufuoka, H.: *A consideration on an anisotropic yield criterion.* Proc. 9th Japan National Congress for Applied Mechanics, 117-120 (1959)

Ottosen, N. S.; Ristinmaa, M.: *The Mechanics of Constitutive Modeling.* Elsevier, Amsterdam (2005)

Owen, D. R.: *A mechanical theory of materials with elastic range*. Arch. Rational Mech. Anal. **37**, 85-110 (1970)

Pach, K.; Frey, T.: *Vector and Tensor Analysis*. Terra, Budapest (1964)

Paglietti, A.: *Plasticity of Cold Worked Metals – a Deductive Approach.* WIT-press (2007)

Palmow, W. A.: *Rheologische Modelle für Materialien bei endlichen Deformationen.* Techn. Mech. **5**,4, 20-31 (1984)

Parma, S.; Plešek, J.; Mareka, R.; Hrubýa, Z.; Feigenbaum, H. P.; Dafalias, Y. F.: *Calibration of a simple directional distortional hardening model for metal plasticity.* Int. J. Solids Structures **143**, 113-124 (2018)

Penn, R. W.: *Volume changes accompanying the extension of rubber*. Trans Soc. Rheol. **14**,4, 509-517 (1970)

Perzyna, P.; Kosiński, W.: *A mathematical theory of materials*. Bull. acad. polonaise sci., Série sci. techn. **21**,12, 647-654 (1973)

Petroski, H. J.; Carlson, D. E.: *Controllable states of elastic heat conductors.* Arch. Rational Mech. Anal. **31**,2, 127-150 (1968)

Petroski, H. J.; Carlson, D. E.: *Some exact solutions to the equations of nonlinear thermoelasticty*. J. Appl. Mech. **37**, 1151-1154 (1970)

Petroski, H. J.: *On the insufficiency of controllable states to characterize a class of rigid heat conductors*. Z. Angew. Math. Mech. **51**, 481-482 (1971)

Petryk, H.: *Macroscopic rate-variables in solids undergoing phase transformation.* J. Mech. Phys. Solids **46**,5, 873-894 (1998)

Petryk, H.: *On the micro-macro transition and hardening moduli in plasticity.* In: Proc. IUTAM Symposium on *Micro- and Macrostructural Aspects of Thermoplasticity*. Bochum 1997, Eds.: O. T. Bruhns, E. Stein, Kluwer, Dordrecht 219-230 (1999)

Phillips, A.; Liu, C. S.; Justusson, J. W.: *An experimental investigation of yield surfaces at elevated temperatures*. Acta Mech. **14**, 119-146 (1972)

Phillips, A.: *The foundations of thermoplasticity - experiments and theory*. In: *Topics in Applied Continuum Mechanics*. Eds.: J. L. Zeman, F. Ziegler, 1-21 (1974)

Pietryga, M. P.; Vladimorov, I. N.; Reese, S.: *A finite deformation model for evolving flow anisotropy with distortional hardening inlcuding experimental validation.* Mech. Materials **44**, 163-173 (2012)

Piola, G.: *Intorno alle equazioni fondamentali del movimento di corpi qualsivogliono, considerati secondo la naturale loro forma e costituzione.* Memorie di Matematica e Fisica della Società Italiana delle Scienze residente en Modena, **24**, 1-186 (1848)

Plunkett, B.; Lebensohn, R. A.; Cazacu, O.; Barlat, F.: *Anisotropic yield function of hexagonal materials taking into account texture development and anisotropic hardening.* Acta Mater. **54**, 4159-4169 (2006)

Podio-Guidugli, P.: *A primer in elasticity*. J. Elasticity **58**,1, 1-103 (2000). In book form: Kluwer Academic Publishers (2000)

Polizzotto, C.: *A nonlocal strain gradient plasticity theroy for finite deformations.* Int. J. Plast. **25**, 1280-1300 (2009)

Podio-Guidugli, P.; Vianello, M.: *On a stress-power-based characterization of second-gradient elastic fluids.* Continuum Mech. Thermodyn. **25**, 399-421 (2013)

Polizzotto, C.: *A second strain gradient elasticity theory with second velocity gradient inertia - Part I: Constitutive equations and quasi-static behavior*. Int. J. Solids Structures **50**,24, 3749-3765 (2013)

Pozdeyev, A. A.; Trusov, P. V.; Nyashin, Y. I.: *Large Elastoplastic Strains* (in Russian). Nauka, Moscow (1986)

Raabe, D.; Roters, F.; Barlat, F.; Chen, L.-Q.: *Continuum Scale Simulations of Engineering Materials.* WILEY-VCH Verlag (2004)

Raghavan, M. L.; Vorp, D. A.: *Toward a biomechanical tool to evaluate rupture potential of abdominal aortic aneurysm: identification of a finite strain constitutive model and evaluation of its applicability.* J. Biomechanics **33**, 475-482 (2000)

Rajagopal, K. R., Srinivasa, A. R.: *Mechanics of the inelastic behavior of materials – Part I Theoretical underpinnings.* Int. J. Plasticity **14**,10-11, 945-968 (1998)

Reddy, J. N.: *An Introduction to Continuum Mechanics*. Cambridge (2008)

Rees, D. W. A.: *The theory of scalar plastic deformation functions*. Z. Angew. Math. Mech. **63**, 217-228 (1983)

Reese, S.; Govindjee, S.: *A theory of finite visoelasticity and numerical aspects*. Int. J. Solids Structures **35**,26-27, 3455-3482 (1998)

Reese, S.: *On material and geometrical instabilities in finite elasticity and elasto-plasticity*. Arch. Mech. **52**,6, 969-999 (2000)

Reiher, J. C.; Bertram, A.: *Finite third-order gradient elasticity and thermoelasticity*. J. Elasticity, **133**, 223-252 (2018)

Reiher, J. C.; Bertram, A.: *Finite third*-order *gradient elastoplasticity and thermoplasticity*. J. Elasticity **138**,2, 169-193 (2020)

Reina, C.; Conti, S.: *Kinematic description of crystal plasticity in the finite kinematic framework: A micromechanical understanding of* $\mathbf{F} = \mathbf{F}^e \mathbf{F}^p$. J. Mech. Phys. Solids **67**, 40-61 (2014)

Reuss, A.: *Vereinfachte Berechnung der plastischen Formänderungsgeschwindigkeiten bei Voraussetzung der Schubspannungsfließbedingung.* Z. Angew. Math. Mech. **13**,5, 356-360 (1930)

Rice, J. R.: *Inelastic constitutive relations for solids: an internal-variable theory and its application to metal plasticity.* J. Mech. Phys. Sol. **19**, 433-455 (1971)

Richter, H.: *Das isotrope Elastizitätsgesetz.* Z. Angew. Math. Mech. **28**,7/8, 205-209 (1948)

Richter, H.: *Zur Elastizitätstheorie endlicher Verformungen.* Math. Nachr. **8**, 65-73 (1952)

Ristinmaa, M.; Wallin, M.; Ottosen, N. S.: *Thermodynamic format and heat generation of isotropic hardening plasticity*. Acta Mech. **194**, 103-121 (2007)

Rivlin, R. S.: *Large elastic deformations of isotropic materials, I. Fundamental concepts.* Phil. Trans. Roy. Soc. Lond. A **240**, 459-490 (1948)

Rivlin, R. S.; Saunders, D. W.: *Large elastic deformations of isotropic materials, VII. Experiments on the deformation of rubber.* Phil. Trans. Roy. Soc. Lond. A **243**, 251-288 (1951)

Rivlin, R. S.; Ericksen, J. L.: *Stress-deformation relations for isotropic materials.* J. Rational Mechan. Anal. **4**, 323-425 (1955)

Rivlin, R. S.: *Material symmetry revisited.* GAMM-Mitteilungen **1/2**, 109-126 (2002)

Rösler, J.; Harders, H.; Bäker, M.: *Mechanisches Verhalten der Werkstoffe*. B. G. Teubner, Stuttgart (2003)

Rougée, P.: *Mécanique des grandes transformations.* Springer, Berlin (1997)

Rubin, M. B.: *Plasticity theory formulated in terms of physically based microstructural variables Part- I. Theory.* Int. J. Solids Structures **31**,19, 2615-2634 (1994)

Rubin, M. B.: *On the treatment of elastic deformation in finite elastic-viscoplastic theory.* Int. J. Plasticity **12**,7, 951-965 (1996)

Rubin, M. B.: *Continuum Mechanics with Eulerian Formulations of Constitutive Equations.* Springer (2021)

Ruiz-Tolosa, J. R.; Castillo, E.: *From Vectors to Tensors*. Spinger, Berlin (2005)

Saccomandi, G.: *On inhomogeneous deformations in finite thermoelasticity.* IMA J. Appl. Math. **63**, 131-148 (1999)

Saccomandi, G.: *Universal solutions and relations in finite elasticity.* In: *Topics in Finite Elasticity.* Eds.: M. Hayes, G. Saccomandi. CISM course 424. Springer, Wien, 95-167 (2001)

Salençon, J.: *Handbook of Continuum Mechanics*. Springer, Berlin (2001)

Sansour, C.: *On the dual variable of the logarithmic strain tensor, the dual variable of the Cauchy stress tensor, and related issues*. Int. J. Solids Structures **38**, 9221-9232 (2001)

Sansour, C.; Karšaj, I.; Sorić, J.: *A formulation of anisotropic continuum elastoplasticity at finite strains. Part I: Modelling*. Int. J. Plasticity **22**, 2346-2365 (2006)

Schade, H.: *Tensoranalysis.* Walter de Gruyter, Berlin (1997)

Saint-Venant, A. J. C. B. de: *Rapport sur un mémoire de M. Maurice Levy, relatif à l'hydromechanique des liquides homogènes, particulièrment à leur écoulement rectiligne et permantent.* C R Acad. Sci. Paris **68**, 582-592 (1869a)

Saint-Venant, A. J. C. B. de: *Note sur les valeurs que prennent les pressions dans un solide élastique isotrope lorsque l'ont tient compte des dérivées d'ordre supérieur des déplacements très-petits que leurs points ont éprouvés.* C. R. Acad. Sci. Paris **68**, 569-571 (1869b)

Scheidler, M.; Wright, T. W.: *A continuum framework for finite viscoplasticity.* Int. J. Plasticity **17**,8, 1033-1085 (2001)

Schmid, E.: *"Streckgrenze" von Kristallen. Schubspannungsgesetz.* Proc. Int. Congr. Appl. Mech., Delft, 342 (1924)

Schmid, E.; Boas, W.: *Kristallplastizität.* Julius Springer, Berlin (1935)

Schmidt-Baldassari, M.: *Numerical concepts for rate-independent single crystal plasticity.* Comp. Methods Appl. Mech. Engrg. **192**, 1261-1280 (2003)

Schreyer, H. L.; Zuo, Q. H.: *Anisotropic yield surfaces based on elastic projection operators.* J. Appl. Mech. **62**, 780-785 (1995)

Schröder, J.; Miehe, C.: *Aspects of computational rate-independent crystal plasticity.* Comp. Material Sci. **9**, 168-176 (1997)

Schurig, M.; Bertram, A.: *A rate independent approach to crystal plasticity with a power law.* Comp. Material Sci. **26**, 154-158 (2003)

Seth, B. R.: *Generalized strain measure with applications to physical problems.* In: *Second-Order Effects in Elasticity, Plasticity and Fluid Dynamics.* Eds.: M. Reiner, D. Abir, Pergamon Press, Oxford, 162-172 (1964)

Shouten, J. A.: *Tensor Analysis for Physicists.* Dover Pub., New York (1990)

Shutov, A. V. ; Panhans, S.; Kreißig, R.: *A phenomenological model of finite strain viscoplasticity with distortional hardening.* Z. Angew. Math. Mech. 1-28 (2011)

Sidoroff, F.: *Quelques réflexions sur le principe d'indifférence materiélle pour un milieu ayant un état relâché.* C. R. Acad. Sc. Paris, Serie A, **271**, 1026-1029 (1970)

Sidoroff, F.: *The geometrical concept of intermediate configuration and elastic-plastic finite strain.* Arch. Mech. **25**,2, 299-308 (1973)

Sievert, R.: *A geometrically nonlinear elasto-viscoplasticity theory of second grade.* Technische Mechanik **31**, 83-111 (2011)

Silber, G.: Eine Systematik nichtlokaler kelvinhafter Fluide vom Grade drei auf der Basis eines klassischen Kontinuumsmodelles. Fortschritt-Berichte VDI Nr. 26 (1986)

Silber, G.: *Aggregate isotroper Tensoren zur Darstellung hyperelastischer anisotroper Stoffe.* Z. angew. Math. Mech., **68**,1, 39-45 (1988)

Silber, G.: *Gradiententheorien in der biomedizinischen Physik.* In: *Beiträge zur Mechanik.* Trostel-Festschrift. Edts. C. Alexandru, G. Gödert, U. Görn, R. Parchem, J. Villwock, 308-324 (1993)

Silber, G.; Trostel, R.; Alizadeh, M.; Benderoth, G.: *A continuum mechanical gradient theory with applications to fluid mechanics.* In: Proceedings 2nd European Mechanics of Materials Conference on Mechanics of Materials with

Intrinsic Length Scale: Physics, Experiments, Modeling and Applications, Pr8-365 (1998)

Šilhavý, M.; Kratochvíl, J.: *A theory of inelastic behavior of materials.* Part I. *Ideal inelastic materials.* Arch. Rational Mech. Anal. **65**,2, 97-129 (1977) Part II. *Inelastic materials.* Arch. Rational Mech. Anal. **65**,2, 131-152 (1977)

Šilhavý, M.: *On transformation laws for plastic deformations of materials with elastic range.* Arch. Rational Mech. Anal. **63**,2, 169-182 (1977)

Šilhavý, M.: *The Mechanics and Thermodynamics of Continuous Media.* Springer, Berlin (1997)

Simo, J. C.; Hughes, T. J. R.: *Computational Inelasticity.* Springer, New York (1998)

Simo, J. C.: *Numerical analysis and simulation of plasticity.* In: *Handbook of Numerical Analysis.* Vol. VI. Eds.: P. G. Ciarlet, J. L. Lions, North-Holland (1998)

Skrzypek, J. J.: *Plasticity and Creep.* CRC Press, Boca Raton (1993)

Sluzalec, A.: *Theory of Metal Forming Plasticity.* Springer, Berlin (2004)

Smith, D. R.: *An Introduction to Continuum Mechanics – after Truesdell and Noll.* Kluwer, Dordrecht (1993)

Sokolnikoff, I. S.: *Tensor Analysis.* John Wiley & Sons, New York (1951, 1964)

Souza Neto, E. D. de; Peric, D.; Owen, D. R.: J.: *Computational Methods for Plasticity.* Wiley (2009)

Speziale, C.G.: *A review of material frame-indifference in mechanics.* Appl. Mech. Rev., **51**,8, 489-504 (1998)

Spitzig, W. A.; Richmond, O.: *The effect of pressure on the flow stress of metals.* Acta Metallurgica **31**,1, 457-463 (1984)

Stahn, O.; Bertram, A.; Müller, W. H.: *The evolution of Hooke's law under finite plastic deformations for fiber reinforced materials* (in preparation 2020)

Steinmann, P.: *On localization analysis in multisurface hyperelasto-plasticity.* J. Mech. Phys. Solids, **44**,10, 1691-1713 (1996)

Steinmann, P.: *Geometrical Foundations of Continuum Mechanics.* Springer (2015)

Suiker, A. S. J.; Chang, C. S.: *Application of higher-order tensor theory for formulating enhanced continuum models.* Acta Mech. **142**, 223-234 (2000)

Svendsen, B.: *A thermodynamic formulation of finite deformation elastoplasticity with hardening based on the concept of material isomorphism.* Int. J. Plasticity **14**,6, 473-488 (1998)

Svendsen, B.; Bertram, A.: *On frame-indifference and form-invariance in constitutive theory.* Acta Mech. **132**, 195-207 (1999)

Svendsen, B.: *On the modelling of anisotropic elastic and inelastic material behaviour at large deformation.* Int. J. Solids Structures **38**, 9579-9599 (2001)

Svendsen, B.; Neff, P.; Menzel, A.: *On constitutive and configurational aspects of models for gradient continua with microstructure.* Z. angew. Math. Mech. **89**,8, 687-697 (2009)

Szczepiński, W.: *Introduction to the Mechanics of Plastic Forming of Metals.* Sijthoff & Noordhoff, Alphen aan den Rijn (1979)

Taber, L. A.: *Nonlinear Theory of Elasticity. Applications in Biomechanics.* World Scientific, New Jersey (2004)

Talbert, S. H.; Avitzur, B.: *Elementary Mechanics of Plastic Flow in Metal Forming.* John Wiley & Sons, Chichester (1996)
Taylor, G. I.; Elam, C. F.: *The plastic extension and fracture of aluminium crystals.* Proc. Roy. Soc. A108, 28-51 (1925)
Taylor, G. I.: *Plastic strain in metals.* J. Inst. Metals **62**, 307-324 (1938)
Teodosiu, C.: *A dynamic theory of dislocations and its applications to the theory of the elastic-plastic continuum.* In: *Fundamental Aspects of Dislocation Theory.* Eds.: J. A. Simmons, R. de Wit, R. Bullough, Nat. Bur. Stand. (U. S.), Spec. Publ. **317**,II, 837-876 (1970)
Teodosiu, C.; Sidoroff, F.: *A theory of finite elastoviscoplasticity of single crystals.* Int. J. Engn. Sci. **14**, 165-176 (1976)
Teodosiu, C.: *The plastic spin: microstructural origin and computational significance.* In: *Computational Plasticity.* Eds.: D. R. J. Owen, E. Hinton, E. Oñate, Pineridge Press, Swansea (1989)
Teodosiu, C.; Raphanel, J. L.; Sidoroff, F. (eds.): *Large Plastic Deformations.* A. A. Balkema, Rotterdam (1993)
Thomson, W.: *Elements of a Mathematical Theory of Elasticity.* London (1856)
Toledano, J.-C.: *Physical Basis of Plasticity in Solids.* World Scientific (2012)
Tong, W.; Tao, H.; Jiang, X.: *Modelling the rotation of orthotropic axes of sheet metals subjected to off-axis uniaxial tension.* J. Appl. Mech. **71**, 521-531 (2004)
Toupin, R. A.: *Elastic materials with couple stresses.* Arch. Rat. Mech. Anal. **11**, 385-414 (1962)
Treloar, L. R. G.: *The Physics of Rubber Elasticity.* Oxford University Press, (1975)
Tresca, H.: *Mémoire sur l'écoulement des corps solides soumis à de fortes pressions.* Comptes Rendus Acad. Sci. Paris **59**, 754 (1864)
Trostel, R.: *Zur Systematik fest-idealplastischer Medien.* In: *Aus Theorie und Praxis der Ingenieurwissenschaften.* Edts. R. Trostel, P. Zimmermann.Verlag von Wilhelm Ernst & Sohn. Berlin, München, Düsseldorf (1971)
Trostel, R.: Gedanken zur Konstruktion mechanischer Theorie. In: Beiträge zu den Ingenieurwissenschaften, Sattler-Festschrift, TU Berlin, 96-134 (1985)
Trostel, R.: *Gedanken zur Konstruktion mechanischer Theorien II.* Forschungsbericht Nr. 7 des 2. Inst. für Mechanik, TU Berlin (1988)
Trostel, R.: *Mathematische Grundlagen the Technischen Mechanik I, Vektor- und Tensoralgebra.* Vieweg, Braunschweig (1993)
Trostel, R.: *Mathematische Grundlagen the Technischen Mechanik II, Vektor- und Tensoranalysis.* Vieweg, Braunschweig (1997)
Trostel, R.: *Mathematische Grundlagen the Technischen Mechanik III, Materialmodelle in the Ingenieurmechanik.* Vieweg, Braunschweig (1999)
Trostel, R.: Theoriekonstruktionen in der Mechanik. Tectum Verlag, Marburg (2010)
Truesdell, C.; Toupin, R. A.: *The Classical Field Theories.* Handbuch der Physik. Vol. III/1. Ed.: S. Flügge, Springer, Berlin (1960)
Truesdell, C. A.: *The nonlinear field theories in mechanics.* In: *Topics in Nonlinear Physics.* Ed.: N. J. Zabusky, Springer, Berlin, 19-215 (1968)
Truesdell, C. A.: *Rational Thermodynamics.* McGraw-Hill, New York (1969)

Truesdell, C. A.; Noll, W.: *The Non-linear Field Theories of Mechanics*. In: Handbuch der Physik. Vol. III/3. Ed.: S. Flügge, Springer, Berlin (1965), 2nd ed. (1992), 3rd ed. by S. Antman (2004)
Tsai, S. W.; Wu, E. M.: *A general theory of strength for anisotropic materials*. J. Composite Materials **5**, 58-80 (1971)
Tsakmakis, C.: *Description of plastic anisotropy effects at large deformations – part I: restrictions imposed by the second law and the postulate of Il'yushin*. Int. J. Plasticity **20**,2, 176-198 (2004)
Valanis, K. C.; Landel, R. F.: *The strain-energy function of a hyperelastic material in terms of the extension ratios*. J. Appl. Phys. **38**,7, 2997-3002 (1967)
Valent, T.: *Boundary Value Problems of Finite Elasticity – Local Theorems on Existence, Uniqueness, and Analytic Dependence on Data*. Springer, New York (1988)
Vallée, C.; He, Q.-C.; Lerintiu, C.: *Convex analysis of the eigenvalues of a 3D second-order symmetric tensor*. J. Elasticity **83**, 191-204 (2006)
Van Houtte, P.: *Models for the prediction of deformation textures*. In: *Advances and Applications of Quantitative Texture Analysis*. Eds.: H. J. Bunge, C. Esling, DGM Informationsgesellschaft-Verlag, Oberursel, 175-198 (1991)
Vegter, H.; van den Boogaard, A. H.: *A plane stress yield function for anisotropic sheet material by interpolation of biaxial stress states*. Int. J. Plasticity **22**, 557-580 (2006)
Vianello, M.: *The representation problem for constrained hyperelastic materials*. Arch. Rational Mech. Anal. **111**,1, 87-98 (1990)
Vidoli, S.; Sciarra, G.: *A model for crystal plasticity based on micro-slip descriptors*. Continuum Mech. Thermodyn. **14**, 425-435 (2002)
Vladimirov, I. N.; Pietryga, M. P.; Reese, S.: *Anisotropic finite elastoplasticity with nonlinear kinematic and isotropic hardening and application to sheet metal forming*. Int. J. Plasticity **26**, 659-687 (2010)
Voce, E.: *A practical strain-hardening function*. Metallurgia, 219-226 (1955)
Voigt, W.: *Lehrbuch der Kristallphysik*. Teubner-Verlag, Leipzig (1910)
Voyiadjis, G. Z.; Foroozesh, M.: *Anisotropic distortional yield model*. J. Appl. Mech. **57**, 537-547 (1990)
Voyiadjis, G. Z.; Thiagarajan, G.; Petrakis, E.: *Constitutive modelling for granular media using an anisotropic distortional yield model*. Acta Mech. **110**, 151-171 (1995)
Wang, C.-C.: *On the stored-energy functions of hyperelastic materials*. Arch. Rational Mech. Anal. **23**,1, 1-14 (1966)
Wang, C.-C.: *On the geometric structures of simple bodies, a mathematical foundation for the theory of continuous distributions of dislocations*. Arch. Rational Mech. Anal. **27**,1, 33-94 (1967)
Wang, C. C.; Truesdell, C. A.: *Introduction to Rational Elasticity*. Noordhoff, Leyden (1973)
Wang, C.-C.; Bloom, F.: *Material uniformity and inhomogeneity in anelastic bodies*. Arch. Rational Mech. Anal. **53**, 246-276 (1974)
Wang, C.-C.: *Global equations of motion for anelastic bodies and bodies with elastic range*. Arch. Rational Mech. Anal. **59**,1, 9-23 (1975)

Weber, M.; Glüge, R.; Altenbach, H.: *Evolution of the stiffness tetrad in fiber-reinforced materials under large plastic strain.* Arch. Appl. Mech., https://doi.org/10.1007/s00419-020-01827-8 (2020)

Wegener, K.: *Zur Berechnung großer plastischer Deformationen mit einem Stoffgesetz vom Überspannungstyp.* Braunschweig Series on Mechnics 2, TU Braunschweig (1991)

Wegener, K.; Schlegel, M.: *Suitability of yield functions for the approximation of subsequent yield surfaces.* Int. J. Plasticity **12**,9, 1151-1177 (1996)

Weissman, S. L.; Sackman, J. L.: *Elastic-plastic multiplicative decomposition with a stressed intermediate configuration.* Comput. Methods Appl. Mech. Engrn. **200**, 1607-1618 (2011)

Weyl, H.: *The classical groups – their invariants and representations.* Princeton University Press (1939)

Wu, P. D.; v. d. Giessen, E.: *On improved network models for rubber elasticity and their applications to orientation hardening in glassy polymers.* J. Mech. Phys. Solids **41**,3, 427-465 (1993)

Wu, H.-C.: *Continuum Mechanics and Plasticity.* Chapman & Hall/CRC, London (2005)

Wilmański, K.: *Thermomechanics of Continua.* Springer, Berlin (1998)

Xiao, H.; Bruhns, O. T.; Meyers, A.: *A new aspect in kinematics of large deformations.* In: Plasticity and Impact Problems. Ed.: N. K. Gupta, New Age, New Delhi, 100-109 (1996)

Xiao, H.; Bruhns, O. T.; Meyers, A.: *Logarithmic strain, logarithmic spin and logarithmic rate.* Acta Mech. **124**, 89-105 (1997)

Xiao, H.; Bruhns, O. T.; Meyers, A.: *The choice of objective rates in finite elastoplasticity: general results on the uniqueness of the logarithmic rate.* Proc. R. Soc. Lond. A **456**, 1865-1882 (2000)

Xiao, H.; Bruhns, O. T.; Meyers, A.: *A consistent finite elastoplasticity theory combining additive and multiplicative decomposition of the stretching and the deformation gradient.* Int. J. Plasticity **16**,2, 143-177 (2000)

Xiao, H.; Bruhns, O. T.; Meyers, A.: *Basic issues concerning finite strain measures and isotropic stress-deformation relations.* J. Elasticity **67**,1, 1-23 (2002)

Xiao, H.; Bruhns, O. T.; Meyers, A.: *Elastoplasticity beyond small deformations.* Acta Mech. **182**, 31-111 (2006)

Yang, W. H. (ed.): *Topics in Plasticity.* A. M. Press, Ann Arbor (1991)

Yang, W.; Lee, W. B.: *Mesoplasticity and its Applications.* Springer, Berlin (1993)

Yeoh, O. H.: *Characterization of elastic properties of carbon-black-filled rubber vulcanizates.* Rubber Chem. Technol. **63**, 792-805 (1990)

Yeoh, O. H.: *Some forms of the strain energy function for rubber.* Rubber Chem. Technol. **66**, 754-771 (1993)

Yu, M.: *Advances in strength theories for materials under complex stress state in the 20th Century.* Appl. Mech. Rev. **55**,3, 169-218 (2002)

Yu, M.-H.: *Generalized Plasticity.* Springer, Berlin (2006)

Zaremba, S.: *Sur une forme perfectionnée de la théorie de la relaxation.* Bull. Int. Acad. Sci. Cracovie 594-614 (1903)

Zheng, Q.-S.; He, Q.-C.; Curnier, A.: *Simple shear decomposition of the deformation gradient.* Acta Mech. **140**, 131-147 (2000)

Ziegler, H.: *An Introduction to Thermomechanics.* North-Holland Pub., Amsterdam (1983)

Życzkowski, M.: *Anisotropic yield conditions.* In: *Handbook of Materials Behavior Models.* Vol. 1. Ed.: J. Lemaitre, 155-165 (2001)

Index

A. Bertram, *Elasticity and Plasticity of Large Deformations*,
https://doi.org/10.1007/978-3-030-72328-6

Printed in the United States
by Baker & Taylor Publisher Services

Printed in the United States
by Baker & Taylor Publisher Services